# Conservation Biology for All

EDITED BY:

**Navjot S. Sodhi**
*Department of Biological Sciences, National University of Singapore* AND *\*Department of Organismic and Evolutionary Biology, Harvard University (\*Address while the book was prepared)*

**Paul R. Ehrlich**
*Department of Biology, Stanford University*

# OXFORD
UNIVERSITY PRESS

Great Clarendon Street, Oxford OX2 6DP

Oxford University Press is a department of the University of Oxford.
It furthers the University's objective of excellence in research, scholarship,
and education by publishing worldwide in

Oxford New York

Auckland Cape Town Dar es Salaam Hong Kong Karachi
Kuala Lumpur Madrid Melbourne Mexico City Nairobi
New Delhi Shanghai Taipei Toronto

With offices in

Argentina Austria Brazil Chile Czech Republic France Greece
Guatemala Hungary Italy Japan Poland Portugal Singapore
South Korea Switzerland Thailand Turkey Ukraine Vietnam

Oxford is a registered trade mark of Oxford University Press
in the UK and in certain other countries

Published in the United States
by Oxford University Press Inc., New York

© Oxford University Press 2010

The moral rights of the author have been asserted
Database right Oxford University Press (maker)

First published 2010
Reprinted with corrections 2010
Reprinted with Foreword 2011

All rights reserved. No part of this publication may be reproduced,
stored in a retrieval system, or transmitted, in any form or by any means,
without the prior permission in writing of Oxford University Press,
or as expressly permitted by law, or under terms agreed with the appropriate
reprographics rights organization. Enquiries concerning reproduction
outside the scope of the above should be sent to the Rights Department,
Oxford University Press, at the address above

You must not circulate this book in any other binding or cover
and you must impose the same condition on any acquirer

British Library Cataloguing in Publication Data
Data available

Library of Congress Cataloging in Publication Data
Data available

Typeset by SPI Publisher Services, Pondicherry, India
Printed and bound by
CPI Group (UK) Ltd,
Croydon, CR0 4YY

ISBN 978–0–19–955423–2 (Hbk.)
ISBN 978–0–19–955424–9 (Pbk.)

7 9 10 8 6

# Contents

| | |
|---|---|
| **Dedication** | ix |
| **Acknowledgements** | x |
| **List of Contributors** | xi |
| *Georgina Mace* | |
| **Foreword** | xv |

**Introduction**
*Navjot S. Sodhi and Paul R. Ehrlich* — 1

    Introduction Box 1: Human population and conservation (*Paul R. Ehrlich*) — 2
    Introduction Box 2: Ecoethics (*Paul R. Ehrlich*) — 3

**1: Conservation biology: past and present**
*Curt Meine* — 7

    1.1 Historical foundations of conservation biology — 7
        Box 1.1: Traditional ecological knowledge and biodiversity conservation (*Fikret Berkes*) — 8
    1.2 Establishing a new interdisciplinary field — 12
    1.3 Consolidation: conservation biology secures its niche — 15
    1.4 Years of growth and evolution — 16
        Box 1.2: Conservation in the Philippines (*Mary Rose C. Posa*) — 19
    1.5 Conservation biology: a work in progress — 21
    Summary — 21
    Suggested reading — 22
    Relevant websites — 22

**2: Biodiversity**
*Kevin J. Gaston* — 27

    2.1 How much biodiversity is there? — 27
    2.2 How has biodiversity changed through time? — 33
    2.3 Where is biodiversity? — 35
    2.4 In conclusion — 39
        Box 2.1: Invaluable biodiversity inventories (*Navjot S. Sodhi*) — 40
    Summary — 41
    Suggested reading — 41
    Revelant websites — 42

**3: Ecosystem functions and services**
*Cagan H. Sekercioglu* — 45

    3.1 Climate and the Biogeochemical Cycles — 45

3.2 Regulation of the Hydrologic Cycle — 48
3.3 Soils and Erosion — 50
3.4 Biodiversity and Ecosystem Function — 51
    Box 3.1: The costs of large-mammal extinctions (*Robert M. Pringle*) — 52
    Box 3.2: Carnivore conservation (*Mark S. Boyce*) — 54
    Box 3.3: Ecosystem services and agroecosystems in a landscape context (*Teja Tscharntke*) — 55
3.5 Mobile Links — 57
    Box 3.4: Conservation of plant-animal mutualisms (*Priya Davidar*) — 58
    Box 3.5: Consequences of pollinator decline for the global food supply (*Claire Kremen*) — 60
3.6 Nature's Cures versus Emerging Diseases — 64
3.7 Valuing Ecosystem Services — 65
Summary — 66
Relevant websites — 67
Acknowledgements — 67

## 4: Habitat destruction: death by a thousand cuts
*William F. Laurance* — 73

4.1 Habitat loss and fragmentation — 73
4.2 Geography of habitat loss — 73
    Box 4.1: The changing drivers of tropical deforestation (*William F. Laurance*) — 75
4.3 Loss of biomes and ecosystems — 76
    Box 4.2: Boreal forest management: harvest, natural disturbance, and climate change (*Ian G. Warkentin*) — 80
4.4 Land-use intensification and abandonment — 82
    Box 4.3: Human impacts on marine ecosystems (*Benjamin S. Halpern, Carrie V. Kappel, Fiorenza Micheli, and Kimberly A. Selkoe*) — 83
Summary — 86
Suggested reading — 86
Relevant websites — 86

## 5: Habitat fragmentation and landscape change
*Andrew F. Bennett and Denis A. Saunders* — 88

5.1 Understanding the effects of landscape change — 88
5.2 Biophysical aspects of landscape change — 90
5.3 Effects of landscape change on species — 92
    Box 5.1: Time lags and extinction debt in fragmented landscapes (*Andrew F. Bennett and Denis A. Saunders*) — 92
5.4 Effects of landscape change on communities — 96
5.5 Temporal change in fragmented landscapes — 99
5.6 Conservation in fragmented landscapes — 99
    Box 5.2: Gondwana Link: a major landscape reconnection project (*Andrew F. Bennett and Denis A. Saunders*) — 101
    Box 5.3: Rewilding (*Paul R. Ehrlich*) — 102
Summary — 104
Suggested reading — 104
Relevant websites — 104

## 6: Overharvesting
*Carlos A. Peres* — 107

|  |  |
|---|---|
| 6.1 A brief history of exploitation | 108 |
| 6.2 Overexploitation in tropical forests | 110 |
| 6.3 Overexploitation in aquatic ecosystems | 113 |
| 6.4 Cascading effects of overexploitation on ecosystems | 115 |
|     Box 6.1: The state of fisheries (*Daniel Pauly*) | 118 |
| 6.5 Managing overexploitation | 120 |
|     Box 6.2: Managing the exploitation of wildlife in tropical forests (*Douglas W. Yu*) | 121 |
| Summary | 126 |
| Relevant websites | 126 |

## 7: Invasive species
### *Daniel Simberloff* — 131

|  |  |
|---|---|
|     Box 7.1: Native invasives (*Daniel Simberloff*) | 131 |
|     Box 7.2: Invasive species in New Zealand (*Daniel Simberloff*) | 132 |
| 7.1 Invasive species impacts | 133 |
| 7.2 Lag times | 143 |
| 7.3 What to do about invasive species | 144 |
| Summary | 148 |
| Suggested reading | 148 |
| Relevant websites | 148 |

## 8: Climate change
### *Thomas E. Lovejoy* — 153

|  |  |
|---|---|
| 8.1 Effects on the physical environment | 153 |
| 8.2 Effects on biodiversity | 154 |
|     Box 8.1: Lowland tropical biodiversity under global warming (*Navjot S. Sodhi*) | 156 |
| 8.3 Effects on biotic interactions | 158 |
| 8.4 Synergies with other biodiversity change drivers | 159 |
| 8.5 Mitigation | 159 |
|     Box 8.2: Derivative threats to biodiversity from climate change (*Paul R. Ehrlich*) | 160 |
| Summary | 161 |
| Suggested reading | 161 |
| Relevant websites | 161 |

## 9: Fire and biodiversity
### *David M. J. S. Bowman and Brett P. Murphy* — 163

|  |  |
|---|---|
| 9.1 What is fire? | 164 |
| 9.2 Evolution and fire in geological time | 164 |
| 9.3 Pyrogeography | 165 |
|     Box 9.1: Fire and the destruction of tropical forests (*David M. J. S. Bowman and Brett P. Murphy*) | 167 |
| 9.4 Vegetation–climate patterns decoupled by fire | 167 |
| 9.5 Humans and their use of fire | 170 |
|     Box 9.2: The grass-fire cycle (*David M. J. S. Bowman and Brett P. Murphy*) | 171 |
|     Box 9.3: Australia's giant fireweeds (*David M. J. S. Bowman and Brett P. Murphy*) | 173 |
| 9.6 Fire and the maintenance of biodiversity | 173 |
| 9.7 Climate change and fire regimes | 176 |
| Summary | 177 |
| Suggested reading | 178 |

| | |
|---|---:|
| Relevant websites | 178 |
| Acknowledgements | 178 |

## 10: Extinctions and the practice of preventing them
### Stuart L. Pimm and Clinton N. Jenkins — **181**

| | |
|---|---:|
| 10.1 Why species extinctions have primacy | 181 |
|     Box 10.1: Population conservation (*Jennifer B.H. Martiny*) | 182 |
| 10.2 How fast are species becoming extinct? | 183 |
| 10.3 Which species become extinct? | 186 |
| 10.4 Where are species becoming extinct? | 187 |
| 10.5 Future extinctions | 192 |
| 10.6 How does all this help prevent extinctions? | 195 |
| Summary | 196 |
| Suggested reading | 196 |
| Relevant websites | 196 |

## 11: Conservation planning and priorities
### Thomas Brooks — **199**

| | |
|---|---:|
| 11.1 Global biodiversity conservation planning and priorities | 199 |
| 11.2 Conservation planning and priorities on the ground | 204 |
|     Box 11.1: Conservation planning for Key Biodiversity Areas in Turkey (*Güven Eken, Murat Ataol, Murat Bozdoğan, Özge Balkız, Süreyya İsfendiyaroğlu, Dicle Tuba Kılıç, and Yıldıray Lise*) | 209 |
| 11.3 Coda: the completion of conservation planning | 213 |
| Summary | 214 |
| Suggested reading | 214 |
| Relevant websites | 214 |
| Acknowledgments | 215 |

## 12: Endangered species management: the US experience
### David. S. Wilcove — **220**

| | |
|---|---:|
| 12.1 Identification | 220 |
|     Box 12.1: Rare and threatened species and conservation planning in Madagascar (*Claire Kremen, Alison Cameron, Tom Allnutt, and Andriamandimbisoa Razafimpahanana*) | 221 |
|     Box 12.2: Flagship species create Pride (*Peter Vaughan*) | 223 |
| 12.2 Protection | 226 |
| 12.3 Recovery | 230 |
| 12.4 Incentives and disincentives | 232 |
| 12.5 Limitations of endangered species programs | 233 |
| Summary | 234 |
| Suggested reading | 234 |
| Relevant websites | 234 |

## 13: Conservation in human-modified landscapes
### Lian Pin Koh and Toby A. Gardner — **236**

| | |
|---|---:|
| 13.1 A history of human modification and the concept of "wild nature" | 236 |
|     Box 13.1: Endocrine disruption and biological diversity (*J. P. Myers*) | 237 |
| 13.2 Conservation in a human-modified world | 240 |

|  |  |  |
|---|---|---|
| 13.3 | Selectively logged forests | 242 |
| 13.4 | Agroforestry systems | 243 |
| 13.5 | Tree plantations | 245 |
|  | Box 13.2: Quantifying the biodiversity value of tropical secondary forests and exotic tree plantations (*Jos Barlow*) | 247 |
| 13.6 | Agricultural land | 248 |
|  | Box 13.3: Conservation in the face of oil palm expansion (*Matthew Struebig, Ben Phalan, and Emily Fitzherbert*) | 249 |
|  | Box 13.4: Countryside biogeography: harmonizing biodiversity and agriculture (*Jai Ranganathan and Gretchen C. Daily*) | 251 |
| 13.7 | Urban areas | 253 |
| 13.8 | Regenerating forests on degraded land | 254 |
| 13.9 | Conservation and human livelihoods in modified landscapes | 255 |
| 13.10 | Conclusion | 256 |
| Summary | | 257 |
| Suggested reading | | 257 |
| Relevant websites | | 258 |

## 14: The roles of people in conservation
### *C. Anne Claus, Kai M. A. Chan, and Terre Satterfield* — 262

| | | |
|---|---|---|
| 14.1 | A brief history of humanity's influence on ecosystems | 262 |
| 14.2 | A brief history of conservation | 262 |
|  | Box 14.1: Customary management and marine conservation (*C. Anne Claus, Kai M. A. Chan, and Terre Satterfield*) | 264 |
|  | Box 14.2: Historical ecology and conservation effectiveness in West Africa (*C. Anne Claus, Kai M. A. Chan, and Terre Satterfield*) | 265 |
| 14.3 | Common conservation perceptions | 265 |
|  | Box 14.3: Elephants, animal rights, and *Campfire* (*Paul R. Ehrlich*) | 267 |
| 14.4 | Factors mediating human-environment relations | 269 |
|  | Box 14.4: Conservation, biology, and religion (*Kyle S. Van Houtan*) | 270 |
| 14.5 | Biodiversity conservation and local resource use | 273 |
| 14.6 | Equity, resource rights, and conservation | 275 |
|  | Box 14.5: Empowering women: the Chipko movement in India (*Priya Davidar*) | 276 |
| 14.7 | Social research and conservation | 278 |
| Summary | | 281 |
| Relevant websites | | 281 |
| Suggested reading | | 281 |

## 15: From conservation theory to practice: crossing the divide
### *Madhu Rao and Joshua Ginsberg* — 284

|  |  |  |
|---|---|---|
|  | Box 15.1: Swords into Ploughshares: reducing military demand for wildlife products (*Lisa Hickey, Heidi Kretser, Elizabeth Bennett, and McKenzie Johnson*) | 285 |
|  | Box 15.2: The World Bank and biodiversity conservation (*Tony Whitten*) | 286 |
|  | Box 15.3: The Natural Capital Project (*Heather Tallis, Joshua H. Goldstein, and Gretchen C. Daily*) | 288 |
| 15.1 | Integration of Science and Conservation Implementation | 290 |
|  | Box 15.4: Measuring the effectiveness of conservation spending (*Matthew Linkie and Robert J. Smith*) | 291 |

15.2  Looking beyond protected areas ... 292
    Box 15.5:  From managing protected areas to conserving landscapes (*Karl Didier*) ... 293
15.3  Biodiversity and human poverty ... 293
    Box 15.6:  Bird nest protection in the Northern Plains of Cambodia (*Tom Clements*) ... 297
    Box 15.7:  International activities of the Missouri Botanical Garden (*Peter Raven*) ... 301
15.4  Capacity needs for practical conservation in developing countries ... 303
15.5  Beyond the science: reaching out for conservation ... 304
15.6  People making a difference: A Rare approach ... 305
15.7  Pride in the La Amistad Biosphere Reserve, Panama ... 305
15.8  Outreach for policy ... 306
15.9  Monitoring of Biodiversity at Local and Global Scales ... 306
    Box 15.8:  Hunter self-monitoring by the Isoseño-Guaraní in the Bolivian Chaco (*Andrew Noss*) ... 307
Summary ... 310
Suggested reading ... 310
Relevant websites ... 310

## 16: The conservation biologist's toolbox – principles for the design and analysis of conservation studies
*Corey J. A. Bradshaw and Barry W. Brook* ... **313**

16.1  Measuring and comparing 'biodiversity' ... 314
    Box 16.1:  Cost effectiveness of biodiversity monitoring (*Toby Gardner*) ... 314
    Box 16.2:  Working across cultures (*David Bickford*) ... 316
16.2  Mensurative and manipulative experimental design ... 319
    Box 16.3:  Multiple working hypotheses (*Corey J. A. Bradshaw and Barry W. Brook*) ... 321
    Box 16.4:  Bayesian inference (*Corey J. A. Bradshaw and Barry W. Brook*) ... 324
16.3  Abundance Time Series ... 326
16.4  Predicting Risk ... 328
16.5  Genetic Principles and Tools ... 330
    Box 16.5:  Functional genetics and genomics (*Noah K. Whiteman*) ... 331
16.6  Concluding Remarks ... 333
    Box 16.6:  Useful textbook guides (*Corey J. A. Bradshaw and Barry W. Brook*) ... 334
Summary ... 335
Suggested reading ... 335
Relevant websites ... 335
Acknowledgements ... 336

**Index** ... **341**

# Dedication

NSS: To those who have or want to make the difference.
PRE: To my mentors—Charles Birch, Charles Michener, and Robert Sokal.

# Acknowledgements

NSS thanks the Sarah and Daniel Hrdy Fellowship in Conservation Biology (Harvard University) and the National University of Singapore for support while this book was prepared. He also thanks Naomi Pierce for providing him with an office. PRE thanks Peter and Helen Bing, Larry Condon, Wren Wirth, and the Mertz Gilmore Foundation for their support. We thank Mary Rose C. Posa, Pei Xin, Ross McFarland, Hugh Tan, and Peter Ng for their invaluable assistance. We also thank Ian Sherman, Helen Eaton, and Carol Bestley at Oxford University Press for their help/support.

# List of Contributors

**Tom Allnutt**
*Department of Environmental Sciences, Policy and Management, 137 Mulford Hall, University of California, Berkeley, CA 94720-3114, USA.*

**Murat Ataol**
*Doğa Derneği, Hürriyet Cad. 43/12 Dikmen, Ankara, Turkey.*

**Özge Balkız**
*Doğa Derneği, Hürriyet Cad. 43/12 Dikmen, Ankara, Turkey.*

**Jos Barlow**
*Lancaster Environment Centre, Lancaster University, Lancaster, LA1 4YQ, UK.*

**Andrew F. Bennett**
*School of Life and Environmental Sciences, Deakin University, 221 Burwod Highway, Burwood, VIC 3125, Australia.*

**Elizabeth Bennett**
*Wildlife Conservation Society, 2300 Southern Boulevard., Bronx, NY 10464-1099, USA.*

**Fikret Berkes**
*Natural Resources Institute, 70 Dysart Road, University of Manitoba, Winnipeg MB R3T 2N2, Canada.*

**David Bickford**
*Department of Biological Sciences, National University of Singapore, 14 Science Drive 4, Singapore 117543, Republic of Singapore.*

**David M. J. S. Bowman**
*School of Plant Science, University of Tasmania, Private Bag 55, Hobart, TAS 7001, Australia.*

**Mark S. Boyce**
*Department of Biological Sciences, University of Alberta, Edmonton, Alberta T6G 2E9, Canada.*

**Murat Bozdoğan**
*Doğa Derneği, Hürriyet Cad. 43/12 Dikmen, Ankara, Turkey.*

**Corey J. A. Bradshaw**
*Environmental Institute, School of Earth and Environmental Sciences, University of Adelaide, South Australia 5005 AND South Australian Research and Development Institute, P.O. Box 120, Henley Beach, South Australia 5022, Australia.*

**Barry W. Brook**
*Environment Institute, School of Earth and Environmental Sciences, University of Adelaide, South Australia 5005, Australia.*

**Thomas Brooks**
*Center for Applied Biodiversity Science, Conservation International, 2011 Crystal Drive Suite 500, Arlington VA 22202, USA; World Agroforestry Center (ICRAF), University of the Philippines Los Baños, Laguna 4031, Philippines; AND School of Geography and Environmental Studies, University of Tasmania, Hobart TAS 7001, Australia.*

**Alison Cameron**
*Max Planck Institute for Ornithology, Eberhard-Gwinner-Straße, 82319 Seewiesen, Germany.*

**Kai M. A. Chan**
*Institute for Resources, Environment and Sustainability, University of British Columbia, Vancouver, British Columbia V6T 1Z4, Canada.*

**C. Anne Claus**
*Departments of Anthropology and Forestry & Environmental Studies, Yale University,10 Sachem Street, New Haven, CT 06511, USA.*

**Tom Clements**
*Wildlife Conservation Society, Phnom Penh, Cambodia.*

**Gretchen C. Daily**
Center for Conservation Biology, Department of Biology, and Woods Institute, 371 Serra Mall, Stanford University, Stanford, CA 94305-5020, USA.

**Priya Davidar**
School of Life Sciences, Pondicherry University, Kalapet, Pondicherry 605014, India.

**Karl Didier**
Wildlife Conservation Society, 2300 Southern Boulevard, Bronx, NY 10464-1099, USA.

**Paul R. Ehrlich**
Center for Conservation Biology, Department of Biology, Stanford University, Stanford, CA 94305-5020, USA.

**Güven Eken**
Doğa Derneği, Hürriyet Cad. 43/12 Dikmen, Ankara, Turkey.

**Emily Fitzherbert**
Institute of Zoology, Zoological Society of London, Regent's Park, London, NW1 4RY, UK.

**Toby A. Gardner**
Department of Zoology, University of Cambridge, Downing Street, Cambridge, CB2 3EJ, UK AND Departamento de Biologia, Universidade Federal de Lavras, Lavras, Minas Gerais, 37200-000, Brazil.

**Kevin J. Gaston**
Department of Animal & Plant Sciences, University of Sheffield, Sheffield, S10 2TN, UK.

**Joshua Ginsberg**
Wildlife Conservation Society, 2300 Southern Boulevard, Bronx, NY 10464-1099, USA.

**Joshua H. Goldstein**
Human Dimensions of Natural Resources, Warner College of Natural Resources, Colorado State University, Fort Collins, CO 80523-1480, USA.

**Benjamin S. Halpern**
National Center for Ecological Analysis and Synthesis, 735 State Street, Santa Barbara, CA 93101, USA.

**Lisa Hickey**
Wildlife Conservation Society, 2300 Southern Boulevard, Bronx, NY 10464-1099, USA.

**Süreyya İsfendiyaroğlu**
Doğa Derneği, Hürriyet Cad. 43/12 Dikmen, Ankara, Turkey.

**Clinton N. Jenkins**
Nicholas School of the Environment, Duke University, Box 90328, LSRC A201, Durham, NC 27708, USA.

**McKenzie Johnson**
Wildlife Conservation Society, 2300 Southern Boulevard, Bronx, NY 10464-1099, USA.

**Carrie V. Kappel**
National Center for Ecological Analysis and Synthesis, 735 State Street, Santa Barbara, CA 93101, USA.

**Dicle Tuba Kılıç**
Doğa Derneği, Hürriyet Cad. 43/12 Dikmen, Ankara, Turkey.

**Lian Pin Koh**
Institute of Terrestrial Ecosystems, Swiss Federal Institute of Technology (ETH Zürich), CHN G 74.2, Universitätstrasse 16, Zurich 8092, Switzerland.

**Claire Kremen**
Department of Environmental Sciences, Policy and Management, 137 Mulford Hall, University of California, Berkeley, CA 94720-3114, USA.

**Heidi Kretser**
Wildlife Conservation Society, 2300 Southern Boulevard, Bronx, NY 10464-1099, USA.

**William F. Laurance**
Smithsonian Tropical Research Institute, Apartado 0843-03092, Balboa, Ancón, Republic of Panama.

**Matthew Linkie**
Fauna & Flora International, 4th Floor, Jupiter House, Station Road, Cambridge, CB1 2JD, UK.

**Yıldıray Lise**
Doğa Derneği, Hürriyet Cad. 43/12 Dikmen, Ankara, Turkey.

**Thomas E. Lovejoy**
The H. John Heinz III Center for Science, Economics and the Environment, 900 17th Street NW, Suite 700, Washington, DC 20006, USA.

**Jennifer B. H. Martiny**
Department of Ecology and Evolutionary Biology, University of California, Irvine, CA 92697, USA.

## LIST OF CONTRIBUTORS

**Curt Meine**
Aldo Leopold Foundation/International Crane Foundation, P.O. Box 38, Prairie du Sac, WI 53578, USA.

**Fiorenza Micheli**
Hopkins Marine Station, Stanford University, Pacific Grove, CA 93950, USA.

**Brett P. Murphy**
School of Plant Science, University of Tasmania, Private Bag 55, Hobart, TAS 7001, Australia.

**J. P. Myers**
Environmental Health Sciences, 421 E Park Street, Charlottesville VA 22902, USA.

**Andrew Noss**
Proyecto Gestión Integrada de Territorios Indigenas WCS-Ecuador, Av. Eloy Alfaro N37-224 y Coremo Apartado, Postal 17-21-168, Quito, Ecuador.

**Daniel Pauly**
Seas Around Us Project, University of British Columbia, Vancouver, British Columbia, V6T 1Z4, Canada.

**Carlos A. Peres**
School of Environmental Sciences, University of East Anglia, Norwich, NR4 7TJ, UK.

**Ben Phalan**
Conservation Science Group, Department of Zoology, University of Cambridge, Downing Street, Cambridge, CB2 3EJ, UK.

**Stuart L. Pimm**
Nicholas School of the Environment, Duke University, Box 90328, LSRC A201, Durham, NC 27708, USA.

**Mary Rose C. Posa**
Department of Biology, National University of Singapore, 14 Science Drive 4, Singapore 117543, Republic of Singapore.

**Robert M. Pringle**
Department of Biology, Stanford University, Stanford, CA 94305, USA.

**Jai Ranganathan**
National Center for Ecological Analysis and Synthesis, 735 State Street, Suite 300 Santa Barbara, CA 93109, USA.

**Madhu Rao**
Wildlife Conservation Society Asia Program 2300 S. Blvd., Bronx, New York, NY 10460, USA.

**Peter Raven**
Missouri Botanical Garden, Post Office Box 299, St. Louis, MO 63166-0299, USA.

**Andriamandimbisoa Razafimpahanana**
Réseau de la Biodiversité de Madagascar, Wildlife Conservation Society, Villa Ifanomezantsoa, Soavimbahoaka, Boîte Postale 8500, Antananarivo 101, Madagascar.

**Terre Satterfield**
Institute for Resources, Environment and Sustainability, University of British Columbia, Vancouver, British Columbia V6T 1Z4, Canada.

**Denis A. Saunders**
CSIRO Sustainable Ecosystems, GPO Box 284, Canberra, ACT 2601, Australia.

**Cagan H. Sekercioglu**
Center for Conservation Biology, Department of Biology, Stanford University, Stanford, CA 94305-5020, USA.

**Kimberly A. Selkoe**
National Center for Ecological Analysis and Synthesis, 735 State Street, Santa Barbara, CA 93101, USA.

**Daniel Simberloff**
Department of Ecology and Evolutionary Biology, University of Tennessee, Knoxville, TN 37996, USA.

**Robert J. Smith**
Durrell Institute of Conservation and Ecology, University of Kent, Canterbury, Kent, CT2 7NR, UK.

**Navjot S. Sodhi**
Department of Biological Sciences, National University of Singapore, 14 Science Drive 4, Singapore 117543, Republic of Singapore AND Department of Organismic and Evolutionary Biology, Harvard University, Cambridge, MA 02138, USA.

**Matthew Struebig**
School of Biological & Chemical Sciences, Queen Mary, University of London, Mile End Road, London, E1 4NS, UK.

**Heather Tallis**
*The Natural Capital Project, Woods Institute for the Environment, 371 Serra Mall, Stanford University, Stanford, CA 94305-5020, USA.*

**Teja Tscharntke**
*Agroecology, University of Göttingen, Germany.*

**Kyle S. Van Houtan**
*Department of Biology, O W Rollins Research Ctr, 1st Floor, 1510 Clifton Road, Lab# 1112 Emory University AND Center for Ethics, 1531 Dickey Drive, Emory University, Atlanta, GA 30322, USA.*

**Peter Vaughan**
*Rare, 1840 Wilson Boulevard, Suite 204, Arlington, VA 22201, USA.*

**Ian G. Warkentin**
*Environmental Science – Biology, Memorial University of Newfoundland, Corner Brook, Newfoundland and Labrador A2H 6P9, Canada.*

**Noah K. Whiteman**
*Department of Organismic and Evolutionary Biology, Harvard University, Cambridge, MA 02138, USA.*

**Tony Whitten**
*The World Bank, Washington, DC, USA.*

**David Wilcove**
*Department of Ecology and Evolutionary Biology, Princeton University, Princeton, NJ 08544-1003, USA.*

**Douglas W. Yu**
*School of Biological Sciences, University of East Anglia, Norwich, NR4 7TJ, UK.*

# Foreword

2010 was named by the United Nations to be the International Year of Biodiversity, coinciding with major political events that set the stage for a radical review of the way we treat our environment and its biological riches. So far, the reports have been dominated by reconfirmations that people and their lifestyles continue to deplete the earth's biodiversity. We are still vastly overspending our natural capital and thereby depriving future generations. If that were not bad enough news in itself, there are no signs that actions to date have slowed the rate of depletion. In fact, it continues to increase, due largely to growing levels of consumption that provide increasingly unequal benefits to different groups of people.

It is easy to continue to delve into the patterns and processes that lie at the heart of the problem. But it is critical that we also start to do everything we can to reverse all the damaging trends. These actions cannot and should not be just the responsibility of governments and their agencies. It must be the responsibility of all of us, including scientists, wildlife managers, naturalists, and indeed everyone who cares that future generations can have the same choices, and the same opportunity to marvel at and benefit from nature, as our generation has had. We all can be involved in actions to improve matters and making conservation biology relevant to and applicable by all is therefore a key task.

It is in this context that Navjot Sodhi and Paul Ehrlich have contributed this important book. Covering all aspects of conservation biology from the deleterious drivers, through to the impacts on people, and providing tools, techniques and background to practical solutions, the book provides a resource for many different people and contexts. Written by the world's leading experts you will find clear summaries of the latest literature on how to decide what to do, and then how to do it. Presented in clear and accessible text, this book will support the work of many people. There are different kinds of conservation actions, at different scales and affecting different part of the biosphere, all laid out clearly and concisely.

There is something in here for everyone who is, or wishes to be, a conservation biologist. I am sure you will all be inspired and better informed to do something that will improve the prospects for all, so that in a decade or so when the world community next examines the biodiversity accounts, things will definitely be taking a turn for the better!

Georgina Mace CBE FRS
Imperial College London
18 November 2010

# Introduction

Navjot S. Sodhi and Paul R. Ehrlich

Our actions have put humanity into a deep environmental crisis. We have destroyed, degraded, and polluted Earth's natural habitats – indeed, virtually all of them have felt the influence of the dominant species. As a result, the vast majority of populations and species of plants and animals – key working parts of human life support systems – are in decline, and many are already extinct. Increasing human population size and consumption per person (see Introduction Box 1) have precipitated an extinction crisis – the "sixth mass extinction", which is comparable to past extinction events such as the Cretaceous-Tertiary mass extinction 65 million years ago that wiped out all the dinosaurs except for the birds. Unlike the previous extinction events, which were attributed to natural catastrophes including volcanic eruptions, meteorite impact and global cooling, the current mass extinction is exclusively humanity's fault. Estimates indicate that numerous species and populations are currently likely being extinguished every year. But all is not lost – yet.

Being the dominant species on Earth, humans have a moral obligation (see Introduction Box 2) to ensure the long-term persistence of rainforests, coral reefs, and tidepools as well as saguaro cacti, baobab trees, tigers, rhinos, pandas, birds of paradise, morpho butterflies, and a plethora of other creatures. All these landmarks and life make this planet remarkable – our imagination will be bankrupt if wild nature is obliterated – even if civilization could survive the disaster. In addition to moral and aesthetic reasons, we have a selfish reason to preserve nature – it provides society with countless and invaluable goods and absolutely crucial services (e.g. food, medicines, pollination, pest control, and flood protection).

Habitat loss and pollution are particularly acute in developing countries, which are of special concern because these harbor the greatest species diversity and are the richest centers of endemism. Sadly, developing world conservation scientists have found it difficult to afford an authoritative textbook of conservation biology, which is particularly ironic, since it is these countries where the rates of habitat loss are highest and the potential benefits of superior information in the hands of scientists and managers are therefore greatest. There is also now a pressing need to educate the next generation of conservation biologists in developing countries, so that hopefully they are in a better position to protect their natural resources. With this book, we intend to provide cutting-edge but basic conservation science to developing as well as developed country inhabitants. The contents of this book will be freely available on the web twelve months following book publication.

Since our main aim is to make up-to-date conservation knowledge widely available, we have invited many of the top names in conservation biology to write on specific topics. Overall, this book represents a project that the conservation community has deemed worthy of support by donations of time and effort. None of the authors, including ourselves, will gain financially from this project.

It is our hope that this book will be of relevance and use to both undergraduate and graduate students as well as scientists, managers, and personnel in non-governmental organizations. The book should have all the necessary topics to become a required reading for various undergraduate and graduate conservation-related courses. English is

## Introduction Box 1  Human population and conservation
### Paul R. Ehrlich

The size of the human population is approaching 7 billion people, and its most fundamental connection with conservation is simple: people compete with other animals, which unlike green plants cannot make their own food. At present *Homo sapiens* uses, coopts, or destroys close to half of all the food available to the rest of the animal kingdom (see Introduction Box 1 Figure). That means that, in essence, every human being added to the population means fewer individuals can be supported in the remaining fauna.

But human population growth does much more than simply cause a proportional decline in animal biodiversity – since as you know, we degrade nature in many ways besides competing with animals for food. Each additional person will have a disproportionate negative impact on biodiversity in general. The first farmers started farming the richest soils they could find and utilized the richest and most accessible resources first (Ehrlich and Ehrlich 2005). Now much of the soil that people first farmed has been eroded away or paved over, and agriculturalists increasingly are forced to turn to marginal land to grow more food. Equally, deeper and poorer ore deposits must be mined and smelted today, water and petroleum must come from lower quality sources, deeper wells, or (for oil) from deep beneath the ocean and must be transported over longer distances, all at ever-greater environmental cost.

The tasks of conservation biologists are made more difficult by human population growth, as is readily seen in the I=PAT equation (Holdren and Ehrlich 1974; Ehrlich and Ehrlich 1981). Impact (I) on biodiversity is not only a result of population size (P), but of that size multiplied by affluence (A) measured as per capita consumption, and that product multiplied by another factor (T), which summarizes the technologies and socio-political-economic arrangements to service that consumption. More people surrounding a rainforest reserve in a poor nation often means more individuals invading the reserve to gather firewood or bush meat. More people in a rich country may mean more off-road vehicles (ORVs) assaulting the biota – especially if the ORV manufacturers are politically powerful and can successfully fight bans on their use. As poor countries' populations grow and segments of them become more affluent, demand rises for meat and automobiles, with domesticated animals

**Introduction Box 1 Figure** Human beings consuming resources. Photograph by Mary Rose Posa.

*continues*

### Introduction Box 1 (Continued)

competing with or devouring native biota, cars causing all sorts of assaults on biodiversity, and both adding to climate disruption. Globally, as a growing population demands greater quantities of plastics, industrial chemicals, pesticides, fertilizers, cosmetics, and medicines, the toxification of the planet escalates, bringing frightening problems for organisms ranging from polar bears to frogs (to say nothing of people!) (see Box 13.1).

In sum, population growth (along with escalating consumption and the use of environmentally malign technologies) is a major driver of the ongoing destruction of populations, species, and communities that is a salient feature of the Anthropocene (Anonymous 2008). Humanity, as the dominant animal (Ehrlich and Ehrlich 2008), simply out competes other animals for the planet's productivity, and often both plants and animals for its freshwater. While dealing with more limited problems, it therefore behooves every conservation biologist to put part of her time into restraining those drivers, including working to humanely lower birth rates until population growth stops and begins a slow decline toward a sustainable size (Daily et al. 1994).

### REFERENCES

Anonymous. (2008). Welcome to the Anthropocene. *Chemical and Engineering News*, **86**, 3.

Daily, G. C. and Ehrlich, A. H. (1994). Optimum human population size. *Population and Environment*, **15**, 469–475.

Ehrlich, P. R. and Ehrlich, A. H. (1981). *Extinction: the causes and consequences of the disappearance of species*. Random House, New York, NY.

Ehrlich, P. R. and Ehrlich, A. H. (2005). *One with Nineveh: politics, consumption, and the human future*, (with new afterword). Island Press, Washington, DC.

Ehrlich, P. R. and Ehrlich, A. H. (2008). *The Dominant Animal: human evolution and the environment*. Island Press, Washington, DC.

Holdren J. P. and Ehrlich, P. R. (1974). Human population and the global environment. *American Scientist*, **62**, 282–292.

### Introduction Box 2   Ecoethics
### Paul R. Ehrlich

The land ethic simply enlarges the boundaries of the community to include soils, waters, plants, and animals, or collectively: the land.... Aldo Leopold (1949)

As you read this book, you should keep in mind that the problem of conserving biodiversity is replete with issues of practical ethics – agreed-upon notions of the right or wrong of actual behaviors (Singer 1993; Jamieson 2008). If civilization is to maintain the ecosystem services (Chapter 3) that can support a sustainable society and provide virtually everyone with a reasonable quality of life, humanity will need to focus much more on issues with a significant conservation connection, "ecoethics".

Ultimately everything must be examined from common "small-scale" personal ecoethical decisions to the ethics of power wielded by large-scale institutions that try (and sometimes succeed) to control broad aspects of our global civilization. Those institutions include governments, religions, transnational corporations, and the like. To ignore these power relations is, in essence, to ignore the most important large-scale issues, such as conservation in the face of further human population growth and of rapid climate change – issues that demand global ethical discussion.

Small-scale ecoethical dilemmas are commonly faced by conservation biologists. Should we eat shrimp in a restaurant when we can't determine its provenance? Should we become more vegetarian? Is it legitimate to fly around the world in jet aircraft to try and persuade people to change a lifestyle that includes flying around the world in jet aircraft? How should we think about all the trees cut

*continues*

## Introduction Box 2 (Continued)

down to produce the books and articles we've written? These sorts of decisions are poignantly discussed by Bearzi (2009), who calls for conservation biologists to think more carefully about their individual decisions and set a better example where possible. Some personal decisions are not so minor – such as how many children to have. But ironically Bearzi does not discuss child-bearing decisions, even though especially in rich countries these are often the most conservation-significant ethical decisions an individual makes.

Ecotourism is a hotbed of difficult ethical issues, some incredibly complex, as shown in Box 14.3. But perhaps the most vexing ethical questions in conservation concern conflicts between the needs and prerogatives of peoples and non-human organisms. This is seen in issues like protecting reserves from people, where in the extreme some conservation biologists plead for strict exclusion of human beings (e.g. Terborgh 2004), and by the debates over the preservation of endangered organisms and traditional rights to hunt them. The latter is exemplified by complex aboriginal "subsistence" whaling issues (Reeves 2002). While commercial whaling is largely responsible for the collapse of many stocks, aboriginal whaling may threaten some of the remnants. Does one then side with the whales or the people, to whom the hunts may be an important part of their tradition? Preserving the stocks by limiting aboriginal takes seems the ecoethical thing to do, since it allows for traditional hunting to persist, which will not happen if the whales go extinct. Tradition is a tricky thing – coal mining or land development may be family traditions, but ecoethically those occupations should end.

Perhaps most daunting of all is the task of getting broad agreement from diverse cultures on ecoethical issues. It has been suggested that a world-wide *Millennium Assessment of Human Behavior* (MAHB) be established to, among other things, facilitate discussion and debate (Ehrlich and Kennedy 2005). My own views of the basic ecoethical paths that should be pursued follow. Others may differ, but if we don't start debating ecoethics now, the current ethical stasis will likely persist.

- Work hard to humanely bring human population growth to a halt and start a slow decline.
- Reduce overconsumption by the already rich while increasing consumption by the needy poor, while striving to limit aggregate consumption by humanity.
- Start a global World War II type mobilization to shift to more benign energy technologies and thus reduce the chances of a world-wide conservation disaster caused by rapid climate change.
- Judge technologies not just on what they do *for* people but also *to* people and the organisms that are key parts of their life-support systems.
- Educate students, starting in kindergarten, about the crucial need to preserve biodiversity and expand peoples' empathy not just to all human beings but also to the living elements in the natural world.

Most conservation biologists view the task of preserving biodiversity as fundamentally one of ethics (Ehrlich and Ehrlich 1981). Nonetheless, long experience has shown that arguments based on a proposed ethical need to preserve our only known living relatives in the entire universe, the products of incredible evolutionary sequences billions of years in extent, have largely fallen on deaf ears. Most ecologists have therefore switched to admittedly risky instrumental arguments for conservation (Daily 1997). What proportion of conservation effort should be put into promoting instrumental approaches that might backfire or be effective in only the short or middle term is an ethical-tactical issue. One of the best arguments for emphasizing the instrumental is that they can at least buy time for the necessarily slow cultural evolutionary process of changing the norms that favor attention to reproducible capital and property rights to the near exclusion of natural capital. Some day Aldo Leopold's "Land Ethic" may become universal – until then conservation biologists will face many ethical challenges.

### REFERENCES

Bearzi, G. (2009). When swordfish conservation biologists eat swordfish. *Conservation Biology*, 23, 1–2.

Daily, G. C., ed. (1997). *Nature's services: societal dependence on natural ecosystems*. Island Press, Washington, DC.

*continues*

## Introduction Box 2 (Continued)

Ehrlich, P. R. and Ehrlich, A. H. (1981). Extinction: the causes and consequences of the disappearance of species. Random House, New York, NY.

Ehrlich, P. R. and Kennedy, D. (2005). Millennium assessment of human behavior: a challenge to scientists. *Science*, **309**, 562–563.

Jamieson, D. (2008). *Ethics and the environment: an introduction*. Cambridge University Press, Cambridge, UK.

Leopold, A. (1949). *Sand county almanac*. Oxford University Press, New York, NY.

Reeves, R. R. (2002). The origins and character of 'aboriginal subsistence' whaling: a global review. *Mammal Review*, **32**, 71–106.

Singer, P. (1993). Practical ethics. 2nd edn. University Press, Cambridge, UK.

Terborgh, J. (2004). *Requiem for nature*. Island Press, Washington, DC.

---

kept at a level comprehensible to readers for whom English is a second language.

The book contains 16 chapters, which are briefly introduced below:

### Chapter 1. Conservation biology: past and present

In this chapter, Curt Meine introduces the discipline by tracing its history. He also highlights the interdisciplinary nature of conservation science.

### Chapter 2. Biodiversity

Kevin J. Gaston defines biodiversity and lays out the obstacles to its better understanding in this chapter.

### Chapter 3. Ecosystem functioning and services

In this chapter, Cagan H. Sekercioglu recapitulates natural ecosystem functions and services.

### Chapter 4. Habitat destruction: death by a thousand cuts

William F. Laurance provides an overview of contemporary habitat loss in this chapter. He evaluates patterns of habitat destruction geographically and contrasts it in different biomes and ecosystems. He also reviews some of the ultimate and proximate factors causing habitat loss.

### Chapter 5. Habitat fragmentation and landscape change

Conceptual approaches used to understand conservation in fragmented landscapes are summarized in this chapter by Andrew F. Bennett and Denis A. Saunders. They also examine biophysical aspects of landscape change, and how such change affects populations, species, and communities.

### Chapter 6. Overharvesting

Biodiversity is under heavy threat from anthropogenic overexploitation (e.g. harvest for food or decoration or of live animals for the pet trade). For example, bushmeat or wild meat hunting is imperiling many tropical species as expanding human populations in these regions seek new sources of protein and create potentially profitable new avenues for trade at both local and international levels. In this Chapter, Carlos A. Peres highlights the effects of human exploitation of terrestrial and aquatic biomes on biodiversity.

### Chapter 7. Invasive species

Daniel Simberloff presents an overview of invasive species, their impacts and management in this chapter.

### Chapter 8. Climate change

Climate change is quickly emerging as a key issue in the battle to preserve biodiversity. In this chapter, Thomas E. Lovejoy reports on the documented impacts of climate change on biotas.

### Chapter 9. Fire and biodiversity

Evolutionary and ecological principles related to conservation in landscapes subject to regular fires are presented in this chapter by David M. J. S. Bowman and Brett P. Murphy.

### Chapter 10. Extinctions and the practice of preventing them

Stuart L. Pimm and Clinton N. Jenkins explore why extinctions are the critical issue for conservation science. They also list a number of conservation options.

### Chapter 11. Conservation planning and priorities

In this chapter, Thomas Brooks charts the history, state, and prospects of conservation planning and prioritization in terrestrial and aquatic habitats. He focuses on successful conservation implementation planned through the discipline's conceptual framework of vulnerability and irreplaceability.

### Chapter 12. Endangered species management: the US experience

In this chapter, David S. Wilcove focuses on endangered species management, emphasizing the United States of America (US) experience. Because the US has one of the oldest and possibly strongest laws to protect endangered species, it provides an illuminating case history.

### Chapter 13. Conservation in human-modified landscapes

Lian Pin Koh and Toby A. Gardner discuss the challenges of conserving biodiversity in degraded and modified landscapes with a focus on the tropical terrestrial biome in this chapter. They highlight the extent to which human activities have modified natural ecosystems and outline opportunities for conserving biodiversity in human-modified landscapes.

### Chapter 14. The roles of people in conservation

The effective and sustainable protection of biodiversity will require that the sustenance needs of native people are adequately considered. In this chapter, C. Anne Claus, Kai M. A. Chan, and Terre Satterfield highlight that understanding human activities and human roles in conservation is fundamental to effective conservation.

### Chapter 15. From conservation theory to practice: crossing the divide

Madhu Rao and Joshua Ginsberg explore the implementation of conservation science in this chapter.

### Chapter 16. The conservation biologist's toolbox – principles for the design and analysis of conservation studies

In this chapter, Corey J. A. Bradshaw and Barry W. Brook, discuss measures of biodiversity patterns followed by an overview of experimental design and associated statistical paradigms. They also present the analysis of abundance time series, assessments of species' endangerment, and a brief introduction to genetic tools to assess the conservation status of species.

Each chapter includes boxes written by various experts describing additional relevant material, case studies/success stories, or personal perspectives.

# CHAPTER 1

## Conservation biology: past and present[1]

### Curt Meine

Our job is to harmonize the increasing kit of scientific tools and the increasing recklessness in using them with the shrinking biotas to which they are applied. In the nature of things we are mediators and moderators, and unless we can help rewrite the objectives of science we are pre-destined to failure.

—Aldo Leopold (1940; 1991)

Conservation in the old sense, of this or that resource in isolation from all other resources, is not enough. Environmental conservation based on ecological knowledge and social understanding is required.

—Raymond Dasmann (1959)

Conservation biology is a mission-driven discipline comprising both pure and applied science. ...We feel that conservation biology is a new field, or at least a new rallying point for biologists wishing to pool their knowledge and techniques to solve problems.

—Michael E. Soulé and Bruce A. Wilcox (1980)

Conservation biology, though rooted in older scientific, professional, and philosophical traditions, gained its contemporary definition only in the mid-1980s. Anyone seeking to understand the history and growth of conservation biology thus faces inherent challenges. The field has formed too recently to be viewed with historical detachment, and the trends shaping it are still too fluid to be easily traced. Conservation biology's practitioners remain embedded within a process of change that has challenged conservation "in the old sense," even while extending conservation's core commitment to the future of life, human and non-human, on Earth.

There is as yet no comprehensive history of conservation that allows us to understand the causes and context of conservation biology's emergence. Environmental ethicists and historians have provided essential studies of particular conservation ideas, disciplines, institutions, individuals, ecosystems, landscapes, and resources. Yet we still lack a broad, fully integrated account of the dynamic coevolution of conservation science, philosophy, policy, and practice (Meine 2004). The rise of conservation biology marked a new "rallying point" at the intersection of these domains; exactly how, when, and why it did so are still questions awaiting exploration.

## 1.1 Historical foundations of conservation biology

Since conservation biology's emergence, commentary on (and in) the field has rightly emphasized its departure from prior conservation science and practice. However, the main "thread" of the field—the description, explanation, appreciation, protection, and perpetuation of *biological diversity* can be traced much further back through the historical tapestry of the biological sciences and the conservation movement (Mayr 1982;

---

[1] Adapted from Meine, C., Soulé, M., and Noss, R. F. (2006). "A mission-driven discipline": the growth of conservation biology. *Conservation Biology*, **20**, 631–651.

McIntosh 1985; Grumbine 1996; Quammen 1996). That thread weaves through related themes and concepts in conservation, including wilderness protection, sustained yield, wildlife protection and management, the diversity-stability hypothesis, ecological restoration, sustainability, and ecosystem health. By focusing on the thread itself, conservation biology brought the theme of biological diversity to the fore.

In so doing, conservation biology has reconnected conservation to deep sources in Western natural history and science, and to cultural traditions of respect for the natural world both within and beyond the Western experience (see Box 1.1 and Chapter 14). Long before environmentalism began to reshape "conservation in the old sense" in the 1960s—prior even to the Progressive Era conservation movement of the early 1900s—the foundations of conservation biology were being laid over the course of biology's epic advances over the last four centuries. The "discovery of diversity" (to use Ernst Mayr's phrase) was the driving force behind the growth of biological thought. "Hardly any aspect of life is more

---

**Box 1.1 Traditional ecological knowledge and biodiversity conservation**
**Fikret Berkes**

Conservation biology is a discipline of Western science, but there are other traditions of conservation in various parts of the world (see also Chapter 14). These traditions are based on local and indigenous knowledge and practice. Traditional ecological knowledge may be defined as a cumulative body of knowledge, practice and belief, evolving by adaptive processes and handed down through generations by cultural transmission. It is experiential knowledge closely related to a way of life, multi-generational, based on oral transmission rather than book learning, and hence different from science in a number of ways.

Traditional knowledge does not always result in conservation, just as science does not always result in conservation. But there are a number of ways in which traditional knowledge and practice may lead to conservation outcomes. First, sacred groves and other sacred areas are protected through religious practice and enforced by social rules. UNESCO's (the United Nations Educational, Scientific and Cultural Organization) World Heritage Sites network includes many sacred sites, such as Machu Picchu in Peru. Second, many national parks have been established at the sites of former sacred areas, and are based on the legacy of traditional conservation. Alto Fragua Indiwasi National Park in Colombia and Kaz Daglari National Park in Turkey are examples. Third, new protected areas are being established at the request of indigenous peoples as a safeguard against development. One example is the Paakumshumwaau Biodiversity Reserve in James Bay, Quebec, Canada (see Box 1.1 Figure). In the Peruvian Andes, the centre of origin of the potato, the Quetchua people maintain a mosaic of agricultural and natural areas as a biocultural heritage site with some 1200 potato varieties, both cultivated and wild.

**Box 1.1 Figure** Paakumshumwaau Biodiversity Reserve in James Bay, Quebec, Canada, established at the request of the Cree Nation of Wemindji. Photograph by F. Berkes.

In some cases, high biodiversity is explainable in terms of traditional livelihood practices that maintain a diversity of varieties, species and landscapes. For example, Oaxaca State in Mexico exhibits high species richness despite the absence of official protected areas. This may be attributed to the diversity of local and indigenous practices resulting in multi-functional cultural landscapes. In many parts of the world, agroforestry systems that rely on the cultivation of a diversity of crops and trees together (as opposed to modern

*continues*

### Box 1.1 (Continued)

monocultures), seem to harbor high species richness. There are at least three mechanisms that help conserve biodiversity in the use of agroforestry and other traditional practices:

- Land use regimes that maintain forest patches at different successional stages conserve biodiversity because each stage represents a unique community. At the same time, such land use contributes to continued ecosystem renewal.
- The creation of patches, gaps and mosaics enhance biodiversity in a given area. In the study of landscape ecology, the principle is that low and intermediate levels of disturbance often increase biodiversity, as compared to non-disturbed areas.
- Boundaries between ecological zones are characterized by high diversity, and the creation of new edges (ecotones) by disturbance enhances biodiversity, but mostly of "edge-loving" species. Overlaps and mixing of plant and animal species produce dynamic landscapes.

The objective of formal protected areas is biodiversity conservation, whereas traditional conservation is often practiced for livelihood and cultural reasons. Making biodiversity conservation relevant to most of the world requires bridging this gap, with an emphasis on sustainability, equity and a diversity of approaches. There is international interest in community-conserved areas as a class of protected areas. Attention to time-tested practices of traditional conservation can help develop a pluralistic, more inclusive definition of conservation, and build more robust constituencies for conservation.

### SUGGESTED READING

Berkes, F. (2008). *Sacred ecology*, 2nd edn. Routledge, New York, NY.

---

characteristic than its almost unlimited diversity," wrote Mayr (1982:133). "Indeed, there is hardly any biological process or phenomenon where diversity is not involved."

This "discovery" unfolded as colonialism, the Industrial Revolution, human population growth, expansion of capitalist and collectivist economies, and developing trade networks transformed human social, economic, political, and ecological relationships ever more quickly and profoundly (e.g. Crosby 1986; Grove 1995; Diamond 1997). Technological change accelerated humanity's capacity to reshape the world to meet human needs and desires. In so doing, it amplified tensions along basic philosophical fault lines: mechanistic/organic; utilitarian/reverential; imperialist/arcadian; reductionism/holism (Thomas *et al.* 1956; Worster 1985). As recognition of human environmental impacts grew, an array of 19th century philosophers, scientists, naturalists, theologians, artists, writers, and poets began to regard the natural world within an expanded sphere of moral concern (Nash 1989).

For example, Alfred Russel Wallace (1863) warned against the "extinction of the numerous forms of life which the progress of cultivation invariably entails" and urged his scientific colleagues to assume the responsibility for stewardship that came with knowledge of diversity.

The first edition of George Perkins Marsh's *Man and Nature* appeared the following year. In his second chapter, "Transfer, Modification, and Extirpation of Vegetable and of Animal Species," Marsh examined the effect of humans on biotic diversity. Marsh described human beings as a "new geographical force" and surveyed human impacts on "minute organisms," plants, insects, fish, "aquatic animals," reptiles, birds, and "quadrupeds." "All nature," he wrote, "is linked together by invisible bonds, and every organic creature, however low, however feeble, however dependent, is necessary to the well-being of some other among the myriad forms of life with which the Creator has peopled the earth." He concluded his chapter with the hope that people might "learn to put a wiser estimate on the works of creation" (Marsh 1864). Through the veil of 19th century language, modern conservation biologists may recognize Marsh, Wallace, and others as common intellectual ancestors.

Marsh's landmark volume appeared just as the post-Civil War era of rampant resource exploitation commenced in the United States. A generation later, Marsh's book undergirded the Progressive Era reforms that gave conservation in the United States its modern meaning and turned it into a national movement. That movement rode Theodore Roosevelt's presidency into public consciousness and across the American landscape. Conservationists in the Progressive Era were famously split along utilitarian-preservationist lines. The utilitarian Resource Conservation Ethic, realized within new federal conservation agencies, was committed to the efficient, scientifically informed management of natural resources, to provide "the greatest good to the greatest number for the longest time" (Pinchot 1910:48). By contrast, the Romantic-Transcendental Preservation Ethic, overshadowed but persistent through the Progressive Era, celebrated the aesthetic and spiritual value of contact with wild nature, and inspired campaigns for the protection of parklands, refuges, forests, and "wild life."

Callicott (1990) notes that both ethical camps were "essentially human-centered or 'anthropocentric'... (and) regarded human beings or human interests as the only legitimate ends and nonhuman natural entities and nature as a whole as means." Moreover, the science upon which both relied had not yet experienced its 20th century revolutions. Ecology had not yet united the scientific understanding of the abiotic, plant, and animal components of living systems. Evolutionary biology had not yet synthesized knowledge of genetics, population biology, and evolutionary biology. Geology, paleontology, and biogeography were just beginning to provide a coherent narrative of the temporal dynamics and spatial distribution of life on Earth. Although explicitly informed by the natural sciences, conservation in the Progressive Era was primarily economic in its orientation, reductionist in its tendencies, and selective in its application.

New concepts from ecology and evolutionary biology began to filter into conservation and the resource management disciplines during the early 20th century. "Proto-conservation biologists" from this period include Henry C. Cowles, whose pioneering studies of plant succession and the flora of the Indiana Dunes led him into active advocacy for their protection (Engel 1983); Victor Shelford, who prodded his fellow ecologists to become active in establishing biologically representative nature reserves (Croker 1991); Arthur Tansley, who similarly advocated establishment of nature reserves in Britain, and who in 1935 contributed the concept of the "ecosystem" to science (McIntosh 1985; Golley 1993); Charles Elton, whose text *Animal Ecology* (1927) provided the foundations for a more dynamic ecology through his definition of food chains, food webs, trophic levels, the niche, and other basic concepts; Joseph Grinnell, Paul Errington, Olaus Murie, and other field biologists who challenged prevailing notions on the ecological role and value of predators (Dunlap 1988); and biologists who sought to place national park management in the USA on a sound ecological footing (Sellars 1997; Shafer 2001). Importantly, the crisis of the Dust Bowl in North America invited similar ecological critiques of agricultural practices during the 1930s (Worster 1979; Beeman and Pritchard 2001).

By the late 1930s an array of conservation concerns—soil erosion, watershed degradation, urban pollution, deforestation, depletion of fisheries and wildlife populations—brought academic ecologists and resource managers closer together and generated a new awareness of conservation's ecological foundations, in particular the significance of biological diversity. In 1939 Aldo Leopold summarized the point in a speech to a symbolically appropriate joint meeting of the Ecological Society of America and the Society of American Foresters:

> The emergence of ecology has placed the economic biologist in a peculiar dilemma: with one hand he points out the accumulated findings of his search for utility, or lack of utility, in this or that species; with the other he lifts the veil from a biota

so complex, so conditioned by interwoven cooperations and competitions, that no man can say where utility begins or ends. No species can be 'rated' without the tongue in the cheek; the old categories of 'useful' and 'harmful' have validity only as conditioned by time, place, and circumstance. The only sure conclusion is that the biota as a whole is useful, and (the) biota includes not only plants and animals, but soils and waters as well (Leopold 1991:266–67).

With appreciation of "the biota as a whole" came greater appreciation of the functioning of ecological communities and systems (Golley 1993). For Leopold and others, this translated into a redefinition of conservation's aims: away from the narrow goal of sustaining outputs of discrete commodities, and toward the more complex goal of sustaining what we now call ecosystem health and resilience.

As conservation's aims were thus being redefined, its ethical foundations were being reconsidered. The accumulation of revolutionary biological insights, combined with a generation's experience of fragmented policy, short-term economics, and environmental decline, yielded Leopold's assertion of an Evolutionary-Ecological Land Ethic (Callicott 1990). A land ethic, Leopold wrote, "enlarges the boundaries of the community to include soils, waters, plants, and animals, or collectively: the land"; it "changes the role of *Homo sapiens* from conqueror of the land-community to plain member and citizen of it" (Leopold 1949:204). These ethical concepts only slowly gained ground in forestry, fisheries management, wildlife management, and other resource management disciplines; indeed, they are contentious still.

In the years following World War II, as consumer demands increased and technologies evolved, resource development pressures grew. Resource managers responded by expanding their efforts to increase the yields of their particular commodities. Meanwhile, the pace of scientific change accelerated in disciplines across the biological spectrum, from microbiology, genetics, systematics, and population biology to ecology, limnology, marine biology, and biogeography (Mayr 1982). As these advances accrued, maintaining healthy connections between the basic sciences and their application in resource management fields proved challenging. It fell to a diverse cohort of scientific researchers, interpreters, and advocates to enter the public policy fray (including such notable figures as Rachel Carson, Jacques-Yves Cousteau, Ray Dasmann, G. Evelyn Hutchinson, Julian Huxley, Eugene and Howard Odum, and Sir Peter Scott). Many of these had worldwide influence through their writings and students, their collaborations, and their ecological concepts and methodologies. Working from within traditional disciplines, government agencies, and academic seats, they stood at the complicated intersection of conservation science, policy, and practice—a place that would come to define conservation biology.

More pragmatically, new federal legislation in the USA and a growing body of international agreements expanded the role and responsibilities of biologists in conservation. In the USA the National Environmental Policy Act (1970) required analysis of environmental impacts in federal decision-making. The Endangered Species Act (1973) called for an unprecedented degree of scientific involvement in the identification, protection, and recovery of threatened species (see Chapter 12). Other laws that broadened the role of biologists in conservation and environmental protection include the Marine Mammal Protection Act (1972), the Clean Water Act (1972), the Forest and Rangeland Renewable Resources Planning Act (1974), the National Forest Management Act (1976), and the Federal Land Policy Management Act (1976).

At the international level, the responsibilities of biologists were also expanding in response to the adoption of bilateral treaties and multilateral agreements, including the UNESCO (United Nations Educational, Scientific and Cultural Organization) Man and the Biosphere Programme (1970), the Convention on International Trade in Endangered Species of Wild Fauna and Flora (CITES) (1975), and the Convention on Wetlands of International Importance (the "Ramsar Convention") (1975). In 1966 the International Union for the Conservation of Nature (IUCN) published it first "red list" inventories of threatened species. In short, the need for rigorous science input *into* conservation decision-making was increasing, even as the science *of* conservation was changing. This state of affairs challenged the traditional orientation of resource managers and research biologists alike.

## 1.2 Establishing a new interdisciplinary field

In the opening chapter of *Conservation Biology: An Evolutionary-Ecological Perspective*, editors Michael Soulé and Bruce Wilcox (1980) described conservation biology as "a mission-oriented discipline comprising both pure and applied science." The phrase *crisis-oriented* (or *crisis-driven*) was soon added to the list of modifiers describing the emerging field (Soulé 1985). This characterization of conservation biology as a *mission-oriented, crisis-driven, problem-solving* field resonates with echoes of the past. The history of conservation and environmental management demonstrates that the emergence of problem-solving fields (or new emphases within established fields) invariably involves new interdisciplinary connections, new institutions, new research programs, and new practices. Conservation biology would follow this pattern in the 1970s, 1980s, and 1990s.

In 1970 David Ehrenfeld published *Biological Conservation*, an early text in a series of publications that altered the scope, content, and direction of conservation science (e.g. MacArthur and Wilson 1963; MacArthur and Wilson 1967; MacArthur 1972; Soulé and Wilcox 1980; CEQ 1980; Frankel and Soulé 1981; Schonewald-Cox *et al.* 1983; Harris 1984; Caughley and Gunn 1986; Soulé 1986; Soulé 1987a) (The journal *Biological Conservation* had also begun publication a year earlier in England). In his preface Ehrenfeld stated, "Biologists are beginning to forge a discipline in that turbulent and vital area where biology meets the social sciences and humanities". Ehrenfeld recognized that the "acts of conservationists are often motivated by strongly humanistic principles," but cautioned that "the practice of conservation must also have a firm scientific basis or, plainly stated, it is not likely to work". Constructing that "firm scientific basis" required—and attracted—researchers and practitioners from varied disciplines (including Ehrenfeld himself, whose professional background was in medicine and physiological ecology). The common concern that transcended the disciplinary boundaries was *biological diversity*: its extent, role, value, and fate.

By the mid-1970s, the recurring debates within theoretical ecology over the relationship between species diversity and ecosystem stability were intensifying (Pimm 1991; Golley 1993; McCann 2000). Among conservationists the theme of diversity, in eclipse since Leopold's day, began to re-emerge. In 1951, renegade ecologists had created The Nature Conservancy for the purpose of protecting threatened sites of special biological and ecological value. In the 1960s voices for diversity began to be heard within the traditional conservation fields. Ray Dasmann, in *A Different Kind of Country* (1968: vii) lamented "the prevailing trend toward uniformity" and made the case "for the preservation of natural diversity" and for cultural diversity as well. Pimlott (1969) detected "a sudden stirring of interest in diversity ... Not until this decade did the word diversity, as an ecological and genetic concept, begin to enter the vocabulary of the wildlife manager or land-use planner." Hickey (1974) argued that wildlife ecologists and managers should concern themselves with "all living things"; that "a scientifically sound wildlife conservation program" should "encompass the wide spectrum from one-celled plants and animals to the complex species we call birds and mammals." Conservation scientists and advocates of varied backgrounds increasingly framed the fundamental conservation problem in these new and broader terms (Farnham 2002).

As the theme of biological diversity gained traction among conservationists in the 1970s, the key components of conservation biology began to coalesce around it:

• Within the sciences proper, the synthesis of knowledge from island biogeography and population biology greatly expanded understanding of the distribution of species diversity and the phenomena of speciation and extinction.

- The fate of threatened species (both *in situ* and *ex situ*) and the loss of rare breeds and plant germplasm stimulated interest in the heretofore neglected (and occasionally even denigrated) application of genetics in conservation.
- Driven in part by the IUCN red listing process, captive breeding programs grew; zoos, aquaria, and botanical gardens expanded and redefined their role as partners in conservation.
- Wildlife ecologists, community ecologists, and limnologists were gaining greater insight into the role of keystone species and top-down interactions in maintaining species diversity and ecosystem health.
- Within forestry, wildlife management, range management, fisheries management, and other applied disciplines, ecological approaches to resource management gained more advocates.
- Advances in ecosystem ecology, landscape ecology, and remote sensing provided increasingly sophisticated concepts and tools for land use and conservation planning at larger spatial scales.
- As awareness of conservation's social dimensions increased, discussion of the role of values in science became explicit. Interdisciplinary inquiry gave rise to environmental history, environmental ethics, ecological economics, and other hybrid fields.

As these trends unfolded, "keystone individuals" also had special impact. Peter Raven and Paul Ehrlich (to name two) made fundamental contributions to coevolution and population biology in the 1960s before becoming leading proponents of conservation biology. Michael Soulé, a central figure in the emergence of conservation biology, recalls that Ehrlich encouraged his students to speculate across disciplines, and had his students read Thomas Kuhn's *The Structure of Scientific Revolutions* (1962). The intellectual syntheses in *population biology* led Soulé to adopt (around 1976) the term *conservation biology* for his own synthesizing efforts.

For Soulé, that integration especially entailed the merging of genetics and conservation (Soulé 1980). In 1974 Soulé visited Sir Otto Frankel while on sabbatical in Australia. Frankel approached Soulé with the idea of collaborating on a volume on the theme (later published as *Conservation and Evolution*) (Frankel and Soulé 1981). Soulé's work on that volume led to the convening of the First International Conference on Conservation Biology in September 1978. The meeting brought together what looked from the outside like "an odd assortment of academics, zoo-keepers, and wildlife conservationists" (Gibbons 1992). Inside, however, the experience was more personal, among individuals who had come together through important, and often very personal, shifts in professional priorities. The proceedings of the 1978 conference were published as *Conservation Biology: An Evolutionary-Ecological Perspective* (Soulé and Wilcox 1980). The conference and the book initiated a series of meetings and proceedings that defined the field for its growing number of participants, as well as for those outside the immediate circle (Brussard 1985; Gibbons 1992).

Attention to the genetic dimension of conservation continued to gain momentum into the early 1980s (Schonewald-Cox *et al.* 1983). Meanwhile, awareness of threats to species diversity and causes of extinction was reaching a broader professional and public audience (e.g. Ziswiler 1967; Iltis 1972; Terborgh 1974; Ehrlich and Ehrlich 1981). In particular, the impact of international development policies on the world's species-rich, humid tropical forests was emerging as a global concern. Field biologists, ecologists, and taxonomists, alarmed by the rapid conversion of the rainforests—and witnesses themselves to the loss of research sites and study organisms—began to sound alarms (e.g. Gómez-Pompa *et al.* 1972; Janzen 1972). By the early 1980s, the issue of rainforest destruction was highlighted through a surge of books, articles, and scientific reports (e.g. Myers 1979, 1980; NAS 1980; NRC 1982; see also Chapter 4).

During these years, recognition of the needs of the world's poor and the developing world was prompting new approaches to integrating conservation and development. This movement was embodied in a series of international programs, meetings, and reports, including the Man and the Biosphere Programme (1970), the United Nations Conference on the Human Environment held in Stockholm (1972), and the World Conservation Strategy (IUCN 1980). These approaches eventually came together under the banner of *sustainable development*, especially as defined in the report of the World Commission on Environment and Development (the "Brundtland Report") (WCED 1987). The complex relationship between development and conservation created tensions within conservation biology from the outset, but also drove the search for deeper consensus and innovation (Meine 2004).

A Second International Conference on Conservation Biology convened at the University of Michigan in May 1985 (Soulé 1986). Prior to the meeting, the organizers formed two committees to consider establishing a new professional society and a new journal. A motion to organize the Society for Conservation Biology (SCB) was approved at the end of the meeting (Soulé 1987b). One of the Society's first acts was to appoint David Ehrenfeld editor of the new journal *Conservation Biology* (Ehrenfeld 2000).

The founding of SCB coincided with planning for the National Forum on BioDiversity, held September 21–24, 1986 in Washington, DC. The forum, broadcast via satellite to a national and international audience, was organized by the US National Academy of Sciences and the Smithsonian Institution. Although arranged independently of the process that led to SCB's creation, the forum represented a convergence of conservation concern, scientific expertise, and interdisciplinary commitment. In planning the event, Walter Rosen, a program officer with the National Research Council, began using a contracted form of the phrase *biological diversity*. The abridged form *biodiversity* began its etymological career.

The forum's proceedings were published as *Biodiversity* (Wilson and Peter 1988). The wide impact of the forum and the book assured that the landscape of conservation science, policy, and action would never be the same. For some, conservation biology appeared as a new, unproven, and unwelcome kid on the conservation block. Its adherents, however, saw it as the culmination of trends long latent within ecology and conservation, and as a necessary adaptation to new knowledge and a gathering crisis. Conservation biology quickly gained its footing within academia, zoos and botanical gardens, non-profit conservation groups, resource management agencies, and international development organizations (Soulé 1987b).

In retrospect, the rapid growth of conservation biology reflected essential qualities that set it apart from predecessor and affiliated fields:

• Conservation biology rests upon a scientific foundation in systematics, genetics, ecology, and evolutionary biology. As the Modern Synthesis rearranged the building blocks of biology, and new insights emerged from population genetics, developmental genetics (heritability studies), and island biogeography in the 1960s, the application of biology in conservation was bound to shift as well. This found expression in conservation biology's primary focus on the conservation of genetic, species, and ecosystem diversity (rather than those ecosystem components with obvious or direct economic value).

• Conservation biology paid attention to the entire biota; to diversity at all levels of biological organization; to patterns of diversity at various temporal and spatial scales; and to the evolutionary and ecological processes that maintain diversity. In particular, emerging insights from ecosystem ecology, disturbance ecology, and landscape ecology in the 1980s shifted the perspective of ecologists and conservationists, placing greater emphasis on the dynamic nature of ecosystems and landscapes (e.g. Pickett and White 1985; Forman 1995).

• Conservation biology was an interdisciplinary, systems-oriented, and inclusive response to conservation dilemmas exacerbated by approaches that were too narrowly focused, fragmented, and exclusive (Soulé 1985; Noss and Cooperrider 1994). It provided an interdisciplinary home for those in established disciplines who sought new ways to organize and use scientific information, and who followed broader ethical imperatives. It also reached beyond its own core scientific disciplines to incorporate insights from the social sciences and humanities, from the empirical experience of resource managers, and from diverse cultural sources (Grumbine 1992; Knight and Bates 1995).

• Conservation biology acknowledged its status as an inherently "value-laden" field. Soulé (1985) asserted that "ethical norms are a genuine part of conservation biology." Noss (1999) regarded this as a distinguishing characteristic, noting an "overarching normative assumption in conservation biology...that biodiversity is good and ought to be preserved." Leopold's land ethic and related appeals to intergenerational responsibilities and the intrinsic value of non-human life motivated growing numbers of conservation scientists and environmental ethicists (Ehrenfeld 1981; Samson and Knopf 1982; Devall and Sessions 1985; Nash 1989). This explicit recognition of conservation biology's ethical content stood in contrast to the usual avoidance of such considerations within the sciences historically (McIntosh 1980; Barbour 1995; Barry and Oelschlaeger 1996).

• Conservation biology recognized a "close linkage" between biodiversity conservation and economic development and sought new ways to improve that relationship. As *sustainability* became the catch-all term for development that sought to blend environmental, social, and economic goals, conservation biology provided a new venue at the intersection of ecology, ethics, and economics (Daly and Cobb 1989). To achieve its goals, conservation biology had to reach beyond the sciences and generate conversations with economists, advocates, policy-makers, ethicists, educators, the private sector, and community-based conservationists.

Conservation biology thus emerged in response to both increasing knowledge and expanding demands. In harnessing that knowledge and meeting those demands, it offered a new, integrative, and interdisciplinary approach to conservation science.

## 1.3 Consolidation: conservation biology secures its niche

In June 1987 more than 200 people attended the first annual meeting of the Society for Conservation Biology in Bozeman, Montana, USA. The rapid growth of the new organization's membership served as an index to the expansion of the field generally. SCB tapped into the burgeoning interest in interdisciplinary conservation science among younger students, faculty, and conservation practitioners. Universities established new courses, seminars, and graduate programs. Scientific organizations and foundations adjusted their funding priorities and encouraged those interested in the new field. A steady agenda of conferences on biodiversity conservation brought together academics, agency officials, resource managers, business representatives, international aid agencies, and non-governmental organizations. In remarkably rapid order, conservation biology gained legitimacy and secured a professional foothold.

Not, however, without resistance, skepticism, and occasional ridicule. As the field grew, complaints came from various quarters. Conservation biology was caricatured as a passing fad, a response to trendy environmental ideas (and momentarily available funds). Its detractors regarded it as too theoretical, amorphous, and eclectic; too promiscuously interdisciplinary; too enamored of models; and too technique-deficient and data-poor to have any practical application (Gibbons 1992). Conservation biologists in North America were accused of being indifferent to the conservation traditions of other nations and regions. Some saw conservation biology as merely putting "old wine in a new bottle" and dismissing the rich experience of foresters, wildlife managers, and other resource managers (Teer 1988; Jensen and Krausman 1993). *Biodiversity* itself was just too broad, or confusing, or "thorny" a term (Udall 1991; Takacs 1996).

Such complaints made headlines within the scientific journals and reflected real tensions within resource agencies, academic departments, and conservation organizations. Conservation biology had indeed challenged prevalent paradigms, and such responses were to be expected. Defending the new field, Ehrenfeld (1992: 1625) wrote, "Conservation biology is not defined by a discipline but by its goal—to halt or repair the undeniable, massive damage that is being done to ecosystems, species, and the relationships of humans to the environment.... Many specialists in a host of fields find it difficult, even hypocritical, to continue business as usual, blinders firmly in place, in a world that is falling apart."

Meanwhile, a spate of new and complex conservation issues were drawing increased attention to biodiversity conservation. In North America, the Northern Spotted Owl (*Strix occidentalis caurina*) became the poster creature in deeply contentious debates over the fate of remaining old-growth forests and alternative approaches to forest management; the Exxon Valdez oil spill and its aftermath put pollution threats and energy policies on the front page; the anti-environmental, anti-regulatory "Wise Use" movement gained in political power and influence; arguments over livestock grazing practices and federal rangeland policies pitted environmentalists against ranchers; perennial attempts to allow oil development within the Arctic National Wildlife Refuge continued; and moratoria were placed on commercial fishing of depleted stocks of northern cod (Alverson et al. 1994; Yaffee 1994; Myers et al. 1997; Knight et al. 2002; Jacobs 2003).

At the international level, attention focused on the discovery of the hole in the stratospheric ozone layer over Antarctica; the growing scientific consensus about the threat of global warming (the Intergovernmental Panel on Climate Change was formed in 1988 and issued its first assessment report in 1990); the environmental legacy of communism in the former Soviet bloc; and the environmental impacts of international aid and development programs. In 1992, 172 nations gathered in Rio de Janeiro at the United Nations Conference on Environment and Development (the "Earth Summit"). Among the products of the summit was the Convention on Biological Diversity. In a few short years, the scope of biodiversity conservation, science, and policy had expanded dramatically (e.g. McNeely et al. 1990; Lubchenco et al. 1991).

To some degree, conservation biology had defined its own niche by synthesizing scientific disciplines, proclaiming its special mission, and gathering together a core group of leading scientists, students, and conservation practitioners. However, the field was also filling a niche that was rapidly opening around it. It provided a meeting ground for those with converging interests in the conservation of biological diversity. It was not alone in gaining ground for interdisciplinary conservation research and practice. It joined restoration ecology, landscape ecology, agroecology, ecological economics, and other new fields in seeking solutions across traditional academic and intellectual boundaries.

Amid the flush of excitement in establishing conservation biology, it was sometimes easy to overlook the challenges inherent in the effort. Ehrenfeld (2000) noted that the nascent field was "controversy-rich." Friction was inherent not only in conservation biology's relationship to related fields, but within the field itself. Some of this was simply a result of high energy applied to a new endeavor. Often, however, this reflected deeper tensions in conservation: between sustainable use and protection; between public and private resources; between the immediate needs of people, and obligations to future generations and other life forms. Conservation biology would be the latest stage on which these long-standing tensions would express themselves.

Other tensions reflected the special role that conservation biology carved out for itself. Conservation biology was largely a product of American institutions and individuals, yet sought to address a problem of global proportions (Meffe 2002). Effective biodiversity conservation entailed work at scales from the global to the local, and on levels from the genetic to the species to the community; yet actions at these different scales and levels required different types of information, skills, and partnerships (Noss 1990). Professionals in the new field had to be firmly grounded within particular professional specialties, yet conversant across disciplines (Trombulak 1994; Noss 1997). Success in the *practice* of biodiversity conservation was measured by on-the-ground impact, yet the *science* of conservation biology was obliged (as are all sciences) to undertake rigorous research and to define uncertainty (Noss 2000). Conservation biology was a "value-laden" field adhering to explicit ethical norms, yet sought to advance conservation through careful scientific analysis (Barry and Oelschlager 1996). These tensions within conservation biology were present at birth. They continue to present important challenges to conservation biologists. They also give the field its creativity and vitality.

## 1.4 Years of growth and evolution

Although conservation biology has been an organized field only since the mid-1980s, it is possible to identify and summarize at least several salient trends that have shaped it since.

## 1.4.1 Implementation and transformation

Conservation biologists now work in a much more elaborate field than existed at the time of its founding. Much of the early energy—and debate—in conservation biology focused on questions of the genetics and demographics of small populations, population and habitat viability, landscape fragmentation, reserve design, and management of natural areas and endangered species. These topics remain close to the core of conservation biology, but the field has grown around them. Conservation biologists now tend to work more flexibly, at varied scales and in varied ways. In recent years, for example, more attention has focused on landscape permeability and connectivity, the role of strongly interacting species in top-down ecosystem regulation, and the impacts of global warming on biodiversity (Hudson 1991; Lovejoy and Peters 1994; Soulé and Terborgh 1999; Ripple and Beschta 2005; Pringle et al. 2007; Pringle 2008; see Chapters 5 and 8).

Innovative techniques and technologies (such as computer modeling and geographic information systems) have obviously played an important role in the growth of conservation biology. The most revolutionary changes, however, have involved the reconceptualizing of science's role in conservation. The principles of conservation biology have spawned creative applications among conservation visionaries, practitioners, planners, and policy-makers (Noss et al. 1997; Adams 2005). To safeguard biological diversity, larger-scale and longer-term thinking and planning had to take hold. It has done so under many rubrics, including: adaptation of the biosphere reserve concept (Batisse 1986); the development of gap analysis (Scott et al. 1993); the movement toward ecosystem management and adaptive management (Grumbine 1994b; Salafsky et al. 2001; Meffe et al. 2002); ecoregional planning and analogous efforts at other scales (Redford et al. 2003); and the establishment of marine protected areas and networks (Roberts et al. 2001).

Even as conservation biologists have honed tools for designing protected area networks and managing protected areas more effectively (see Chapter 11), they have looked beyond reserve boundary lines to the matrix of surrounding lands (Knight and Landres 1998). Conservation biologists play increasingly important roles in defining the biodiversity values of aquatic ecosystems, private lands, and agroecosystems. The result is much greater attention to private land conservation, more research and demonstration at the interface of agriculture and biodiversity conservation, and a growing watershed- and community-based conservation movement. Conservation biologists are now active across the entire landscape continuum, from wildlands to agricultural lands and from suburbs to cities, where conservation planning now meets urban design and green infrastructure mapping (e.g. Wang and Moskovits 2001; CNT and Openlands Project 2004).

## 1.4.2 Adoption and integration

Since the emergence of conservation biology, the conceptual boundaries between it and other fields have become increasingly porous. Researchers and practitioners from other fields have come into conservation biology's circle, adopting and applying its core concepts while contributing in turn to its further development. Botanists, ecosystem ecologists, marine biologists, and agricultural scientists (among other groups) were underrepresented in the field's early years. The role of the social sciences in conservation biology has also expanded within the field (Mascia et al. 2003). Meanwhile, conservation biology's concepts, approaches, and findings have filtered into other fields. This "permeation" (Noss 1999) is reflected in the number of biodiversity conservation-related articles appearing in the general science journals such as *Science* and *Nature*, and in more specialized ecological and resource management journals. Since 1986 several new journals with related content have appeared, including *Ecological Applications* (1991), the *Journal of Applied Ecology* (1998), the on-line journal *Conservation Ecology* (1997) (now called

*Ecology and Society*), *Frontiers in Ecology and the Environment* (2003), and *Conservation Letters* (2008).

The influence of conservation biology is even more broadly evident in environmental design, planning, and decision-making. Conservation biologists are now routinely involved in land-use and urban planning, ecological design, landscape architecture, and agriculture (e.g. Soulé 1991; Nassauer 1997; Babbitt 1999; Jackson and Jackson 2002; Miller and Hobbs 2002; Imhoff and Carra 2003; Orr 2004). Conservation biology has spurred activity within such emerging areas of interest as conservation psychology (Saunders 2003) and conservation medicine (Grifo and Rosenthal 1997; Pokras *et al.* 1997; Tabor *et al.* 2001; Aguirre *et al.* 2002). Lidicker (1998) noted that "conservation needs conservation biologists for sure, but it also needs conservation sociologists, conservation political scientists, conservation chemists, conservation economists, conservation psychologists, and conservation humanitarians." Conservation biology has helped to meet this need by catalyzing communication and action among colleagues across a wide spectrum of disciplines.

### 1.4.3 Marine and freshwater conservation biology

Conservation biology's "permeation" has been especially notable with regard to aquatic ecosystems and marine environments. In response to long-standing concerns over "maximum sustained yield" fisheries management, protection of marine mammals, depletion of salmon stocks, degradation of coral reef systems, and other issues, marine conservation biology has emerged as a distinct focus area (Norse 1993; Boersma 1996; Bohnsack and Ault 1996; Safina 1998; Thorne-Miller 1998; Norse and Crowder 2005). The application of conservation biology in marine environments has been pursued by a number of non-governmental organizations, including SCB's Marine Section, the Ocean Conservancy, the Marine Conservation Biology Institute, the Center for Marine Biodiversity and Conservation at the Scripps Institution of Oceanography, the Blue Ocean Institute, and the Pew Institute for Ocean Science.

Interest in freshwater conservation biology has also increased as intensified human demands continue to affect water quality, quantity, distribution, and use. Conservationists have come to appreciate even more deeply the essential hydrological connections between groundwater, surface waters, and atmospheric waters, and the impact of human land use on the health and biological diversity of aquatic ecosystems (Leopold 1990; Baron *et al.* 2002; Glennon 2002; Hunt and Wilcox 2003; Postel and Richter 2003). Conservation biologists have become vital partners in interdisciplinary efforts, often at the watershed level, to steward freshwater as both an essential ecosystem component and a basic human need.

### 1.4.4 Building capacity

At the time of its founding, conservation biology was little known beyond the core group of scientists and conservationists who had created it. Now the field is broadly accepted and well represented as a distinct body of interdisciplinary knowledge worldwide. Several textbooks appeared soon after conservation biology gained its footing (Primack 1993; Meffe and Carroll 1994; Hunter 1996). These are now into their second and third editions. Additional textbooks have been published in more specialized subject areas, including insect conservation biology (Samways 1994), conservation of plant biodiversity (Frankel *et al.* 1995), forest biodiversity (Hunter and Seymour 1999), conservation genetics (Frankham *et al.* 2002), marine conservation biology (Norse and Crowder 2005), and tropical conservation biology (Sodhi *et al.* 2007).

Academic training programs in conservation biology have expanded and now exist around the world (Jacobson 1990; Jacobson *et al.* 1995; Rodríguez *et al.* 2005). The interdisciplinary skills of conservation biologists have found acceptance within universities, agencies, non-governmental organizations, and the private sector. Funders have likewise helped build conservation biology's capacity through support for students, academic programs, and basic research and field projects. Despite such growth, most conservation biologists would likely agree that the capacity does not nearly meet the need, given the urgent problems in biodiversity conservation. Even the existing support is highly vulnerable to budget cutbacks, changing priorities, and political pressures.

### 1.4.5 Internationalization

Conservation biology has greatly expanded its international reach (Meffe 2002; Meffe 2003). The scientific roots of biodiversity conservation are obviously not limited to one nation or continent (see Box 1.2). Although the international conservation movement dates back more than a century, the history of the science from an international perspective has been inadequately studied (Blandin 2004). This has occasionally led to healthy debate over the origins and development of conservation biology. Such debates, however, have not hindered the trend toward greater international collaboration and representation within the field (e.g. Medellín 1998).

---

**Box 1.2 Conservation in the Philippines**
**Mary Rose C. Posa**

Conservation biology has been referred to as a "discipline with a deadline" (Wilson 2000). As the rapid loss and degradation of ecosystems accelerates across the globe, some scientists suggest a strategy of triage—in effect, writing off countries that are beyond help (Terborgh 1999). But are there any truly lost causes in conservation?

The Philippines is a mega-biodiversity country with exceptionally high levels of endemism (~50% of terrestrial vertebrates and 45–60% of vascular plants; Heaney and Mittermeier 1997). However, centuries of exploitation and negligence have pushed its ecosystems to their limit, reducing primary forest cover [less than 3% remaining; FAO (Food and Agriculture Organization of the United Nations) 2005], decimating mangroves (>90% lost; Primavera 2000), and severely damaging coral reefs (~5% retaining 75–100% live cover; Gomez et al. 1994), leading to a high number of species at risk of extinction [~21% of vertebrates assessed; IUCN (International Union for Conservation of Nature and Natural Resources) 2006]. Environmental degradation has also brought the loss of soil fertility, pollution, and diminished fisheries productivity, affecting the livelihood of millions of rural inhabitants. Efforts to preserve biodiversity and implement sound environmental policies are hampered by entrenched corruption, weak governance and opposition by small but powerful interest groups. In addition, remaining natural resources are under tremendous pressure from a burgeoning human population. The Philippines has thus been pegged as a top conservation "hotspot" for terrestrial and marine ecosystems, and there are fears that it could be the site of the first major extinction spasm (Heaney and Mittermeier 1997; Myers et al. 2000; Roberts et al. 2002). Remarkably, and despite this precarious situation, there is evidence that hope exists for biodiversity conservation in the Philippines.

Indication of the growing valuation of biodiversity, sustainable development and environmental protection can be seen in different sectors of Philippine society. Stirrings of grassroots environmental consciousness began in the 1970s, when marginalized communities actively opposed unsustainable commercial developments, blocking logging trucks, and protesting the construction of large dams (Broad and Cavanagh 1993). After the 1986 overthrow of dictator Ferdinand Marcos, a revived democracy fostered the emergence of civil society groups focused on environmental issues. The devolution of authority over natural resources from central to local governments also empowered communities to create and enforce regulations on the use of local resources. There are now laudable examples where efforts by communities and non-governmental organizations (NGOs) have made direct impacts on conserving endangered species and habitats (Posa et al. 2008).

Driven in part by public advocacy, there has also been considerable progress in environmental legislation. In particular, the

*continues*

**Box 1.2 (Continued)**

National Integrated Protected Areas System Act provides for stakeholder involvement in protected area management, which has been a key element of success for various reserves. Perhaps the best examples of where people-centered resource use and conservation have come together are marine protected areas (MPAs) managed by coastal communities across the country—a survey of 156 MPAs reported that 44.2% had good to excellent management (Alcala and Russ 2006).

Last, but not least, there has been renewed interest in biodiversity research in academia, increasing the amount and quality of biodiversity information (see Box 1.2 Figure). Labors of field researchers result in hundreds of additional species yet to be described, and some rediscoveries of species thought to be extinct (e.g. Cebu flowerpecker *Dicaeum quadricolor*; Dutson et al. 1993). There are increasing synergies and networks among conservation workers, politicians, community leaders, park rangers, researchers, local people, and international NGOs, as seen from the growth of the Wildlife Conservation Society of the Philippines, which has a diverse membership from all these sectors.

complacency, that positive progress has been made in the Philippines—a conservation "worst case scenario"—suggests that there are grounds for optimism for biodiversity conservation in tropical countries worldwide.

## REFERENCES

Alcala, A. C. and Russ, G. R. (2006). No-take marine reserves and reef fisheries management in the Philippines: a new people power revolution. *Ambio*, **35**, 245–254.

Broad, R. and Cavanagh, J. (1993). *Plundering paradise: the struggle for the environment in the Philippines*. University of California Press, Berkeley, CA.

Dutson, G. C. L., Magsalay, P. M., and Timmins, R. J. (1993). The rediscovery of the Cebu Flowerpecker *Dicaeum quadricolor*, with notes on other forest birds on Cebu, Philippines. *Bird Conservation International*, **3**, 235–243.

FAO (Food and Agriculture Organization of the United Nations) (2005). *Global forest resources assessment 2005, Country report 202: Philippines*. Forestry Department, FAO, Rome, Italy.

Gomez, E. D., Aliño, P. M., Yap, H. T., Licuanan, W. Y. (1994). A review of the status of Philippine reefs. *Marine Pollution Bulletin*, **29**, 62–68.

Heaney, L. and Mittermeier, R. A. (1997). The Philippines. In R. A. Mittermeier, G. P. Robles, and C. G. Mittermeier, eds *Megadiversity: earth's biologically wealthiest nations*, pp. 236–255. CEMEX, Monterrey, Mexico.

IUCN (International Union for Conservation of Nature and Natural Resources) (2006). *2006 IUCN Red List of threatened species*. www.iucnredlist.org.

Myers, N., Mittermeier, R. A., Mittermeier, C. G., da Fonseca, G. A. B., and Kent, J. (2000). Biodiversity hotspots for conservation priorities. *Nature*, **403**, 853–858.

Posa, M. R. C., Diesmos, A. C., Sodhi, N. S., and Brooks, T. M. (2008). Hope for threatened biodiversity: lessons from the Philippines. *BioScience*, **58**, 231–240.

Primavera, J. H. (2000). Development and conservation of Philippine mangroves: Institutional issues. *Ecological Economics*, **35**, 91–106.

Roberts, C. M., McClean, C. J., Veron, J. E. N., et al. (2002). Marine biodiversity hotspots and conservation priorities for tropical reefs. *Science*, **295**, 1280–1284.

Terborgh, J. (1999). *Requiem for nature*. Island Press, Washington, DC.

Wilson, E. O. (2000). On the future of conservation biology. *Conservation Biology*, **14**, 1–3.

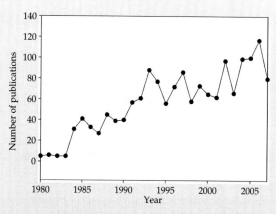

**Box 1.2 Figure** Steady increase in the number of publications on Philippine biodiversity and conservation, obtained from searching three ISI Web of Knowledge databases for the period 1980–2007.

While many daunting challenges remain especially in the area of conservation of populations (Chapter 10) and ecosystems services (Chapter 3), and there is no room for

This growth is reflected in the expanding institutional and membership base of the Society for Conservation Biology. The need to reach across national boundaries was recognized by the founders of the SCB. From its initial issue *Conservation Biology* included Spanish translations of article abstracts. The Society has diversified its editorial board, recognized the accomplishments of leading conservation biologists from around the world, and regularly convened its meetings outside the USA. A significant move toward greater international participation in the SCB came when, in 2000, the SCB began to develop its regional sections.

### 1.4.6 Seeking a policy voice

Conservation biology has long sought to define an appropriate and effective role for itself in shaping public policy (Grumbine 1994a). Most who call themselves conservation biologists feel obligated to be advocates for biodiversity (Odenbaugh 2003). How that obligation ought to be fulfilled has been a source of continuing debate within the field. Some scientists are wary of playing an active advocacy or policy role, lest their objectivity be called into question. Conversely, biodiversity advocates have responded to the effect that "if you don't use your science to shape policy, we will."

Conservation biology's inherent mix of science and ethics all but invited such debate. Far from avoiding controversy, *Conservation Biology*'s founding editor David Ehrenfeld built dialogue on conservation issues and policy into the journal at the outset. *Conservation Biology* has regularly published letters and editorials on the question of values, advocacy, and the role of science in shaping policy. Conservation biologists have not achieved final resolution on the matter. Perhaps in the end it is irresolvable, a matter of personal judgment involving a mixture of scientific confidence levels, uncertainty, and individual conscience and responsibility. "Responsibility" is the key word, as all parties to the debate seem to agree that advocacy, to be responsible, must rest on a foundation of solid science and must be undertaken with honesty and integrity (Noss 1999).

## 1.5 Conservation biology: a work in progress

These trends (and no doubt others) raise important questions for the future. Conservation biology has grown quickly in a few brief decades, yet most conservation biologists would assert that growth for growth's sake is hardly justified. As disciplines and organizations become more structured, they are liable to equate mere expansion with progress in meeting their missions (Ehrenfeld 2000). Can conservation biology sustain its own creativity, freshness, and vision? In its collective research agenda, is the field asking, and answering, the appropriate questions? Is it performing its core function—providing reliable and useful scientific information on biological diversity and its conservation—in the most effective manner possible? Is that information making a difference? What "constituencies" need to be more fully involved and engaged?

While continuing to ponder such questions, conservation biologists cannot claim to have turned back the threats to life's diversity. Yet the field has contributed essential knowledge at a time when those threats have continued to mount. It has focused attention on the full spectrum of biological diversity, on the ecological processes that maintain it, on the ways we value it, and on steps that can be taken to conserve it. It has brought scientific knowledge, long-range perspectives, and a conservation ethic into the public and professional arenas in new ways. It has organized scientific information to inform decisions affecting biodiversity at all levels and scales. In so doing, it has helped to reframe fundamentally the relationship between conservation philosophy, science, and practice.

### Summary

- Conservation biology emerged in the mid-1980s as a new field focused on understanding, protecting, and perpetuating biological diversity at all scales and all levels of biological organization.
- Conservation biology has deep roots in the growth of biology over several centuries, but its

emergence reflects more recent developments in an array of biological sciences (ecology, genetics, evolutionary biology, etc.) and natural resource management fields (forestry, wildlife and fisheries management, etc.).

- Conservation biology was conceived as a "mission-oriented" field based in the biological sciences, but with an explicit interdisciplinary approach that incorporated insights from the social sciences, humanities, and ethics.
- Since its founding, conservation biology has greatly elaborated its research agenda; built stronger connections with other fields and disciplines; extended its reach especially into aquatic and marine environments; developed its professional capacity for training, research, and field application; become an increasingly international field; and become increasingly active at the interface of conservation science and policy.

## Suggested reading

- Farnham, T. J. (2007). *Saving Nature's Legacy: Origins of the Idea of Biological Diversity*. Yale University Press, New Haven.
- Quammen, D. (1996). *The Song of the Dodo: Island Biogeography in an Age of Extinctions*. Simon and Schuster, New York.
- Meine, C. (2004). *Correction Lines: Essays on Land, Leopold, and Conservation*. Island Press, Washington, DC.
- Minteer, B. A. and Manning, R. E. (2003). *Reconstructing Conservation: Finding Common Ground*. Island Press, Washington, DC.

## Relevant website

- Society for Conservation Biology: http://www.conbio.org/

## REFERENCES

Adams, J. S. (2005). *The future of the wild: radical conservation for a crowded world*. Beacon Press, Boston, Massachusetts.

Aguirre, A. A., Ostfeld, R. S., Tabor, G. M., House, C., and Pearl, M. C. (2002). *Conservation medicine: ecological health in practice*. Oxford University Press, New York.

Alverson, W. S., Kuhlman, W., and Waller, D. M. (1994). *Wild forests: conservation biology and public policy*. Island Press, Washington, DC.

Babbitt, B. (1999). Noah's mandate and the birth of urban bioplanning. *Conservation Biology*, **13**, 677–678.

Barbour, M. G. (1995). Ecological fragmentation in the Fifties. In W. Cronon, ed. *Uncommon ground: toward reinventing nature*, pp. 233–255. W. W. Norton, New York.

Baron, J. S., Poff, N. L., Angermeier, P. L., et al. (2002). Meeting ecological and societal needs for freshwater. *Ecological Applications*, **12**, 1247–1260.

Barry, D. and Oelschlaeger, M. (1996). A science for survival: values and conservation biology. *Conservation Biology*, **10**, 905–911.

Batisse, M. (1986). Developing and focusing the Biosphere Reserve concept. *Nature and Resources*, **22**, 2–11.

Beeman, R. S. and Pritchard, J. A. (2001). *A green and permanent land: ecology and agriculture in the twentieth century*. University Press of Kansas, Lawrence, Kansas.

Blandin, P. (2004). Biodiversity, between science and ethics. In S. H. Shakir, and W. Z. A. Mikhail, eds *Soil zoology for sustainable development in the 21st Century*, pp. 17–49. Eigenverlag, Cairo, Egypt.

Boersma, P. D. (1996). Maine conservation: protecting the exploited commons. *Society for Conservation Biology Newsletter*, **3**, 1–6.

Boersma, P. D., Kareiva, P., Fagan, W. F., Clark, J. A., and Hoekstra, J. M. (2001). How good are endangered species recovery plans? *BioScience*, **51**, 643–650.

Bohnsack, J. and Ault, J. (1996). Management strategies to conserve marine biodiversity. *Oceanography*, **9**, 73–81.

Brussard, P. (1985). The current status of conservation biology. *Bulletin of the Ecological Society of America*, **66**, 9–11.

Callicott, J. B. (1990). Whither conservation ethics? *Conservation Biology*, **4**, 15–20.

Caughley, G. and Gunn, A. (1986). *Conservation biology in theory and practice*. Blackwell Science, Cambridge, Massachusetts.

CNT (Center For Neighborhood Technologies) and Openlands Project (2004). *Natural connections: green infrastructure in Wisconsin, Illinois, and Indiana*. (Online) Available at http://www.greenmapping.org. (Accessed February 2006).

CEQ (Council On Environmental Quality) (1980). *Environmental quality—1980: the eleventh annual report of the CEQ*. US Government Printing Office, Washington, DC.

Croker, R. A. (1991). *Pioneer ecologist: the life and work of Victor Ernest Shelford*. Smithsonian Institution Press, Washington, DC.

Crosby, A. W. (1986). *Ecological imperialism: the biological expansion of Europe, 900–1900*. Cambridge University Press, New York.

Daly, H. E. and Cobb, J. B. Jr. (1989). *For the common good: redirecting the economy toward community, the environment, and a sustainable future.* Beacon Press, Boston, Massachusetts.

Dasmann, R. (1959). *Environmental conservation.* John Wiley and Sons, New York.

Dasmann, R. (1968). *A different kind of country.* Macmillan, New York.

Devall, B. and Sessions, G. (1985). *Deep ecology: living as if nature mattered.* Peregrine Smith Books, Salt Lake City, Utah.

Diamond, J. M. (1997). *Guns, germs, and steel: the fates of human societies.* Norton, New York.

Dunlap, T. R. (1988). *Saving America's wildlife: ecology and the American mind, 1850–1990.* Princeton University Press, Princeton, New Jersey.

Ehrenfeld, D. W. (1970). *Biological conservation.* Holt, Rinehard, and Winston, New York.

Ehrenfeld, D. W. (1981). *The arrogance of humanism.* Oxford University Press, New York.

Ehrenfeld, D. W. (1992). Conservation biology: its origins and definition. *Science*, **255**, 1625–26.

Ehrenfeld, D. (2000). War and peace and conservation biology. *Conservation Biology*, **14**, 105–112.

Ehrlich, P. and Ehrlich, A. (1981). *Extinction: the causes and consequences of the disappearance of species.* Random House, New York.

Elton, C. (1927). *Animal Ecology.* University of Chicago Press, Chicago, Illinois.

Engel, R. J. (1983). *Sacred sands: the struggle for community in the Indiana Dunes.* Wesleyan University Press, Middletown, Connecticut.

Farnham, T. J. (2002). The concept of biological diversity: the evolution of a conservation paradigm. Ph.D. dissertation, Yale University.

Forman, R. T. T. (1995). *Land mosaics: the ecology of landscapes and regions.* Cambridge University Press, Cambridge, UK.

Frankel, O. H. and Soulé, M. E. (1981). *Conservation and evolution.* Cambridge University Press, Cambridge, UK.

Frankel, O. H., Brown, A. H. D., and Burdon, J. J. (1995). *The conservation of plant biodiversity.* Cambridge University Press, New York.

Frankham, R., Briscoe, D. A., and Ballou, J. D. (2002). *Introduction to conservation genetics.* Cambridge University Press, Cambridge, UK.

Gibbons, A. (1992). Conservation biology in the fast lane. *Science*, **255**, 20–22.

Glennon, R. (2002). *Water follies: groundwater pumping and the fate of America's fresh waters.* Island Press, Washington, DC.

Golley, F. B. (1993). *A history of the ecosystem concept in ecology: more than the sum of the parts.* Yale University Press, New Haven, Connecticut.

Gómez-Pompa, A., Vázquez-Yanes, C., and Guevara, S. (1972). The tropical rainforest: a non-renewable resource. *Science*, **177**, 762–765.

Grifo, F. T. and Rosenthal, J., eds (1997). *Biodiversity and human health.* Island Press, Washington, DC.

Grove, R. H. (1995). *Green imperialism: colonial expansion, tropical island Edens, and the origins of environmentalism, 1600–1860.* Cambridge University Press, Cambridge, UK.

Grumbine, R. E. (1992). *Ghost bears: exploring the biodiversity crisis.* Island Press, Washington, DC.

Grumbine, R. E. (1994a). *Environmental policy and biodiversity.* Island Press, Washington, DC.

Grumbine, R. E. (1994b). What is ecosystem management? *Conservation Biology*, **8**, 27–38.

Grumbine, R. E. (1996). Using biodiversity as a justification for nature protection in the U. S. *Wild Earth*, **6**, 71–80.

Harris, L. D. (1984). *The fragmented forest: island biogeography theory and the preservation of biotic diversity.* University of Chicago Press, Chicago, Illinois.

Hickey, J. J. (1974). Some historical phases in wildlife conservation. *Wildlife Society Bulletin*, **2**, 164–170.

Hudson, W., ed. (1991). *Landscape linkages and biodiversity.* Island Press, Washington, DC.

Hunt, R. J. and Wilcox, D. A. (2003). Ecohydrology: why hydrologists should care. *Ground Water*, **41**, 289.

Hunter, M. (1996). *Fundamentals of conservation biology.* Blackwell Science, Cambridge, Massachusetts.

Hunter, M. L. and Seymour, R. S. (1999). *Maintaining biodiversity in forest ecosystems.* Cambridge University Press, Cambridge, UK.

Iltis, H. H. (1972). The extinction of species in the destruction of ecosystems. *American Biology Teacher*, **34**, 201–05.

Imhoff, D. and Carra, R. (2003). *Farming with the wild: a new vision for conservation-based agriculture.* Sierra Club Books, San Francisco, California.

IUCN (International Union For The Conservation Of Nature) (1980). *World conservation strategy: living resource conservation for sustainable development.* International Union for the Conservation of Nature, World Resources Institute, World Wildlife Fund, Gland, Switzerland.

Jackson, D. L. and Jackson, L. L. (2002). *The farm as natural habitat: reconnecting food systems with ecosystems.* Island Press, Washington, DC.

Jacobs, H. M. (2003). The politics of property rights at the national level: signals and trends. *Journal of the American Planning Association*, **69**, 181–189.

Jacobson, S. K. (1990). Graduate education in conservation biology. *Conservation Biology*, **4**, 431–440.

Jacobson, S. K., Vaughan, E., and Miller, S. (1995). New directions in conservation biology: graduate programs. *Conservation Biology*, **9**, 5–17.

Janzen, D. H. (1972). The uncertain future of the tropics. *Natural History*, **81**, 80–90.

Jensen, M. N. and Krausman, P. R. (1993). Conservation biology's literature: new wine or just a new bottle? *Wildlife Society Bulletin*, **21**, 199–203.

Knight, R. L. and Bates, S. F., eds (1995). *A new century for natural resources management*. Island Press, Washington, DC.

Knight, R. L. and Landres, P., eds (1998). *Stewardship across boundaries*. Island Press, Washington, DC.

Knight, R. L., Gilgert, W. C., and Marston, E., eds (2002). *Ranching west of the 100th meridian: culture, ecology, and economics*. Island Press, Washington, DC.

Kuhn, T. (1962). *The structure of scientific revolutions*. University of Chicago Press, Chicago, Illinois.

Leopold, A. (1940). The state of the profession. In S. L. Flader, and J. B. Callicott, eds *The river of the mother of God and other essays by Aldo Leopold*, pp. 276–280. University of Wisconsin Press, Madison, Wisconsin.

Leopold, A. (1949). *A sand county almanac and sketches here and there*. Oxford University Press, New York.

Leopold, A. (1991). *The river of the mother of God and other essays by Aldo Leopold*. S. L. Flader and J. B. Callicott, eds University of Wisconsin Press, Madison, Wisconsin.

Leopold, L. B. (1990). Ethos, equity, and the water resource. *Environment*, **32**, 16–20, 37–42.

Lidicker, W. Z., Jr. (1998). Revisiting the human dimension in conservation biology. *Conservation Biology*, **12**, 1170–1171.

Lovejoy, T. E. and Peters, R. L. (1994). *Global warming and biological diversity*. Yale University Press, New Haven, Connecticut.

Lubchenco, J., Olson, A. M., Brubaker, L. B., *et al*. (1991). The sustainable biosphere initiative: an ecological research agenda. *Ecology*, **72**, 371–412.

MacArthur, R. (1972). *Geographical ecology: patterns in the distribution of species*. Princeton University Press, Princeton, New Jersey.

MacArthur, R. H. and Wilson, E. O. (1963). An equilibrium theory of insular zoogeography. *Evolution*, **17**, 373–387.

MacArthur, R. H. and Wilson, E. O. (1967). *The theory of island biogeography*. Princeton University Press, Princeton, New Jersey.

Marsh, G. P. (1864). *Man and nature: or, physical geography as modified by human action*. D. Lowenthal, D., ed. 1965 reprint. Harvard University Press, Cambridge, Massachusetts.

Mascia, M. B., Brosius, J. P., Dobson, T. A., *et al*. (2003). Conservation and the social sciences. *Conservation Biology*, **17**, 649–650.

Mayr, E. (1982). *The growth of biological thought: diversity, evolution, and inheritance*. Harvard University Press, Cambridge, Massachusetts.

McCann, K. S. (2000). The diversity-stability debate. *Nature*, **405**, 228–233.

McIntosh, R. P. (1980). The background and some current problems of theoretical ecology. *Synthese*, **43**, 195–255.

McIntosh, R. P. (1985). *The background of ecology: concept and theory*. Cambridge University Press, Cambridge, UK.

McNeely, J., Miller, K. R., Reid, W. V., Mittermeier, R. A., and Werner, T. B. (1990). *Conserving the World's Biological Diversity*. International Union for the Conservation of Nature, World Resources Institute, World Wildlife Fund, Gland, Switzerland.

Medellín, R. A. (1998). True international collaboration: now or never. *Conservation Biology*, **12**, 939–940.

Meffe, G. K. (2002). Going international. *Conservation Biology*, **16**, 573–574.

Meffe, G. K. (2003). Toward further internationalization. *Conservation Biology*, **17**, 1197–1199.

Meffe, G. K. and Carroll, R. (1994). *Principles of conservation biology*. Sinauer Associates, Sunderland, Massachusetts.

Meffe, G. K., Nielsen, A. A., Knight, R. L., and Schenborn, D. A. (2002). *Ecosystem management: adaptive, community-based conservation*. Island Press, Washington, DC.

Meine, C. (2004). *Correction lines: essays on land, Leopold, and conservation*. Island Press, Washington, DC.

Miller, J. R., and Hobbs, R. J. (2002). Conservation where people live and work. *Conservation Biology*, **16**, 330–337.

Myers, N. (1979). *The sinking ark: a new look at the problem of disappearing species*. Pergamon Press, Oxford, England.

Myers, N. (1980). *Conversion of tropical moist forests*. National Academy of Sciences, Washington, DC.

Myers, R. A., Hutchings, J. A., and Barrowman, N. J. (1997). Why do fish stocks collapse? The example of cod in Atlantic Canada. *Ecological Applications*, **7**, 91–106.

Nash, R. (1989). *The rights of nature: a history of environmental ethics*. University of Wisconsin Press, Madison, Wisconsin.

Nassauer, J. (1997). *Placing nature: culture and landscape ecology*. Island Press, Washington, DC.

NAS (National Academy Of Sciences) (1980). *Research priorities in tropical biology*. National Academy of Sciences, Washington, DC.

NRC (National Research Council) (1982). *Ecological aspects of development in the humid tropics*. National Academy of Sciences, Washington, DC.

Norse, E. A., ed. (1993). *Global marine biological diversity: a strategy for building conservation into decision making*. Island Press, Washington, DC.

Norse, E. A. and Crowder, L. B. (2005). *Marine conservation biology: the science of maintaining the sea's biodiversity*. Island Press, Washington, DC.

Noss, R. F. (1990). Indicators for monitoring biodiversity: a hierarchical approach. *Conservation Biology*, **4**, 355–364.

Noss, R. F. (1997). The failure of universities to produce conservation biologists. *Conservation Biology*, **11**, 1267–1269.

Noss, R. F. (1999). Is there a special conservation biology? *Ecography*, **22**, 113–122.

Noss, R. F. (2000). Science on the bridge. *Conservation Biology*, **14**, 333–335.

Noss, R. F. and Cooperrider, A. Y. (1994). *Saving nature's legacy: protecting and restoring biodiversity*. Island Press, Washington, DC.

Noss, R. F., O'Connell, M. A., and Murphy, D. D. (1997). *The science of conservation planning*. Island Press, Washington, DC.

Odenbaugh, J. (2003). Values, advocacy, and conservation biology. *Environmental Values*, **12**, 55–69.

Orr, D. W. (2004). *The nature of design: ecology, culture, and human intention*. Oxford University Press, New York.

Pickett, S. T. A., and White, P. S. (1985). *The ecology of natural disturbance and patch dynamics*. Academic Press, Orlando, Florida.

Pimlott, D. H. (1969). The value of diversity. *Transactions of the North American Wildlife and Natural Resources Conference*, **34**, 265–280.

Pimm, S. L. (1991). *The balance of nature? Ecological issues in the conservation of species and communities*. University of Chicago Press, Chicago, Illinois.

Pinchot, G. (1910). *The fight for conservation*. Doubleday, Page, and Company, New York.

Pokras, M., Tabor, G., Pearl, M., Sherman, D., and Epstein, P. (1997). Conservation medicine: an emerging field. In P. H. Raven, ed. *Nature and human society: the quest for a sustainable world*, pp. 551–556. National Academy Press, Washington, DC.

Postel, S. and Richter, B. (2003). *Rivers for life: managing water for people and nature*. Island Press, Washington, DC.

Primack, R. (1993). *Essentials of conservation biology*. Sinauer Associates, Sunderland, Massachusetts.

Pringle, R. M. (2008). Elephants as agents of habitat creation for small vertebrates at patch scale. *Ecology*, **89**, 26–33.

Pringle, R. M., Young, T. P., Rubenstein, D. I., and Mccauly, D. J. (2007). Herbivore-initiated interaction cascades and their modulation by primary productivity in an African savanna. *Proceedings of the National Academy of Sciences of the United States of America*, **104**, 193–197.

Quammen, D. (1996). *The song of the dodo: island biogeography in an age of extinctions*. Simon and Schuster, New York.

Redford, K. H., Coppolillo, P., Sanderson, E. W., et al. (2003). Mapping the conservation landscape. *Conservation Biology*, **17**, 116–131.

Ripple, W. J. and Beschta, R. L. (2005). Linking wolves and plants: Aldo Leopold on trophic cascades. *BioScience*, **55**, 613–621.

Roberts, C. M., Halpern, B., Palumbi, S. R., and Warner, R. R. (2001). Designing marine reserve networks: why small, isolated protected areas are not enough. *Conservation Biology in Practice*, **2**, 10–17.

Rodríguez, J. P., Simonetti, J. A., Premoli, A., and Marini, M. Â. (2005). Conservation in Austral and Neotropical America: building scientific capacity equal to the challenges. *Conservation Biology*, **19**, 969–972.

Safina, C. (1998). *Song for the blue ocean*. Henry Holt and Company, New York.

Salafsky, N., Margoluis, R., and Redford, K. (2001). *Adaptive management: a tool for conservation practitioners*. Biodiversity Support Program, Washington, DC.

Samson, F. B. and Knopf, F. L. (1982). In search of a diversity ethic for wildlife management. *Transactions of the North American Wildlife and Natural Resources Conference*, **47**, 421–431.

Samways, M. J. (1994). *Insect conservation biology*. Chapman and Hall, New York.

Saunders, C. (2003). The emerging field of conservation psychology. *Human Ecology Review*, **10**, 137–149.

Schonewald-Cox, C., Chambers, S. M., Macbryde, B., and Thomas, L., eds (1983). *Genetics and conservation: a reference for managing wild animal and plant populations*. Benjamin-Cummings, Menlo Park, California.

Scott, M. J., Davis, F., Cusuti, B., et al. (1993). GAP analysis: a geographic approach to protection of biological diversity. *Wildlife Monographs*, **123**, 1–41.

Sellars, R. W. (1997). *Preserving nature in the national parks: a history*. Yale University Press, New Haven, Connecticut.

Shafer, C. L. (2001). Conservation biology trailblazers: George Wright, Ben Thompson, and Joseph Dixon. *Conservation Biology*, **15**, 332–344.

Sodhi, N. S., Brook, B. W., and Bradshaw, C. J. A. (2007). *Tropical conservation biology*. Blackwell, Oxford, UK.

Soulé, M. E. (1980). Thresholds for survival: criteria for maintenance of fitness and evolutionary potential. In M. E. Soulé, and B. A. Wilcox, eds *Conservation biology: an evolutionary-ecological perspective*, pp. 151–170. Sinauer Associates, Sunderland, Massachusetts.

Soulé, M. E. (1985). What is conservation biology? *BioScience*, **35**, 727–734.

Soulé, M. E., ed. (1986). *Conservation biology: the science of scarcity and diversity*. Sinauer Associates, Sunderland, Massachusetts.

Soulé, M. E., ed. (1987a). *Viable populations for conservation*. Cambridge University Press. Cambridge, UK.

Soulé, M. E. (1987b). History of the Society for Conservation Biology: how and why we got here. *Conservation Biology*, **1**, 4–5.

Soulé, M. E. (1991). Land use planning and wildlife maintenance: guidelines for preserving wildlife in an urban landscape. *Journal of the American Planning Association*, **57**, 313–323.

Soulé, M. E. and Terborgh, J., eds (1999). *Continental conservation: scientific foundations of regional reserve networks*. Washington, DC., Island Press.

Soulé, M. E. and Wilcox, B. A., eds (1980). *Conservation biology: an evolutionary-ecological perspective*. Sinauer, Sunderland, Massachusetts.

Tabor, G. M., Ostfeld, R. S., Poss, M., Dobson, A. P., and Aguirre, A. A. (2001). Conservation biology and the health sciences. In M. E. Soulé and G. H. Orians, eds *Conservation biology: research priorities for the next decade*, pp. 155–173. Island Press, Washington, DC.

Takacs, D. (1996). *The idea of biodiversity: philosophies of paradise*. Johns Hopkins University Press, Baltimore, Maryland.

Teer, J. G. (1988). Review of Conservation biology: the science of scarcity and diversity. *Journal of Wildlife Management*, **52**, 570–572.

Terborgh, J. (1974). Preservation of natural diversity: the problem of extinction prone species. *BioScience*, **24**, 715–722.

Thomas, W. L. Jr., Sauer, C. O., Bates, M., and Mumford, L. (1956). *Man's role in changing the face of the Earth*. University of Chicago Press, Chicago, Illinois.

Thorne-Miller, B. (1998). *The Living Ocean: Understanding and Protecting Marine Biodiversity*, 2nd ed. Island Press, Washington, DC.

Trombulak, S. C. (1994). Undergraduate education and the next generation of conservation biologists. *Conservation Biology*, **8**, 589–591.

Udall, J. R. (1991). Launching the natural ark. *Sierra*, **7**, 80–89.

Wallace, A. R. (1863). On the physical geography of the Malay Archipelago. *Journal of the Royal Geographical Society of London*, **33**, 217–234.

Wang, Y. and Moskovits, D. K. (2001). Tracking fragmentation of natural communities and changes in land cover: applications of Landsat data for conservation in an urban landscape (Chicago Wilderness). *Conservation Biology*, **15**, 835–843.

Wilson, E. O. and Peter, F. M., eds (1988). *Biodiversity*. National Academy Press, Washington, DC.

WCED (World Commission On Environment and Development) (1987). *Our common future*. Oxford University Press, Oxford, UK.

Worster, D. (1979). *Dust bowl: the southern plains in the 1930s*. Oxford University Press, New York.

Worster, D. (1985). *Nature's economy: a history of ecological ideas*. Cambridge University Press, Cambridge, UK.

Yaffee, S. L. (1994). *The wisdom of the spotted owl: policy lessons for a new century*. Island Press, Washington, DC.

Ziswiler, V. (1967). *Extinct and vanishing animals: a biology of extinction and survival*. Springer-Verlag, New York.

# CHAPTER 2

# Biodiversity

Kevin J. Gaston

Biological diversity or biodiversity (the latter term is simply a contraction of the former) is *the variety of life*, in all of its many manifestations. It is a broad unifying concept, encompassing all forms, levels and combinations of natural variation, at all levels of biological organization (Gaston and Spicer 2004). A rather longer and more formal definition is given in the international Convention on Biological Diversity (CBD; the definition is provided in Article 2), which states that "'Biological diversity' means the variability among living organisms from all sources including, *inter alia*, terrestrial, marine and other aquatic ecosystems and the ecological complexes of which they are part; this includes diversity within species, between species and of ecosystems". Whichever definition is preferred, one can, for example, speak equally of the biodiversity of some given area or volume (be it large or small) of the land or sea, of the biodiversity of a continent or an ocean basin, or of the biodiversity of the entire Earth. Likewise, one can speak of biodiversity at present, at a given time or period in the past or in the future, or over the entire history of life on Earth.

The scale of the variety of life is difficult, and perhaps impossible, for any of us truly to visualize or comprehend. In this chapter I first attempt to give some sense of the magnitude of biodiversity by distinguishing between different key elements and what is known about their variation. Second, I consider how the variety of life has changed through time, and third and finally how it varies in space. In short, the chapter will, inevitably in highly summarized form, address the three key issues of how much biodiversity there is, how it arose, and where it can be found.

## 2.1 How much biodiversity is there?

Some understanding of what the variety of life comprises can be obtained by distinguishing between different key elements. These are the basic building blocks of biodiversity. For convenience, they can be divided into three groups: genetic diversity, organismal diversity, and ecological diversity (Table 2.1). Within each, the elements are organized in nested hierarchies, with those higher order elements comprising lower order

**Table 2.1** Elements of biodiversity (focusing on those levels that are most commonly used). Modified from Heywood and Baste (1995).

| Ecological diversity | | Organismal diversity |
|---|---|---|
| Biogeographic realms | | Domains or Kingdoms |
| Biomes | | Phyla |
| Provinces | | Families |
| Ecoregions | | Genera |
| Ecosystems | | Species |
| Habitats | Genetic diversity | Subspecies |
| Populations | Populations | Populations |
| | Individuals | Individuals |
| | Chromosomes | |
| | Genes | |
| | Nucleotides | |

ones. The three groups are intimately linked and share some elements in common.

## 2.1.1 Genetic diversity

Genetic diversity encompasses the components of the genetic coding that structures organisms (nucleotides, genes, chromosomes) and variation in the genetic make-up between individuals within a population and between populations. This is the raw material on which evolutionary processes act. Perhaps the most basic measure of genetic diversity is genome size—the amount of DNA (Deoxyribonucleic acid) in one copy of a species' chromosomes (also called the C-value). This can vary enormously, with published eukaryote genome sizes ranging between 0.0023 pg (picograms) in the parasitic microsporidium *Encephalitozoon intestinalis* and 1400 pg in the free-living amoeba *Chaos chaos* (Gregory 2008). These translate into estimates of 2.2 million and 1369 billion base pairs (the nucleotides on opposing DNA strands), respectively. Thus, even at this level the scale of biodiversity is daunting. Cell size tends to increase with genome size. Humans have a genome size of 3.5 pg (3.4 billion base pairs).

Much of genome size comprises non-coding DNA, and there is usually no correlation between genome size and the number of genes coded. The genomes of more than 180 species have been completely sequenced and it is estimated that, for example, there are around 1750 genes for the bacteria *Haemophilus influenzae* and 3200 for *Escherichia coli*, 6000 for the yeast *Saccharomyces cerevisiae*, 19 000 for the nematode *Caenorhabditis elegans*, 13 500 for the fruit fly *Drosophila melanogaster*, and ~25 000 for the plant *Arabidopsis thaliana*, the mouse *Mus musculus*, brown rat *Rattus norvegicus* and human *Homo sapiens*. There is strong conservatism of some genes across much of the diversity of life. The differences in genetic composition of species give us indications of their relatedness, and thus important information as to how the history and variety of life developed.

Genes are packaged into chromosomes. The number of chromosomes per somatic cell thus far observed varies between 2 for the jumper ant *Myrmecia pilosula* and 1260 for the adders-tongue fern *Ophioglossum reticulatum*. The ant species reproduces by haplodiploidy, in which fertilized eggs (diploid) develop into females and unfertilized eggs (haploid) become males, hence the latter have the minimal achievable single chromosome in their cells (Gould 1991). Humans have 46 chromosomes (22 pairs of autosomes, and one pair of sex chromosomes).

Within a species, genetic diversity is commonly measured in terms of allelic diversity (average number of alleles per locus), gene diversity (heterozygosity across loci), or nucleotide differences. Large populations tend to have more genetic diversity than small ones, more stable populations more than those that wildly fluctuate, and populations at the center of a species' geographic range often have more genetic diversity than those at the periphery. Such variation can have a variety of population-level influences, including on productivity/biomass, fitness components, behavior, and responses to disturbance, as well as influences on species diversity and ecosystem processes (Hughes *et al.* 2008).

## 2.1.2 Organismal diversity

Organismal diversity encompasses the full taxonomic hierarchy and its components, from individuals upwards to populations, subspecies and species, genera, families, phyla, and beyond to kingdoms and domains. Measures of organismal diversity thus include some of the most familiar expressions of biodiversity, such as the numbers of species (i.e. species richness). Others should be better studied and more routinely employed than they have been thus far.

Starting at the lowest level of organismal diversity, little is known about how many individual organisms there are at any one time, although this is arguably an important measure of the quantity and variety of life (given that, even if sometimes only in small ways, most individuals differ from one another). Nonetheless, the numbers must be extraordinary. The global number of prokaryotes has been estimated to be $4-6 \times 10^{30}$ cells—many million times more than there are stars in the visible universe (Copley 2002)—with a production rate of $1.7 \times 10^{30}$ cells per annum (Whitman *et al.* 1998). The numbers of protists is estimated at $10^4-10^7$ individuals per $m^2$ (Finlay 2004).

Impoverished habitats have been estimated to have $10^5$ individual nematodes per m$^2$, and more productive habitats $10^6$–$10^7$ per m$^2$, possibly with an upper limit of $10^8$ per m$^2$; $10^{19}$ has been suggested as a conservative estimate of the global number of individuals of free-living nematodes (Lambshead 2004). By contrast, it has been estimated that globally there may be less than $10^{11}$ breeding birds at any one time, fewer than 17 for every person on the planet (Gaston et al. 2003).

Individual organisms can be grouped into relatively independent populations of a species on the basis of limited gene flow and some level of genetic differentiation (as well as on ecological criteria). The population is a particularly important element of biodiversity. First, it provides an important link between the different groups of elements of biodiversity (Table 2.1). Second, it is the scale at which it is perhaps most sensible to consider linkages between biodiversity and the provision of ecosystem services (supporting services—e.g. nutrient cycling, soil formation, primary production; provisioning services—e.g. food, freshwater, timber and fiber, fuel; regulating services—e.g. climate regulation, flood regulation, disease regulation, water purification; cultural services—e.g. aesthetic, spiritual, educational, recreational; MEA 2005). Estimates of the density of such populations and the average geographic range sizes of species suggest a total of about 220 distinct populations per eukaryote species (Hughes et al. 1997). Multiplying this by a range of estimates of the extant numbers of species, gives a global total of 1.1 to 6.6 x $10^9$ populations (Hughes et al. 1997), one or fewer for every person on the planet. The accuracy of this figure is essentially unknown, with major uncertainties at each step of the calculation, but the ease with which populations can be eradicated (e.g. through habitat destruction) suggests that the total is being eroded at a rapid rate.

People have long pondered one of the important contributors to the calculation of the total number of populations, namely how many different species of organisms there might be. Greatest uncertainty continues to surround the richness of prokaryotes, and in consequence they are often ignored in global totals of species numbers. This is in part variously because of difficulties in applying standard species concepts, in culturing the vast majority of these organisms and thereby applying classical identification techniques, and by the vast numbers of individuals. Indeed, depending on the approach taken, the numbers of prokaryotic species estimated to occur even in very small areas can vary by a few orders of magnitude (Curtis et al. 2002; Ward 2002). The rate of reassociation of denatured (i.e. single stranded) DNA has revealed that in pristine soils and sediments with high organic content samples of 30 to 100 cm$^3$ correspond to c. 3000 to 11 000 different genomes, and may contain $10^4$ different prokaryotic species of equivalent abundances (Torsvik et al. 2002). Samples from the intestinal microbial flora of just three adult humans contained representatives of 395 bacterial operational taxonomic units (groups without formal designation of taxonomic rank, but thought here to be roughly equivalent to species), of which 244 were previously unknown, and 80% were from species that have not been cultured (Eckburg et al. 2005). Likewise, samples from leaves were estimated to harbor at least 95 to 671 bacterial species from each of nine tropical tree species, with only 0.5% common to all the tree species, and almost all of the bacterial species being undescribed (Lambais et al. 2006). On the basis of such findings, global prokaryote diversity has been argued to comprise possibly millions of species, and some have suggested it may be many orders of magnitude more than that (Fuhrman and Campbell 1998; Dykhuizen 1998; Torsvik et al. 2002; Venter et al. 2004).

Although much more certainty surrounds estimates of the numbers of eukaryotic than prokaryotic species, this is true only in a relative and not an absolute sense. Numbers of eukaryotic species are still poorly understood. A wide variety of approaches have been employed to estimate the global numbers in large taxonomic groups and, by summation of these estimates, how many extant species there are overall. These approaches include extrapolations based on counting species, canvassing taxonomic experts, temporal patterns of species description, proportions of undescribed species in samples, well-studied areas, well-studied groups, species-abundance distributions, species-body size distributions, and trophic relations (Gaston 2008). One recent summary for eukaryotes gives lower and upper estimates of 3.5 and 108 million species, respectively, and a working figure of around 8 million species (Table 2.2). Based on current information the two extremes seem rather unlikely, but the working figure at least seems tenable. However, major uncertainties surround global numbers of eukaryotic species in particular environments which have been poorly sampled (e.g. deep sea, soils, tropical forest canopies), in higher taxa which are extremely species rich or with species which are very difficult to discriminate (e.g. nematodes, arthropods), and in particular functional groups which are less readily studied (e.g. parasites). A wide array of techniques is now being employed to gain access to some of the environments that have been less well explored, including rope climbing techniques, aerial walkways, cranes and balloons for tropical forest canopies, and remotely operated vehicles, bottom landers, submarines, sonar, and video for the deep ocean. Molecular and better imaging techniques are also improving species discrimination. Perhaps most significantly, however, it seems highly probable that the majority of species are parasites, and yet few people tend to think about biodiversity from this viewpoint.

How many of the total numbers of species have been taxonomically described remains surprisingly uncertain, in the continued absence of a single unified, complete and maintained database of valid formal names. However, probably about 2 million extant species are regarded as being known to science (MEA 2005). Importantly, this total hides two kinds of error. First, there are instances in which the same species is known under more than one name (synonymy). This is more frequent amongst widespread species, which may show marked geographic variation in morphology, and may be described anew repeatedly in different regions. Second, one name may actually encompass multiple species (homonymy). This typically occurs because these species are very closely related, and look very similar (cryptic species), and molecular analyses may be required to recognize or confirm their differences. Levels of as yet unresolved synonymy are undoubtedly high in many taxonomic groups. Indeed, the actual levels have proven to be a key issue in, for example, attempts to estimate the global species richness of plants, with the highly variable synonymy rate amongst the few groups that have been well studied in this regard making difficult the assessment of the overall level of synonymy across all the known species. Equally, however, it is apparent that cryptic species abound, with, for example, one species of neotropical skipper butterfly recently having been shown actually to be a complex of ten species (Hebert *et al.* 2004).

New species are being described at a rate of about 13 000 per annum (Hawksworth and

**Table 2.2** Estimates (in thousands), by different taxonomic groups, of the overall global numbers of extant eukaryote species. Modified from Hawksworth and Kalin-Arroyo (1995) and May (2000).

|  | Overall species | | | |
| --- | --- | --- | --- | --- |
|  | High | Low | Working figure | Accuracy of working figure |
| 'Protozoa' | 200 | 60 | 100 | very poor |
| 'Algae' | 1000 | 150 | 300 | very poor |
| Plants | 500 | 300 | 320 | good |
| Fungi | 2700 | 200 | 1500 | moderate |
| Nematodes | 1000 | 100 | 500 | very poor |
| Arthropods | 101 200 | 2375 | 4650 | moderate |
| Molluscs | 200 | 100 | 120 | moderate |
| Chordates | 55 | 50 | 50 | good |
| Others | 800 | 200 | 250 | moderate |
| **Totals** | 107 655 | 3535 | 7790 | very poor |

Kalin-Arroyo 1995), or about 36 species on the average day. Given even the lower estimates of overall species numbers this means that there is little immediate prospect of greatly reducing the numbers that remain unknown to science. This is particularly problematic because the described species are a highly biased sample of the extant biota rather than the random one that might enable more ready extrapolation of its properties to all extant species. On average, described species tend to be larger bodied, more abundant and more widespread, and disproportionately from temperate regions. Nonetheless, new species continue to be discovered in even otherwise relatively well-known taxonomic groups. New extant fish species are described at the rate of about 130–160 each year (Berra 1997), amphibian species at about 95 each year (from data in Frost 2004), bird species at about 6–7 each year (Van Rootselaar 1999, 2002), and terrestrial mammals at 25–30 each year (Ceballos and Ehrlich 2009). Recently discovered mammals include marsupials, whales and dolphins, a sloth, an elephant, primates, rodents, bats and ungulates.

Given the high proportion of species that have yet to be discovered, it seems highly likely that there are entire major taxonomic groups of organisms still to be found. That is, new examples of higher level elements of organismal diversity. This is supported by recent discoveries of possible new phyla (e.g. Nanoarchaeota), new orders (e.g. Mantophasmatodea), new families (e.g. Aspidytidae) and new subfamilies (e.g. Martialinae). Discoveries at the highest taxonomic levels have particularly served to highlight the much greater phyletic diversity of microorganisms compared with macroorganisms. Under one classification 60% of living phyla consist entirely or largely of unicellular species (Cavalier-Smith 2004). Again, this perspective on the variety of life is not well reflected in much of the literature on biodiversity.

### 2.1.3 Ecological diversity

The third group of elements of biodiversity encompasses the scales of ecological differences from populations, through habitats, to ecosystems, ecoregions, provinces, and on up to biomes and biogeographic realms (Table 2.1). This is an important dimension to biodiversity not readily captured by genetic or organismal diversity, and in many ways is that which is most immediately apparent to us, giving the structure of the natural and semi-natural world in which we live. However, ecological diversity is arguably also the least satisfactory of the groups of elements of biodiversity. There are two reasons. First, whilst these elements clearly constitute useful ways of breaking up continua of phenomena, they are difficult to distinguish without recourse to what ultimately constitute some essentially arbitrary rules. For example, whilst it is helpful to be able to label different habitat types, it is not always obvious precisely where one should end and another begin, because no such beginnings and endings really exist. In consequence, numerous schemes have been developed for distinguishing between many elements of ecological diversity, often with wide variation in the numbers of entities recognized for a given element. Second, some of the elements of ecological diversity clearly have both abiotic and biotic components (e.g. ecosystems, ecoregions, biomes), and yet biodiversity is defined as the variety of *life*.

Much recent interest has focused particularly on delineating ecoregions and biomes, principally for the purposes of spatial conservation planning (see Chapter 11), and there has thus been a growing sense of standardization of the schemes used. Ecoregions are large areal units containing geographically distinct species assemblages and experiencing geographically distinct environmental conditions. Careful mapping schemes have identified 867 terrestrial ecoregions (Figure 2.1 and Plate 1; Olson *et al.* 2001), 426 freshwater ecoregions (Abell *et al.* 2008), and 232 marine coastal & shelf area ecoregions (Spalding *et al.* 2007). Ecoregions can in turn be grouped into biomes, global-scale biogeographic regions distinguished by unique collections of species assemblages and ecosystems. Olson *et al.* (2001) distinguish 14 terrestrial biomes, some of which at least will be very familiar wherever in the world one resides (tropical & subtropical moist broadleaf forests;

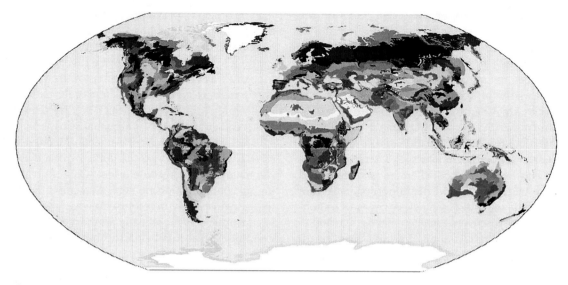

**Figure 2.1** The terrestrial ecoregions. Reprinted from Olson et al. (2001).

tropical & subtropical dry broadleaf forests; tropical & subtropical coniferous forests; temperate broadleaf & mixed forests; temperate coniferous forests; boreal forest/taiga; tropical & subtropical grasslands, savannas & shrublands; temperate grasslands, savannas & shrublands; flooded grasslands & savannas; montane grasslands & shrublands; tundra; Mediterranean forests, woodlands & scrub; deserts & xeric shrublands; mangroves).

At a yet coarser spatial resolution, terrestrial and aquatic systems can be divided into biogeographic realms. Terrestrially, eight such realms are typically recognized, Australasia, Antarctic, Afrotropic, Indo-Malaya, Nearctic, Neotropic, Oceania and Palearctic (Olson et al. 2001). Marine coastal & shelf areas have been divided into 12 realms (Arctic, Temperate North Atlantic, Temperate Northern Pacific, Tropical Atlantic, Western Indo-Pacific, Central Indo-Pacific, Eastern Indo-Pacific, Tropical Eastern Pacific, Temperate South America, Temperate Southern Africa, Temperate Australasia, and Southern Ocean; Spalding et al. 2007). There is no strictly equivalent scheme for the pelagic open ocean, although one has divided the oceans into four primary units (Polar, Westerlies, Trades and Coastal boundary), which are then subdivided, on the basis principally of biogeochemical features, into a further 12 biomes (Antarctic Polar, Antarctic Westerly Winds, Atlantic Coastal, Atlantic Polar, Atlantic Trade Wind, Atlantic Westerly Winds, Indian Ocean Coastal, Indian Ocean Trade Wind, Pacific Coastal, Pacific Polar, Pacific Trade Wind, Pacific Westerly Winds), and then into a finer 51 units (Longhurst 1998).

### 2.1.4 Measuring biodiversity

Given the multiple dimensions and the complexity of the variety of life, it should be obvious that there can be no single measure of biodiversity (see Chapter 16). Analyses and discussions of biodiversity have almost invariably to be framed in terms of particular elements or groups of elements, although this may not always be apparent from the terminology being employed (the term 'biodiversity' is used widely and without explicit qualification to refer to only some subset of the variety of life). Moreover, they have to be framed in terms either of "number" or of "heterogeneity" measures of biodiversity, with the former disregarding the degrees of difference between the occurrences of an element of biodiversity and the latter explicitly incorporating such differences. For example, organismal diversity could be expressed in terms of species richness, which is a number measure, or using an index of diversity that incorporates differences in the abundances of the species, which is a heterogeneity measure. The two approaches constitute different responses to the question of whether biodiversity is similar or different in an assemblage in which a small proportion of the species comprise most of the individuals, and therefore would predominantly be obtained in a small sample of individuals, or in an assemblage of the same total number of species in which abundances are more evenly distributed, and thus more species would occur in a small sample of individuals (Purvis and Hector 2000). The distinction between number and heterogeneity measures is also captured in answers to questions that reflect taxonomic heterogeneity, for example whether the above-mentioned group of 10 skipper butterflies is as biodiverse as a group of five skipper species and five swallowtail species (e.g. Hendrickson and Ehrlich 1971).

In practice, biodiversity tends most commonly to be expressed in terms of number measures of organismal diversity, often the numbers of a given taxonomic level, and particularly the numbers of species. This is in large part a pragmatic choice. Organismal diversity is better documented and often more readily estimated than is genetic diversity, and more finely and consistently resolved than much of ecological diversity. Organismal diversity, however, is problematic inasmuch as the majority of it remains unknown (and thus studies have to be based on subsets), and precisely how naturally and well many taxonomic groups are themselves delimited remains in dispute. Perhaps most importantly it also remains but one, and arguably a quite narrow, perspective on biodiversity.

Whilst accepting the limitations of measuring biodiversity principally in terms of organismal diversity, the following sections on temporal and spatial variation in biodiversity will follow this course, focusing in many cases on species richness.

## 2.2 How has biodiversity changed through time?

The Earth is estimated to have formed, by the accretion through large and violent impacts of numerous bodies, approximately 4.5 billion years ago (Ga). Traditionally, habitable worlds are considered to be those on which liquid water is stable at the surface. On Earth, both the atmosphere and the oceans may well have started to form as the planet itself did so. Certainly, life is thought to have originated on Earth quite early in its history, probably after about 3.8–4.0 Ga, when impacts from large bodies from space are likely to have declined or ceased. It may have originated in a shallow marine pool, experiencing intense radiation, or possibly in the environment of a deeper water hydrothermal vent. Because of the subsequent recrystallisation and deformation of the oldest sediments on Earth, evidence for early life must be found in its metabolic interaction with the environment. The earliest, and highly controversial, evidence of life, from such indirect geochemical data, is from more than 3.83 billion years ago (Dauphas *et al.* 2004). Relatively unambiguous fossil evidence of life dates to 2.7 Ga (López-García *et al.* 2006). Either way, life has thus been present throughout much of the Earth's existence. Although inevitably attention tends to fall on more immediate concerns, it is perhaps worth occasionally recalling this deep heritage in the face of the conservation challenges of today. For much of this time, however, life comprised Precambrian chemosynthetic and photosynthetic prokaryotes, with oxygen-producing cyanobacteria being particularly important (Labandeira 2005). Indeed, the evolution of oxygenic photosynthesis, followed by oxygen becoming a major component of the atmosphere, brought about a dramatic transformation of the environment on Earth. Geochemical data has been argued to suggest that oxygenic photosynthesis evolved before 3.7 Ga (Rosing and Frei 2004), although others have proposed that it could not have arisen before c.2.9 Ga (Kopp *et al.* 2005).

These cyanobacteria were initially responsible for the accumulation of atmospheric oxygen. This in turn enabled the emergence of aerobically metabolizing eukaryotes. At an early stage, eukaryotes incorporated within their structure aerobically metabolizing bacteria, giving rise to eukaryotic cells with mitochondria; all anaerobically metabolizing eukaryotes that have been studied in detail have thus far been found to have had aerobic ancestors, making it highly likely that the ancestral eukaryote was aerobic (Cavalier-Smith 2004). This was a fundamentally important event, leading to heterotrophic microorganisms and sexual means of reproduction. Such endosymbiosis occurred serially, by simpler and more complex routes, enabling eukaryotes to diversify in a variety of ways. Thus, the inclusion of photosynthesizing cyanobacteria into a eukaryote cell that already contained a mitochondrion gave rise to eukaryotic cells with plastids and capable of photosynthesis. This event alone would lead to dramatic alterations in the Earth's ecosystems.

Precisely when eukaryotes originated, when they diversified, and how congruent was the diversification of different groups remains unclear, with analyses giving a very wide range of dates (Simpson and Roger 2004). The uncertainty, which is particularly acute when attempting to understand evolutionary events in deep time, results principally from the inadequacy of the fossil record (which, because of the low probabilities of fossilization and fossil recovery, will always tend to underestimate the ages of taxa) and the difficulties of correctly calibrating molecular clocks so as to use the information embodied in genetic sequences to date these events. Nonetheless, there is increasing convergence on the idea that most known eukaryotes can be placed in one of five or six major clades—Unikonts (Opisthokonts and Amoebozoa), Plantae, Chromalveolates, Rhizaria and Excavata (Keeling *et al.* 2005; Roger and Hug 2006).

Focusing on the last 600 million years, attention shifts somewhat from the timing of key diversification events (which becomes less controversial) to how diversity *per se* has changed through time (which becomes more measurable). Arguably the critical issue is how well the known fossil record reflects the actual patterns of change that took place and how this record can best be analyzed to address its associated biases to determine those actual patterns. The best fossil data are for marine invertebrates and it was long thought that these principally demonstrated a dramatic rise in diversity, albeit punctuated by significant periods of stasis and mass extinction events. However, analyses based on standardized sampling have markedly altered this picture (Figure 2.2). They identify the key features of change in the numbers of genera (widely assumed to correlate with species richness) as comprising: (i) a rise in richness from the Cambrian through to the mid-Devonian (~525–400 million years ago, Ma); (ii) a large extinction in the mid-Devonian with no clear recovery until the Permian (~400–300 Ma); (iii) a large extinction in the late-Permian and again in the late-Triassic (~250–200 Ma); and (iv) a rise in richness through the late-Triassic to the present (~200–0 Ma; Alroy *et al.* 2008).

Whatever the detailed pattern of change in diversity through time, most of the species that have ever existed are extinct. Across a variety of groups (both terrestrial and marine), the best present estimate based on fossil evidence is that the average species has had a lifespan (from its appearance in the fossil record until the time it disappeared) of perhaps around 1–10 Myr (McKinney 1997; May 2000). However, the variability both within and between groups is very marked, making estimation of what is the overall average difficult. The longest-lived species that is well documented is a bryozoan that persisted from the early Cretaceous to the present, a period of approximately 85 million years (May 2000). If the fossil record spans 600 million years, total species numbers were to have been roughly constant over this period, and the average life span of individual species were 1–10 million years, then at any specific instant the extant species would have represented 0.2–2% of those that have ever lived (May 2000). If this were true of the present time then, if the number of extant eukaryote species numbers 8 million, 400 million might once have existed.

The frequency distribution of the numbers of time periods with different levels of extinction is markedly right-skewed, with most periods having relatively low levels of extinction and a

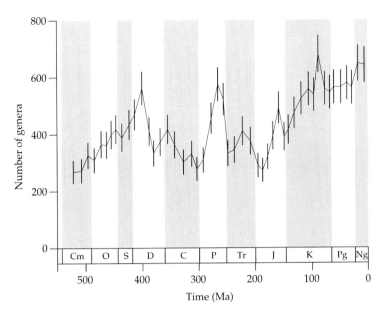

**Figure 2.2** Changes in generic richness of marine invertebrates over the last 600 million years based on a sampling-standardized analysis of the fossil record. Ma, million years ago. Reprinted from Alroy et al. (2008) with permission from AAAS (American Association for the Advancement of Science).

minority having very high levels (Raup 1994). The latter are the periods of mass extinction when 75–95% of species that were extant are estimated to have become extinct. Their significance lies not, however, in the overall numbers of extinctions for which they account (over the last 500 Myr this has been rather small), but in the hugely disruptive effect they have had on the development of biodiversity. Clearly neither terrestrial nor marine biotas are infinitely resilient to environmental stresses. Rather, when pushed beyond their limits they can experience dramatic collapses in genetic, organismal and ecological diversity (Erwin 2008). This is highly significant given the intensity and range of pressures that have been exerted on biodiversity by humankind, and which have drastically reshaped the natural world over a sufficiently long period in respect to available data that we have rather little concept of what a truly natural system should look like (Jackson 2008). Recovery from past mass extinction events has invariably taken place. But, whilst this may have been rapid in geological terms, it has nonetheless taken of the order of a few million years (Erwin 1998), and the resultant assemblages have invariably had a markedly different composition from those that preceded a mass extinction, with groups which were previously highly successful in terms of species richness being lost entirely or persisting at reduced numbers.

## 2.3 Where is biodiversity?

Just as biodiversity has varied markedly through time, so it also varies across space. Indeed, one can think of it as forming a richly textured land and seascape, with peaks (hotspots) and troughs (coldspots), and extensive plains in between (Figure 2.3 and Plate 2, and 2.4 and Plate 3; Gaston 2000). Even locally, and just for particular groups, the numbers of species can be impressive, with for example c.900 species of fungal fruiting bodies recorded from 13 plots totaling just 14.7 ha (hectare) near Vienna, Austria (Straatsma and Krisai-Greilhuber 2003), 173 species of lichens on a single tree in Papua New Guinea (Aptroot

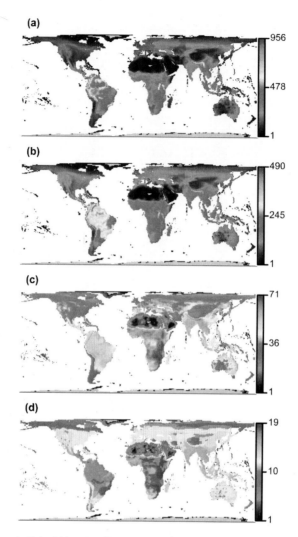

**Figure 2.3** Global richness patterns for birds of (a) species, (b) genera, (c) families, and (d) orders. Reprinted from Thomas et al. (2008).

1997), 814 species of trees from a 50 ha study plot in Peninsular Malaysia (Manokaran et al. 1992), 850 species of invertebrates estimated to occur at a sandy beach site in the North Sea (Armonies and Reise 2000), 245 resident species of birds recorded holding territories on a 97 ha plot in Peru (Terborgh et al. 1990), and >200 species of mammals occurring at some sites in the Amazonian rain forest (Voss and Emmons 1996).

Although it remains the case that for no even moderately sized area do we have a comprehensive inventory of all of the species that are present (microorganisms typically remain insufficiently documented even in otherwise well studied areas), knowledge of the basic patterns has been developing rapidly. Although long constrained to data on higher vertebrates, the breadth of organisms for which information is available has been growing, with much recent work particularly attempting to determine whether microorganisms show the same geographic patterns as do other groups.

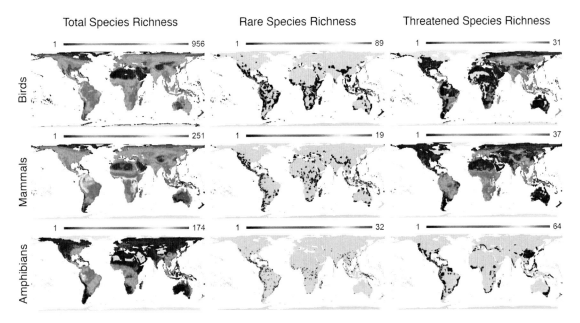

**Figure 2.4** Global species richness patterns of birds, mammals, and amphibians, for total, rare (those in the lower quartile of range size for each group) and threatened (according to the IUCN criteria) species. Reprinted from Grenyer *et al.* (2006).

## 2.3.1 Land and water

The oceans cover ~340.1 million km² (67%), the land ~170.3 million km² (33%), and freshwaters (lakes and rivers) ~1.5 million km² (0.3%; with another 16 million km² under ice and permanent snow, and 2.6 million km² as wetlands, soil water and permafrost) of the Earth's surface. It would therefore seem reasonable to predict that the oceans would be most biodiverse, followed by the land and then freshwaters. In terms of numbers of higher taxa, there is indeed some evidence that marine systems are especially diverse. For example, of the 96 phyla recognized by Margulis and Schwartz (1998), about 69 have marine representatives, 55 have terrestrial ones, and 60 have freshwater representatives. However, of the species described to date only about 15% are marine and 6% are freshwater. The fact that life began in the sea seems likely to have played an important role in explaining why there are larger numbers of higher taxa in marine systems than in terrestrial ones. The heterogeneity and fragmentation of the land masses (particularly that associated with the breakup of the "supercontinent" of Gondwana from ~180 Ma) is important in explaining why there are more species in terrestrial systems than in marine ones. Finally, the extreme fragmentation and isolation of freshwater bodies seems key to why these are so diverse for their area.

## 2.3.2 Biogeographic realms and ecoregions

Of the terrestrial realms, the Neotropics is generally regarded as overall being the most biodiverse, followed by the Afrotropics and Indo-Malaya, although the precise ranking of these tropical regions depends on the way in which organismal diversity is measured. For example, for species the richest realm is the Neotropics for amphibians, reptiles, birds and mammals, but for families it is the Afrotropics for amphibians and mammals, the Neotropics for reptiles, and the Indo-Malayan for birds (MEA 2005). In parts, these differences reflect variation in the histories of the realms (especially mountain uplift and climate changes) and the interaction with the emergence and spread of the groups, albeit perhaps

**Table 2.3** The five most species rich terrestrial ecoregions for each of four vertebrate groups. AT – Afrotropic, IM – Indo-Malaya, NA – Nearctic, and NT–Neotropic. Data from Olson et al. (2001).

| | Amphibians | Reptiles | Birds | Mammals |
|---|---|---|---|---|
| 1 | Northwestern Andean montane forests (NT) | Peten-Veracruz moist forests (NT) | Northern Indochina subtropical forests (IM) | Sierra Madre de Oaxaca pine-oak forests (NT) |
| 2 | Eastern Cordillera real montane forests (NT) | Southwest Amazon moist forests (NT) | Southwest Amazon moist forests (NT) | Northern Indochina subtropical forests (IM) |
| 3 | Napo moist forests (NT) | Napo moist forests (NT) | Albertine Rift montane forests (AT) | Sierra Madre Oriental pine-oak forests (NA) |
| 4 | Southwest Amazon moist forests (NT) | Southern Pacific dry forests (NT) | Central Zambezian Miombo woodlands (AT) | Southwest Amazon moist forests (NT) |
| 5 | Choco-Darien moist forests (NT) | Central American pine-oak forests (NT) | Northern Acacia-Commiphora bushlands & thickets (AT) | Central Zambezian Miombo woodlands (AT) |

complicated by issues of geographic consistency in the definition of higher taxonomic groupings.

The Western Indo-Pacific and Central Indo-Pacific realms have been argued to be a center for the evolutionary radiation of many groups, and are thought to be perhaps the global hotspot of marine species richness and endemism (Briggs 1999; Roberts et al. 2002). With a shelf area of 6 570 000 km$^2$, which is considered to be a significant influence, it has more than 6000 species of molluscs, 800 species of echinoderms, 500 species of hermatypic (reef forming) corals, and 4000 species of fish (Briggs 1999).

At the scale of terrestrial ecoregions, the most speciose for amphibians and reptiles are in the Neotropics, for birds in Indo-Malaya, Neotropics and Afrotropics, and for mammals in the Neotropics, Indo-Malaya, Nearctic, and Afrotropics (Table 2.3). Amongst the freshwater ecoregions, those with globally high richness of freshwater fish include the Brahmaputra, Ganges, and Yangtze basins in Asia, and large portions of the Mekong, Chao Phraya, and Sitang and Irrawaddy; the lower Guinea in Africa; and the Paraná and Orinoco in South America (Abell et al. 2008).

## 2.3.3 Latitude

Perhaps the best known of all spatial patterns in biodiversity is the general increase in species richness (and some other elements of organismal diversity) towards lower (tropical) latitudes. Several features of this gradient are of note: (i) it is exhibited in marine, terrestrial and freshwaters, and by virtually all major taxonomic groups, including microbes, plants, invertebrates and vertebrates (Hillebrand 2004; Fuhrman et al. 2008); (ii) it is typically manifest whether biodiversity is determined at local sites, across large regions, or across entire latitudinal bands; (iii) it has been a persistent feature of much of the history of life on Earth (Crane and Lidgard 1989; Alroy et al. 2008); (iv) the peak of diversity is seldom at the equator itself, but seems often to be displaced somewhat further north (often at ~20–30°N); (v) it is commonly, though far from universally, asymmetrical about the equator, increasing rapidly from northern regions to the equator and declining slowly from the equator to southern regions; and (vi) it varies markedly in steepness for different major taxonomic groups with, for example, butterflies being more tropical than birds.

Although it attracts much attention in its own right, it is important to see the latitudinal pattern in species richness as a component of broader spatial patterns of richness. As such, the mechanisms that give rise to it are also those that give rise to those broader patterns. Ultimately, higher species richness has to be generated by some combination of greater levels of speciation (a cradle of diversity), lower levels of extinction (a museum of diversity) or greater net movements of geographic ranges. It is likely that their relative importance in giving rise to latitudinal gradients varies with taxon and region. This said, greater levels of speciation at low latitudes and range expansion of lineages from lower to higher latitudes seem to be particularly important (Jablonski *et al.* 2006; Martin *et al.* 2007). More proximally, key constraints on speciation and extinction rates and range movements are thought to be levels of: (i) productive energy, which influence the numbers of individuals that can be supported, thereby limiting the numbers of species that can be maintained in viable populations; (ii) ambient energy, which influences mutation rates and thus speciation rates; (iii) climatic variation, which on ecological time scales influences the breadth of physiological tolerances and dispersal abilities and thus the potential for population divergence and speciation, and on evolutionary time scales influences extinctions (e.g. through glacial cycles) and recolonizations; and (iv) topographic variation, which enhances the likelihood of population isolation and thus speciation (Gaston 2000; Evans *et al.* 2005; Clarke and Gaston 2006; Davies *et al.* 2007).

### 2.3.4 Altitude and Depth

Variations in depth in marine systems and altitude in terrestrial ones are small relative to the areal coverage of these systems. The oceans average c.3.8 km in depth but reach down to 10.9 km (Challenger Deep), and land averages 0.84 km in elevation and reaches up to 8.85 km (Mt. Everest). Nonetheless, there are profound changes in organismal diversity both with depth and altitude. This is in large part because of the environmental differences (but also the effects of area and isolation), with some of those changes in depth or altitude of a few hundred meters being similar to those experienced over latitudinal distances of several hundred kilometers (e.g. temperature).

In both terrestrial and marine (pelagic and benthic) systems, species richness across a wide variety of taxonomic groups has been found progressively to decrease with distance from sea level (above or below) and to show a pronounced hump-shaped pattern in which it first increases and then declines (Angel 1994; Rahbek 1995; Bryant *et al.* 2008). The latter pattern tends to become more apparent when the effects of variation in area have been accounted for, and is probably the more general, although in either case richness tends to be lowest at the most extreme elevations or depths.

Microbial assemblages can be found at considerable depths (in some instances up to a few kilometers) below the terrestrial land surface and the seafloor, often exhibiting unusual metabolic capabilities (White *et al.* 1998; D'Hondt *et al.* 2004). Knowledge of these assemblages remains, however, extremely poor, given the physical challenges of sampling and of doing so without contamination from other sources.

## 2.4 In conclusion

Understanding of the nature and scale of biodiversity, of how it has changed through time, and of how it varies spatially has developed immeasurably in recent decades. Improvements in the levels of interest, the resources invested and the application of technology have all helped. Indeed, it seems likely that the basic principles are in the main well established. However, much remains to be learnt. The obstacles are fourfold. First, the sheer magnitude and complexity of biodiversity constitute a huge challenge to addressing perhaps the majority of questions that are posed about it, and one that is unlikely to be resolved in the near future. Second, the biases of the fossil record and the apparent variability in rates of molecular evolution continue to thwart a better understanding of the history of biodiversity. Third, knowledge of the spatial patterning of biodiversity is limited by the relative paucity of quantitative sampling of biodiversity over much of the planet. Finally, the levels and patterns of biodiversity are being profoundly altered by human activities (see Box 2.1 and Chapter 10).

## Box 2.1 Invaluable biodiversity inventories
### Navjot S. Sodhi

This chapter defines biodiversity. Due to massive loss of native habitats around the globe (Chapter 4), biodiversity is rapidly being eroded (Chapter 10). Therefore, it is critical to understand which species will survive human onslaught and which will not. We also need to comprehend the composition of new communities that arise after the loss or disturbance of native habitats. Such a determination needs a "peek" into the past. That is, which species were present before the habitat was disturbed. Perhaps naturalists in the 19th and early 20th centuries did not realize that they were doing a great service to future conservation biologists by publishing species inventories. These historic inventories are treasure troves—they can be used as baselines for current (and future) species loss and turnover assessments.

Singapore represents a worst-case scenario in tropical deforestation. This island (540 km$^2$) has lost over 95% of its primary forests since 1819. Comparing historic and modern inventories, Brook et al. (2003) could determine losses in vascular plants, freshwater decapod crustaceans, phasmids, butterflies, freshwater fish, amphibians, reptiles, birds, and mammals. They found that overall, 28% of original species were lost in Singapore, probably due to deforestation. Extinctions were higher (34–43%) in butterflies, freshwater fish, birds, and mammals. Due to low endemism in Singapore, all of these extinctions likely represented population than species extinctions (see Box 10.1). Using extinction data from Singapore, Brook et al. (2003) also projected that if the current levels of deforestation in Southeast Asia continue, between 13–42% of regional populations could be lost by 2100. Half of these extinctions could represent global species losses.

Fragments are becoming a prevalent feature in most landscapes around the globe (Chapter 5). Very little is known about whether fragments can sustain forest biodiversity over the long-term. Using an old species inventory, Sodhi et al. (2005) studied the avifaunal change over 100 years (1898–1998) in a four hectare patch of rain forest in Singapore (Singapore Botanic Gardens). Over this period, many forest species (e.g. green broadbill (*Calyptomena viridis*); Box 2.1 Figure) were lost, and replaced with introduced species such as the house crow (*Corvus splendens*). By 1998, 20% of individuals observed belonged to introduced species, with more native species expected to be extirpated from the site in the future through competition and predation. This study shows that small fragments decline in their value for forest birds over time.

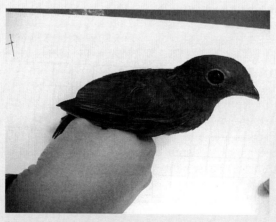

**Box 2.1 Figure** Green broadbill. Photograph by Haw Chuan Lim.

*continues*

### Box 2.1 (Continued)

The old species inventories not only help in understanding species losses but also help determine the characteristics of species that are vulnerable to habitat perturbations. Koh *et al.* (2004) compared ecological traits (e.g. body size) between extinct and extant butterflies in Singapore. They found that butterflies species restricted to forests and those which had high larval host plant specificity were particularly vulnerable to extirpation. In a similar study, but on angiosperms, Sodhi *et al.* (2008) found that plant species susceptible to habitat disturbance possessed traits such as dependence on forests and pollination by mammals. These trait comparison studies may assist in understanding underlying mechanisms that make species vulnerable to extinction and in preemptive identification of species at risk from extinction.

The above highlights the value of species inventories. I urge scientists and amateurs to make species lists every time they visit a site. Data such as species numbers should also be included in these as such can be used to determine the effect of abundance on species persistence. All these checklists should be placed on the web for wide dissemination. Remember, like antiques, species inventories become more valuable with time.

### REFERENCES

Brook, B. W., Sodhi, N. S., and Ng, P. K. L. (2003). Catastrophic extinctions follow deforestation in Singapore. *Nature*, **424**, 420–423.

Koh, L. P., Sodhi, N. S., and Brook, B. W. (2004). Prediction extinction proneness of tropical butterflies. *Conservation Biology*, **18**, 1571–1578.

Sodhi, N.S., Lee, T. M., Koh, L. P., and Dunn, R. R. (2005). A century of avifaunal turnover in a small tropical rainforest fragment. *Animal Conservation*, **8**, 217–222.

Sodhi, N. S., Koh, L. P., Peh, K. S.-H. *et al.* (2008). Correlates of extinction proneness in tropical angiosperms. *Diversity and Distributions*, **14**, 1–10.

## Summary

- Biodiversity is the variety of life in all of its many manifestations.
- This variety can usefully be thought of in terms of three hierarchical sets of elements, which capture different facets: genetic diversity, organismal diversity, and ecological diversity.
- There is by definition no single measure of biodiversity, although two different kinds of measures (number and heterogeneity) can be distinguished.
- Pragmatically, and rather restrictively, biodiversity tends in the main to be measured in terms of number measures of organismal diversity, and especially species richness.
- Biodiversity has been present for much of the history of the Earth, but the levels have changed dramatically and have proven challenging to document reliably.
- Biodiversity is variably distributed across the Earth, although some marked spatial gradients seem common to numerous higher taxonomic groups.
- The obstacles to an improved understanding of biodiversity are: (i) its sheer magnitude and complexity; (ii) the biases of the fossil record and the apparent variability in rates of molecular evolution; (iii) the relative paucity of quantitative sampling over much of the planet; and (iv) that levels and patterns of biodiversity are being profoundly altered by human activities.

## Suggested reading

- Gaston, K. J. and Spicer, J. I. (2004). *Biodiversity: an introduction*, 2nd edition. Blackwell Publishing, Oxford, UK.

- Groombridge, B. and Jenkins, M. D. (2002). *World atlas of biodiversity: earth's living resources in the 21st century*. University of California Press, London, UK.
- Levin, S. A., ed. (2001). *Encyclopedia of biodiversity, Vols. 1–5*. Academic Press, London, UK.
- MEA (millennium Ecosystem Assessment) (2005). *Ecosystems and human well-being: current state and trends, Volume 1*. Island Press, Washington, DC.
- Wilson, E. O. (2001). *The diversity of life*, 2nd edition. Penguin, London, UK.

## Relevant website

- Convention on Biological Diversity: http://www.cbd.int/

## REFERENCES

Abell, R., Thieme, M. L., Revenga, C., *et al.* (2008). Freshwater ecoregions of the world: a new map of biogeographic units for freshwater biodiversity conservation. *BioScience*, **58**, 403–414.

Alroy, J., Aberhan, M., Bottjer, D. J., *et al.* (2008). Phanerozoic trends in the global diversity of marine invertebrates. *Science*, **321**, 97–100.

Angel, M. V. (1994). Spatial distribution of marine organisms: patterns and processes. In P. J. Edwards, R. M. May and N. R. Webb, eds *Large-scale ecology and conservation biology*, pp. 59–109. Blackwell Scientific, Oxford.

Aptroot, A. (1997). Species diversity in tropical rainforest ascomycetes: lichenized *versus* non-lichenized; folicolous *versus* corticolous. *Abstracta Botanica*, **21**, 37–44.

Armonies, W. and Reise, K. (2000). Faunal diversity across a sandy shore. *Marine Ecology Progress Series*, **196**, 49–57.

Berra, T. M. (1997). Some 20th century fish discoveries. *Environmental Biology of Fishes*, **50**, 1–12.

Briggs, J. C. (1999). Coincident biogeographic patterns: Indo-west Pacific ocean. *Evolution*, **53**, 326–335.

Bryant, J. A., Lamanna, C., Morlon, H., Kerkhoff, A. J., Enquist, B. J., and Green, J. L. (2008). Microbes on mountainsides: contrasting elevational patterns of bacterial and plant diversity. *Proceedings of the National Academy of Sciences of the United States of America*, **105**, 11505–11511.

Cavalier-Smith, T. (2004). Only six kingdoms of life. *Proceedings of the Royal Society of London B*, **271**, 1251–1262.

Ceballos, G. and Ehrlich, P. R. (2009). Discoveries of new mammal species and their implications for conservation and ecosystem services. *Proceedings of the National Academy of Sciences of the United States of America*, **106**, 3841–3846.

Clarke, A. and Gaston, K. J. (2006). Climate, energy and diversity. *Proceedings of the Royal Society of London Series B*, **273**, 2257–2266.

Copley, J. (2002). All at sea. *Nature*, **415**, 572–574.

Crane, P. R. and Lidgard, S. (1989). Angiosperm diversification and paleolatitudinal gradients in Cretaceous floristic diversity. *Science*, **246**, 675–678.

Curtis, T. P., Sloan, W. T., and Scannell, J. W. (2002). Estimating prokaryotic diversity and its limits. *Proceedings of the National Academy of Sciences of the United States of America*, **99**, 10494–10499.

Dauphas, N., van Zuilen, M., Wadhwa, M., Davis, A. M., Marty, B., and Janney, P. E. (2004). Clues from Fe isotope variations on the origin of early archaen BIFs from Greenland. *Science*, **306**, 2077–2080.

Davies, R. G., Orme, C. D. L., Storch, D., *et al.* (2007). Topography, energy and the global distribution of bird species richness. *Proceedings of the Royal Society of London B*, **274**, 1189–1197.

D'Hondt, S., Jørgensen, B. B., Miller, D. J., *et al.* (2004). Distributions of microbial activities in deep subseafloor sediments. *Science*, **306**, 2216–2221.

Dykhuizen, D. E. (1998). Santa Rosalia revisited: Why are there so many species of bacteria? *Antonie van Leeuwenhoek*, **73**, 25–33.

Eckburg, P. B., Bik, E. M., Bernstein, C. N., *et al.* (2005). Diversity of the human intestinal microbial flora. *Science*, **308**, 1635–1638.

Erwin, D. H. (1998). The end and the beginning: recoveries from mass extinctions. *Trends in Ecology and Evolution*, **13**, 344–349.

Erwin, D. H. (2008). Extinction as the loss of evolutionary history. *Proceedings of the National Academy of Sciences of the United States of America*, **105** (Suppl. 1), 11520–11527.

Evans, K. L., Warren, P. H., and Gaston, K. J. (2005). Species-energy relationships at the macroecological scale: a review of the mechanisms. *Biological Reviews*, **80**, 1–25.

Finlay, B. J. (2004). Protist taxonomy: an ecological perspective. *Philosophical Transactions of the Royal Society of London B*, **359**, 599–610.

Frost, D. R. (2004). *Amphibian species of the world: an online reference*. [Online database] http://research.amnh.org/herpetology/amphibia/index.php. Version 3.0 [22 August 2004]. American Museum of Natural History, New York.

Fuhrman, J. A. and Campbell, L. (1998). Microbial microdiversity. *Nature*, **393**, 410–411.

Fuhrman, J. A., Steele, J. A., Schwalbach, M. S., Brown, M. V., Green, J. L., and Brown, J. H. (2008). A latitudinal diversity gradient in planktonic marine bacteria. *Proceedings of the National Academy of Sciences of the United States of America*, **105**, 7774–7778.

Gaston, K. J. (2000). Global patterns in biodiversity. *Nature*, **405**, 220–227.

Gaston, K. J. (2008). Global species richness. In S.A. Levin, ed. *Encyclopedia of biodiversity*. Academic Press, San Diego, California.

Gaston, K. J., Blackburn, T. M., and Klein Goldewijk, K. (2003). Habitat conversion and global avian biodiversity loss. *Proceedings of the Royal Society of London B*, **270**, 1293–1300.

Gaston, K. J. and Spicer, J. I. (2004). *Biodiversity: an introduction*. 2nd edn. Blackwell Publishing, Oxford, UK.

Gould, S. J. (1991). *Bully for brontosaurus: reflections in natural history*. Hutchinson Radius, London, UK.

Gregory, T. R. (2008). *Animal genome size database*. [Online] http://www.genomesize.com.

Grenyer, R., Orme, C. D. L., Jackson, S. F. *et al.* (2006). The global distribution and conservation of rare and threatened vertebrates. *Nature*, **444**, 93–96.

Hawksworth, D. L. and Kalin-Arroyo, M. T. (1995). Magnitude and distribution of biodiversity. In V. H. Heywood, ed. *Global biodiversity assessment*, pp. 107–199. Cambridge University Press, Cambridge, UK.

Hebert, P. D. N., Penton, E. H., Burns, J. M., Janzen, D. H., and Hallwachs, W. (2004). Ten species in one: DNA barcoding reveals cryptic species in the neotropical skipper butterfly *Astraptes fulgerator*. *Proceedings of the National Academy of Sciences of the United States of America*, **101**, 14812–14817.

Hendrickson, J. A. and Ehrlich, P. R. (1971). An expanded concept of "species diversity". *Notulae Naturae*, **439**: 1–6.

Heywood, V. H. and Baste, I. (1995). Introduction. In V. H. Heywood, ed. *Global biodiversity assessment*, pp. 1–19. Cambridge University Press, Cambridge, UK.

Hillebrand, H. (2004). On the generality of the latitudinal diversity gradient. *American Naturalist*, **163**, 192–211.

Hughes, A. R., Inouye, B. D., Johnson, M. T. J., Underwood, N., and Vellend, M. (2008). Ecological consequences of genetic diversity. *Ecology Letters*, **11**, 609–623.

Hughes, J. B., Daily, G. C., and Ehrlich, P. R. (1997). Population diversity: its extent and extinction. *Science*, **278**, 689–692.

Jablonski, D., Roy, K., and Valentine, J. W. (2006). Out of the tropics: evolutionary dynamics of the latitudinal diversity gradient. *Science*, **314**, 102–106.

Jackson, J. B. C. (2008). Ecological extinction and evolution in the brave new ocean. *Proceedings of the National Academy of Sciences of the United States of America*, **105** (Suppl. 1), 11458–11465.

Keeling, P. J., Burger, G., Durnford, D. G., *et al.* (2005). The tree of eukaryotes. *Trends in Ecology and Evolution*, **20**, 670–676.

Kopp, R. E., Kirschvink, J. L., Hilburn, I. A., and Nash, C. Z. (2005). The Paleoproterozoic snowball Earth: A climate disaster triggered by the evolution of oxygenic photosynthesis. *Proceedings of the National Academy of Sciences of the United States of America*, **102**, 11131–11136.

Labandeira, C. C. (2005). Invasion of the continents: cyanobacterial crusts to tree-inhabiting arthropods. *Trends in Ecology and Evolution*, **20**, 253–262.

Lambais, M. R., Crowley, D. E., Cury, J. C., Büll, R. C., and Rodrigues, R. R. (2006). Bacterial diversity in tree canopies of the Atlantic Forest. *Science*, **312**, 1917.

Lambshead, P. J. D. (2004). Marine nematode biodiversity. In Z. X. Chen, S. Y. Chen and D. W. Dickson, eds *Nematology: advances and perspectives Vol. 1: Nematode morphology, physiology and ecology*, pp. 436–467. CABI Publishing, Oxfordshire, UK.

Longhurst, A. (1998). *Ecological geography of the sea*. Academic Press, San Diego, California.

López-García, P., Moreira, D., Douzery, E., *et al.* (2006). Ancient fossil record and early evolution (ca. 3.8 to 0.5 Ga). *Earth, Moon and Planets*, **98**, 247–290.

Manokaran, N., La Frankie, J. V., Kochummen, K. M., *et al.* (1992). Stand table and distribution of species in the 50-ha research plot at Pasoh Forest Reserve. *Forest Research Institute Malaysia, Research Data*, **1**, 1–454.

Margulis, L. and Schwartz, K. V. (1998). *Five kingdoms: an illustrated guide to the phyla of life on earth*, 3rd edn W. H. Freeman & Co., New York.

Martin, P. R., Bonier, F., and Tewksbury, J. J. (2007). Revisiting Jablonski (1993): cladogenesis and range expansion explain latitudinal variation in taxonomic richness. *Journal of Evolutionary Biology*, **20**, 930–936.

May, R. M. (2000). The dimensions of life on earth. In P. H. Raven and T. Williams, eds *Nature and Human Society*, pp. 30–45. National Academy Press, Washington, DC.

McKinney, M. L. (1997). Extinction vulnerability and selectivity: combining ecological and paleontological views. *Annual Review of Ecology and Systematics*, **28**, 495–516.

MEA (Millennium Ecosystem Assessment) (2005). *Ecosystems and human well-being: current state and trends, Volume 1*. Island Press, Washington, DC.

Olson, D. M., Dinerstein, E., Wikramanayake, E. D., *et al.* (2001). Terrestrial ecoregions of the world: a new map of life on earth. *BioScience*, **51**, 933–938.

Purvis, A. and Hector, A. (2000). Getting the measure of biodiversity. *Nature*, **405**, 212–219.

Rahbek, C. (1995). The elevational gradient of species richness: a uniform pattern? *Ecography*, **18**, 200–205.

Raup, D. M. (1994). The role of extinction in evolution. *Proceedings of the National Academy of Sciences of the United States of America*, **91**, 6758–6763.

Roberts, C. M., McClean, C. J., Veron, J. E. N., *et al.* (2002) Marine biodiversity hotspots and conservation priorities for tropical reefs. *Science*, **295**, 1280–1284.

Roger, A. J. and Hug, L. A. (2006). The origin and diversification of eukaryotes: problems with molecular phylogenies and molecular clock estimation. *Philosophical Transactions of the Royal Society of London B*, **361**, 1039–1054.

Rosing, M. T. and Frei, R. (2004). U-rich Archaean sea-floor sediments from Greenland - indications of >3700 Ma oxygenic photosynthesis. *Earth and Planetary Science Letters*, **217**, 237–244.

Simpson, A. G. B. and Roger, A. J. (2004). The real 'kingdoms' of eukaryotes. *Current Biology*, **14**, R693–R696.

Spalding, M. D., Fox, H. E., Allen, G. R., *et al.* (2007). Marine ecoregions of the world: a bioregionalisation of coastal and shelf areas. *BioScience*, **57**, 573–583.

Straatsma, G. and Krisai-Greilhuber, I. (2003). Assemblage structure, species richness, abundance and distribution of fungal fruit bodies in a seven year plot-based survey near Vienna. *Mycological Research*, **107**, 632–640.

Terborgh, J., Robinson, S. K., Parker, T. A. III, Munn, C. A., and Pierpont, N. (1990). Structure and organization of an Amazonian forest bird community. *Ecological Monographs*, **60**, 213–238.

Thomas, G. H., Orme, C. D., Davies, R. G., *et al.* (2008). Regional variation in the historical components of global avian species richness. *Global Ecology and Biogeography*, **17**, 340–351.

Torsvik, V., Øvreås, L., and Thingstad, T. F. (2002). Prokaryotic diversity-magnitude, dynamics, and controlling factors. *Science*, **296**, 1064–1066.

van Rootselaar, O. (1999). New birds for the world: species discovered during 1980–1999. *Birding World*, **12**, 286–293.

van Rootselaar, O. (2002). New birds for the world: species described during 1999–2002. *Birding World*, **15**, 428–431.

Venter, J. C., Remington, K., Heidelberg, J. F., *et al.* (2004). Environment genome shotgun sequencing of the Sargasso Sea. *Science*, **304**, 66–74.

Voss, R. S. and Emmons, L. H. (1996). Mammalian diversity in Neotropical lowland rainforests: a preliminary assessment. *Bulletin of the American Museum of Natural History*, **230**, 1–115.

Ward, B. B. (2002). How many species of prokaryotes are there? *Proceedings of the National Academy of Sciences of the United States of America*, **99**, 10234–10236.

White, D. C., Phelps, T. J., and Onstott, T. C. (1998). What's up down there? *Current Opinion in Microbiology*, **1**, 286–290.

Whitman, W. B., Coleman, D. C., and Wiebe, W. J. (1998). Prokaryotes: the unseen majority. *Proceedings of the National Academy of Sciences of the United States of America*, **95**, 6578–6583.

# CHAPTER 3

# Ecosystem functions and services

Cagan H. Sekercioglu

In our increasingly technological society, people give little thought to how dependent they are on the proper functioning of ecosystems and the crucial services for humanity that flow from them. Ecosystem services are "the conditions and processes through which natural ecosystems, and the species that make them up, sustain and fulfill human life" (Daily 1997); in other words, "the set of ecosystem functions that are useful to humans" (Kremen 2005). Although people have been long aware that natural ecosystems help support human societies, the explicit recognition of "ecosystem services" is relatively recent (Ehrlich and Ehrlich 1981a; Mooney and Ehrlich 1997).

Since the entire planet is a vast network of integrated ecosystems, ecosystem services range from global to microscopic in scale (Table 3.1; Millennium Ecosystem Assessment 2005a). Ecosystems purify the air and water, generate oxygen, and stabilize our climate. Earth would not be fit for our survival if it were not for plants that have created and maintained a suitable atmosphere. Organisms decompose and detoxify detritus, preventing our civilization from being buried under its own waste. Other species help to create the soils on which we grow our food, and recycle the nutrients essential to agriculture. Myriad creatures maintain these soils, play key roles in recycling nutrients, and by so doing help to mitigate erosion and floods. Thousands of animal species pollinate and fertilize plants, protect them from pests, and disperse their seeds. And of course, humans use and trade thousands of plant, animal and microorganism species for food, shelter, medicinal, cultural, aesthetic and many other purposes. Although most people may not know what an ecosystem is, the proper functioning of the world's ecosystems is critical to human survival, and understanding the basics of ecosystem services is essential. Entire volumes have been written on ecosystem services (National Research Council 2005; Daily 1997), culminating in a formal, in-depth, and global overview by hundreds of scientists: the *Millennium Ecosystem Assessment* (2005a). It is virtually impossible to list all the ecosystem services let alone the natural products that people directly consume, so this discussion presents a brief introduction to ecosystem function and an overview of critical ecosystem services.

## 3.1 Climate and the Biogeochemical Cycles

Ecosystem services start at the most fundamental level: the creation of the air we breathe and the supply and distribution of water we drink. Through photosynthesis by bacteria, algae, plankton, and plants, atmospheric oxygen is mostly generated and maintained by ecosystems and their constituent species, allowing humans and innumerable other oxygen-dependent organisms to survive. Oxygen also enables the atmosphere to "clean" itself via the oxidation of compounds such as carbon monoxide (Sodhi *et al.* 2007) and another form of oxygen in the ozone layer, protects life from the sun's carcinogenic, ultraviolet (UV) rays.

Global biogeochemical cycles consist of "the transport and transformation of substances in the environment through life, air, sea, land, and ice" (Alexander *et al.* 1997). Through these cycles, the planet's climate, ecosystems, and creatures

**Table 3.1** Ecosystem services, classified according to the Millennium Ecosystem Assessment (2003), and their ecosystem service providers. 'Functional units' refer to the unit of study for assessing functional contributions ($f_{ik}$) of ecosystem service providers; spatial scale indicates the scale(s) of operation of the service. Assessment of the potential to apply this conceptual framework to the service is purposefully conservative and is based on the degree to which the contributions of individual species or communities can currently be quantified (Kremen 2005).

| Service | Ecosystem service providers/ trophic level | Functional units | Spatial scale | Potential to apply this conceptual framework for ecological study |
|---|---|---|---|---|
| Aesthetic, cultural | All biodiversity | Populations, species, communities, ecosystems | Local-global | Low |
| Ecosystem goods | Diverse species | Populations, species, communities, ecosystems | Local-global | Medium |
| UV protection | Biogeochemical cycles, micro-organisms, plants | Biogeochemical cycles, functional groups | Global | Low |
| Purification of air | Micro-organisms, plants | Biogeochemical cycles, populations, species, functional groups | Regional-global | Medium (plants) |
| Flood mitigation | Vegetation | Communities, habitats | Local-regional | Medium |
| Drought mitigation | Vegetation | Communities, habitats | Local-regional | Medium |
| Climate stability | Vegetation | Communities, habitats | Local-global | Medium |
| Pollination | Insects, birds, mammals | Populations, species, functional groups | Local | High |
| Pest control | Invertebrate parasitoids and predators and vertebrate predators | Populations, species, functional groups | Local | High |
| Purification of water | Vegetation, soil micro-organisms, aquatic micro-organisms, aquatic invertebrates | Populations, species, functional groups, communities, habitats | Local-regional | Medium to high* |
| Detoxification and decomposition of wastes | Leaf litter and soil invertebrates, soil micro-organisms, aquatic micro-organisms | Populations, species, functional groups, communities, habitats | Local-regional | Medium |
| Soil generation and soil fertility | Leaf litter and soil invertebrates, soil micro-organisms, nitrogen-fixing plants, plant and animal production of waste products | Populations, species, functional groups | Local | Medium |
| Seed dispersal | Ants, birds, mammals | Populations, species, functional groups | Local | High |

* Waste-water engineers 'design' microbial communities; in turn, wastewater treatments provide ideal replicated experiments for ecological work (Graham and Smith 2004 in Kremen 2005).

are tightly linked. Changes in one component can have drastic effects on another, as exemplified by the effects of deforestation on climatic change (Phat *et al.* 2004). The hydrologic cycle is one that most immediately affects our lives and it is treated separately below.

As carbon-based life forms, every single organism on our planet is a part of the global carbon cycle. This cycle takes place between the four main reservoirs of carbon: carbon dioxide ($CO_2$) in the atmosphere; organic carbon compounds within organisms; dissolved carbon in water bodies; and carbon compounds inside the earth as part of soil, limestone (calcium carbonate), and buried organic matter like coal, natural gas, peat, and petroleum (Alexander *et al.* 1997). Plants play a major role in fixing atmospheric $CO_2$ through photosynthesis and most terrestrial carbon storage occurs in forest trees (Falkowski *et al.* 2000). The global carbon cycle has been disturbed by about 13% compared to the pre-industrial era, as opposed to 100% or more for nitrogen, phosphorous, and sulfur cycles (Falkowski *et al.* 2000). Given the dominance of carbon in shaping life and in regulating climate, however, this perturbation has already been enough to lead to significant climate change with worse likely to come in the future [IPCC (Intergovernmental Panel on Climate Change) 2007].

Because gases like $CO_2$, methane ($CH_4$), and nitrous oxide ($N_2O$) trap the sun's heat, especially the long-wave infrared radiation that's emitted by the warmed planet, the atmosphere creates a natural "greenhouse" (Houghton 2004). Without this greenhouse effect, humans and most other organisms would be unable to survive, as the global mean surface temperature would drop from the current 14° C to –19° C (IPCC 2007). Ironically, the ever-rising consumption of fossil fuels during the industrial age and the resultant increasing emission of greenhouse gases have created the opposite problem, leading to an increase in the magnitude of the greenhouse effect and a consequent rise in global temperatures (IPCC 2007). Since 1750, atmospheric $CO_2$ concentrations have increased by 34% (Millennium Ecosystem Assessment 2005a) and by the end of this century, average global temperature is projected to rise by 1.8°–6.4° C (IPCC 2007). Increasing deforestation and warming both exacerbate the problem as forest ecosystems switch from being major carbon sinks to being carbon sources (Phat *et al.* 2004; IPCC 2007). If fossil fuel consumption and deforestation continue unabated, global $CO_2$ emissions are expected to be about 2–4 times higher than at present by the year 2100 (IPCC 2007). As climate and life have coevolved for billions of years and interact with each other through various feedback mechanisms (Schneider and Londer 1984), rapid climate change would have major consequences for the planet's life-support systems. There are now plans under way for developed nations to finance the conservation of tropical forests in the developing world so that these forests can continue to provide the ecosystem service of acting as carbon sinks (Butler 2008).

Changes in ecosystems affect nitrogen, phosphorus, and sulfur cycles as well (Alexander *et al.* 1997; Millennium Ecosystem Assessment 2005b; Vitousek *et al.* 1997). Although nitrogen in its gaseous form ($N_2$) makes up 80% of the atmosphere, it is only made available to organisms through nitrogen fixation by cyanobacteria in aquatic systems and on land by bacteria and algae that live in the root nodules of lichens and legumes (Alexander *et al.* 1997). Eighty million tons of nitrogen every year are fixed artificially by industry to be used as fertilizer (Millennium Ecosystem Assessment 2005b). However, the excessive use of nitrogen fertilizers can lead to nutrient overload, eutrophication, and elimination of oxygen in water bodies. Nitrogen oxides, regularly produced as a result of fossil fuel combustion, are potent greenhouse gases that increase global warming and also lead to smog, breakdown of the ozone layer, and acid rain (Alexander *et al.* 1997). Similarly, although sulfur is an essential element in proteins, excessive sulfur emissions from human activities lead to sulfuric acid smog and acid rain that harms people and ecosystems alike (Alexander *et al.* 1997).

Phosphorous (P) scarcity limits biological nitrogen fixation (Smith 1992). In many terrestrial ecosystems, where P is scarce, specialized symbiotic fungi (mycorrhizae) facilitate P uptake by plants (Millennium Ecosystem Assessment 2005b). Even though P is among the least naturally available of major nutrients, use of phosphorous in artificial fertilizers and runoff from animal husbandry often also leads to eutrophication in aquatic systems (Millenium Ecosystem Assessment 2005b). The mining of phosphate deposits and their addition to terrestrial ecosystems as fertilizers represents a six fold increase over the natural rate of mobilization of P by the weathering of phosphate rock and by plant activity (Reeburgh 1997). P enters aquatic ecosystems mainly through erosion, but no-till agriculture and the use of hedgerows can substantially reduce the rate of this process (Millenium Ecosystem Assessment 2005a).

## 3.2 Regulation of the Hydrologic Cycle

One of the most vital and immediate services of ecosystems, particularly of forests, rivers and wetlands, is the provisioning and regulation of water resources. These services provide a vast range of benefits from spiritual to life-saving, illustrated by the classification of hydrologic services into five broad categories by Brauman et al. (2007): improvement of extractive water supply, improvement of in-stream water supply, water damage mitigation, provision of water-related cultural services, and water-associated supporting services (Figure 3.1). Although 71% of the planet is covered by water, most of this is seawater unfit for drinking or agriculture (Postel et al. 1996). Fresh water not locked away in glaciers and icecaps constitutes 0.77% of the planet's water (Shiklomanov 1993). To provide sufficient fresh water to meet human needs via industrial desalination (removing the salt from seawater) would cost US$3 000 billion per year (Postel and Carpenter 1997).

Quantity, quality, location, and timing of water provision determine the scale and impact of hydrologic services (Brauman et al. 2007). These attributes can make the difference between water as a blessing (e.g. drinking water) or a curse (e.g. floods). Water is constantly redistributed through the hydrologic cycle. Fresh water comes down as precipitation, collects in water bodies or is absorbed by the soil and plants. Some of the water flows unutilized into the sea or seeps into

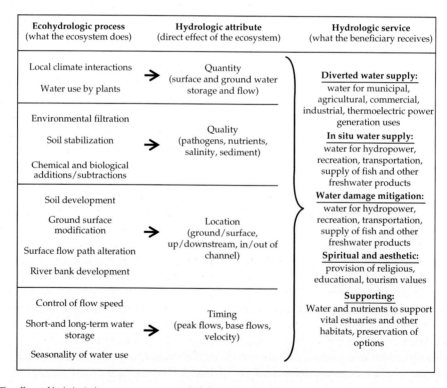

**Figure 3.1** The effects of hydrological ecosystem processes on hydrological services. Reprinted from Brauman et al. (2007).

underground aquifers where it can remain for millennia unless extracted by people; mining this "fossil" groundwater is often unsustainable and is a serious problem in desert regions like Libya (Millennium Ecosystem Assessment 2005c). The cycle is completed when water vapor is released back into the atmosphere either through evaporation from land and water bodies or by being released from plants (transpiration) and other organisms. Rising environmental temperatures are expected to increase evaporation and consequent precipitation in some places and raise the likelihood of droughts and fires in other places, both scenarios that would have major consequences for the world's vegetation (Wright 2005). These changes in turn can lead to further climatic problems, affecting agriculture and communities worldwide. Ecosystems, particularly forests, play major roles in the regulation of the hydrologic cycle and also have the potential to moderate the effects of climate change. Tropical forests act as heat and humidity pumps, transferring heat from the tropics to the temperate zones and releasing water vapor that comes back as rain (Sodhi *et al.* 2007). Extensive tropical deforestation is expected to lead to higher temperatures, reduced precipitation, and increased frequency of droughts and fires, all of which are likely to reduce tropical forest cover in a positive feedback loop (Sodhi *et al.* 2007).

Forest ecosystems alone are thought to regulate approximately a third of the planet's watersheds on which nearly five billion people rely (Millennium Ecosystem Assessment 2005c). With increasing human population and consequent water pollution, fresh water is becoming an increasingly precious resource, especially in arid areas like the Middle East, where the scarcity of water is likely to lead to increasing local conflicts in the 21$^{st}$ century (Klare 2001; Selby 2005). Aquatic ecosystems, in addition to being vital sources of water, fish, waterfowl, reeds, and other resources, also moderate the local climate and can act as buffers for floods, tsunamis, and other water incursions (Figure 3.1). For example, the flooding following Hurricane Katrina would have done less damage if the coastal wetlands surrounding New Orleans had had their original extent (Day *et al.* 2007). The impact of the 24 December 2004 tsunami in Southeast Asia would have been reduced if some of the hardest-hit areas had not been stripped of their mangrove forests (Dahdouh-guebas *et al.* 2005; Danielsen *et al.* 2005). These observations support analytical models in which thirty "waru" trees (*Hibiscus tiliaceus*) planted along a 100 m by 1 meter band reduced the impact of a tsunami by 90% (Hiraishi and Harada 2003), a solution more effective and cheaper than artificial barriers.

Hydrologic regulation by ecosystems begins with the first drop of rain. Vegetation layers, especially trees, intercept raindrops, which gradually descend into the soil, rather than hitting it directly and leading to erosion and floods. By intercepting rainfall and promoting soil development, vegetation can modulate the timing of flows and potentially reduce flooding. Flood mitigation is particularly crucial in tropical areas where downpours can rapidly deposit enormous amounts of water that can lead to increased erosion, floods, and deaths if there is little natural forest to absorb the rainfall (Bradshaw *et al.* 2007). Studies of some watersheds have shown that native forests reduced flood risks only at small scales, leading some hydrologists to question directly connecting forest cover to flood reduction (Calder and Aylward 2006). However, in the first global-scale empirical demonstration that forests are correlated with flood risk and severity in developing countries, Bradshaw *et al.* (2007) estimated that a 10% decrease in natural forest area would lead to a flood frequency increase between 4% and 28%, and to a 4–8% increase in total flood duration at the country scale. Compared to natural forests, however, afforestation programs or forest plantations may not reduce floods, or may even increase flood volume due to road construction, soil compaction, and changes in drainage regimes (Calder and Aylward 2006). Non-native plantations can do more harm than good, particularly when they reduce dry season water flows (Scott *et al.* 2005).

Despite covering only 6% of the planet's surface, tropical forests receive nearly half of the world's rainfall, which can be as much as 22 500 mm during five months of monsoon season in

India (Myers 1997). In Southeast Asia, an intact old-growth dipterocarp forest intercepts at least 35% of the rainfall, while a logged forest intercepts less than 20%, and an oil palm (*Elaeis* spp.) plantation intercepts only 12% (Ba 1977). As a consequence, primary forest can moderate seasonal extremes in water flow and availability better than more intensive land uses like plantation forestry and agriculture. For example, primary forest in Ivory Coast releases three to five times as much water at the end of the dry season compared to a coffee plantation (Dosso 1981). However, it is difficult to make generalizations about hydrologic response in the tropics. For example, local soil and rainfall patterns can result in a 65-fold variation in tropical natural sedimentation rates (Bruijnzeel 2004). This underlines the importance of site-specific studies in the tropics, but most hydrologic studies of ecosystems have taken place in temperate ecosystems (Brauman *et al.* 2007).

## 3.3 Soils and Erosion

Without forest cover, erosion rates skyrocket, and many countries, especially in the tropics, lose astounding amounts of soil to erosion. Worldwide, 11 million km$^2$ of land (the area of USA and Mexico combined) are affected by high rates of erosion (Millennium Ecosystem Assessment 2005b). Every year about 75 billion tons of soil are thought to be eroded from terrestrial ecosystems, at rates 13–40 times faster than the average rate of soil formation (Pimentel and Kounang 1998). Pimentel *et al.* (1995) estimated that in the second half of the 20$^{th}$ century about a third of the world's arable land was lost to erosion. This means losing vital harvests and income (Myers 1997), not to mention losing lives to malnutrition and starvation. Soil is one of the most critical but also most underappreciated and abused elements of natural capital, one that can take a few years to lose and millennia to replace. A soil's character is determined by six factors: topography, the nature of the parent material, the age of the soil, soil organisms and plants, climate, and human activity (Daily *et al.* 1997). For example, in the tropics, farming can result in the loss of half the soil nutrients in less than a decade (Bolin and Cook 1983), a loss that can take centuries to restore. In arid areas, the replacement of native deep-rooted plants with shallow-rooted crop plants can lead to a rise in the water table, which can bring soil salts to the surface (salinization), cause waterlogging, and consequently result in crop losses (Lefroy *et al.* 1993).

Soil provides six major ecosystem services (Daily *et al.* 1997):

- Moderating the hydrologic cycle.
- Physical support of plants.
- Retention and delivery of nutrients to plants.
- Disposal of wastes and dead organic matter.
- Renewal of soil fertility.
- Regulation of major element cycles.

Every year enough rain falls to cover the planet with one meter of water (Shiklomanov 1993), but thanks to soil's enormous water retention capacity, most of this water is absorbed and gradually released to feed plants, underground aquifers, and rivers. However, intensive cultivation, by lowering soil's organic matter content, can reduce this capacity, leading to floods, erosion, pollution, and further loss of organic matter (Pimentel *et al.* 1995).

Soil particles usually carry a negative charge, which plays a critical role in delivering nutrient cations (positively-charged ions) like $Ca2+$, $K+$, $Na+$, $NH4+$, and $Mg2+$ to plants (Daily *et al.* 1997). To deliver these nutrients without soil would be exceedingly expensive as modern hydroponic (water-based) systems cost more than US$250 000 per ha (Canada's Office of Urban Agriculture 2008; Avinash 2008). Soil is also critical in filtering and purifying water by removing contaminants, bacteria, and other impurities (Fujii *et al.* 2001). Soils harbor an astounding diversity of microorganisms, including thousands of species of protozoa, antibiotic-producing bacteria (which produce streptomycin) and fungi (producing penicillin), as well as myriad invertebrates, worms and algae (Daily *et al.* 1997). These organisms play fundamental roles in decomposing dead matter, neutralizing deadly pathogens, and recycling waste into valuable nutrients. Just the nitrogen fixed by soil organisms like

*Rhizobium* bacteria amounts to about 100 million metric tons per year (Schlesinger 1991). It would cost at least US$320 billon/year to replace natural nitrogen fertilization with fertilizers (Daily *et al.* 1997).

As the accelerating release of $CO_2$, $N_2O$ (Nitrous Oxide), methane and other greenhouse gases increasingly modifies climate (IPCC 2007), the soil's capacity to store these molecules is becoming even more vital. Per area, soil stores 1.8 times the carbon and 18 times the nitrogen that plants alone can store (Schlesinger 1991). For peatlands, soil carbon storage can be 10 times greater than that stored by the plants growing on it and peatland fires release massive amounts of $CO_2$ into the atmosphere (Page and Rieley 1998).

Despite soil's vital importance, 17% of the Earth's vegetated land surface (Oldeman 1998) or 23% of all land used for food production [FAO (Food and Agriculture Organization of the United Nations) 1990] has experienced soil degradation since 1945. Erosion is the best-known example of the disruption of the sedimentary cycle. Although erosion is responsible for releasing nutrients from bedrock and making them available to plants, excessive wind and water erosion results in the removal of top soil, the loss of valuable nutrients, and desertification. The direct costs of erosion total about US$250 billion per year and the indirect costs (e.g. siltation, obsolescence of dams, water quality declines) approximately $150 billion per year (Pimentel *et al.* 1995). Sufficient preventive measures would cost only 19% of this total (Pimentel *et al.* 1995).

The loss of vegetative cover increases the erosional impact of rain. In intact forests, most rain water does not hit the ground directly and tree roots hold the soil together against being washed away (Brauman *et al.* 2007), better than in logged forest or plantations (Myers 1997) where roads can increase erosion rates (Bruijnzeel 2004). The expansion of farming and deforestation have doubled the amount of sediment discharged into the oceans. Coral reefs can experience high mortality after being buried by sediment discharge (Pandolfi *et al.* 2003; Bruno and Selig 2007). Wind erosion can be particularly severe in desert ecosystems, where even small increases in vegetative cover (Hupy 2004) and reduced tillage practices (Gomes *et al.* 2003) can lessen wind erosion substantially. Montane areas are especially prone to rapid erosion (Milliman and Syvitski 1992), and revegetation programs are critical in such ecosystems (Vanacker *et al.* 2007). Interestingly, soil carbon buried in deposits resulting from erosion, can produce carbon sinks that can offset up to 10% of the global fossil fuel emissions of $CO_2$ (Berhe *et al.* 2007). However, erosion also lowers soil productivity and reduces the organic carbon returned to soil as plant residue (Gregorich *et al.* 1998). Increasing soil carbon capacity by 5–15% through soil-friendly tillage practices not only offsets fossil-fuel carbon emissions by a roughly equal amount but also increases crop yields and enhances food security (Lal 2004). An increase of one ton of soil carbon pool in degraded cropland soils may increase crop yield by 20 to 40 kilograms per ha (kg/ha) for wheat, 10 to 20 kg/ha for maize, and 0.5 to 1 kg/ha for cowpeas (Lal 2004).

## 3.4 Biodiversity and Ecosystem Function

The role of biodiversity in providing ecosystem services is actively debated in ecology. The diversity of functional groups (groups of ecologically equivalent species (Naeem and Li 1997)), is as important as species diversity, if not more so (Kremen 2005), and in most services a few dominant species seem to play the major role (Hooper *et al.* 2005). However, many other species are critical for ecosystem functioning and provide "insurance" against disturbance, environmental change, and the decline of the dominant species (Tilman 1997; Ricketts *et al.* 2004; Hobbs *et al.* 2007). As for many other ecological processes, it was Charles Darwin who first wrote of this, noting that several distinct genera of grasses grown together would produce more plants and more herbage than a single species growing alone (Darwin 1872). Many studies have confirmed that increased biodiversity improves ecosystem functioning in plant communities (Naeem and Li 1997; Tilman 1997). Different plant species capture different resources, leading to greater efficiency and higher productivity (Tilman *et al.* 1996). Due to the

## Box 3.1 The costs of large-mammal extinctions
**Robert M. Pringle**

When humans alter ecosystems, large mammals are typically the first species to disappear. They are hunted for meat, hides, and horns; they are harassed and killed if they pose a threat; they require expansive habitat; and they are susceptible to diseases, such as anthrax, rinderpest, and distemper, that are spread by domestic animals. Ten thousand years ago, humans played at least a supporting, if not leading, role in extinguishing most of the large mammals in the Americas and Australia. Over the last 30 years, we have extinguished many large-mammal populations (and currently threaten many more) in Africa and Asia—the two continents that still support diverse assemblages of these charismatic creatures.

**Box 3.1 Figure 1** White-footed mice (*Peromyscus leucopus*, shown with an engorged tick on its ear) are highly competent reservoirs for Lyme disease. When larger mammals disappear, mice often thrive, increasing disease risk. Photograph courtesy of Richard Ostfeld Laboratory.

The ecological and economic consequences of losing large-mammal populations vary depending on the location and the ecological role of the species lost. The loss of carnivores has induced trophic cascades: in the absence of top predators, herbivores can multiply and deplete the plants, which in turn drives down the density and the diversity of other species (Ripple and Beschta 2006). Losing large herbivores and their predators can have the opposite effect, releasing plants and producing compensatory increases in the populations of smaller herbivores (e.g. rodents: Keesing 2000) and their predators (e.g. snakes: McCauley *et al*. 2006). Such increases, while not necessarily detrimental themselves, can have unpleasant consequences (see below).

Many species depend on the activities of particular large mammal species. Certain trees produce large fruits and seeds apparently adapted for dispersal by large browsers (Guimarães *et al*. 2008). Defecation by large mammals deposits these seeds and provides food for many dung beetles of varying degrees of specialization. In East Africa, the disturbance caused by browsing elephants creates habitat for tree-dwelling lizards (Pringle 2008), while the total loss of large herbivores dramatically altered the character of an ant-plant symbiosis via a complex string of species interactions (Palmer *et al*. 2008).

**Box 3.1 Figure 2** Ecotourists gather around a pair of lions in Tanzania's Ngorongoro Crater. Ecotourism is one of the most powerful driving forces for biodiversity conservation, especially in tropical regions where money is short. But tourists must be managed in such a way that they do not damage or deplete the very resources they have traveled to visit. Photograph by Robert M. Pringle

These examples and others suggest that the loss of large mammals may precipitate extinctions of other taxa and the relationships among them, thus decreasing the diversity of both species and interactions. Conversely, protecting the large areas needed to conserve large mammals may often serve to conserve the greater diversity of smaller organisms—the so-called umbrella effect.

The potential economic costs of losing large mammals also vary from place to place. Because

*continues*

### Box 3.1 (Continued)

cattle do not eat many species of woody plants, the loss of wildlife from rangelands can result in bush encroachment and decreased pastoral profitability. Because some rodents and their parasites are reservoirs and vectors of various human diseases, increases in rodent densities may increase disease transmission (Ostfeld and Mills 2007; Box 3.1 Figure 1). Perhaps most importantly, because large mammals form the basis of an enormous tourism industry, the loss of these species deprives regions of an important source of future revenue and foreign exchange (Box 3.1 Figure 2).

Arguably, the most profound cost of losing large mammals is the toll that it takes on our ability to relate to nature. Being large mammals ourselves, we find it easier to identify and sympathize with similar species—they behave in familiar ways, hence the term "charismatic megafauna." While only a handful of large mammal species have gone globally extinct in the past century, we are dismantling many species population by population, pushing them towards extinction. At a time when we desperately need to mobilize popular support for conservation, the loss over the next 50 years of even a few emblematic species—great apes in central Africa, polar bears in the arctic, rhinoceroses in Asia—could deal a crippling blow to efforts to salvage the greater portion of biodiversity.

### REFERENCES

Guimarães, P. R. J., Galleti, M., and Jordano, P. (2008). Seed dispersal anachronisms: rethinking the fruits extinct megafauna ate. *PloS One*, **3**, e1745.

Keesing, F. (2000). Cryptic consumers and the ecology of an African savanna. *BioScience*, **50**, 205–215.

McCauley, D. J., Keesing, F., Young, T. P., Allan, B. F., and Pringle, R. M. (2006). Indirect effects of large herbivores on snakes in an African savanna. *Ecology*, **87**, 2657–2663.

Ostfeld, R. S., and Mills, J. N. (2007). Social behavior, demography, and rodent-borne pathogens. In J. O. Wolff and P. W. Sherman, eds *Rodent societies*, pp. 478–486. University of Chicago Press, Chicago, IL.

Palmer, T. M., Stanton, M. L., Young, T. P., *et al.* (2008). Breakdown of an ant-plant mutualism follows the loss of large mammals from an African savanna. *Science*, **319**, 192–195.

Pringle, R. M. (2008). Elephants as agents of habitat creation for small vertebrates at the patch scale. Ecology, **89**, 26–33.

Ripple, W. J. and Beschta, R. L. (2006). Linking a cougar decline, trophic cascade, and catastrophic regime shift in Zion National Park. *Biological Conservation*, **133**, 397–408.

---

"sampling-competition effect" the presence of more species increases the probability of having a particularly productive species in any given environment (Tilman 1997). Furthermore, different species' ecologies lead to complementary resource use, where each species grows best under a specific range of environmental conditions, and different species can improve environmental conditions for other species (facilitation effect; Hooper *et al.* 2005). Consequently, the more complex an ecosystem is, the more biodiversity will increase ecosystem function, as more species are needed to fully exploit the many combinations of environmental variables (Tilman 1997). More biodiverse ecosystems are also likely to be more stable and more efficient due to the presence of more pathways for energy flow and nutrient recycling (Macarthur 1955; Hooper *et al.* 2005; Vitousek and Hooper 1993; Worm *et al.* 2006).

Greenhouse and field experiments have confirmed that biodiversity does increase ecosystem productivity, while reducing fluctuations in productivity (Naeem *et al.* 1995; Tilman *et al.* 1996). Although increased diversity can increase the population fluctuations of individual species, diversity is thought to stabilize overall ecosystem functioning (Chapin *et al.* 2000; Tilman 1996) and make the ecosystem more resistant to perturbations (Pimm 1984). These hypotheses have been confirmed in field experiments, where species-rich plots showed less yearly variation in productivity (Tilman 1996) and their productivity during a drought year declined much less than species-poor plots (Tilman and Downing 1994). Because

## Box 3.2 Carnivore conservation
### Mark S. Boyce

Predation by carnivores can alter prey population abundance and distribution, and these predator effects have been shown to influence many aspects of community ecology. Examples include the effect of sea otters that kill and eat sea urchins reducing their abundance and herbivory on the kelp forests that sustain diverse near-shore marine communities of the North Pacific. Likewise, subsequent to wolf (see Box 3.2 Figure) recovery in Yellowstone National Park (USA), elk have become preferred prey of wolves resulting in shifts in the distribution and abundance of elk that has released vegetation from ungulate herbivory with associated increases in beavers, song birds, and other plants and animals.

Yet, carnivore conservation can be very challenging because the actions of carnivores often are resented by humans. Carnivores depredate livestock or reduce abundance of wildlife valued by hunters thereby coming into direct conflict with humans. Some larger species of carnivores can prey on humans. Every year, people are killed by lions in Africa, children are killed by wolves in India, and people are killed or mauled by cougars and bears in western North America (see also Box 14.3). Retaliation is invariably swift and involves killing those individuals responsible for the depredation, but furthermore such incidents of human predation usually result in fear-driven management actions that seldom consider the ecological significance of the carnivores in question.

Another consideration that often plays a major role in carnivore conservation is public opinion. Draconian methods for predator control, including aerial gunning and poisoning of wolves by government agencies, typically meets with fierce public opposition. Yet, some livestock ranchers and hunters lobby to have the carnivores eradicated. Rural people who are at risk of depredation losses from carnivores usually want the animals controlled or eliminated, whereas tourists and broader publics usually push for protection of the carnivores.

Most insightful are programs that change human management practices to reduce the probability of conflict. Bringing cattle into areas where they can be watched during calving can reduce the probability that bears or wolves will kill the calves. Ensuring that garbage is unavailable to bears and other large carnivores reduces the risk that carnivores will become habituated to humans and consequently come into conflict. Livestock ranchers can monitor their animals in back-country areas and can dispose of dead animal carcasses to reduce the risk of depredation. Killing those individuals that are known to depredate livestock can be an effective approach because individuals sometimes learn to kill livestock whereas most carnivores in the population take only wild prey. Managing recreational access to selected trails and roads can be an effective tool for reducing conflicts between large carnivores and people. Finding socially acceptable methods of predator control whilst learning to live in proximity with large carnivores is the key challenge for carnivore conservation.

**Box 3.2 Figure** Grey wolf (*Canis lupus*). Photograph from www.all-about-wolves.com.

more species do better at utilizing and recycling nutrients, in the long-term, species-rich plots are better at reducing nutrient losses and maintaining soil fertility (Tilman *et al.* 1996; Vitousek and Hooper 1993).

Although it makes intuitive sense that the species that dominate in number and/or biomass are more likely to be important for ecosystem function (Raffaelli 2004; Smith *et al.* 2004), in some cases, even rare species can have a role, for

---

**Box 3.3 Ecosystem services and agroecosystems in a landscape context**
**Teja Tscharntke**

Agroecosystems result from the transformation of natural ecosystems to promote ecosystem services, which are defined as benefits people obtain from ecosystems (MEA 2005). Major challenges in managing ecosystem services are that they are not independent of each other and attempts to optimize a single service (e.g. reforestation) lead to losses in other services (e.g. food production; Rodriguez *et al.* 2006). Agroecosystems such as arable fields and grasslands are typically extremely open ecosystems, characterized by high levels of input (e.g. labour, agrochemicals) and output (e.g. food resources), while agricultural management reduces structural complexity and associated biodiversity.

The world's agroecosystems deliver a number of key goods and services valued by society such as food, feed, fibre, water, functional biodiversity, and carbon storage. These services may directly contribute to human well-being, for example through food production, or just indirectly through ecosystem processes such as natural biological control of crop pests (Tscharntke *et al.* 2007) or pollination of crops (Klein *et al.* 2007). Farmers are mostly interested in the privately owned, marketable goods and services, while they may also produce public goods such as aesthetic landscapes or regulated water levels. Finding win-win solutions that serve both private economic gains in agroecosystems and public long-term conservation in agricultural landscapes is often difficult (but see Steffan-Dewenter *et al.* 2007). The goal of long-lasting ecosystem services providing sustainable human well-being may become compromised by the short-term interest of farmers in increasing marketable services, but incentives may encourage environment friendly agriculture. This is why governments implement payment-for-ecosystem service programs such as the agri-environment schemes in the European Community or the Chinese programs motivated by large floods on the Yangtze River (Tallis *et al.* 2008).

In addition, conservation of most services needs a landscape perspective. Agricultural land use is often focused on few species and local processes, but in dynamic, human-dominated landscapes, only a diversity of insurance species may guarantee resilience, i.e. the capacity to re-organize after disturbances (see Box 3.3 Figure). Biodiversity and associated ecosystem services can be maintained only in complex landscapes with a minimum of near-natural habitat (in central Europe roughly 20%) supporting a minimum number of species dispersing across natural and managed systems (Tscharntke *et al.* 2005). For example, pollen beetles causing economically meaningful damage in oilseed rape (canola) are naturally controlled by parasitic wasps in complex but not in simplified landscapes. Similarly, high levels of pollination and yield in coffee and pumpkin depend on a high diversity of bee species, which is only available in heterogeneous environments. The landscape context may be even more important for local biodiversity and associated ecosystem services than differences in local management, for example between organic and conventional farming or between crop fields with or without near-natural field margins, because the organisms immigrating into agroecosystems from the landscape-wide species pool may compensate for agricultural intensification at a local scale (Tscharntke *et al.* 2005).

*continues*

### Box 3.3 (Continued)

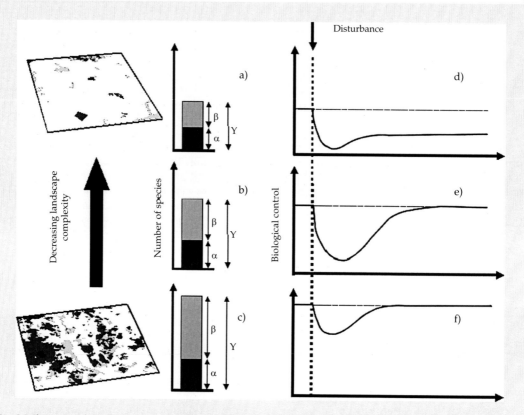

**Box 3.3 Figure** Hypothesized responses to disturbance on ecosystem services such as biological control and pollination by native natural enemies and pollinators in different landscapes, showing how beta diversity (a-c) and recover of biological control and pollination after disturbance (d-f) change with landscape heterogeneity. Adapted from Tscharntke et al. (2007). a) and d) Intensely used monotonous landscape with a small available species pool, giving a low general level of ecosystem services, a greater dip in the service after a disturbance and an ecosystem that is unable to recover. b) and e) Intermediate landscape harboring slightly higher species richness, rendering deeper dip and slower return from a somewhat lower maximum level of biological control or pollination after a disturbance. c) and f) Heterogeneous landscape with large species richness, mainly due to the higher beta diversity, rendering high maximum level of the service, and low dip and quick return after a disturbance.

The turnover of species among patches (the dissimilarity of communities creating high beta diversity, in contrast to the local, patch-level alpha diversity) is the dominant driver of landscape-wide biodiversity. Beta diversity reflects the high spatial and temporal heterogeneity experienced by communities at a landscape scale. Pollinator or biocontrol species that do not contribute to the service in one patch may be important in other patches, providing spatial insurance through complementary resource use (see Box 3.3 Figure). Sustaining ecosystem services in landscapes depends on a high beta diversity coping with the spatial and temporal heterogeneity in a real world under Global Change.

### REFERENCES

Klein, A.-M., Vaissière, B. E., Cane. J. H., et al. (2007). Importance of pollinators in changing landscapes for world crops. *Proceedings of the Royal Society of London B*, **274**, 303–313.

MEA (2005). Millenium Ecosystem Assessment. Island Press, Washington, DC.

*continues*

## Box 3.3 (Continued)

Rodriguez, J. J., Beard, T. D. Jr, Bennett, E. M., et al. (2006). Trade-offs across space, time, and ecosystem services. *Ecology and Society*, **11**, 28 (online).

Steffan-Dewenter, I., Kessler, M., Barkmann, J., et al. (2007). Tradeoffs between income, biodiversity, and ecosystem functioning during tropical rainforest conversion and agroforestry intensification. *Proceedings of the National Academy of Sciences of the United States of America*, **104**, 4973–4978

Tallis, H., Kareiva, P., Marvier, M., and Chang, A. (2008). An ecosystem services framework to support both practical conservation and economic development. *Proceedings of the National Academy of Sciences of the United States of America*, **105**, 9457–9464.

Tscharntke, T., Klein, A.-M., Kruess, A., Steffan-Dewenter, I, and Thies, C. (2005). Landscape perspectives on agricultural intensification and biodiversity - ecosystem service management. *Ecology Letters*, **8**, 857–874.

Tscharntke, T., Bommarco, R., Clough, Y., et al. (2007). Conservation biological control and enemy diversity on a landscape scale. *Biological Control*, **43**, 294–309.

---

example, in increasing resistance to invasion (Lyons and Schwartz 2001). A keystone species is one that has an ecosystem impact that is disproportionately large in relation to its abundance (Hooper et al. 2005; Power et al. 1996; see Boxes 3.1, 3.2, and 5.3). Species that are not thought of as "typical" keystones can turn out to be so, sometimes in more ways than one (Daily et al. 1993). Even though in many communities only a few species have strong effects, the weak effects of many species can add up to a substantial stabilizing effect and seemingly "weak" effects over broad scales can be strong at the local level (Berlow 1999). Increased species richness can "insure" against sudden change, which is now a global phenomenon (Parmesan and Yohe 2003; Root et al. 2003). Even though a few species may make up most of the biomass of most functional groups, this does not mean that other species are unnecessary (Walker et al. 1999). Species may act like the rivets in an airplane wing, the loss of each unnoticed until a catastrophic threshold is passed (Ehrlich and Ehrlich 1981b).

As humanity's footprint on the planet increases and formerly stable ecosystems experience constant disruptions in the form of introduced species (Chapter 7), pollution (Box 13.1), climate change (Chapter 8), excessive nutrient loads, fires (Chapter 9), and many other perturbations, the insurance value of biodiversity has become increasingly vital over the entire range of habitats and systems, from diverse forest stands sequestering $CO_2$ better in the long-term (Bolker et al. 1995; Hooper et al. 2005; but see Tallis and Kareiva 2006) to forest-dwelling native bees' coffee pollination services increasing coffee production in Costa Rica (Ricketts et al. 2004; also see Box 3.3). With accelerating losses of unique species, humanity, far from hedging its bets, is moving ever closer to the day when we will run out of options on an increasingly unstable planet.

## 3.5 Mobile Links

"Mobile links" are animal species that provide critical ecosystem services and increase ecosystem resilience by connecting habitats and ecosystems as they move between them (Gilbert 1980; Lundberg and Moberg 2003; Box 3.4). Mobile links are crucial for maintaining ecosystem function, memory, and resilience (Nystrm and Folke 2001). The three main types of mobile links: genetic, process, and resource links (Lundberg and Moberg 2003), encompass many fundamental ecosystem services (Sekercioglu 2006a, 2006b). Pollinating nectarivores and seed dispersing frugivores are genetic links that carry genetic material from an individual plant to another plant or to a habitat suitable for regeneration, respectively

## Box 3.4 Conservation of plant-animal mutualisms
## Priya Davidar

Plant-animal mutualisms such as pollination and seed dispersal link plant productivity and ecosystem functioning, and maintain gene flow in plant populations. Insects, particularly bees, are the major pollinators of wild and crop plants worldwide, whereas vertebrates such as birds and mammals contribute disproportionately to dispersal of seeds. About 1200 vertebrate and 100 000 invertebrate species are involved in pollination (Roubik 1995; Buchmann and Nabhan 1996). Pollinators are estimated to be responsible for 35% of global crop production (Klein et al. 2007) and for 60–90% of the reproduction of wild plants (Kremen et al. 2007). It is estimated that feral and managed honey bee colonies have declined by 25% in the USA since the 1990s, and globally about 200 species of wild vertebrate pollinators might be on the verge of extinction (Allen-Wardell et al. 1998). The widespread decline of pollinators and consequently pollination services is a cause for concern and is expected to reduce crop productivity and contribute towards loss of biodiversity in natural ecosystems (Buchmann and Nabhan 1996; Kevan and Viana 2003). Habitat loss, modification and the indiscriminate use of pesticides are cited as major reasons for pollinator loss (Kevan and Viana 2003). This alarming trend has led to the creation of an "International Initiative for the Conservation and Sustainable use of Pollinators" as a key element under the Convention on Biodiversity, and the International Union for the Conservation of Nature has a task force on declining pollination in the Survival Service Commission.

Frugivores tend to be less specialized than pollinators since many animals include some fruit in their diet (Wheelwright and Orians 1982). Decline of frugivores from overhunting and loss of habitat, can affect forest regeneration (Wright et al. 2007a). Hunting pressure differentially affects recruitment of species, where seeds dispersed by game animals decrease, and small non-game animals and by abiotic means increase in the community (Wright et al. 2007b).

Habitat fragmentation is another process that can disrupt mutualistic interactions by reducing the diversity and abundance of pollinators and seed dispersal agents, and creating barriers to pollen and seed dispersal (Cordeiro and Howe 2001, 2003; Aguilar et al. 2006).

Plant-animal mutualisms form webs or networks that contribute to the maintenance of biodiversity. Specialized interactions tend to be nested within generalized interactions where generalists interact more with each other than by chance, whereas specialists interact with generalists (Bascompte and Jordano 2006). Interactions are usually asymmetric, where one partner is more dependent on the other than vice-versa. These characteristics allow for the persistence of rare specialist species. Habitat loss and fragmentation (Chapters 4 and 5), hunting (Chapter 6) and other factors can disrupt mutualistic networks and result in loss of biodiversity. Models suggest that structured networks are less resilient to habitat loss than randomly generated communities (Fortuna and Bascompte 2006).

Therefore maintenance of contiguous forests and intact functioning ecosystems is needed to sustain mutualistic interactions such as pollination and seed dispersal. For agricultural production, wild biodiversity needs to be preserved in the surrounding matrix to promote native pollinators.

### REFERENCES

Aguilar, R., Ashworth, L., Galetto, L., and Aizen, M. A. (2006). Plant reproductive susceptibility to habitat fragmentation: review and synthesis through a meta-analysis. *Ecology Letters* 9, 968–980.

Allen-Wardell, G., Bernhardt, P., Bitner, R., et al. (1998). The potential consequences of pollinator declines on the conservation of biodiversity and stability of food crop yields. *Conservation Biology*, 12, 8–17.

Bascompte, J. and Jordano, P. (2006). The structure of plant-animal mutualistic networks. In M. Pascual and

*continues*

### Box 3.4 (Continued)

J. Dunne, eds *Ecological networks*, pp. 143–159. Oxford University Press, Oxford, UK.

Buchmann, S. L. and Nabhan, G. P. (1996). *The forgotten pollinators*. Island Press, Washington, DC.

Cordeiro, N. J. and Howe, H. F. (2001). Low recruitment of trees dispersed by animals in African forest fragments. *Conservation Biology*, **15**, 1733–1741.

Cordeiro, N. J. and Howe, H. F. (2003). Forest fragmentation severs mutualism between seed dispersers and an endemic African tree. *Proceedings of the National Academy of Sciences of the United States of America*, **100**, 14052–14056.

Fortuna, M. A. and Bascompte, J. (2006). Habitat loss and the structure of plant-animal mutualistic networks. *Ecology Letters*, **9**, 281–286.

Kevan, P. G. and Viana, B. F. (2003). The global decline of pollination services. *Biodiversity*, **4**, 3–8.

Klein, A-M., Vaissiere, B. E., Cane, J. H., *et al.* (2007). Importance of pollinators in changing landscapes for world crops. *Proceedings of the Royal Society of London B*, **274**, 303–313.

Kremen, C., Williams, N. M., Aizen, M. A., *et al.* (2007). Pollination and other ecosystem services produced by mobile organisms: a conceptual framework for the effects of land-use change. *Ecology Letters*, **10**, 299–314.

Roubik, D. W. (1995). *Pollination of cultivated plants in the tropics*. Bulletin 118. FAO, Rome, Italy.

Wheelwright, N. T. and Orians, G. H. (1982). Seed dispersal by animals: contrasts with pollen dispersal, problems of terminology, and constraints on coevolution. *American Naturalist*, **119**, 402–413.

Wright, S. J., Hernandez, A., and Condit, R. (2007a). The bushmeat harvest alters seedling banks by favoring lianas, large seeds and seeds dispersed by bats, birds and wind. *Biotropica*, **39**, 363–371.

Wright, S. J., Stoner, K. E., Beckman, N., *et al.* (2007b). The plight of large animals in tropical forests and the consequences for plant regeneration. *Biotropica*, **39**, 289–291.

---

(Box 3.4). Trophic process links are grazers, such as antelopes, and predators, such as lions, bats, and birds of prey that influence the populations of plant, invertebrate, and vertebrate prey (Boxes 3.1 and 3.2). Scavengers, such as vultures, are crucial process links that hasten the decomposition of potentially disease-carrying carcasses (Houston 1994). Predators often provide natural pest control (Holmes *et al.* 1979). Many animals, such as fish-eating birds that nest in colonies, are resource links that transport nutrients in their droppings and often contribute significant resources to nutrient-deprived ecosystems (Anderson and Polis 1999). Some organisms like woodpeckers or beavers act as physical process linkers or "ecosystem engineers" (Jones *et al.* 1994). By building dams and flooding large areas, beavers engineer ecosystems, create new wetlands, and lead to major changes in species composition (see Chapter 6). In addition to consuming insects (trophic linkers), many woodpeckers also engineer their environment and build nest holes later used by a variety of other species (Daily *et al.* 1993). Through mobile links, distant ecosystems and habitats are linked to and influence one another (Lundberg and Moberg 2003). The long-distance migrations of many species, such as African antelopes, songbirds, waterfowl, and gray whales (*Eschrichtius robustus*) are particularly important examples of critical mobile links. However, many major migrations are disappearing (Wilcove 2008) and nearly two hundred migratory bird species are threatened or near threatened with extinction (Sekercioglu 2007).

Dispersing seeds is among the most important functions of mobile links. Vertebrates are the main seed vectors for flowering plants (Regal 1977; Tiffney and Mazer 1995), particularly woody species (Howe and Smallwood 1982; Levey *et al.* 1994; Jordano 2000). This is especially true in the tropics where bird seed dispersal may have led to the emergence of flowering plant dominance (Regal 1977; Tiffney and Mazer 1995). Seed dispersal is thought to benefit plants in three major ways (Howe and Smallwood 1982):

- Escape from density-dependent mortality caused by pathogens, seed predators, competitors, and herbivores (Janzen-Connell escape hypothesis).

- Chance colonization of favorable but unpredictable sites via wide dissemination of seeds.
- Directed dispersal to specific sites that are particularly favorable for establishment and survival.

Although most seeds are dispersed over short distances, long-distance dispersal is crucial (Cain et al. 2000), especially over geological time scales during which some plant species have been calculated to achieve colonization distances 20 times higher than would be possible without vertebrate seed dispersers (Cain et al. 2000). Seed dispersers play critical roles in the regeneration and restoration of disturbed and degraded ecosystems (Wunderle 1997; Chapter 6), including newly-formed volcanic soils (Nishi and Tsuyuzaki 2004).

Plant reproduction is particularly pollination-limited in the tropics relative to the temperate zone (Vamosi et al. 2006) due to the tropics greater biodiversity, and up to 98% of tropical rainforest trees are pollinated by animals (Bawa 1990). Pollination is a critical ecosystem function for the continued persistence of the most biodiverse terrestrial habitats on Earth. Nabhan and Buchmann (1997) estimated that more than 1200 vertebrate and about 289 000 invertebrate species are involved in pollinating over 90% of flowering plant species (angiosperms) and 95% of food crops. Bees, which pollinate about two thirds of the world's flowering plant species and three quarters of food crops (Nabhan and Buchmann 1997), are the most important group of pollinators (Box 3.3). In California alone, their services are estimated to be worth $4.2 billion (Brauman and Daily 2008). However, bee numbers worldwide are declining (Nabhan and Buchmann 1997) (Box 3.5). In addition to the ubiquitous European honeybee (*Apis mellifera*), native bee species that depend on natural habitats also provide valuable services to farmers, exemplified by Costa Rican forest bees whose activities increase coffee yield by 20% near forest fragments (Ricketts et al. 2004).

Some plant species mostly depend on a single (Parra et al. 1993) or a few (Rathcke 2000) pollinator species. Plants are more likely to be pollinator-limited than disperser-limited (Kelly et al. 2004) and a survey of pollination experiments for 186 species showed that about half were pollinator-limited (Burd 1994). Compared to seed dispersal, pollination is more demanding due to the faster ripening rates and shorter lives of flowers (Kelly et al. 2004). Seed disperser and pollinator limitation are often more important in island ecosystems with fewer species, tighter linkages, and higher vulnerability to disturbance and introduced species. Island plant species are more vulnerable to the extinctions of their pollinators since many island plants have lost

---

**Box 3.5 Consequences of pollinator decline for the global food supply**
**Claire Kremen**

Both wild and managed pollinators have suffered significant declines in recent years. Managed *Apis mellifera*, the most important source of pollination services for crops around the world, have been diminishing around the globe (NRC 2006), particularly in the US where colony numbers are now at < 50% of their 1950 levels. In addition, major and extensive colony losses have occurred over the past several years in North America and Europe, possibly due to diseases as well as other factors (Cox-Foster et al. 2007; Stokstad 2007), causing shortages and rapid increases in the price of pollination services (Sumner and Boriss 2006). These recent trends in honey bee health illustrate the extreme risk of relying on a single pollinator to provide services for the world's crop species. Seventy-five percent of globally important crops rely on animal pollinators, providing up to 35% of crop production (Klein et al. 2007).

At the same time, although records are sorely lacking for most regions, comparisons of recent with historical (pre-1980) records have indicated significant regional declines in species richness of major pollinator groups (bees and hoverflies in Britain; bees alone in the Netherlands) (Biesmeijer et al. 2006). Large reductions in species richness and abundance of bees have also been documented in regions of high agricultural intensity in California's

*continues*

## Box 3.5 (Continued)

Central Valley (Kremen et al. 2002; Klein and Kremen unpublished data). Traits associated with bee, bumble bee and hoverfly declines in Europe included floral specialization, slower (univoltine) development and lower dispersal (non-migratory) species (Biesmeijer et al. 2006; Goulson et al. 2008). Specialization is also indicated as a possible correlate of local extinction in pollinator communities studied across a disturbance gradient in Canada; communities in disturbed habitat contained significantly more generalized species than those associated with pristine habitats (Taki and Kevan 2007). Large-bodied bees were more sensitive to increasing agricultural intensification in California's Central Valley, and ominously, bees with the highest per-visit pollination efficiencies were also most likely to go locally extinct with agricultural intensification (Larsen et al. 2005).

Thus, in highly intensive farming regions, such as California's Central Valley, that contribute comparatively large amounts to global food production (e.g. 50% of the world supply of almonds), the supply of native bee pollinators is lowest in exactly the regions where the demand for pollination services is highest. Published (Kremen et al. 2002) and recent studies (Klein et al. unpublished data) clearly show that the services provided by wild bee pollinators are not sufficient to meet the demand for pollinators in these intensive regions; such regions are instead entirely reliant on managed honey bees for pollination services. If trends towards increased agricultural intensification continue elsewhere (e.g. as in Brazil, Morton et al. 2006), then pollination services from wild pollinators are highly likely to decline in other regions (Ricketts et al. 2008). At the same time, global food production is shifting increasingly towards production of pollinator-dependent foods (Aizen et al. 2008), increasing our need for managed and wild pollinators yet further. Global warming, which could cause mismatches between pollinators and the plants they feed upon, may exacerbate pollinator decline (Memmott et al. 2007). For these reasons, we may indeed face more serious shortages of pollinators in the future.

A recent, carefully analyzed, global assessment of the economic impact of pollinator loss (e.g. total loss of pollinators worldwide) estimates our vulnerability (loss of economic value) at Euro 153 billion or 10% of the total economic value of annual crop production (Gallai et al. 2009). Although total loss of pollination services is both unlikely to occur and to cause widespread famine if it were to occur, it potentially has both serious economic and human health consequences. For example, some regions of the world produce large proportions of the world's pollinator-dependent crops—such regions would experience more severe economic consequences from the loss of pollinators, although growers and industries would undoubtedly quickly respond to these changes in a variety of ways passing the principle economic burden on to consumers globally (Southwick and Southwick 1992; Gallai et al. 2009). Measures of the impacts on consumers (consumer surplus) are of the same order of magnitude (Euro 195–310 billion based on reasonable estimates for price elasticities, Gallai et al. 2009) as the impact on total economic value of crop production. Nutritional consequences may be more fixed and more serious than economic consequences, due to the likely plasticity of responses to economic change. Pollinator-dependent crop species supply not only up to 35% of crop production by weight (Klein et al. 2007), but also provide essential vitamins, nutrients and fiber for a healthy diet and provide diet diversity (Gallai et al. 2009; Kremen et al. 2007). The nutritional consequences of total pollinator loss for human health have yet to be quantified; however food recommendations for minimal daily portions of fruits and vegetables are well-known and already often not met in diets of both developed and underdeveloped countries.

### REFERENCES

Aizen, M. A., Garibaldi, L. A., Cunningham, S. A., and Klein, A. M. (2008). Long-term global trends in crop yield and production reveal no current pollination

*continues*

### Box 3.5 (Continued)

shortage but increasing pollinator dependency. *Current Biology*, **18**, 1572–1575.

Biesmeijer, J. C., Roberts, S. P. M., Reemer, M., *et al.* (2006). Parallel declines in pollinators and insect-pollinated plants in Britain and the Netherlands. *Science*, **313**, 351–354.

Cox-Foster, D. L., Conlan, S., Holmes, E. C., *et al.* (2007). A metagenomic survey of microbes in honey bee colony collapse disorder. *Science*, **318**, 283–287.

Gallai, N., Salles, J.-M., Settele, J., and Vaissière, B. E. (2009). Economic valuation of the vulnerability of world agriculture confronted with pollinator decline. *Ecological Economics*, **68**, 810–821.

Goulson, D., Lye, G. C., and Darvill, B. (2008). Decline and conservation of bumblebees. *Annual Review of Entomology*, **53**, 191–208.

Klein, A. M., Vaissièrie, B., Cane, J. H., *et al.* (2007). Importance of crop pollinators in changing landscapes for world crops. *Proceedings of the Royal Society of London B*, **274**, 303–313.

Kremen, C., Williams, N. M., and Thorp, R. W. (2002). Crop pollination from native bees at risk from agricultural intensification. *Proceedings of the National Academy of Sciences of the United States of America*, **99**, 16812–16816.

Kremen, C., Williams, N. M., Aizen, M. A., *et al.* (2007). Pollination and other ecosystem services produced by mobile organisms: a conceptual framework for the effects of land-use change. *Ecology Letters*, **10**, 299314.

Larsen, T. H., Williams, N. M., and Kremen, C. (2005). Extinction order and altered community structure rapidly disrupt ecosystem functioning. *Ecology Letters*, **8**, 538–547.

Memmott, J., Craze, P. G., Waser, N. M., and Price, M. V. (2007). Global warming and the disruption of plant-pollinator interactions. *Ecology Letters*, **10**, 710–717.

Morton, D. C., DeFries, R. S., Shimabukuro, Y. E., *et al.*, (2006). Cropland expansion changes deforestation dynamics in the southern Brazilian Amazon. *Proceedings of the National Academy of Sciences of the United States of America*, **103**, 14637–14641.

NRC (National Research Council of the National Academies) (2006). *Status of Pollinators in North America*. National Academy Press, Washington, DC.

Ricketts, T. H., Regetz, J., Steffan-Dewenter, I., *et al.* (2008) Landscape effects on crop pollination services: are there general patterns? *Ecology Letters*, **11**, 499–515.

Southwick, E. E. and Southwick, L. Jr. (1992). Estimating the economic value of honey bees (Hymenoptera: Apidae) as agricultural pollinators in the United States. *Journal of Economic Entomology*, **85**, 621–633.

Stokstad, E. (2007). The case of the empty hives. *Science*, **316**, 970–972.

Sumner, D. A. and Boriss, H. (2006). Bee-conomics and the leap in pollination fees. *Giannini Foundation of Agricultural Economics Update*, **9**, 9–11.

Taki, H. and Kevan, P. G. (2007). Does habitat loss affect the communities of plants and insects equally in plant-pollinator interactions? Preliminary findings. *Biodiversity and Conservation*, **16**, 3147–3161.

---

their ability to self-pollinate and have become completely dependent on endemic pollinators (Cox and Elmqvist 2000). Pollination limitation due to the reduced species richness of pollinators on islands like New Zealand and Madagascar (Farwig *et al.* 2004) can significantly reduce fruit sets and probably decrease the reproductive success of dioecious plant species.

Predators are important trophic process links and can control the populations of pest species. For millennia, agricultural pests have been competing with people for the food and fiber plants that feed and clothe humanity. Pests, particularly herbivorous insects, consume 25–50% of humanity's crops every year (Pimentel *et al.* 1989). In the US alone, despite the US$25 billion spent on pesticides annually (Naylor and Ehrlich 1997), pests destroy 37% of the potential crop yield (Pimentel *et al.* 1997). However, many pests have evolved resistance to the millions of tons of synthetic pesticide sprayed each year (Pimentel and Lehman 1993), largely due to insects' short generation times and their experience with millions of years of coevolution with plant toxins (Ehrlich and Raven 1964). Consequently, these chemicals poison the environment (Carson 1962), lead to thousands of wildlife fatalities every year, and by killing pests' natural enemies faster than the pests themselves, often lead to the emergence of new pest populations (Naylor and Ehrlich 1997). As a result, the value of natural pest control has been increasingly recognized worldwide, some major successes have been achieved, and natural controls now form a core component of "integrated pest management" (IPM) that aims to restore the natural pest-predator balance in agricultural ecosystems (Naylor and Ehrlich 1997).

Species that provide natural pest control range from bacteria and viruses to invertebrate and vertebrate predators feeding on insect and rodent pests (Polis *et al.* 2000; Perfecto *et al.* 2004; Sekercioglu 2006b). For example, a review by Holmes (1990) showed that reductions in moth and butterfly populations due to temperate forest birds was mostly between 40–70% at low insect densities, 20–60% at intermediate densities, and 0–10% at high densities. Although birds are not usually thought of as important control agents, avian control of insect herbivores and consequent reductions in plant damage can have important economic value (Mols and Visser 2002). Takekawa and Garton (1984) calculated avian control of western spruce budworm in northern Washington State to be worth at least US$1820/km$^2$/year. To make Beijing greener for the 2008 Olympics without using chemicals, entomologists reared four billion parasitic wasps to get rid of the defoliating moths in less than three months (Rayner 2008). Collectively, natural enemies of crop pests may save humanity at least US$54 billion per year, not to mention the critical importance of natural controls for food security and human survival (Naylor and Ehrlich 1997). Promoting natural predators and preserving their native habitat patches like hedgerows and forests may increase crop yields, improve food security, and lead to a healthier environment.

Often underappreciated are the scavenging and nutrient deposition services of mobile links. Scavengers like vultures rapidly get rid of rotting carcasses, recycle nutrients, and lead other animals to carcasses (Sekercioglu 2006a). Besides their ecological significance, vultures are particularly important in many tropical developing countries where sanitary waste and carcass disposal programs may be limited or non-existent (Prakash *et al.* 2003) and where vultures contribute to human and ecosystem health by getting rid of refuse (Pomeroy 1975), feces (Negro *et al.* 2002), and dead animals (Prakash *et al.* 2003).

Mobile links also transport nutrients from one habitat to another. Some important examples are geese transporting terrestrial nutrients to wetlands (Post *et al.* 1998) and seabirds transferring marine productivity to terrestrial ecosystems, especially in coastal areas and unproductive island systems (Sanchez-pinero and Polis 2000). Seabird droppings (guano) are enriched in important plant nutrients such as calcium, magnesium, nitrogen, phosphorous, and potassium (Gillham 1956). Murphy (1981) estimated that seabirds around the world transfer $10^4$ to $10^5$ tons of phosphorous from sea to land every year. Guano also provides an important source of fertilizer and income to many people living near seabird colonies.

Scavengers and seabirds provide good examples of how the population declines of ecosystem service providers lead to reductions in their services (Hughes *et al.* 1997). Scavenging and fish-eating birds comprise the most threatened avian functional groups, with about 40% and 33%, respectively, of these species being threatened or near threatened with extinction (Sekercioglu *et al.* 2004). The large declines in the populations of many scavenging and fish-eating species mean that even if none of these species go extinct, their services are declining substantially. Seabird losses can trigger trophic cascades and ecosystem shifts (Croll *et al.* 2005). Vulture declines can lead to the emergence of public health problems. In India, *Gyps* vulture populations declined as much as 99% in the 1990s (Prakash *et al.* 2003). Vultures compete with feral dogs, which often carry rabies. As the vultures declined between 1992 and 2001, the numbers of feral dogs increased 20-fold at a garbage dump in India (Prakash *et al.* 2003). Most of world's rabies deaths take place in India (World Health Organization 1998) and feral dogs replacing vultures is likely to aggravate this problem.

Mobile links, however, can be double-edged swords and can harm ecosystems and human populations, particularly in concert with human related poor land-use practices, climate change, and introduced species. Invasive plants can spread via native and introduced seed dispersers (Larosa et al. 1985; Cordeiro et al. 2004). Land use change can increase the numbers of mobile links that damage distant areas, such as when geese overload wetlands with excessive nutrients (Post et al. 1998). Climate change can lead to asynchronies in insect emergence and their predators timing of breeding (Both et al. 2006), and in flowering and their pollinators lifecycles (Harrington et al. 1999) (Chapter 8).

Mobile links are often critical to ecosystem functioning as sources of "external memory" that promote the resilience of ecosystems (Scheffer et al. 2001). More attention needs to be paid to mobile links in ecosystem management and biodiversity conservation (Lundberg and Moberg 2003). This is especially the case for migrating species that face countless challenges during their annual migrations that sometimes cover more than 20 000 kilometers (Wilcove 2008). Some of the characteristics that make mobile links important for ecosystems, such as high mobility and specialized diets, also make them more vulnerable to human impact. Protecting pollinators, seed dispersers, predators, scavengers, nutrient depositors, and other mobile links must be a top conservation priority to prevent collapses in ecosystem services provided by these vital organisms (Boxes 3.1–3.5).

## 3.6 Nature's Cures versus Emerging Diseases

While many people know about how plants prevent erosion, protect water supplies, and "clean the air", how bees pollinate plants or how owls reduce rodent activity, many lesser-known organisms not only have crucial ecological roles, but also produce unique chemicals and pharmaceuticals that can literally save people's lives. Thousands of plant species are used medically by traditional, indigenous communities worldwide. These peoples' ethnobotanical knowledge has led to the patenting, by pharmaceutical companies, of more than a quarter of all medicines (Posey 1999), although the indigenous communities rarely benefit from these patents (Mgbeoji 2006). Furthermore, the eroding of traditions worldwide, increasing emigration from traditional, rural communities to urban areas, and disappearing cultures and languages mean that the priceless ethnobotanical knowledge of many cultures is rapidly disappearing in parallel with the impending extinctions of many medicinal plants due to habitat loss and overharvesting (Millennium Ecosystem Assessment 2005a). Some of the rainforest areas that are being deforested fastest, like the island of Borneo, harbor plant species that produce active anti-HIV (Human Immunodeficiency Virus) agents (Chung 1996; Jassim and Naji 2003). Doubtlessly, thousands more useful and vital plant compounds await discovery in the forests of the world, particularly in the biodiverse tropics (Laurance 1999; Sodhi et al. 2007). However, without an effective strategy that integrates community-based habitat conservation, rewarding of local ethnobotanical knowledge, and scientific research on these compounds, many species, the local knowledge of them, and the priceless cures they offer will disappear before scientists discover them.

As with many of nature's services, there is a flip side to the medicinal benefits of biodiversity, namely, emerging diseases (Jones et al. 2008). The planet's organisms also include countless diseases, many of which are making the transition to humans as people increasingly invade the habitats of the hosts of these diseases and consume the hosts themselves. Three quarters of human diseases are thought to have their origins in domestic or wild animals and new diseases are emerging as humans increase their presence in formerly wilderness areas (Daily and Ehrlich 1996; Foley et al. 2005). Some of the deadliest diseases, such as monkeypox, malaria, HIV and Ebola, are thought to have initially crossed from central African primates to the people who hunted, butchered, and consumed them (Hahn et al. 2000; Wolfe et al. 2005; Rich et al. 2009). Some diseases emerge in ways that show the difficulty of predicting the consequences of disturbing ecosystems. The extensive smoke from the massive 1997–1998 forest fires in Southeast Asia is thought to have led to the fruiting failure of many forest trees, forcing frugivorous bats to switch to fruit trees in pig farms. The bats, which host the Nipah virus, likely passed it to the pigs, from which the virus made the jump to people (Chivian 2002). Another classic example from Southeast Asia is the Severe Acute Respiratory Syndrome (SARS). So far having killed 774 people, the SARS coronavirus has been recently discovered in wild animals like the masked palm civet (*Paguma larvata*) and raccoon dog (*Nyctereuteus procyonoides*) that are frequently consumed by people in the region (Guan *et al.* 2003). SARS-like coronaviruses have been discovered in bats (Li *et al.* 2005) and the virus was probably passed to civets and other animals as they ate fruits partially eaten and dropped by those bats (Jamie H. Jones, personal communication). It is probable that SARS made the final jump to people through such animals bought for food in wildlife markets.

The recent emergence of the deadly avian influenza strain H5N1 provides another good example. Even though there are known to be at least 144 strains of avian flu, only a few strains kill people. However, some of the deadliest pandemics have been among these strains, including H1N1, H2N, and H3N2 (Cox and Subbarao 2000). H5N1, the cause of the recent bird flu panic, has a 50% fatality rate and may cause another human pandemic. At low host densities, viruses that become too deadly, fail to spread. It is likely that raising domestic birds in increasingly higher densities led to the evolution of higher virulence in H5N1, as it became easier for the virus to jump to another host before it killed its original host. There is also a possibility that increased invasion of wilderness areas by people led to the jump of H5N1 from wild birds to domestic birds, but that is yet to be proven.

Malaria, recently shown to have jumped from chimpanzees to humans (Rich *et al.* 2009), is perhaps the best example of a resurging disease that increases as a result of tropical deforestation (Singer and Castro 2001; Foley *et al.* 2005; Yasuoka and Levins 2007). Pearson (2003) calculated that every 1% increase in deforestation in the Amazon leads to an 8% increase in the population of the malaria vector mosquito (*Anopheles darlingi*). In addition, some immigrants colonizing deforested areas brought new sources of malaria (Moran 1988) whereas other immigrants come from malaria-free areas and thus become ideal hosts with no immunity (Aiken and Leigh 1992).

Collectively, the conditions leading to and resulting from tropical deforestation, combined with climate change, human migration, agricultural intensification, and animal trafficking create the perfect storm for the emergence of new diseases as well as the resurgence of old ones. In the face of rapid global change, ecologically intact and relatively stable communities may be our best weapon against the emergence of new diseases.

## 3.7 Valuing Ecosystem Services

Ecosystems and their constituent species provide an endless stream of products, functions, and services that keep our world running and make our existence possible. To many, even the thought of putting a price tag on services like photosynthesis, purification of water, and pollination of food crops may seem like hubris, as these are truly priceless services without which not only humans, but most of life would perish. A distinguished economist put it best in response to a seminar at the USA Federal Trade Commission, where the speaker downplayed the impact of global warming by saying agriculture and forestry "accounted for only three percent of the US gross national product". The economist's response was: "What does this genius think we're going to eat?" (Naylor and Ehrlich 1997).

Nevertheless, in our financially-driven world, we need to quantify the trade-offs involved in land use scenarios that maximize biodiversity conservation and ecosystem services versus scenarios that maximize profit from a single commodity. Without such assessments, special interests representing single objectives dominate the debate and sideline the integration of ecosystem services into the decision-making process (Nelson *et al.* 2009). Valuing ecosystem services is not an end in itself, but is the first step towards integrating these services into public decision-making and ensuring the continuity of ecosystems that provide the services (Goulder and Kennedy 1997; National Research Council 2005; Daily *et al.* 2009). Historically, ecosystem services have been mostly thought of as free public goods, an approach which has too frequently led to the "tragedy of the commons" where vital ecosystem goods like clean water have been degraded and consumed to extinction (Daily 1997). Too often, ecosystem services have been valued, if at all, based on "marginal utility" (Brauman and Daily 2008). When the service (like clean water) is abundant, the marginal utility of one additional unit can be as low as zero. However, as the service becomes more scarce, the marginal utility of each additional unit becomes increasingly valuable (Goulder and Kennedy 1997). Using the marginal value for a service when it is abundant drastically underestimates the value of the service as it becomes scarcer. As Benjamin Franklin wryly observed, "When the well's dry, we know the worth of water."

As the societal importance of ecosystem services becomes increasingly appreciated, there has been a growing realization that successful application of this concept requires a skilful combination of biological, physical, and social sciences, as well as the creation of new programs and institutions. The scientific community needs to help develop the necessary quantitative tools to calculate the value of ecosystem services and to present them to the decision makers (Daily *et al.* 2009). A promising example is the InVEST (Integrated Valuation of Ecosystem Services and Tradeoffs) system (Daily *et al.* 2009; Nelson 2009) developed by the Natural Capital Project (www.naturalcapital.org; see Box 15.3). However, good tools are valuable only if they are used. A more difficult goal is convincing the private and public sectors to incorporate ecosystem services into their decision-making processes (Daily *et al.* 2009). Nevertheless, with the socio-economic impacts and human costs of environmental catastrophes, such as Hurricane Katrina, getting bigger and more visible, and with climate change and related carbon sequestration schemes having reached a prominent place in the public consciousness, the value of these services and the necessity of maintaining them has become increasingly mainstream.

Recent market-based approaches such as payments for Costa Rican ecosystem services, wetland mitigation banks, and the Chicago Climate Exchange have proven useful in the valuation of ecosystem services (Brauman and Daily 2008). Even though the planet's ecosystems, the biodiversity they harbor, and the services they collectively provide are truly priceless, market-based and other quantitative approaches for valuing ecosystem services will raise the profile of nature's services in the public consciousness, integrate these services into decision-making, and help ensure the continuity of ecosystem contributions to the healthy functioning of our planet and its residents.

## Summary

- Ecosystem services are the set of ecosystem functions that are useful to humans.
- These services make the planet inhabitable by supplying and purifying the air we breathe and the water we drink.
- Water, carbon, nitrogen, phosphorus, and sulfur are the major global biogeochemical cycles. Disruptions of these cycles can lead to floods, droughts, climate change, pollution, acid rain, and many other environmental problems.
- Soils provide critical ecosystem services, especially for sustaining ecosystems and growing food crops, but soil erosion and degradation are serious problems worldwide.
- Higher biodiversity usually increases ecosystem efficiency and productivity, stabilizes overall ecosystem functioning, and makes ecosystems more resistant to perturbations.
- Mobile link animal species provide critical ecosystem functions and increase ecosystem resilience by connecting habitats and ecosystems through their movements. Their services include pollination, seed dispersal, nutrient deposition, pest control, and scavenging.
- Thousands of species that are the components of ecosystems harbor unique chemicals and pharmaceuticals that can save people's lives, but traditional knowledge of medicinal plants is disappearing and many potentially valuable species are threatened with extinction.
- Increasing habitat loss, climate change, settlement of wild areas, and wildlife consumption facilitate the transition of diseases of animals to humans, and other ecosystem alterations are increasing the prevalence of other diseases.

- Valuation of ecosystem services and tradeoffs helps integrate these services into public decision-making and can ensure the continuity of ecosystems that provide the services.

## Relevant websites

- Millennium Ecosystem Assessment: http://www.millenniumassessment.org/
- Intergovernmental Panel on Climate Change: http://www.ipcc.ch/
- Ecosystem Marketplace: http://www.ecosystemmarketplace.com/
- United States Department of Agriculture, Forest Service Website on Ecosystem Services: http://www.fs.fed.us/ecosystemservices/
- Ecosystem Services Project: http://www.ecosystemservicesproject.org/index.htm
- Natural Capital Project: http://www.naturalcapitalproject.org
- Carbon Trading: http://www.carbontrading.com/

## Acknowledgements

I am grateful to Karim Al-Khafaji, Berry Brosi, Paul R. Ehrlich, Jamie H. Jones, Stephen Schneider, Navjot Sodhi, Tanya Williams, and especially Kate Brauman for their valuable comments. I thank the Christensen Fund for their support of my conservation and ecology work.

## REFERENCES

Aiken, S. R. and Leigh, C. H. (1992). *Vanishing rain forests*. Clarendon Press, Oxford, UK.

Alexander, S. E., Schneider, S. H., and Lagerquist, K. (1997). The interaction of climate and life. In G. C. Daily, ed. *Nature's Services*, pp. 71–92. Island Press, Washington DC.

Anderson, W. B. and Polis, G. A. (1999). Nutrient fluxes from water to land: seabirds affect plant nutrient status on Gulf of California islands. *Oecologia*, **118**, 324–32.

Avinash, N. (2008). Soil no bar: Gujarat farmers going hi-tech. *The Economic Times*, **24 July**.

Ba, L. K. (1977). *Bio-economics of trees in native Malayan forest*. Department of Botany, University of Malaya, Kuala Lumpur.

Bawa, K. S. (1990). Plant-pollinator interactions in Tropical Rain-Forests. *Annual Review of Ecology and Systematics*, **21**, 399–422.

Berhe, A. A., Harte, J., Harden, J. W., and Torn, M. S. (2007). The significance of the erosion-induced terrestrial carbon sink. *BioScience*, **57**, 337–46.

Berlow, E. L. (1999). Strong effects of weak interactions in ecological communities. *Nature*, **398**, 330–34.

Bolin, B. and Cook, R. B., eds (1983). *The major biogeochemical cycles and their interactions*. Wiley, New York.

Bolker, B. M., Pacala, S. W., Bazzaz, F. A., Canham, C. D., and Levin, S. A. (1995). Species diversity and ecosystem response to carbon dioxide fertilization: conclusions from a temperate forest model. *Global Change Biology*, **1**, 373–381.

Both, C., Bouwhuis, S., Lessells, C. M., and Visser, M. E. (2006). Climate change and population declines in a long-distance migratory bird. *Nature*, **441**, 81–83.

Bradshaw, C. J. A., Sodhi, N. S., Peh, K. S.-H., and Brook, B. W. (2007). Global evidence that deforestation amplifies flood risk and severity in the developing world. *Global Change Biology*, **13**, 2379–2395.

Brauman, K. A., and G. C. Daily. (2008). Ecosystem services. In S. E. Jorgensen and B. D. Fath, ed. *Human Ecology*, pp. 1148–1154. Elsevier, Oxford, UK.

Brauman, K. A., Daily, G. C., Duarte, T. K., and Mooney, H. A. (2007). The nature and value of ecosystem services: an overview highlighting hydrologic services. *Annual Review of Environment and Resources*, **32**, 67–98.

Bruijnzeel, L. A. (2004). Hydrological functions of tropical forests: not seeing the soil for the trees? *Agriculture Ecosystems and Environment*, **104**, 185–228.

Bruno, J. F. and Selig, E. R. (2007). Regional decline of coral cover in the Indo-Pacific: timing, extent, and subregional comparisons. *PLoS One*, **2**, e711.

Burd, M. (1994). Bateman's principle and plant reproduction: the role of pollen limitation in fruit and seed set. *Botanical Review*, **60**, 81–109.

Butler, R. (2008). Despite financial chaos, donors pledge $100M for rainforest conservation. http://news.mongabay.com/2008/1023-fcpf.html.

Cain, M. L., Milligan, B. G., and Strand, A. E. (2000). Long-distance seed dispersal in plant populations. *American Journal of Botany*, **87**, 1217–1227.

Calder, I. R. and Aylward, B. (2006). Forest and floods: Moving to an evidence-based approach to watershed and integrated flood management. *Water International*, **31**, 87–99.

Canada's Office of Urban Agriculture (2008). *Urban agriculture notes*. City Farmer, Vancouver, Canada.

Carson, R. (1962). *Silent Spring*. Houghton Mifflin, Boston, MA.

Chapin, F. S., Zavaleta, E. S., Eviner, V. T., et al. (2000). Consequences of changing biodiversity. *Nature*, **405**, 234–242.

Chivian, E. (2002). *Biodiversity: its importance to human health*. Center for Health and the Global Environment, Harvard Medical School, Cambridge, MA.

Chung, F. J. (1996). Interests and policies of the state of Sarawak, Malaysia regarding intellectual property rights for plant derived drugs. *Journal of Ethnopharmacology*, **51**, 201–204.

Cordeiro, N. J., D. A. G. Patrick, B. Munisi, and V. Gupta. (2004). Role of dispersal in the invasion of an exotic tree in an East African submontane forest. *Journal of Tropical Ecology*, **20**: 449–457.

National Research Council (2005). *Valuing ecosystem service: toward better environmental decision-making*. The National Academies Press, Washington, DC.

Cox, N.J. and Subbarao, K. (2000). Global epidemiology of influenza: past and present.

Cox, P. A. and Elmqvist, T. (2000). Pollinator extinction in the Pacific islands. *Conservation Biology*, **14**, 1237–1239.

Croll, D. A., Maron, J. L., Estes, J. A., et al. (2005). Introduced predators transform subarctic islands from grassland to tundra. *Science*, **307**, 1959–1961.

Dahdouh-Guebas, F., Jayatissa, L. P., Di Nitto, D., et al. (2005). How effective were mangroves as a defence against the recent tsunami? *Current Biology*, **15**, R443–R447.

Daily, G. C., ed. (1997). *Nature's services: societal dependence on natural ecosystems*. Island Press, Washington, DC.

Daily, G. C. and Ehrlich, P. R. (1996). Global change and human susceptibility to disease. *Annual Review of Energy Environment*, **21**, 125–144.

Daily, G. C., Ehrlich, P. R., and Haddad, N. M. (1993). Double keystone bird in a keystone species complex. *Proceedings of the National Academy of Sciences of the United States of America*, **90**, 592–594.

Daily, G. C., Matson, P. A., and Vitousek, P. M. (1997). Ecosystem services supplied by soil. In G. C. Daily, ed. *Nature Services: societal dependence on natural ecosystems*, pp. 113–132. Island Press, Washington, DC.

Daily, G. C., Polasky, S., Goldstein, J., et al. (2009). Ecosystem services in decision-making: time to deliver. *Frontiers in Ecology and the Environment*, **7**, 21–28.

Danielsen, F., Sørensen, M. K., Olwig, M. F., et al. (2005). The Asian tsunami: a protective role for coastal vegetation. *Science*, **310**, 643.

Darwin, C. R. (1872). *The origin of species*. 6th London edn Thompson & Thomas, Chicago, IL.

Day, J. W., Boesch, D. F., Clairain, E. J., et al. (2007). Restoration of the Mississippi Delta: Lessons from Hurricanes Katrina and Rita. *Science*, **315**, 1679–1684.

Dosso, H. (1981). The Tai Project: land use problems in a tropical forest. *Ambio*, **10**, 120–125.

Ehrlich, P. and Ehrlich, A. (1981a). *Extinction: the causes and consequences of the disappearance of species*. Ballantine Books, New York, NY.

Ehrlich, P. R. and Ehrlich, A. H. (1981b). The rivet poppers. *Not Man Apart*, **2**, 15.

Ehrlich, P. R. and Mooney, H. M. (1983). Extinction, substitution and ecosystem services. *BioScience*, **33**, 248–254.

Ehrlich, P. R. and Raven, P. H. (1964). Butterflies and plants: a study in coevolution. *Evolution*, **18**, 586–608.

Falkowski, P., Scholes, R. J., Boyle, E., et al. (2000). The global carbon cycle: a test of our knowledge of earth as a system. *Science*, **290**, 291–296.

FAO. (1990). *Soilless culture for horticultural crop production*. FAO, Rome, Italy.

Farwig, N., Randrianirina, E. F., Voigt, F. A., et al. (2004). Pollination ecology of the dioecious tree *Commiphora guillauminii* in Madagascar. *Journal of Tropical Ecology*, **20**, 307–16.

Foley, J. A., Defries, R., Asner, G. P., et al. (2005). Global consequences of land use. *Science*, **309**, 570–574.

Fujii, S., Somiya, I., Nagare, H., and Serizawa, S. (2001). Water quality characteristics of forest rivers around Lake Biwa. *Water Science and Technology*, **43**, 183–192.

Gilbert, L. E. (1980). Food web organization and the conservation of Neotropical diversity. In M. E. Soulé, and B. A. Wilcox, eds *Conservation Biology: an evolutionary-ecological perspective*, pp. 11–33. Sinauer Associates, Sunderland, MA.

Gillham, M. E. (1956). The ecology of the Pembrokeshire Islands V: manuring by the colonial seabirds and mammals with a note on seed distribution by gulls. *Journal of Ecology*, **44**, 428–454.

Gomes, L., Arrue, J. L., Lopez, M. V., et al. (2003). Wind erosion in a semiarid agricultural area of Spain: the WELSONS project. *Catena*, **52**, 235–256.

Goulder, L. H. and Kennedy, D. (1997). Valuing ecosystem services: philosophical bases and empirical methods. In G. C. Daily, ed *Nature's Services: societal dependence on natural ecosystems*, pp. 23–47. Island Press, Washington, DC.

Gregorich, E. G., Greer, K. J., Anderson, D. W., and Liang, B. C. (1998). Carbon distribution and losses: erosion and deposition effects. *Soil and Tillage Research* **47**, 291–02.

Guan, Y., Zheng, B. J., He, Y. Q., et al. (2003). Isolation and characterization of viruses related to the SARS coronavirus from animals in Southern China. *Science*, **302**, 276–278.

Hahn, B. H., G. M. Shaw, K. M. De Cock, and P. M. Sharp. (2000). AIDS as a zoonosis: scientific and public health implications. *Science*, **287**, 607–614.

Harrington, R., Woiwod, I., and Sparks, T. (1999). Climate change and trophic interactions. *Trends in Ecology and Evolution*, **14**, 146–150.

Hiraishi, T. and Harada, K. (2003). Greenbelt tsunami prevention in South-Pacific Region. *Report of the Port and Airport Research Institute*, **42**, 3–5, 7–25.

Hobbs, R. J., Yates, S. and Mooney, H. A. (2007). Long-term data reveal complex dynamics in grassland in relation to climate and disturbance. *Ecological Monographs*, **77**, 545–568.

Holmes, R. T. (1990). Ecological and evolutionary impacts of bird predation on forest insects: an overview. In M. L. Morrison, C. J. Ralph, J. Verner, and J. R. Jehl, Jr, eds *Avian Foraging: theory, methodology, and applications*, pp. 6–13. Allen Press, Lawrence, KS.

Holmes, R. T., Schultz, J. C., and Nothnagle, P. (1979). Bird predation on forest insects - an exclosure experiment. *Science*, **206**, 462–463.

Hooper, D. U., Chapin, F. S., Ewel, J. J., et al. (2005). Effects of biodiversity on ecosystem functioning: a consensus of current knowledge. *Ecological Monographs*, **75**, 3–5.

Houghton, J. (2004). *Global Warming: the complete briefing*. Cambridge University Press, Cambridge, UK.

Houston, D. C. (1994). Family Cathartidae (New World vultures). In J. del Hoyo, A. Elliott, and J. Sargatal, eds *Handbook of the Birds of the World (Vol. 2): New World Vultures to Guineafowl*, pp. 24–41. Lynx Ediciones, Barcelona, Spain.

Howe, H. F. and J. Smallwood. (1982). Ecology of seed dispersal. *Annual Review of Ecology and Systematics*, **13**, 201–228.

Hughes, J. B., Daily, G. C., and Ehrlich, P. R. (1997). Population diversity, its extent and extinction. *Science*, **278**, 689–692.

Hupy, J. P. (2004). Influence of vegetation cover and crust type on wind-blown sediment in a semi-arid climate. *Journal of Arid Environments*, **58**, 167–179.

IPCC (2007). *Fourth Assessment Report: Climate Change 2007, The Physical Science Basis*. Cambridge University Press, Cambridge, UK.

Jassim, S. A. A. and Naji, M. A. (2003). Novel antiviral agents: a medicinal plant perspective. *Journal of Applied Microbiology*, **95**, 412–427.

Jones, C. G., Lawton, J. H., and Shachak, M. (1994). Organisms as ecosystem engineers. *Oikos*, **69**, 373–86.

Jones, K. E., Patel, N. G., Levy, M. A., et al. (2008). Global trends in emerging infectious diseases. *Nature*, **451**, 990–994.

Jordano, P. (2000). Fruits and frugivory. In M. Fenner, ed. *Seeds: the ecology of regeneration in plant communities*, pp. 125–165. CAB International, New York, NY.

Kelly, D., Ladley, J. J., and Robertson, A. W. (2004). Is dispersal easier than pollination? Two tests in new Zealand Loranthaceae. *New Zealand Journal of Botany*, **42**, 89–103.

Klare M. T. (2001). *Resource wars: the new landscape of global conflict*. New York, NY. Henry Holt.

Kremen, C. (2005). Managing ecosystem services: what do we need to know about their ecology? *Ecology Letters*, **8**, 468–479.

Lal, R. (2004). Soil carbon sequestration impacts on global climate change and food security. *Science*, **304**, 1623–1627.

Larosa, A. M., Smith, C. W., and Gardner, D. E. (1985). Role of alien and native birds in the dissemination of firetree (Myrica-faya Ait-Myricaceae) and associated plants in Hawaii. *Pacific Science*, **39**, 372–378.

Laurance, W. F. (1999). Reflections on the tropical deforestation crisis. *Biological Conservation*, **91**, 109–117.

Lefroy, E., Hobbs, R., and Scheltma, M. (1993). Reconciling agriculture and nature conservation: toward a restoration strategy for the Western Australia wheatbelt. In D. Saunders, R. Hobbs, and P. Ehrlich, eds *Reconstruction of fragmented ecosystems: global and regional perspectives*, pp. 243–257. Surrey Beattie & Sons, Chipping Norton, NSW, Australia.

Levey, D. J., Moermond, T. C., and Denslow, J. S. (1994). Frugivory: an overview. In L. A. McDade, K. S. Bawa, H. A. Hespenheide, and G. S. Hartshorn, eds *La Selva: ecology and natural history of a Neotropical Rain Forest*, pp.282–294. University of Chicago Press, Chicago.

Li, W. D., Shi, Z. L., Yu, M., et al. (2005). Bats are natural reservoirs of SARS-like coronaviruses. *Science*, **310**, 676–679.

Lundberg, J. and Moberg, F. (2003). Mobile link organisms and ecosystem functioning: implications for ecosystem resilience and management. *Ecosystems*, **6**, 87–98.

Lyons, K. G. and Schwartz, M. W. (2001). Rare species loss alters ecosystem function - invasion resistance. *Ecology Letters*, **4**, 358–65.

MacArthur, R. H. (1955). Fluctuations of animal populations and a measure of community stability. *Ecology*, **36**, 533–536.

Mgbeoji, I. (2006). *Global biopiracy: patents, plants, and indigenous knowledge*. Cornell University Press, Ithaca, NY.

Millennium Ecosystem Assessment (2005a). *Ecosystems and human well-being: synthesis*. Island Press, Washington, DC.

Millennium Ecosystem Assessment. (2005b). *Nutrient cycling*. World Resources Institute, Washington, DC.

Millennium Ecosystem Assessment. (2005c). *Fresh water*. World Resources Institute, Washington, DC.

Milliman, J. D. and Syvitski, J. P. M. (1992). Geomorphic tectonic control of sediment discharge to the ocean - the importance of small mountainous rivers. *Journal of Geology*, **100**, 525–544.

Mols, C. M. M. and Visser, M. E. (2002). Great tits can reduce caterpillar damage in apple orchards. *Journal of Applied Ecology*, **39**, 888–899.

Mooney, H. A. and Ehrlich, P. R. (1997). Ecosystem services: a fragmentary history. In G. C. Daily, ed. *Nature's Services: societal dependence on natural ecosystems*, pp. 11–19. Island Press, Washington, DC.

Moran, E. F. (1988). Following the Amazonian Highways. In J. S. D. A. C. Padoch, ed. *People of the rain forest*, pp. 155–162. University of California Press, Berkeley, CA.

Murphy, G. I. (1981). Guano and the anchovetta fishery. *Research Management Environment Uncertainty*, **11**, 81–106.

Myers, N. (1997). The world's forests and their ecosystem services. In G. C. Daily, ed. *Nature's Services: societal dependence on natural ecosystems*, pp. 215–235. Island Press, Washington, DC.

Nabhan, G. P. and Buchmann, S. L. (1997). Services provided by pollinators. In G. C. Daily, ed. *Nature's Services: societal dependence on natural ecosystems*, pp. 133–150. Island Press, Washington, DC.

Naeem, S. and Li, S. (1997). Biodiversity enhances ecosystem reliability. *Nature*, **390**, 507–509.

Naeem, S., Thompson, L. J., Lawler, S. P., Lawton, J. H., and Woodfin, R. M. (1995). Empirical evidence that declining species diversity may alter the performance of terrestrial ecosystems. *Philosophical Transactions of the Royal Society B*, **347**, 249–262.

Naylor, R. and Ehrlich, P. R. (1997). Natural pest control services and agriculture. In G. C. Daily, ed. *Nature's Services: societal dependence on natural ecosystems*, pp. 151–174. Island Press, Washington, DC.

Negro, J. J., Grande, J. M., Tella, J. L., *et al.* (2002). Coprophagy: An unusual source of essential carotenoids - a yellow-faced vulture includes ungulate faeces in its diet for cosmetic purposes. *Nature*, **416**, 807–808.

Nelson, E., Mendoza, G., Regetz, J., *et al.* (2009). Modeling multiple ecosystem services and tradeoffs at landscape scales. *Frontiers in Ecology and the Environment*, **7**, 4–11.

Nishi, H. and Tsuyuzaki, S. (2004). Seed dispersal and seedling establishment of *Rhus trichocarpa* promoted by a crow (*Corvus macrorhynchos*) on a volcano in Japan. *Ecography*, **27**, 311–22.

Nystrm, M. and Folke, C. (2001). Spatial resilience of coral reefs. *Ecosystems*, **4**, 406–417.

Oldeman, L. R. (1998). *Soil degradation: a threat to food security?* International Soil Reference and Information Center (ISRIC), Wageningen, Netherlands.

Page, S. E. and Rieley, J. O. (1998). Tropical peatlands: A review of their natural resource functions with particular reference to Southwest Asia. *International Peat Journal*, **8**, 95–106.

Pandolfi, J. M., Bradbury, R. H., Sala, E., *et al.* (2003). Global trajectories of the long-term decline of coral reef ecosystems. *Science*, **301**, 955–958.

Parmesan, C. and Yohe, G. (2003). A globally coherent fingerprint of climate change impacts across natural systems. *Nature*, **421**, 37–42.

Parra, V., Vargas, C. F., and Eguiarte, L. E. (1993). Reproductive biology, pollen and seed dispersal, and neighborhood size in the hummingbird-pollinated *Echeveria gibbiflora* (Crassulaceae). *American Journal of Botany*, **80**, 153–159.

Pearson, H. (2003). Lost forest fuels malaria. Nature Science Update 28 November.

Perfecto, I., Vandermeer, J. H., Bautista, G. L., *et al.* (2004). Greater predation in shaded coffee farms: the role of resident Neotropical birds. *Ecology*, **85**, 2677–2681.

Phat, N. K., Knorr, W., and Kim, S. (2004). Appropriate measures for conservation of terrestrial carbon stocks - analysis of trends of forest management in Southeast Asia. *Forest Ecology and Management*, **191**, 283–299.

Pimentel, D. and Kounang, N. (1998). Ecology of soil erosion in ecosystems. *Ecosystems*, **1**, 416–426.

Pimentel, D. and Lehman, H. eds. (1993). *The pesticide question: environment, economics, and ethics*. Chapman and Hall, New York, NY.

Pimentel, D., Mclaughlin, L., Zepp, A., *et al.* (1989). Environmental and economic impacts of reducing U.S. agricultural pesticide use. *Handbook of Pest Management in Agriculture*, **4**, 223–278.

Pimentel, D., Harvey, C., Resosudarmo, P., *et al.* (1995). Environmental and economic costs of soil erosion and conservation benefits. *Science*, **267**, 1117–1123.

Pimentel, D., Wilson, C., McCullum, C., *et al.* (1997). Economic and environmental benefits of biodiversity. *BioScience*, **47**, 747–757.

Pimm, S. L. (1984). The complexity and stability of ecosystems. *Nature*, **307**, 321–26.

Polis, G. A., Sears, A. L. W., Huxel, G. R., Strong, D. R., and Maron, J. (2000). When is a trophic cascade a trophic cascade? *Trends in Ecology and Evolution*, **15**, 473–475.

Pomeroy, D. E. (1975). Birds as scavengers of refuse in Uganda. *Ibis*, **117**, 69–81.

Posey, D. A. E. (1999). *Cultural and spiritual values of biodiversity*. Intermediate Technology Publications, United Nations Environment Programme, London, UK.

Post, D. M., Taylor, J. P., Kitchell, J. F., *et al.* (1998). The role of migratory waterfowl as nutrient vectors in a managed wetland. *Conservation Biology*, **12**, 910–920.

Postel, S. and Carpenter, S. (1997). Freshwater ecosystem services. In G. C. Daily, ed. *Nature's Services: societal dependence on natural ecosystems*, pp.195–214. Island Press, Washington, DC.

Postel, S. L., Daily, G. C., and Ehrlich, P. R. (1996). Human appropriation of renewable fresh water. *Science*, **271**, 785–788.

Power, M. E., Tilman, D., Estes, J. A., et al. (1996). Challenges in the quest for keystones. *BioScience*, **46**, 609–620.

Prakash, V., Pain, D. J., Cunningham, A. A., et al. (2003). Catastrophic collapse of Indian White-backed *Gyps bengalensis* and Long-billed *Gyps indicus* vulture populations. *Biological Conservation*, **109**, 381–90.

Raffaelli, D. (2004). How extinction patterns affect ecosystems. *Science*, **306**, 1141–1142.

Rathcke, B. J. (2000). Hurricane causes resource and pollination limitation of fruit set in a bird-pollinated shrub. *Ecology*, **81**, 1951–1958.

Rayner, R. (2008). *Bug Wars*. New Yorker. Conde Nast, New York.

Reeburgh, W. (1997). Figures summarizing the global cycles of biogeochemically important elements. *Bulletin of Ecological Society of America*, **78**, 260–267.

Regal, P. J. (1977). Ecology and evolution of flowering plant dominance. *Science*, **196**, 622–629.

Rich, S. M., Leendertz, F. H., Xu, G., et al. (2009). The origin of malignant malaria. *Proceedings of the National Academy of Sciences of the United States of America*, in press.

Ricketts, T. H., Daily, G. C., Ehrlich, P. R., and Michener, C. D. (2004). Economic value of tropical forest to coffee production. *Proceedings of the National Academy of Sciences of the United States of America*, **101**, 12579–12582.

Root, T. L., Price, J. T., Hall, K. R., et al. (2003). Fingerprints of global warming on wild animals and plants. *Nature*, **421**, 57–60.

Sanchez-Pinero, F. and Polis, G. A. (2000). Bottom-up dynamics of allochthonous input: direct and indirect effects of seabirds on islands. *Ecology*, **81**, 3117–132.

Scheffer, M., Carpenter, S., Foley, J. A., Folke, C., and Walker, B. (2001). Catastrophic shifts in ecosystems. *Nature*, **413**, 591–596.

Schlesinger, W. H. (1991). *Biogeochemistry: an analysis of global change*. Academic Press, San Diego, CA.

Schneider, S. H. and Londer, R. (1984). *The coevolution of climate and life*. Sierra Club Books, San Francisco, CA.

Scott, D., Bruijnzeel, L., and Mackensen, J. (2005). The hydrological and soil impacts of forestation in the tropics. In M. Bonnell and L. Bruijnzeel, eds *Forests, water and people in the humid tropics*, pp. 622–651. Cambridge University Press, Cambridge, UK.

Sekercioglu, C. H. (2006a). Ecological significance of bird populations. In J. del Hoyo, A. Elliott, and D. A. Christie, eds *Handbook of the Birds of the World, volume 11*, pp. 15–51. Lynx Edicions, Barcelona, Spain.

Sekercioglu, C. H. (2006b). Increasing awareness of avian ecological function. *Trends in Ecology and Evolution*, **21**, 464–471.

Sekercioglu, C. H. (2007). Conservation ecology: area trumps mobility in fragment bird extinctions. *Current Biology*, **17**, R283–R286.

Sekercioglu, C. H., Daily, G. C., and Ehrlich, P. R. (2004). Ecosystem consequences of bird declines. *Proceedings of the National Academy of Sciences of the United States of America*, **101**, 18042–18047.

Selby, J. (2005). The geopolitics of water in the Middle East: fantasies and realities. *Third World Quarterly*, **26**, 329–349.

Shiklomanov, I. A. (1993). World's freshwater resources. In P. E. Gleick, ed. *Water in crisis: a guide to the world's freshwater resources*, pp. 13–24. Oxford University Press, New York, NY.

Singer, B. H., and M. C. De Castro. (2001). Agricultural colonization and malaria on the Amazon frontier. *Annals of the New York Academy of Sciences*, **954**, 184–222.

Smith, M. D., Wilcox, J. C., Kelly, T., and Knapp, A. K. (2004). Dominance not richness determines invasibility of tallgrass prairie. *Oikos*, **106**, 253–262.

Smith, V. H. (1992). Effects of nitrogen:phosphorus supply ratios on nitrogen fixation in agricultural and pastoral systems. *Biogeochemistry*, **18**, 19–35.

Sodhi, N. S., Brook, B. W., and Bradshaw, C. J. A. (2007). *Tropical conservation biology*. Wiley-Blackwell, Boston, MA.

Takekawa, J. Y. and Garton, E. O. (1984). How much is an evening grosbeak worth? *Journal of Forestry*, **82**, 426–428.

Tallis, H. M. and Kareiva, P. (2006). Shaping global environmental decisions using socio-ecological models. *Trends in Ecology and Evolution*, **21**, 562–568.

Tiffney, B. H. and Mazer, S. J. (1995). Angiosperm growth habit, dispersal and diversification reconsidered. *Evolutionary Ecology*, **9**, 93–117.

Tilman, D. (1996). Biodiversity: population versus ecosystem stability. *Ecology*, **77**, 350–63.

Tilman, D. (1997). Biodiversity and ecosystem functioning. In G. C., Daily ed. *Nature's Services: societal dependence on natural ecosystems*, pp.93–112. Island Press, Washington DC.

Tilman, D. and Downing, J. A. (1994). Biodiversity and stability of grasslands. *Nature*, **367**, 363–65.

Tilman, D., Wedin, D., and Knops, J. (1996). Productivity and sustainability influenced by biodiversity in grassland ecosystems. *Nature*, **379**, 718–720.

Vamosi, J. C., Knight, T. M., Steets, J. A., et al. (2006). Pollination decays in biodiversity hotspots. *Proceedings of the National Academy of Sciences of the United States of America*, **103**, 956–961.

Vanacker, V., Von Blanckenburg, F., Govers, G., et al. (2007). Restoring dense vegetation can slow mountain erosion to near natural benchmark levels. *Geology*, **35**, 303–06.

Vitousek, P. M. and Hooper, D. U. (1993). Biological diversity and terrestrial ecosystem productivity. In D. E. Schulze, and H. A. Mooney, eds *Biodiversity and ecosystem function*, pp. 3–14. Springer-Verlag, Berlin, Germany.

Vitousek, P. M., Aber, J. D., Howard, R. W., *et al.* (1997). Human alteration of the global nitrogen cycle: sources and consequences. *Ecological Applications*, **7**, 737–750.

Walker, B., Kinzig, A., and Langridge, J. (1999). Plant attribute diversity, resilience, and ecosystem function: the nature and significance of dominant and minor species. *Ecosystems*, **2**, 95–113.

Wilcove, D. S. (2008). *No way home: the decline of the world's great animal migrations*. Island Press, Washington, DC.

Wolfe, N. D., Daszak, P., Kilpatrick, A. M., and Burke, D. S. (2005). Bushmeat hunting deforestation, and prediction of zoonoses emergence. *Emerging Infectious Diseases*, **11**, 1822–1827.

World Health Organization. (1998). *World Survey for Rabies No. 34*. World Health Organization, Geneva, Switzerland.

Worm, B., Barbier, E. B., Beaumont, N., *et al.* (2006). Impacts of biodiversity loss on ocean ecosystem services. *Science*, **314**, 787–790.

Wright, S. J. (2005). Tropical forests in a changing environment. *Trends in Ecology and Evolution*, **20**, 553–560.

Wunderle, J. M. (1997). The role of animal seed dispersal in accelerating native forest regeneration on degraded tropical lands. *Forest Ecology and Management*, **99**, 223–235.

Yasuoka, J., and R. Levins. (2007). Impact of deforestation and agricultural development on Anopheline ecology and malaria epidemiology. *American Journal of Tropical Medicine and Hygiene*, **76**, 450–460.

# CHAPTER 4

# Habitat destruction: death by a thousand cuts

## William F. Laurance

Humankind has dramatically transformed much of the Earth's surface and its natural ecosystems. This process is not new—it has been ongoing for millennia—but it has accelerated sharply over the last two centuries, and especially in the last several decades.

Today, the loss and degradation of natural habitats can be likened to a war of attrition. Many natural ecosystems are being progressively razed, bulldozed, and felled by axes or chainsaws, until only small scraps of their original extent survive. Forests have been hit especially hard: the global area of forests has been reduced by roughly half over the past three centuries. Twenty-five nations have lost virtually all of their forest cover, and another 29 more than nine-tenths of their forest (MEA 2005). Tropical forests are disappearing at up to 130 000 $km^2$ a year (Figure 4.1)—roughly 50 football fields a minute. Other ecosystems are less imperiled, and a few are even recovering somewhat following past centuries of overexploitation.

Here I provide an overview of contemporary habitat loss. Other chapters in this book describe the many additional ways that ecosystems are being threatened—by overhunting (Chapter 6), habitat fragmentation (Chapter 5), and climate change (Chapter 8), among other causes—but my emphasis here is on habitat destruction *per se*. I evaluate patterns of habitat destruction geographically and draw comparisons among different biomes and ecosystems. I then consider some of the ultimate and proximate factors that drive habitat loss, and how they are changing today.

## 4.1 Habitat loss and fragmentation

Habitat destruction occurs when a natural habitat, such as a forest or wetland, is altered so dramatically that it no longer supports the species it originally sustained. Plant and animal populations are destroyed or displaced, leading to a loss of biodiversity (see Chapter 10). Habitat destruction is considered the most important driver of species extinction worldwide (Pimm and Raven 2000).

Few habitats are destroyed entirely. Very often, habitats are reduced in extent and simultaneously fragmented, leaving small pieces of original habitat persisting like islands in a sea of degraded land. In concert with habitat loss, habitat fragmentation is a grave threat to species survival (Laurance *et al.* 2002; Sekercioglu *et al.* 2002; Chapter 5).

Globally, agriculture is the biggest cause of habitat destruction (Figure 4.2). Other human activities, such as mining, clear-cut logging, trawling, and urban sprawl, also destroy or severely degrade habitats. In developing nations, where most habitat loss is now occurring, the drivers of environmental change have shifted fundamentally in recent decades. Instead of being caused mostly by small-scale farmers and rural residents, habitat loss, especially in the tropics, is now substantially driven by globalization promoting intensive agriculture and other industrial activities (see Box 4.1).

## 4.2 Geography of habitat loss

Some regions of the Earth are far more affected by habitat destruction than others. Among the most

**Figure 4.1** The aftermath of slash-and-burn farming in central Amazonia. Photograph by W. F. Laurance.

imperiled are the so-called "biodiversity hotspots", which contain high species diversity, many locally endemic species (those whose entire geographic range is confined to a small area), and which have lost at least 70% of their native vegetation (Myers *et al.* 2000). Many hotspots are in the tropics. The Atlantic forests of Brazil and rainforests of West Africa, both of which have been severely reduced and degraded, are examples of biodiversity hotspots. Despite encompassing just a small fraction (<2%) of the Earth's land surface, hotspots may sustain over half of the world's terrestrial species (Myers *et al.* 2000).

Many islands have also suffered heavy habitat loss. For instance, most of the original natural habitat has already been lost in Japan, New

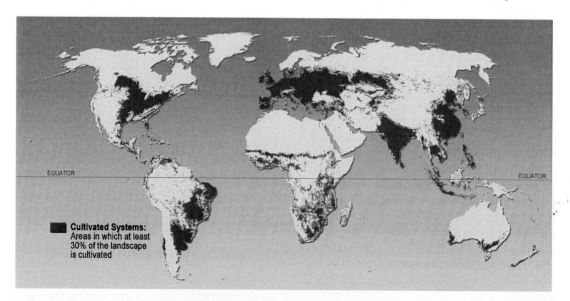

**Figure 4.2** Extent of land area cultivated globally by the year 2000. Reprinted from MEA (2005).

## Box 4.1 The changing drivers of tropical deforestation
### William F. Laurance

Tropical forests are being lost today at an alarming pace. However, the fundamental drivers of tropical forest destruction have changed in recent years (Rudel 2005; Butler and Laurance 2008). Prior to the late 1980s, deforestation was generally caused by rapid human population growth in developing nations, in concert with government policies for rural development. These included agricultural loans, tax incentives, and road construction. Such initiatives, especially evident in countries such as Brazil and Indonesia, promoted large influxes of colonists into frontier areas and often caused dramatic forest loss.

More recently, however, the impacts of rural peoples on tropical forests seem to be stabilizing (see Box 4.1 Figure). Although many tropical nations still have considerable population growth, strong urbanization trends (except in Sub-Saharan Africa) mean that rural populations are growing more slowly, and are even declining in some areas. The popularity of large-scale frontier-colonization programs has also waned. If such trends continue, they could begin to alleviate some pressures on forests from small-scale farming, hunting, and fuel-wood gathering (Wright and Muller-landau 2006).

**Box 4.1 Figure** Changing drivers of deforestation: Small-scale cultivators (a) versus industrial road construction (b) in Gabon, central Africa. Photograph by W. F. Laurance.

At the same time, globalized financial markets and a worldwide commodity boom are creating a highly attractive environment for the private sector. Under these conditions, large-scale agriculture—crops, livestock, and tree plantations—by corporations and wealthy landowners is increasingly emerging as the biggest direct cause of tropical deforestation (Butler and Laurance 2008). Surging demand for grains and edible oils, driven by the global thirst for biofuels and rising standards of living in developing countries, is also spurring this trend. In Brazilian Amazonia, for instance, large-scale ranching has exploded in recent years, with the number of cattle more than tripling (from 22 to 74 million head) since 1990 (Smeraldi and May 2008), while industrial soy farming has also grown dramatically.

Other industrial activities, especially logging, mining, and petroleum development, are also playing a critical but indirect role in forest destruction (Asner et al. 2006; Finer et al. 2008). These provide a key economic impetus for forest road-building (see Box 4.1 Figure), which in turn allows influxes of colonists, hunters, and miners into frontier areas, often leading to rapid forest disruption and cycles of land speculation.

*continues*

> **Box 4.1 (Continued)**
>
> **REFERENCES**
>
> Asner, G. P., Broadbent, E., Oliveira, P., Keller, M., Knapp, D., and Silva, J. (2006). Condition and fate of logged forests in the Amazon. *Proceedings of the National Academy of Sciences of the United States of America*. **103**, 12947–12950.
>
> Butler, R. A. and Laurance, W. F. (2008). New strategies for conserving tropical forests. *Trends in Ecology and Evolution*, **23**, 469–72.
>
> Finer, M., Jenkins, C., Pimm, S., Kean, B., and Rossi, C. (2008). Oil and gas projects in the western Amazon: threats to wilderness, biodiversity, and indigenous peoples. *PLoS One*, doi:10.1371/journal.pone.0002932.
>
> Rudel, T. K. (2005). Changing agents of deforestation: from state-initiated to enterprise driven processes, 1970–2000. *Land Use Policy*, **24**, 35–41.
>
> Smeraldi, R. and May, P. H. (2008). *The cattle realm: a new phase in the livestock colonization of Brazilian Amazonia*. Amigos da Terra, Amazônia Brasileira, São Paulo, Brazil.
>
> Wright, S. J. and Muller-Landau, H. C. (2006). The future of tropical forest species. *Biotropica*, **38**, 287–301.

Zealand, Madagascar, the Philippines, and Java (WRI 2003). Other islands, such as Borneo, Sumatra, and New Guinea, still retain some original habitat but are losing it at alarming rates (Curran *et al.* 2004; MacKinnon 2006).

Most areas of high human population density have suffered heavy habitat destruction. Such areas include much of Europe, eastern North America, South and Southeast Asia, the Middle East, West Africa, Central America, and the Caribbean region, among others. Most of the biodiversity hotspots occur in areas with high human density (Figure 4.3) and many still have rapid population growth (Cincotta *et al.* 2000). Human populations are often densest in coastal areas, many of which have experienced considerable losses of both terrestrial habitats and nearby coral reefs. Among others, coastal zones in Asia, northern South America, the Caribbean, Europe, and eastern North America have all suffered severe habitat loss (MEA 2005).

Finally, habitat destruction can occur swiftly in areas with limited human densities but rapidly expanding agriculture. Large expanses of the Amazon, for example, are currently being cleared for large-scale cattle ranching and industrial soy farming, despite having low population densities (Laurance *et al.* 2001). Likewise, in some relatively sparsely populated areas of Southeast Asia, such as Borneo, Sumatra, and New Guinea, forests are being rapidly felled to establish oil-palm or rubber plantations (MacKinnon 2006; Laurance 2007; Koh and Wilcove 2008; see Box 13.3). Older agricultural frontiers, such as those in Europe, eastern China, the Indian Subcontinent, and eastern and midwestern North America, often have very little native vegetation remaining (Figure 4.2).

## 4.3 Loss of biomes and ecosystems

### 4.3.1 Tropical and subtropical forests

A second way to assess habitat loss is by contrasting major biomes or ecosystem types (Figure 4.4). Today, tropical rainforests (also termed tropical moist and humid forests) are receiving the greatest attention, because they are being destroyed so rapidly and because they are the most biologically diverse of all terrestrial biomes. Of the roughly 16 million $km^2$ of tropical rainforest that originally existed worldwide, less than 9 million $km^2$ remains today (Whitmore 1997; MEA 2005). The current rate of rainforest loss is debated, with different estimates ranging from around 60 000 $km^2$ (Achard *et al.* 2002) to 130 000 $km^2$ per year (FAO 2000). Regardless of which estimate one adheres to, rates of rainforest loss are alarmingly high.

Rates of rainforest destruction vary considerably among geographic regions. Of the world's three major tropical regions, Southeast Asian forests are disappearing most rapidly in relative

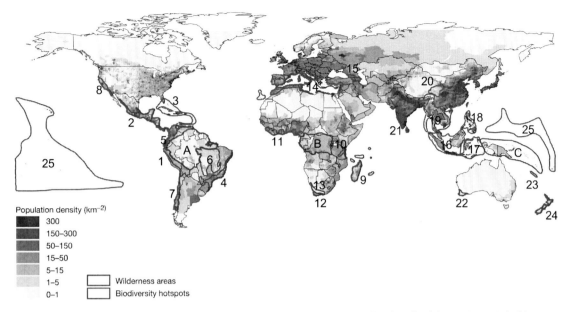

**Figure 4.3** Human population density in 1995 within 25 recognized biodiversity hotspots (numbered 1-25) and three major tropical wildernesses (labeled A-C). Reprinted from Cincotta et al. 2000 © Nature Publishing Group.

terms (Figure 4.5), while the African and New World tropics have somewhat lower rates of percent-annual forest loss (Sodhi et al. 2004). Such averages, however, disguise important smaller-scale variation. In the New World tropics, for example, the Caribbean, MesoAmerican, and Andean regions are all suffering severe rainforest loss, but the relative deforestation rate for the region as a whole is buffered by the vastness of the Amazon. Likewise, in tropical Africa, forest loss is severe in West Africa, montane areas of East Africa, and Madagascar, but substantial forest still survives in the Congo Basin (Laurance 1999).

Other tropical and subtropical biomes have suffered even more heavily than rainforests (Figure 4.4). Tropical dry forests (also known as

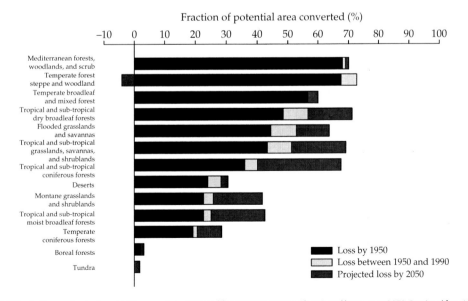

**Figure 4.4** Estimated losses of major terrestrial biomes prior to 1950 and from 1950 to 1990, with projected losses up to 2050. Reprinted from MEA (2005).

**Figure 4.5** Tropical rainforests in Southeast Asia are severely imperiled, as illustrated by this timber operation in Indonesian Borneo. Photograph by W. F. Laurance.

monsoonal or deciduous forests) have been severely reduced, in part because they are easier to clear and burn than rainforests. For instance, along Central America's Pacific coast, much less than 1% of the original dry forest survives. Losses of dry forest have been nearly as severe in Madagascar and parts of Southeast Asia (Laurance 1999; Mayaux et al. 2005).

Mangrove forests, salt-tolerant ecosystems that grow in tropical and subtropical intertidal zones, have also been seriously reduced. Based on countries for which data exist, more than a third of all mangroves were lost in the last few decades of the 20$^{th}$ century (MEA 2005). From 1990 to 2000, over 1% of all mangrove forests were lost annually, with rates of loss especially high in Southeast Asia (Mayaux et al. 2005). Such losses are alarming given the high primary productivity of mangroves, their key role as spawning and rearing areas for economically important fish and shrimp species, and their importance for sheltering coastal areas from destructive storms and tsunamis (Danielsen et al. 2005).

### 4.3.2 Temperate forests and woodlands

Some ecosystems have suffered even worse destruction than tropical forests. Mediterranean forests and woodlands, temperate broadleaf and mixed forests, and temperate forest-steppe and woodlands have all suffered very heavy losses (Figure 4.4), given the long history of human settlement in many temperate regions. By 1990 more than two-thirds of Mediterranean forests and woodlands were lost, usually because they were converted to agriculture (MEA 2005). In the eastern USA and Europe (excluding Russia), old-growth broadleaf forests (>100 years old) have nearly disappeared (Matthews et al. 2000), although forest cover is now regenerating in many areas as former agricultural lands are abandoned and their formerly rural, farming-based populations become increasingly urbanized.

In the cool temperate zone, coniferous forests have been less severely reduced than broadleaf and mixed forests, with only about a fifth being lost by 1990 (Figure 4.4). However, vast expanses of coniferous forest in northwestern North America, northern Europe, and southern Siberia are being clear-felled for timber or pulp production. As a result, these semi-natural forests are converted from old-growth to timber-production forests, which have a much-simplified stand structure and species composition. Large expanses of coniferous forest are also burned each year (Matthews et al. 2000).

**Figure 4.6** African savannas are threatened by livestock overgrazing and conversion to farmland. Photograph by W. F. Laurance.

### 4.3.3 Grasslands and deserts

Grasslands and desert areas have generally suffered to a lesser extent than forests (Figure 4.4). Just 10–20% of all grasslands, which include the savannas of Africa (Figure 4.6), the *llano* and *cerrado* ecosystems of South America, the steppes of Central Asia, the prairies of North America, and the spinifex grasslands of Australia, have been permanently destroyed for agriculture (White *et al.* 2000; Kauffman and Pyke 2001). About a third of the world's deserts have been converted to other land uses (Figure 4.4). Included in this figure is the roughly 9 million km$^2$ of seasonally dry lands, such as the vast Sahel region of Africa, that have been severely degraded via desertification (Primack 2006).

Although deserts and grasslands have not fared as badly as some other biomes, certain regions have suffered very heavily. For instance, less than 3% of the tallgrass prairies of North America survive, with the remainder having been converted to farmland (White *et al.* 2000). In southern Africa, large expanses of dryland are being progressively desertified from overgrazing by livestock (MEA 2005). In South America, more than half of the biologically-rich *cerrado* savannas, which formerly spanned over 2 million km$^2$, have been converted into soy fields and cattle pastures in recent decades, and rates of loss remain very high (Klink and Machado 2005).

### 4.3.4 Boreal and alpine regions

Boreal forests are mainly found in broad continental belts at the higher latitudes of North America and Eurasia. They are vast in Siberia, the largest contiguous forest area in the world, as well as in northern Canada. They also occur at high elevations in more southerly areas, such as the European Alps and Rocky Mountains of North America. Dominated by evergreen conifers, boreal forests are confined to cold, moist climates and are especially rich in soil carbon, because low temperatures and waterlogged soils inhibit decomposition of organic material (Matthews *et al.* 2000).

Habitat loss in boreal forests has historically been low (Figure 4.4; Box 4.2). In Russia, however, legal and illegal logging activity has grown rapidly, with Siberia now a major source of timber exports to China, the world's largest timber importer. In Canada, nearly half of the boreal forest is under tenure for wood production. In addition, fire incidence is high in the boreal zone, with perhaps 100 000 km$^2$ of boreal forest burning each year (Matthews *et al.* 2000).

Like boreal forests, tundra is a vast ecosystem (spanning 9–13 million km² globally) that has been little exploited historically (Figure 4.4) (White et al. 2000). Unlike permafrost areas, tundra ecosystems thaw seasonally on their surface, becoming important wetland habitats for waterfowl and other wildlife. Other boreal habitats, such as taiga grasslands (Figure 4.7), have also suffered little loss. However, all boreal ecosystems are vulnerable to global warming (see Chapter 8; Box 4.2). Boreal forests, in particular, could decline if climatic conditions become significantly warmer or drier, leading to an increased frequency or severity of forest fires (see Box 4.2, Chapter 9).

### Box 4.2 Boreal forest management: harvest, natural disturbance, and climate change
### Ian G. Warkentin

Until recently, the boreal biome has largely been ignored in discussions regarding the global impacts of habitat loss through diminishing forest cover. Events in tropical regions during the past four decades were far more critical due to the high losses of forest and associated species (Dirzo and Raven 2003). While there are ongoing concerns about tropical forest harvest, the implications of increasing boreal forest exploitation now also need to be assessed, particularly in the context of climate change. (Bradshaw et al. 2009) Warnings suggest that forest managers should not overlook the services provided by the boreal ecosystem, especially carbon storage (Odling-Smee 2005). Ranging across northern Eurasia and North America, the boreal biome constitutes one third of all current forest cover on Earth and is home to nearly half of the remaining tracts of extensive, intact forests. Nearly 30% of the Earth's terrestrial stored carbon is held here, and the boreal may well have more influence on mean annual global temperature than any other biome due to its sunlight reflectivity (albedo) properties and evapotranspiration rates (Snyder et al. 2004).

Conversion of North America's boreal forest to other land cover types has been limited (e.g. <3% in Canada; Smith and Lee 2000). In Finland and Sweden forest cover has expanded during recent decades, but historic activities extensively reduced and modified the region's boreal forests for commercial purposes, leaving only a small proportion as natural stands (Imbeau et al. 2001; see Box 4.2 Figure). Conversely, there has been a rapid expansion of harvest across boreal Russia during the past 10–15 years leading to broad shifts from forest to other land cover types (MEA 2005). Forest cover loss across European Russia is associated with intensive harvest, mineral exploitation and urbanization, while in Siberian Russia the combination of logging and a sharp rise in human-ignited fires has led to a 2.3% annual decrease in forest cover (Achard et al. 2006, 2008).

Box 4.2 Figure An example of harvesting in the Boreal forest. Photograph by Greg Mitchell.

The biggest challenge for boreal managers may come from the warmer and drier weather, with a longer growing season, that climate change models predict for upper-latitude ecosystems (IPCC 2001). The two major drivers of boreal disturbance dynamics (fire and insect infestation) are closely associated with weather conditions (Soja et al. 2007) and predicted to be both more frequent and intense over the next century (Kurz et al. 2008); more human-ignited fires are also predicted as access to the forest expands (Achard et al. 2008). Increased harvest, fire and insect infestations will raise the rates of carbon loss to the atmosphere, but climate models also suggest that changes to albedo and evapotranspiration due to these disturbances will offset the lost carbon stores (Bala et al. 2007)—maintaining large non-forested boreal sites potentially may cool the global climate more than the carbon storage resulting from reforestation at those sites. However, to manage the boreal forest based solely on one

*continues*

**Box 4.2 (Continued)**

ecosystem service would be reckless. For example, many migratory songbirds that depend upon intact boreal forest stands for breeding also provide critical services such as insect predation, pollen transport and seed dispersal (Sekercioglu 2006) in habitats extending from boreal breeding grounds, to migratory stopovers and their winter homes in sub-tropical and tropical regions. Thus boreal forest managers attempting to meet climate change objectives (or any other single goal) must also consider the potential costs for biodiversity and the multiple services at risk due to natural and human-associated change.

### REFERENCES

Achard, F., Mollicone, D., Stibig, H.-J., et al. (2006). Areas of rapid forest-cover change in boreal Eurasia. *Forest Ecology and Management*, **237**, 322–334.

Achard, F. D., Eva, H. D., Mollicone, D., and Beuchle, R. (2008). The effect of climate anomalies and human ignition factor on wildfires in Russian boreal forests. *Philosophical Transactions of the Royal Society of London B*, **363**, 2331–2339.

Bala, G., Caldeira, K., Wickett, M., et al. (2007). Combined climate and carbon-cycle effects of large-scale deforestation. *Proceedings of the National Academy of Sciences of the United States of America*, **104**, 6550–6555.

Bradshaw, C. J. A., Warkentin, I. G., and Sodhi, N. S. (2009). Urgent preservation of boreal carbon stocks and biodiversity. *Trends in Ecology and Evolution*, in press.

Dirzo, R. and Raven, P. H. (2003). Global state of biodiversity and loss. *Annual Review of Environment and Resources*, **28**, 137–167

Imbeau, L., Mönkkönen, M., and Desrochers, A. (2001). Long-term effects of forestry on birds of the eastern Canadian boreal forests: a comparison with Fennoscandia. *Conservation Biology*, **15**, 1151–1162.

IPCC (Intergovernmental Panel on Climate Change) (2001). *Climate change 2001: the scientific basis*. Contribution of Working Group I to the Third Assessment Report of the Intergovernmental Panel on Climate Change. Cambridge University Press, New York, NY.

Kurz, W. A., Stinson, G., Rampley, G. J., Dymond, C. C., and Neilson, E. T. (2008). Risk of natural disturbances makes future contribution of Canada's forests to the global carbon cycle highly uncertain. *Proceedings of the National Academy of Sciences of the United States of America*, **105**, 1551–1555.

MEA (Millennium Ecosystem Assessment) (2005). *Ecosystems and human well-being: synthesis*. Island Press, Washington, DC.

Odling-Smee, L. (2005). Dollars and sense. *Nature*, **437**, 614–616.

Sekercioglu, C. H. (2006). Increasing awareness of avian ecological function. *Trends in Ecology and Evolution*, **21**, 464–471.

Smith, W. and Lee, P., eds (2000). *Canada's forests at a crossroads: an assessment in the year 2000*. World Resources Institute, Washington, DC.

Snyder, P. K., Delire, C., and Foley, J. A. (2004). Evaluating the influence of different vegetation biomes on the global climate. *Climate Dynamics*, **23**, 279–302.

Soja, A. J., Tchebakova, N. M., French, N. H. F., et al. (2007). Climate-induced boreal forest change: Predictions versus current observations. *Global and Planetary Change*, **56**, 274–296.

---

In addition, tundra areas will shrink as boreal forests spread north.

### 4.3.5 Wetlands

Although they do not fall into any single biome type, wetlands have endured intense habitat destruction in many parts of the world. In the USA, for instance, over half of all wetlands have been destroyed in the last two centuries (Stein *et al.* 2000). From 60–70% of all European wetlands have been destroyed outright (Ravenga *et al.* 2000). Many developing nations are now suffering similarly high levels of wetland loss, particularly as development in coastal zones accelerates. As discussed above, losses of mangrove forests, which are physiologically specialized for the intertidal zone, are also very high.

**Figure 4.7** Boreal ecosystems, such as this alpine grassland in New Zealand, have suffered relatively little habitat loss but are particularly vulnerable to global warming. Photograph by W. F. Laurance.

## 4.4 Land-use intensification and abandonment

Humans have transformed a large fraction of the Earth's land surface (Figure 4.2). Over the past three centuries, the global extent of cropland has risen sharply, from around 2.7 to 15 million km$^2$, mostly at the expense of forest habitats (Turner *et al.* 1990). Permanent pasturelands are even more extensive, reaching around 34 million km$^2$ by the mid-1990s (Wood *et al.* 2000). The rate of land conversion has accelerated over time: for instance, more land was converted to cropland from 1950 to 1980 than from 1700 to 1850 (MEA 2005).

Globally, the rate of conversion of natural habitats has finally begun to slow, because land readily convertible to new arable use is now in increasingly short supply and because, in temperate and boreal regions, ecosystems are recovering somewhat. Forest cover is now increasing in eastern and western North America, Alaska, western and northern Europe, eastern China, and Japan (Matthews *et al.* 2000; MEA 2005, Figure 4.4). During the 1990s, for instance, forest cover rose by around 29 000 km$^2$ annually in the temperate and boreal zones, although roughly 40% of this increase comprised forest plantations of mostly non-native tree species (MEA 2005). Despite partial recovery of forest cover in some regions (Wright and Muller-Landau 2006), conversion rates for many ecosystems, such as tropical and subtropical forests and South American *cerrado* savanna-woodlands, remain very high.

Because arable land is becoming scarce while agricultural demands for food and biofuel feedstocks are still rising markedly (Koh and Ghazoul 2008), agriculture is becoming increasingly intensified in much of the world. Within agricultural regions, a greater fraction of the available land is actually being cultivated, the intensity of cultivation is increasing, and fallow periods are decreasing (MEA 2005). Cultivated systems (where over 30% of the landscape is in croplands, shifting cultivation, confined-livestock production, or freshwater aquaculture) covered 24% of the global land surface by the year 2000 (Figure 4.2).

Thus, vast expanses of the earth have been altered by human activities. Old-growth forests have diminished greatly in extent in many regions, especially in the temperate zones; for instance, at least 94% of temperate broadleaf forests have been disturbed by farming and logging (Primack 2006). Other ecosystems, such as coniferous forests, are being rapidly converted from old-growth to semi-natural production forests with a simplified stand structure and species composition. Forest cover is increasing in parts of the temperate and boreal zones, but the new forests are secondary and differ from old-growth forests in species composition, structure, and carbon storage. Yet other ecosystems, particularly in the tropics, are being rapidly destroyed and degraded. For example, marine ecosystems have been heavily impacted by human activities (see Box 4.3).

The large-scale transformations of land cover described here consider only habitat loss *per se*. Of the surviving habitat, much is being

## Box 4.3 Human Impacts on marine ecosystems
### Benjamin S. Halpern, Carrie V. Kappel, Fiorenza Micheli, and Kimberly A. Selkoe

The oceans cover 71% of the planet. This vastness has led people to assume ocean resources are inexhaustible, yet evidence to the contrary has recently accumulated (see Box 4.3 Figure 1 and Plate 4). Populations of large fish, mammals, and sea turtles have collapsed due to intense fishing pressure, putting some species at risk of extinction, and fishing gear such as bottom trawls not only catch target fish but also destroy vast swaths of habitat (see Box 6.1). Pollution, sedimentation, and nutrient enrichment have caused die-offs of fish and corals, blooms of jellyfish and algae, and "dead zones" of oxygen-depleted waters around the world. Coastal development has removed much of the world's mangroves, sea grass beds and salt marshes. Effects from climate change, such as rising sea levels and temperatures and ocean acidification, are observed with increasing frequency around the world. Global commerce, aquaculture and the aquarium trade have caused the introduction of thousands of non-native species, many of which become ecologically and economically destructive in their new environment. These human-caused stresses on ocean ecosystems are the most intense and widespread, but many other human activities impact the ocean where they are concentrated, such as shipping, aquaculture, and oil and gas extraction, and many new uses such as wave and wind energy farms are just emerging.

**Box 4.3 Figure 1** A few of the many human threats to marine ecosystems around the world. (A) The seafloor before and after bottom trawl fishing occurred [courtesy CSIRO (Australian Commonwealth Scientific and Research Organization) Marine Research], (B) coastal development in Long Beach, California (courtesy California Coastal Records Project), (C) shrimp farms in coastal Ecuador remove coastal habitat (courtesy Google Earth), and (D) commercial shipping and ports produce pollution and introduce non-native species (courtesy public commons).

There are clear challenges in reducing the impacts of any single human activity on marine ecosystems. These challenges are particularly stark in areas where dozens of activities co-occur because each species and each ecosystem may respond uniquely to each set of human activities, and there may be hard-to-predict synergisms among stressors that can amplify impacts. For example, excess nutrient input combined with overfishing of herbivorous fish on coral reefs can lead to algal proliferation and loss of coral with little chance of recovery, while each stressor alone may not lead to such an outcome. The majority of oceans are subject

*continues*

**Box 4.3 (Continued)**

to at least three different overlapping human stressors, with most coastal areas experiencing over a dozen, especially near centers of commerce like the ports of Los Angeles and Singapore.

The first comprehensive map of the impacts of 17 different types of human uses on the global oceans provides information on where cumulative human impacts to marine ecosystems are most intense (Halpern et al. 2008; see Box 4.3 Figure 2 and Plate 5). The map shows that over 40% of the oceans are heavily impacted and less than 4% are relatively pristine (see Box 4.3 Table). The heaviest impacts occur in the North Sea and East and South China Seas, where industry, dense human population, and a long history of ocean use come together. The least impacted areas are small and scattered throughout the globe, with the largest patches at the poles and the Torres Strait north of Australia. Several of the countries whose seas are significantly impacted, including the United States and China, have huge territorial holdings, suggesting both a responsibility and an opportunity to make a significant difference in improving ocean health.

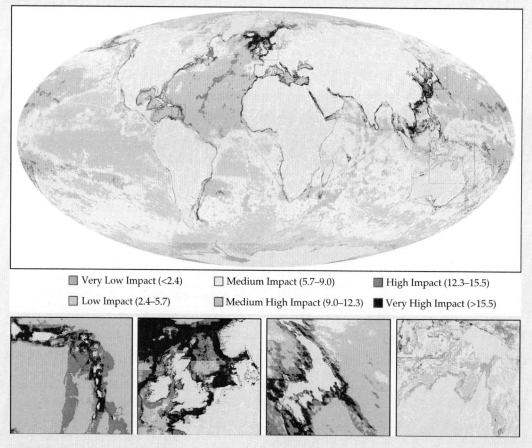

☐ Very Low Impact (<2.4)   ☐ Medium Impact (5.7–9.0)   ■ High Impact (12.3–15.5)
☐ Low Impact (2.4–5.7)     ☐ Medium High Impact (9.0–12.3)   ■ Very High Impact (>15.5)

**Box 4.3 Figure 2** Global map of the cumulative human impact on marine ecosystems, based on 20 ecosystem types and 17 different human activities. Grayscale colors correspond to overall condition of the ocean as indicated in the legend, with cumulative impact score cutoff values for each category of ocean condition indicated.

*continues*

## Box 4.3 (Continued)

**Box 4.3 Table** The amount of marine area within the Exclusive Economic Zone (EEZ) of countries that is heavily impacted. Countries are listed in order of total marine area within a country's EEZ (including territorial waters) and includes a selection of countries chosen for illustrative purposes. Global statistics are provided for comparison. Data are drawn from Halpern et al. (2008).

| | | Impact Category | | | | | |
|---|---|---|---|---|---|---|---|
| | | Very Low | Low | Medium | Medium-High | High | Very High |
| GLOBAL | 100% | 3.7% | 24.5% | 31.3% | 38.2% | 1.8% | 0.5% |
| Largest EEZs | | | | | | | |
| United States | 3.3% | 2.0% | 9.1% | 21.5% | 62.1% | 4.4% | 0.7% |
| France | 2.8% | 0.2% | 36.7% | 40.1% | 21.6% | 0.9% | 0.4% |
| Australia | 2.5% | 3.7% | 26.4% | 42.3% | 26.3% | 1.0% | 0.3% |
| Russia | 2.1% | 22.5% | 30.8% | 32.3% | 13.5% | 0.6% | 0.3% |
| United Kingdom | 1.9% | 0.3% | 25.2% | 36.0% | 29.0% | 6.5% | 3.0% |
| Indonesia | 1.7% | 2.3% | 32.0% | 42.4% | 18.0% | 3.0% | 2.1% |
| Canada | 1.5% | 22.8% | 18.4% | 25.8% | 26.5% | 5.5% | 1.0% |
| Japan | 1.1% | 0.0% | 0.9% | 9.7% | 76.2% | 9.9% | 3.2% |
| Brazil | 1.0% | 3.1% | 17.1% | 32.4% | 44.8% | 2.1% | 0.5% |
| Mexico | 0.9% | 1.2% | 29.1% | 32.7% | 35.5% | 1.2% | 0.3% |
| India | 0.6% | 0.1% | 7.3% | 32.9% | 51.4% | 6.8% | 1.5% |
| China | 0.2% | 0.0% | 24.5% | 5.7% | 27.2% | 20.1% | 22.5% |
| SMALLER EEZs | | | | | | | |
| Germany | 0.02% | 0.4% | 43.7% | 2.4% | 34.6% | 14.4% | 4.4% |
| Iceland | 0.21% | 0.0% | 0.4% | 10.1% | 58.4% | 26.3% | 4.8% |
| Ireland | 0.11% | 0.0% | 0.2% | 2.2% | 50.3% | 40.8% | 6.6% |
| Italy | 0.15% | 0.0% | 3.5% | 15.5% | 64.5% | 11.8% | 4.7% |
| Netherlands | 0.04% | 0.0% | 18.6% | 3.6% | 68.8% | 7.7% | 1.5% |
| Sri Lanka | 0.15% | 0.0% | 2.5% | 8.6% | 45.0% | 37.2% | 6.7% |
| Thailand | 0.08% | 0.4% | 21.9% | 42.6% | 22.1% | 9.6% | 3.4% |
| Vietnam | 0.18% | 1.1% | 21.0% | 26.7% | 35.7% | 10.2% | 5.4% |

Complex but feasible management approaches are needed to address the cumulative impacts of human activities on the oceans. Comprehensive spatial planning of activities affecting marine ecosystems, or ocean zoning, has already been adopted and implemented in Australia's Great Barrier Reef and parts of the North Sea, with the goal of minimizing the overlap and potential synergies of multiple stressors. Many countries, including the United States, are beginning to adopt Ecosystem-Based Management (EBM) approaches that explicitly address cumulative impacts and seek to balance sustainable use of the oceans with conservation and restoration of marine ecosystems. Ultimately, it is now clear that marine resources are not inexhaustible and that precautionary, multi-sector planning of their use is needed to ensure long-term sustainability of marine ecosystems and the crucial services they provide.

## REFERENCES

Halpern, B. S., Walbridge, S., Selkoe, K. A., et al. (2008). A global map of human impact on marine ecosystems. *Science*, **319**, 948–952.

degraded in various ways—such as by habitat fragmentation, increased edge effects, selective logging, pollution, overhunting, altered fire regimes, and climate change. These forms of environmental degradation, as well as the important environmental services these ecosystems provide, are discussed in detail in subsequent chapters.

## Summary

- Vast amounts of habitat destruction have already occurred. For instance, about half of all global forest cover has been lost, and forests have virtually vanished in over 50 nations worldwide.
- Habitat destruction has been highly uneven among different ecosystems. From a geographic perspective, islands, coastal areas, wetlands, regions with large or growing human populations, and emerging agricultural frontiers are all sustaining rapid habitat loss.
- From a biome perspective, habitat loss has been very high in Mediterranean forests, temperate forest-steppe and woodland, temperate broadleaf forests, and tropical coniferous forests. Other ecosystems, particularly tropical rainforests, are now disappearing rapidly.
- Habitat destruction in the temperate zone peaked in the 19th and early 20th centuries. Although considerable habitat loss is occurring in some temperate ecosystems, overall forest cover is now increasing from forest regeneration and plantation establishment in some temperate regions.
- Primary (old-growth) habitats are rapidly diminishing across much of the earth. In their place, a variety of semi-natural or intensively managed ecosystems are being established. For example, although just two-tenths of the temperate coniferous forests have disappeared, vast areas are being converted from old-growth to timber-production forests, with a greatly simplified stand structure and species composition.
- Boreal ecosystems have suffered relatively limited reductions to date but are especially vulnerable to global warming. Boreal forests could become increasingly vulnerable to destructive fires if future conditions become warmer or drier.

## Suggested reading

- Sanderson, E. W., Jaiteh, M., Levy, M., Redford, K., Wannebo, A., and Woolmer, G. (2002). The human footprint and the last of the wild. *BioScience*, **52**, 891–904.
- Sodhi, N.S., Koh, L P., Brook, B.W., and Ng, P. (2004). Southeast Asian biodiversity: an impending catastrophe. *Trends in Ecology and Evolution*, **19**, 654–660.
- MEA. (2005). *Millennium Ecosystem Assessment. Ecosystems and Human Well-Being: Synthesis*. Island Press, Washington, DC.
- Laurance, W.F. and Peres, C. A., eds. (2006). *Emerging Threats to Tropical Forests*. University of Chicago Press, Chicago.

## Relevant websites

- Mongabay: http://www.mongabay.com.
- Forest Protection Portal: http://www.forests.org.
- The Millennium Ecosystem Assessment synthesis reports: http://www.MAweb.org.

## REFERENCES

Achard, F., Eva, H., Stibig, H., Mayaux, P., Gallego, J., Richards, T., and Malingreau, J.-P. (2002). Determination of deforestation rates of the world's humid tropical forests. *Science*, **297**, 999–1002.

Cincotta, R. P., Wisnewski, J., and Engelman, R. (2000). Human population in the biodiversity hotspots. *Nature*, **404**, 990–2.

Curran, L. M., Trigg, S., McDonald, A., *et al.* (2004). Lowland forest loss in protected areas of Indonesian Borneo. *Science*, **303**, 1000–1003.

Danielsen, F., Sørensen, M. K., Olwig, M. F. *et al.* (2005). The Asian Tunami: A protective role for coastal vegetation. *Science*, **310**, 643.

FAO (Food and Agriculture Organization Of The United Nations) (2000). *Global forest resource assessment 2000—main report*. FAO, New York.

Kauffman, J. B. and Pyke, D. A. (2001). Range ecology, global livestock influences. In S. Levin, ed. *Encyclopedia of biodiversity 5*, pp. 33–52. Academic Press, San Diego, California.

Klink, C. A. and Machado, R. B. (2005). Conservation of the Brazilian cerrado. *Conservation Biology*, **19**, 707–713.

Koh, L. P. and Ghazoul, J. (2008). Biofuels, biodiversity, and people: understanding the conflicts and finding opportunities. *Biological Conservation*, **141**, 2450–2460.

Koh, L. P. and Wilcove, D. S. (2008). Is oil palm agriculture really destroying biodiversity. *Conservation Letters*, **1**, 60–64.

Laurance, W. F. (1999). Reflections on the tropical deforestation crisis. *Biological Conservation*, **91**, 109–117.

Laurance, W. F. (2007). Forest destruction in tropical Asia. *Current Science*, **93**, 1544–1550.

Laurance, W. F., Albernaz, A., and Da Costa, C. (2001). Is deforestation accelerating in the Brazilian Amazon? *Environmental Conservation*, **28**, 305–11.

Laurance, W. F., Lovejoy, T., Vasconcelos, H., *et al.* (2002). Ecosystem decay of Amazonian forest fragments: a 22-year investigation. *Conservation Biology*, **16**, 605–618.

MacKinnon, K. (2006). Megadiversity in crisis: politics, policies, and governance in Indonesia's forests. In W. F. Laurance and C. A. Peres, eds *Emerging threats to tropical forests*, pp. 291–305. University of Chicago Press, Chicago, Illinois.

Matthews, E., Rohweder, M., Payne, R., and Murray, S. (2000). *Pilot Analysis of Global Ecosystems: Forest Ecosystems*. World Resources Institute, Washington, DC.

Mayaux, P., Holmgren, P., Achard, F., Eva, H., Stibig, H.-J., and Branthomme, A. (2005). Tropical forest cover change in the 1990s and options for future monitoring. *Philosophical Transactions of the Royal Society of London B*, **360**, 373–384.

MEA (Millenium Ecosystem Assessment) (2005). *Ecosystems and human well-being: synthesis*. Island Press, Washington, DC.

Myers, N., Mittermeier, R., Mittermeier, C., Fonseca, G., and Kent, J. (2000). Biodiversity hotspots for conservation priorities. *Nature*, **403**, 853–858.

Pimm, S. L. and Raven, P. (2000). Biodiversity: Extinction by numbers. *Nature*, **403**, 843–845.

Primack, R. B. (2006). *Essentials of conservation biology*, 4th edn. Sinauer Associates, Sunderland, Massachusetts.

Ravenga, C., Brunner, J., Henninger, N., Kassem, K., and Payne, R. (2000). *Pilot analysis of global ecosystems: wetland ecosystems*. World Resources Institute, Washington, DC.

Sekercioglu C. H., Ehrlich, P. R., Daily, G. C., Aygen, D., Goehring, D., and Sandi, R. (2002). Disappearance of insectivorous birds from tropical forest fragments. *Proceedings of the National Academy of Sciences of the United States of America*, **99**, 263–267.

Sodhi, N. S., Koh, L. P., Brook, B. W., and Ng, P. (2004). Southeast Asian biodiversity: an impending disaster. *Trends in Ecology and Evolution*, **19**, 654–660.

Stein, B. A., Kutner, L. and Adams, J., eds (2000). *Precious heritage: the status of biodiversity in the United States*. Oxford University Press, New York.

Turner, B. L., Clark, W. C., Kates, R., Richards, J., Mathews J., and Meyer, W., eds (1990). *The earth as transformed by human action: global and regional change in the biosphere over the past 300 years*. Cambridge University Press, Cambridge, UK.

White, R. P., Murray, S., and Rohweder, M. (2000) *Pilot analysis of global ecosystems: grassland ecosystems*. World Resources Institute, Washington, DC.

Whitmore, T. C. (1997). Tropical forest disturbance, disappearance, and species loss. In Laurance, W. F. and R. O. Bierregaard, eds *Tropical forest Remnants: ecology, management, and conservation of fragmented communities*, pp. 3–12. University of Chicago Press, Chicago, Illinois.

Wood, S., Sebastian, K., and Scherr, S. J. (2000). *Pilot analysis of global ecosystems: agroecosystems*. World Resources Institute, Washington, DC.

WRI (World Resources Institute) (2003). *World resources 2002–2004: decisions for the earth: balance, voice, and power*. World Resources Institute, Washington, DC.

Wright, S. J. and Muller-Landau, H. C. (2006). The future of tropical forest species. *Biotropica*, **38**, 287–301.

# CHAPTER 5

# Habitat fragmentation and landscape change

Andrew F. Bennett and Denis A. Saunders

Broad-scale destruction and fragmentation of native vegetation is a highly visible result of human land-use throughout the world (Chapter 4). From the Atlantic Forests of South America to the tropical forests of Southeast Asia, and in many other regions on Earth, much of the original vegetation now remains only as fragments amidst expanses of land committed to feeding and housing human beings. Destruction and fragmentation of habitats are major factors in the global decline of populations and species (Chapter 10), the modification of native plant and animal communities and the alteration of ecosystem processes (Chapter 3). Dealing with these changes is among the greatest challenges facing the "mission-orientated crisis discipline" of conservation biology (Soulé 1986; see Chapter 1).

Habitat fragmentation, by definition, is the "breaking apart" of continuous habitat, such as tropical forest or semi-arid shrubland, into distinct pieces. When this occurs, three interrelated processes take place: a reduction in the total amount of the original vegetation (i.e. habitat loss); subdivision of the remaining vegetation into fragments, remnants or patches (i.e. habitat fragmentation); and introduction of new forms of land-use to replace vegetation that is lost. These three processes are closely intertwined such that it is often difficult to separate the relative effect of each on the species or community of concern. Indeed, many studies have not distinguished between these components, leading to concerns that "habitat fragmentation" is an ambiguous, or even meaningless, concept (Lindenmayer and Fischer 2006). Consequently, we use "landscape change" to refer to these combined processes and "habitat fragmentation" for issues directly associated with the subdivision of vegetation and its ecological consequences.

This chapter begins by summarizing the conceptual approaches used to understand conservation in fragmented landscapes. We then examine the biophysical aspects of landscape change, and how such change affects species and communities, posing two main questions: (i) what are the implications for the *patterns* of occurrence of species and communities?; and (ii) how does landscape change affect *processes* that influence the distribution and viability of species and communities? The chapter concludes by identifying the kinds of actions that will enhance the conservation of biota in fragmented landscapes.

## 5.1 Understanding the effects of landscape change

### 5.1.1 Conceptual approaches

The theory of island biogeography (MacArthur and Wilson 1967) had a seminal influence in stimulating ecological and conservation interest in fragmented landscapes. This simple, elegant model highlighted the relationship between the number of species on an island and the island's area and isolation. It predicted that species richness on an island represents a dynamic balance between the rate of colonization of new species to the island and the rate of extinction of species already present. It was quickly perceived that habitat isolates, such as forest fragments, could also be considered as "islands" in a "sea" of developed land and that this theory provided a quantitative approach for studying their biota. This stimulated many studies in which species richness in fragments was related to the area and isolation of the fragment, the primary factors in island biogeographic theory.

The development of landscape ecology contributed new ways of thinking about habitat fragments and landscape change. The concept of patches and connecting corridors set within a matrix (i.e. the background ecosystem or land-use type) became an influential paradigm (Forman and Godron 1986). It recognized the importance of the spatial context of fragments. The environment surrounding fragments is greatly modified during landscape changes associated with fragmentation. Thus, in contrast to islands, fragments and their biota are strongly influenced by physical and biological processes in the wider landscape, and the isolation of fragments depends not only on their distance from a similar habitat but also on their position in the landscape, the types of surrounding land-uses and how they influence the movements of organisms (Saunders *et al.* 1991; Ricketts 2001).

The influence of physical processes and disturbance regimes on fragments means that following habitat destruction and fragmentation, habitat modification also occurs. Mcintyre and Hobbs (1999) incorporated this complexity into a conceptual model by outlining four stages along a trajectory of landscape change. These were: (i) intact landscapes, in which most original vegetation remains with little or no modification; (ii) variegated landscapes, dominated by the original vegetation, but with marked gradients of habitat modification; (iii) fragmented landscapes, in which fragments are a minor component in a landscape dominated by other land uses; and (iv) relict landscapes with little (<10%) cover of original vegetation, set within highly modified surroundings. This framework emphasizes the dynamics of landscape change. Different stages along the trajectory pose different kinds of challenges for conservation management.

Many species are not confined solely to fragments, but also occur in other land uses in modified landscapes. In Nicaragua, for example, riparian forests, secondary forests, forest fallows, live fences, and pastures with dispersed trees each support diverse assemblages of birds, bats, dung beetles and butterflies (Harvey *et al.* 2006). To these species, the landscape represents a mosaic of land uses of differing quality, rather than a contrast between "habitat" and "non-habitat". Recognizing landscapes as mosaics emphasizes the need to appreciate all types of elements in the landscape. This perspective is particularly relevant in regions where cultural habitats, derived from centuries of human land-use, have important conservation values.

Different species have different ecological attributes, such as their scale of movement, life-history stages, longevity, and what constitutes habitat. These each influence how a species "perceives" a landscape, as well as its ability to survive in a modified landscape. Consequently, the same landscape may be perceived by different taxa as having a different structure and different suitability, and quite differently from the way that humans describe the landscape. A "species-centered" view of a landscape can be obtained by mapping contours of habitat suitability for any given species (Fischer *et al.* 2004).

### 5.1.2 Fragment vs landscape perspective

Habitat fragmentation is a landscape-level process. Fragmented landscapes differ in the size and shape of fragments and in their spatial configuration. Most "habitat fragmentation" studies have been undertaken at the fragment level, with individual fragments as the unit of study. However, to draw inferences about the consequences of landscape change and habitat fragmentation, it is necessary to compare "whole" landscapes that differ in their patterns of fragmentation (McGarigal and Cushman 2002). Comparisons of landscapes are also important because: (i) landscapes have properties that differ from those of fragments (Figure 5.1); (ii) many species move between and use multiple patches in the landscape; and (iii) conservation managers must manage entire landscapes (not just individual fragments) and therefore require an understanding of the desirable properties of whole landscapes. Consequently, it is valuable to consider

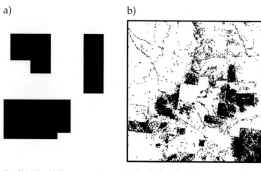

**Figure 5.1** Comparison of the types of attributes of a) individual fragments and b) whole landscapes.

a) Individual fragments
size
shape
core area
vegetation type
isolation

b) Whole landscapes
compositional gradients
diversity of land-uses
number of fragments
aggregation
structural connectivity

the consequences of landscape change at both the fragment and landscape levels.

## 5.2 Biophysical aspects of landscape change

### 5.2.1 Change in landscape pattern

Landscape change is a dynamic process. A series of "snapshots" at intervals through time (Figure 5.2) illustrates the pattern of change to the original vegetation. Characteristic changes along a time trajectory include: (i) a decline in the total area of fragments; (ii) a decrease in the size of many fragments (large tracts become scarce, small fragments predominate); (iii) increased isolation of fragments from similar habitat; and (iv) the shapes of fragments increasingly become dominated by straight edges compared with the curvilinear boundaries of natural features such as rivers. For small fragments and linear features such as fencerows, roadside vegetation, and riparian strips, the ratio of perimeter length to area is high, resulting in a large proportion of "edge" habitat. An increase in the overall proportion of edge habitat is a highly influential consequence of habitat fragmentation.

At the landscape level, a variety of indices have been developed to quantify spatial patterns, but many of these are intercorrelated, especially with the total amount of habitat remaining in the landscape (Fahrig 2003). Several aspects of the spatial configuration of fragments that usefully distinguish between different landscapes include: (i) the degree of subdivision (i.e. number of fragments), (ii) the aggregation of habitat, and (iii) the complexity of fragment shapes (Figure 5.3).

Some kinds of changes are not necessarily evident from a time-series sequence. Landscape change is not random: rather, disproportionate change occurs in certain areas. Clearing of vegetation is more common in flatter areas at lower elevations and on the more-productive soils. Such areas are likely to retain fewer, smaller fragments of original vegetation, whereas larger fragments are more likely to persist in areas less suitable for agricultural or urban development, such as on steep slopes, poorer soils, or regularly inundated floodplains. This has important implications for conservation because sites associated with different soil types and elevations typically support different sets of species. Thus, fragments usually represent a biased sample of the former biota of a region. There also is a strong historical influence on landscape change because many fragments, and the disturbance regimes they experience, are a legacy of past land settlement and land-use (Lunt and Spooner 2005). Land-use history can be an effective predictor of the present distribution of fragments and ecosystem condition within fragments. It is necessary to understand ecological processes and changes in the past in order to manage for the future.

### 5.2.2 Changes to ecosystem processes

Removal of large tracts of native vegetation changes physical processes, such as those relating to solar radiation and the fluxes of wind and water (Saunders *et al.* 1991). The greatest impact on fragments occurs at their boundaries; small remnants and those with complex shapes experience the strongest "edge effects". For example, the microclimate at a forest edge adjacent to cleared land differs from that of the forest interior in attributes such as incident light, humidity, ground and air temperature, and wind speed. In turn, these physical changes affect biological processes such as

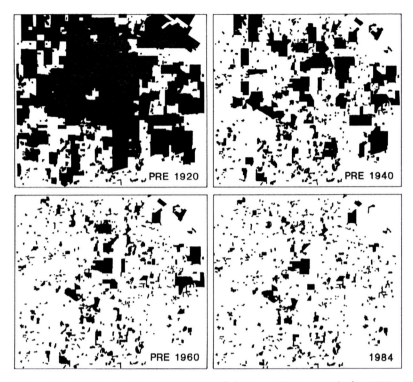

**Figure 5.2** Changes in the extent and pattern of native vegetation in the Kellerberrin area, Western Australia, from 1920 to 1984, illustrating the process of habitat loss and fragmentation. Reprinted from Saunders et al. (1993).

litter decomposition and nutrient cycling, and the structure and composition of vegetation.

Changes to biophysical processes from land use in the surrounding environment, such as the use of fertilizers on farmland, alterations to drainage patterns and water flows, and the presence of exotic plants and animals, also have spill-over effects in fragments. Many native vegetation communities are resistant to invasion by exotic plant species unless they are disturbed. Grazing by domestic stock and altered nutrient levels can facilitate the invasion of exotic species of plants, which markedly alters the vegetation in fragments (Hobbs and Yates 2003) and habitats for animals.

The intensity of edge effects in fragments and the distance over which they act varies between processes and between ecosystems. In tropical forests in the Brazilian Amazon, for example, changes in soil moisture content, vapor pressure deficit, and the number of treefall gaps extend about 50 m into the forest, whereas the invasion of disturbance-adapted butterflies and beetles and elevated tree mortality extend 200 m or more from the forest edge (Laurance 2008). In most situations, changes at edges are generally detrimental to conservation values because they modify formerly intact habitats. However, in some circumstances edges are deliberately managed to achieve specific outcomes. Manipulation of edges is used to enhance the abundance of game species such as deer, pheasants and grouse (see Box 1.1). In England, open linear "rides" in woods may be actively managed to increase incident light and early successional habitat for butterflies and other wildlife (Ferris-Kaan 1995).

Changes to biophysical processes frequently have profound effects for entire landscapes. In highly fragmented landscapes in which most fragments are small or have linear shapes, there may be little interior habitat that is buffered from edge effects. Changes that occur to individual fragments accumulate across the landscape. Changes to biophysical processes such as hydrological regimes can also affect entire landscapes. In the Western Australian wheatbelt (Figure 5.2), massive loss of native vegetation has resulted in a rise in the level of groundwater, bringing stored salt (NaCl) to the surface where it accumulates and reduces agricultural productivity and transforms native vegetation (Hobbs 1993).

## 5.3 Effects of landscape change on species

Species show many kinds of responses to habitat fragmentation: some are advantaged and increase in abundance, while others decline and become locally extinct (see Chapter 10). Understanding these diverse patterns, and the processes underlying them, is an essential foundation for conservation. Those managing fragmented

---

**Box 5.1 Time lags and extinction debt in fragmented landscapes**
**Andrew F. Bennett and Denis A. Saunders**

Habitat destruction and fragmentation result in immediately visible and striking changes to the pattern of habitat in the landscape. However, the effects of these changes on the biota take many years to be expressed: there is a time-lag in experiencing the full consequences of such habitat changes. Long-lived organisms such as trees may persist for many decades before disappearing without replacement; small local populations of animals gradually decline before being lost; and ecological processes in fragments are sensitive to long-term changes in the surroundings. Conservation managers cannot assume that species currently present in fragmented landscapes will persist there. Many fragments and landscapes face impending extinctions, even though there may be no further change in fragment size or the amount of habitat in the landscape. We are still to pay the 'extinction debt' for the consequences of past actions.

Identifying the duration of time-lags and forecasting the size of the extinction debt for fragmented landscapes is difficult. The clearest insights come from long-term studies that document changes in communities. For example, large nocturnal marsupials were surveyed in rainforest fragments in Queensland, Australia, in 1986–87 and again 20 years later in 2006–07 (Laurance et al. 2008). At the time of the first surveys, when fragments had been isolated for 20–50 years, the fauna differed markedly from that in extensive rainforest. Over the subsequent 20 years, even further changes occurred. Notably, the species richness in fragments had

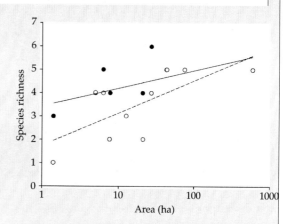

**Box 5.1 Figure** A change in the species-area relationship for mammals in rainforest fragments in Queensland, Australia, between 1986 (filled circles) and 2006 (open circles) illustrates a time-lag in the loss of species following fragmentation. Data from Laurance et al. (2008).

declined further (see Box 5.1 Figure), with most declines in the smaller fragments. By 2006–07, one species, the lemuroid ringtail possum (*Hemibelideus lemuroides*), was almost totally absent from fragments and regrowth forests along streams and its abundance in these habitats was only 0.02% of that in intact forest (Laurance et al. 2008).

### REFERENCES

Laurance, W. F., Laurance, S. G., and Hilbert, D. W. (2008). Long-term dynamics of a fragmented rainforest mammal assemblage. *Conservation Biology*, 22, 1154–1164.

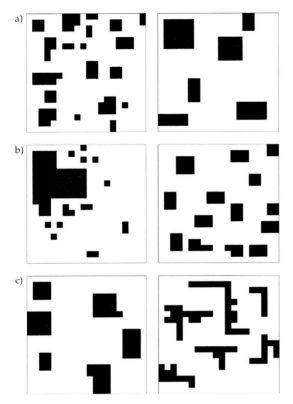

**Figure 5.3** Variation in the spatial configuration of habitat in landscapes with similar cover of native vegetation: a) subdivision (many versus few patches); b) aggregated vs dispersed habitat; and c) compact vs complex shapes. All landscapes have 20% cover (shaded).

landscapes need to know which species are most vulnerable to these processes.

## 5.3.1 Patterns of species occurrence in fragmented landscapes

Many studies have described the occurrence of species in fragments of different sizes, shapes, composition, land-use and context in the landscape. For species that primarily depend on fragmented habitat, particularly animals, fragment size is a key influence on the likelihood of occurrence (Figure 5.4). As fragment size decreases, the frequency of occurrence declines and the species may be absent from many small fragments. Such absences may be because the fragment is smaller than the minimum area required for a single individual or breeding unit, or for a self-sustaining population.

Some species persist in fragmented landscapes by incorporating multiple fragments in their territory or daily foraging movements. In England, the tawny owl (*Strix aluco*) occupies territories of about 26 ha (hectares) in large deciduous woods, but individuals also persist in highly fragmented areas by including several small woods in their territory (Redpath 1995). There is a cost, however: individuals using multiple woods have lower breeding success and there is a higher turnover of territories between years. Species that require different kinds of habitats to meet regular needs (e.g. for foraging and breeding) can be greatly disadvantaged if these habitats become isolated. Individuals may then experience difficulty in moving between different parts of the landscape to obtain their required resources. Amphibians that move between a breeding pond and other habitat, such as overwintering sites in forest, are an example.

Other attributes (in addition to fragment size) that influence the occurrence of species include the type and quality of habitat, fragment shape, land use adjacent to the fragment, and the extent to which the wider landscape isolates populations. In the Iberian region of Spain, for example, the relative abundance of the Eurasian badger (*Meles meles*) in large forest fragments is

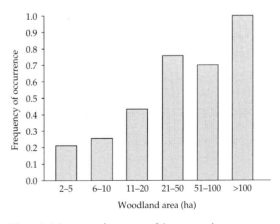

**Figure 5.4** Frequency of occurrence of the common dormouse (*Muscardinus avellanarius*) in ancient semi-natural woods in Herefordshire, England, in relation to increasing size-class of woods. Data from Bright *et al.* (1994).

significantly influenced by habitat quality and forest cover in the wider landscape (Virgos 2001). In areas with less than 20% forest cover, badger abundance in forests was most influenced by isolation (i.e. distance to a potential source area >10 000 ha), whereas in areas with 20–50% cover, badgers were most influenced by the quality of habitat in the forest fragments.

A key issue for conservation is the relative importance of habitat loss versus habitat fragmentation (Fahrig 2003). That is, what is the relative importance of *how much* habitat remains in the landscape versus *how fragmented* it is? Studies of forest birds in landscapes in Canada and Australia suggest that habitat loss and habitat fragmentation are *both* significant influences, although habitat loss generally is a stronger influence for a greater proportion of species (Trczinski *et al.* 1999; Radford and Bennett 2007). Importantly, species respond to landscape pattern in different ways. In southern Australia, the main influence for the eastern yellow robin (*Eopsaltria australis*) was the total amount of wooded cover in the landscape; for the grey shrike-thrush (*Colluricincla harmonica*) it was wooded cover together with its configuration (favoring aggregated habitat); while for the musk lorikeet (*Glossopsitta concinna*) the influential factor was not wooded cover, but the configuration of habitat and diversity of vegetation types (Radford and Bennett 2007).

## 5.3.2 Processes that affect species in fragmented landscapes

The size of any population is determined by the balance between four parameters: births, deaths, immigration, and emigration. Population size is increased by births and immigration of individuals, while deaths and emigration of individuals reduce population size. In fragmented landscapes, these population parameters are influenced by several categories of processes.

*Deterministic processes*
Many factors that affect populations in fragmented landscapes are relatively predictable in their effect. These factors are not necessarily a direct consequence of habitat fragmentation, but arise from land uses typically associated with subdivision. Populations may decline due to deaths of individuals from the use of pesticides, insecticides or other chemicals; hunting by humans; harvesting and removal of plants; and construction of roads with ensuing road kills of animals. For example, in Amazonian forests, subsistence hunting by people compounds the effects of forest fragmentation for large vertebrates such as the lowland tapir (*Tapir terrestris*) and white-lipped peccary (*Tayassu pecari*), and contributes to their local extinction (Peres 2001).

Commonly, populations are also affected by factors such as logging, grazing by domestic stock, or altered disturbance regimes that modify the quality of habitats and affect population growth. For example, in Kibale National Park, an isolated forest in Uganda, logging has resulted in long-term reduction in the density of groups of the blue monkey (*Cercopithecus mitza*) in heavily logged areas: in contrast, populations of black and white colobus (*Colobus guereza*) are higher in regrowth forests than in unlogged forest (Chapman *et al.* 2000). Deterministic processes are particularly important influences on the status of plant species in fragments (Hobbs and Yates 2003).

*Isolation*
Isolation of populations is a fundamental consequence of habitat fragmentation: it affects local populations by restricting immigration and emigration. Isolation is influenced not only by the distance between habitats but also by the effects of human land-use on the ability of organisms to move (or for seeds and spores to be dispersed) through the landscape. Highways, railway lines, and water channels impose barriers to movement, while extensive croplands or urban development create hostile environments for many organisms to move through. Species differ in sensitivity to isolation depending on their type of movement, scale of movement, whether they are nocturnal or diurnal, and their response to landscape change. Populations of one species may be highly isolated, while in the same landscape individuals of another species can move freely.

Isolation affects several types of movements, including: (i) regular movements of individuals between parts of the landscape to obtain different requirements (food, shelter, breeding sites); (ii) seasonal or migratory movements of species at regional, continental or inter-continental scales; and (iii) dispersal movements (immigration, emigration) between fragments, which may supplement population numbers, increase the exchange of genes, or assist recolonization if a local population has disappeared. In Western Australia, dispersal movements of the blue-breasted fairy-wren (*Malurus pulcherrimus*) are affected by the isolation of fragments (Brooker and Brooker 2002). There is greater mortality of individuals during dispersal in poorly connected areas than in well-connected areas, with this difference in survival during dispersal being a key factor determining the persistence of the species in local areas.

For many organisms, detrimental effects of isolation are reduced, at least in part, by habitat components that enhance connectivity in the landscape (Saunders and Hobbs 1991; Bennett 1999). These include continuous "corridors" or "stepping stones" of habitat that assist movements (Haddad *et al.* 2003), or human land-uses (such as coffee-plantations, scattered trees in pasture) that may be relatively benign environments for many species (Daily *et al.* 2003). In tropical regions, one of the strongest influences on the persistence of species in forest fragments is their ability to live in, or move through, modified "countryside" habitats (Gascon *et al.* 1999; Sekercioglu *et al.* 2002).

*Stochastic processes*
When populations become small and isolated, they become vulnerable to a number of stochastic (or chance) processes that may pose little threat to larger populations. Stochastic processes include the following.

• Stochastic variation in demographic parameters such as birth rate, death rate and the sex ratio of offspring.
• Loss of genetic variation, which may occur due to inbreeding, genetic drift, or a founder effect from a small initial population size. A decline in genetic diversity may make a population more vulnerable to recessive lethal alleles or to changing environmental conditions.
• Fluctuations in the environment, such as variation in rainfall and food sources, which affect birth and death rates in populations.
• Small isolated populations are particularly vulnerable to catastrophic events such as flood, fire, drought or hurricanes. A wildfire, for example, may eliminate a small local population whereas in extensive habitats some individuals survive and provide a source for recolonization.

### 5.3.3 Metapopulations and the conservation of subdivided populations

Small populations are vulnerable to local extinction, but a species has a greater likelihood of persistence where there are a number of local populations interconnected by occasional movements of individuals among them. Such a set of subdivided populations is often termed a "metapopulation" (Hanski 1999). Two main kinds of metapopulation have been described (Figure 5.5). A mainland-island model is where a large mainland population (such as a conservation reserve) provides a source of emigrants that disperse to nearby small populations. The mainland population has a low likelihood of extinction, whereas the small populations become extinct relatively frequently. Emigration from the mainland supplements the small populations, introduces new genetic material and allows recolonization should local extinction occur. A second kind of metapopulation is where the set of interacting populations are relatively similar in size and all have a likelihood of experiencing extinction (Figure 5.5b). Although colonization and extinction may occur regularly, the overall population persists through time.

The silver-spotted skipper (*Hesperia comma*), a rare butterfly in the UK, appears to function as a metapopulation (Hill *et al.* 1996). In 1982, butterflies occupied 48 of 69 patches of suitable grassland on the North Downs, Surrey. Over the next 9 years, 12 patches were colonized and seven populations went extinct. Those more susceptible

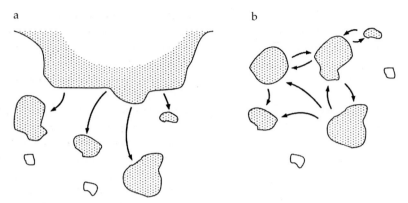

**Figure 5.5** Diagrammatic representation of two main types of metapopulation models: a) a mainland-island metapopulation and b) metapopulation with similar-sized populations. Habitats occupied by a species are shaded, unoccupied habitat fragments are clear, and the arrows indicate typical movements. Reprinted from Bennett (1999).

to extinction were small isolated populations, whereas the patches more likely to be colonized were relatively large and close to other large occupied patches.

The conservation management of patchily-distributed species is likely to be more effective by taking a metapopulation approach than by focusing on individual populations. However, "real world" populations differ from theoretical models. Factors such as the quality of habitat patches and the nature of the land mosaic through which movements occur are seldom considered in theoretical models, which emphasize spatial attributes (patch area, isolation). For example, in a metapopulation of the Bay checkerspot butterfly (*Euphydryas editha bayensis*) in California, USA, populations in topographically heterogeneous fragments were less likely to go extinct than those that were in topographically uniform ones. The heterogeneity provided some areas of suitable topoclimate each year over a wide range of local climates (Ehrlich and Hanski 2004).

There also is much variation in the structure of subdivided populations depending on the frequency of movements between them. At one end of a gradient is a dysfunctional metapopulation where little or no movement occurs; while at the other extreme, movements are so frequent that it is essentially a single patchy population.

## 5.4 Effects of landscape change on communities

### 5.4.1 Patterns of community structure in fragmented landscapes

For many taxa—birds, butterflies, rodents, reptiles, vascular plants, and more—species richness in habitat fragments is positively correlated with fragment size. This is widely known as the species-area relationship (Figure 5.6a). Thus, when habitats are fragmented into smaller pieces, species are lost; and the likely extent of this loss can be predicted from the species-area relationship. Further, species richness in a fragment typically is less than in an area of similar size within continuous habitat, evidence that the fragmentation process itself is a cause of local extinction. However, the species-area relationship does not reveal which particular species will be lost.

Three explanations given for the species-area relationship (Connor and McCoy 1979) are that small areas: (i) have a lower diversity of habitats; (ii) support smaller population sizes and therefore fewer species can maintain viable populations; and (iii) represent a smaller sample of the original habitat and so by chance are likely to have fewer species than a larger sample. While it is difficult to distinguish between these mechanisms, the message is clear: when habitats are fragmented into smaller pieces, species are lost.

specialized ecological requirements are those lost from communities in fragments. In several tropical regions, birds that follow trails of army ants and feed on insects flushed by the ants include specialized ant-following species and others that forage opportunistically in this way. In rainforest in Kenya, comparisons of flocks of ant-following birds between a main forest and forest fragments revealed marked differences (Peters *et al.* 2008). The species richness and number of individuals in ant-following flocks were lower in fragments, and the composition of flocks more variable in small fragments and degraded forest, than in the main forest. This was a consequence of a strong decline in abundance of five species of specialized ant-followers in fragments, whereas the many opportunistic followers (51 species) were little affected by fragmentation (Peters *et al.* 2008).

The way in which fragments are managed is a particularly important influence on the composition of plant communities. In eastern Australia, for example, grassy woodlands dominated by white box (*Eucalyptus albens*) formerly covered several million hectares, but now occur as small fragments surrounded by cropland or agricultural pastures. The species richness of native understory plants increases with fragment size, as expected, but tree clearing and grazing by domestic stock are also strong influences (Prober and Thiele 1995). The history of stock grazing has the strongest influence on the floristic composition in woodland fragments: grazed sites have a greater invasion by weeds and a more depauperate native flora.

The composition of animal communities in fragments commonly shows systematic changes in relation to fragment size. Species-poor communities in small fragments usually support a subset of the species present in larger, richer fragments (Table 5.1). That is, there is a relatively predictable change in composition with species "dropping out" in an ordered sequence in successively smaller fragments (Patterson and Atmar 1986). Typically, rare and less common species occur in larger fragments, whereas those present in smaller fragments are mainly widespread and common. This kind of "nested subset" pattern

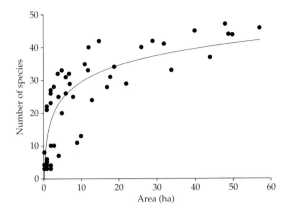

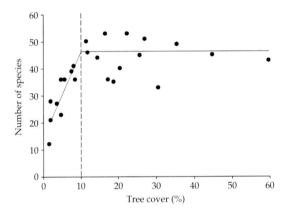

**Figure 5.6** Species-area relationships for forest birds: a) in forest fragments of different sizes in eastern Victoria, Australia (data from Loyn 1997); b) in 24 landscapes (each 100 km$^2$) with differing extent of remnant wooded vegetation, in central Victoria, Australia (data from Radford *et al.* 2005). The piecewise regression highlights a threshold response of species richness to total extent of wooded cover.

Factors other than area, such as the spatial and temporal isolation of fragments, land management or habitat quality may also be significant predictors of the richness of communities in fragments. In Tanzania, for example, the number of forest-understory bird species in forest fragments (from 0.1 to 30 ha in size) was strongly related to fragment size, as predicted by the species-area relationship (Newmark 1991). After taking fragment size into account, further variation in species richness was explained by the isolation distance of each fragment from a large source area of forest.

Species show differential vulnerability to fragmentation. Frequently, species with more-

**Table 5.1** A diagrammatic example of a nested subset pattern of distribution of species (A–J) within habitat fragments (1–9).

| Species | Fragments | | | | | | | | |
|---|---|---|---|---|---|---|---|---|---|
| | 1 | 2 | 3 | 4 | 5 | 6 | 7 | 8 | 9 |
| A | + | + | + | + |   | + | + | + | + |
| B | + | + | + |   | + | + | + | + |   |
| C | + |   | + | + | + | + | + |   | + |
| D | + | + | + | + | + |   |   | + |   |
| E | + | + | + | + | + | + |   |   |   |
| F |   | + |   | + |   |   |   |   |   |
| G | + |   | + | + |   |   |   |   |   |
| H | + | + |   |   |   |   |   |   |   |
| I | + | + | + |   |   |   |   |   |   |
| J | + |   |   |   |   |   |   |   |   |

has been widely observed: for example, in butterfly communities in fragments of lowland rainforest in Borneo (Benedick et al. 2006).

At the landscape level, species richness has frequently been correlated with heterogeneity in the landscape. This relationship is particularly relevant in regions, such as Europe, where human land-use has contributed to cultural habitats that complement fragmented natural or semi-natural habitats. In the Madrid region of Spain, the overall richness of assemblages of birds, amphibians, reptiles and butterflies in 100 km$^2$ landscapes is strongly correlated with the number of different land-uses in the landscape (Atauri and De Lucio 2001). However, where the focus is on the community associated with a particular habitat type (e.g. rainforest butterflies) rather than the entire assemblage of that taxon, the strongest influence on richness is the total amount of habitat in the landscape. For example, the richness of woodland-dependent birds in fragmented landscapes in southern Australia was most strongly influenced by the total extent of wooded cover in each 100 km$^2$ landscape, with a marked threshold around 10% cover below which species richness declined rapidly (Figure 5.6b) (Radford et al. 2005).

### 5.4.2 Processes that affect community structure

Interactions between species, such as predation, competition, parasitism, and an array of mutualisms, have a profound influence on the structure of communities. The loss of a species or a change in its abundance, particularly for species that interact with many others, can have a marked effect on ecological processes throughout fragmented landscapes.

Changes to predator-prey relationships, for example, have been revealed by studies of the level of predation on birds' nests in fragmented landscapes (Wilcove 1985). An increase in the amount of forest edge, a direct consequence of fragmentation, increases the opportunity for generalist predators associated with edges or modified land-uses to prey on birds that nest in forest fragments. In Sweden, elevated levels of nest predation (on artificial eggs in experimental nests) were recorded in agricultural land and at forest edges compared with the interior of forests (Andrén and Angelstam 1988). Approximately 45% of nests at the forest edge were preyed upon compared with less than 10% at distances >200 m into the forest. At the landscape scale, nest predation occurred at a greater rate in agricultural and fragmented forest landscapes than in largely forested landscapes (Andrén 1992). The relative abundance of different corvid species, the main nest predators, varied in relation to landscape composition. The hooded crow (*Corvus corone cornix*) occurred in greatest abundance in heavily cleared landscapes and was primarily responsible for the greater predation pressure recorded at forest edges.

Many mutualisms involve interactions between plants and animals, such as occurs in the pollination of flowering plants by invertebrates, birds or mammals. A change in the occurrence or abundance of animal vectors, as a consequence of fragmentation, can disrupt this process. For many plant species, habitat fragmentation has a negative effect on reproductive success, measured in terms of seed or fruit production, although the relative impact varies among species (Aguilar et al. 2006). Plants that are self-incompatible (i.e. that depend on pollen transfer from another plant) are more susceptible to reduced reproductive success than are self-compatible species. This difference is consistent with an expectation that pollination by animals will be less effective in small and isolated fragments. However, pollinators are a diverse group and they respond to fragmentation in a variety of ways (Hobbs and Yates 2003).

Changes in ecological processes in fragments and throughout fragmented landscapes are complex and poorly understood. Disrupted interactions between species may have flow-on effects to many other species at other trophic levels. However, the kinds of changes to species interactions and ecological processes vary between ecosystems and regions because they depend on the particular sets of species that occur. In parts of North America, nest parasitism by the brown-headed cowbird (*Molothrus ater*) has a marked effect on bird communities in fragments (Brittingham and Temple 1983); while in eastern Australia, bird communities in small fragments may be greatly affected by aggressive competition from the noisy miner (*Manorina melanocephala*) (Grey *et al.* 1997). Both of these examples are idiosyncratic to their region. They illustrate the difficulty of generalizing the effects of habitat fragmentation, and highlight the importance of understanding the consequences of landscape change in relation to the environment, context and biota of a particular region.

## 5.5 Temporal change in fragmented landscapes

Habitat loss and fragmentation do not occur in a single event, but typically extend over many decades. Incremental changes occur year by year as remaining habitats are destroyed, reduced in size, or further fragmented (Figure 5.2). Landscapes are also modified through time as the human population increases, associated settlements expand, and new forms of land use are introduced.

In addition to such changes in spatial pattern, habitat fragmentation sets in motion ongoing changes within fragments and in the interactions between fragments and their surroundings. When a fragment is first isolated, species richness does not immediately fall to a level commensurate with its long-term carrying capacity; rather, a gradual loss of species occurs over time—termed "species relaxation". That is, there is a time-lag in experiencing the full effects of fragmentation (see Box 5.1). The rate of change is most rapid in smaller fragments, a likely consequence of the smaller population sizes of species and the greater vulnerability of such fragments to external disturbances. For example, based on a sequence of surveys of understory birds in tropical forest fragments at Manaus, Brazil, an estimate of the time taken for fragments to lose half their species was approximately 5 years for 1 ha fragments, 8 years for 10 ha fragments, and 12 years for a 100 ha fragment (Ferraz *et al.* 2003).

Ecological processes within fragments also experience ongoing changes in the years after isolation because of altered species interactions and incremental responses to biophysical changes. One example comes from small fragments of tropical dry forest that were isolated by rising water in a large hydroelectric impoundment in Venezuela (Terborgh *et al.* 2001). On small (< 1 ha) and medium (8–12 ha) fragments, isolation resulted in a loss of large predators typical of extensive forest. Seed predators (small rodents) and herbivores (howler monkeys *Alouatta seniculus*, iguanas *Iguana iguana*, and leaf-cutter ants) became hyperabundant in these fragments, with cascading effects on the vegetation. Compared with extensive forest, fragments experienced reduced recruitment of forest trees, changes in vegetation composition, and dramatically modified faunal communities, collectively termed an "ecological meltdown" (Terborgh *et al.* 2001).

## 5.6 Conservation in fragmented landscapes

Conservation of biota in fragmented landscapes is critical to the future success of biodiversity conservation and to the well-being of humans. National parks and dedicated conservation reserves are of great value, but on their own are too few, too small, and not sufficiently representative to conserve all species. The future status of a large portion of Earth's biota depends on how effectively plants and animals can be maintained in fragmented landscapes dominated by agricultural and urban land-uses. Further, the persistence of many species of plants and animals in these landscapes is central to maintaining ecosystem services that sustain food production, clean water, and a sustainable living environment for humans. Outlined below are six kinds of actions necessary for a strategic approach to conservation in fragmented landscapes.

### 5.6.1 Protect and expand the amount of habitat

Many indicators of conservation status, such as population sizes, species richness, and the occurrence of rare species, are positively correlated with the size of individual fragments or the total amount of habitat in the landscape. Consequently, activities that protect and expand natural and semi-natural habitats are critical priorities in maintaining plant and animal assemblages (see also Chapter 11). These include measures that:

- Prevent further destruction and fragmentation of habitats.
- Increase the size of existing fragments and the total amount of habitat in the landscape.
- Increase the area specifically managed for conservation.
- Give priority to protecting large fragments.

All fragments contribute to the overall amount and pattern of habitat in a landscape; consequently, incremental loss, even of small fragments, has a wider impact.

### 5.6.2 Enhance the quality of habitats

Measures that enhance the quality of existing habitats and maintain or restore ecological processes are beneficial. Such management actions must be directed toward specific goals relevant to the ecosystems and biota of concern. These include actions that:

- Control degrading processes, such as the invasion of exotic plants and animals.
- Manage the extent and impact of harvesting natural resources (e.g. timber, firewood, bushmeat).
- Maintain natural disturbance regimes and the conditions suitable for regeneration and establishment of plants.
- Provide specific habitat features required by particular species (e.g. tree hollows, rock crevices, "specimen" rainforest trees used by rainforest birds in agricultural countryside).

### 5.6.3 Manage across entire landscapes

Managing individual fragments is rarely effective because even well managed habitats can be degraded by land uses in the surrounding environment. Further, many species use resources from different parts of the landscape; and the pattern and composition of land uses affect the capacity of species to move throughout the landscape. Two broad kinds of actions relating to the wider landscape are required:

- Manage specific issues that have degrading impacts across the boundaries of fragments, such as pest plants or animals, soil erosion, sources of pollution or nutrient addition, and human recreational pressure.
- Address issues that affect the physical environment and composition of the land mosaic across broad scales, such as altered hydrological regimes and the density of roads and other barriers.

### 5.6.4 Increase landscape connectivity

Measures that enhance connectivity and create linked networks of habitat will benefit the conservation of biota in fragmented landscapes. Connectivity can be increased by providing specific linkages, such as continuous corridors or stepping stones, or by managing the entire mosaic to allow movements of organisms. Actions that enhance connectivity include:

- Protecting connecting features already present, such as streamside vegetation, hedges and live fences.
- Filling gaps in links or restoring missing connections.
- Maintaining stepping-stone habitats for mobile species (such as migratory species).
- Retaining broad habitat links between conservation reserves.
- Developing regional and continental networks of habitat (see Boxes 5.2 and 5.3).

## Box 5.2 Gondwana Link: a major landscape reconnection project
### Denis A. Saunders and Andrew F. Bennett

In many locations throughout the world, conservation organizations and community groups are working together to protect and restore habitats as ecological links between otherwise-isolated areas. These actions are a practical response to the threats posed by habitat destruction and fragmentation and are undertaken at a range of scales, from local to continental. Gondwana Link, in south-western Australia, is one such example of an ambitious plan to restore ecological connectivity and enhance nature conservation across a large geographic region.

The southwest region of Australia is one of the world's 34 biodiversity "hotspots". It is particularly rich in endemic plant species. The region has undergone massive changes over the past 150 years as a result of development for broadscale agricultural cropping and raising of livestock. Over 70% of the area of native vegetation has been removed. The remaining native vegetation consists of thousands of fragments, most of which are less than 100 ha. Many areas within the region have less than 5–10% of their original vegetation remaining.

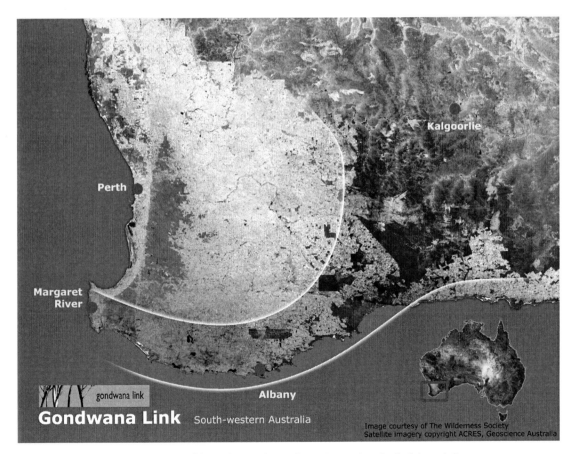

**Box 5.2 Figure** Diagrammatic representation of the Gondwana Link in south-west Western Australia. Shaded areas indicate remnant native vegetation.

*continues*

### Box 5.2 (Continued)

This massive removal of native vegetation has led to a series of changes to ecological processes, producing a wide range of problems that must be addressed. Without some form of remedial action, over 6 million hectares of land (30% of the region's cleared land) will become salinized over the next 50 years, over 50% of vegetation on nature reserves will be destroyed, around 450 endemic species of plant will become extinct, over half of all bird species from the region will be adversely affected, and no potable surface water will be available in the region because of water pollution by salt.

Addressing the detrimental ecological consequences involves the revegetation, with deep-rooted trees and shrubs, of up to 40% of cleared land in the region. Gondwana Link is an ambitious conservation project involving individuals, local, regional and national groups addressing these detrimental ecological consequences. The objective of Gondwana Link is to restore ecological connectivity across south-western Australia. The aim is to provide ecological connections from the tall wet forests of the southwest corner of the state to the dry woodland in the arid interior. This will involve protecting and replanting native vegetation along a "living link" that stretches over 1000 km from the wettest corner of Western Australia into the arid zone (see Box 5.2 Figure and Plate 6). It also involves protecting and managing the fragments of native vegetation that they are reconnecting.

The groups believe that by increasing connectivity and restoring key habitats they will enable more mobile species that are dependent on native vegetation to move safely between isolated populations. This should reduce the localized extinctions of species from isolated fragments of native vegetation that is happening at present. Gondwana Link should also allow species to move as climatic conditions change over time. The revegetation should also have an impact on the hydrological regime by decreasing the amount of water entering the ground water, and reduce the quantity of sediment and pollution from agriculture entering the river and estuarine systems.

In addition to addressing environmental issues the project is speeding up the development of new cultural and economic ways for the region's human population to exist sustainably.

### Relevant website

- Gondwana Link: http://gondwanalink.org/index.html.

### Box 5.3 Rewilding
Paul R. Ehrlich

Some conservation scientists believe that the ultimate cure for habitat loss and fragmentation that is now spreading like ecological smallpox over Earth is a radical form of restoration, called rewilding in North America. The objective of rewilding is to restore resilience and biodiversity by re-connecting severed habitats over large scales and by facilitating the recovery of strongly interactive species, including predators. Rewilding is the goal of the "Wildlands Network," an effort led by Michael Soulé and Dave Foreman (Foreman 2004). The plan is to re-connect relatively undisturbed, but isolated areas of North America, into extensive networks in which large mammals such as bears, mountain lions,

*continues*

### Box 5.3 (Continued)

wolves, elk, and even horses and elephants (which disappeared from North America only 11 000 years ago) can roam free and resume their important ecological roles in ecosystems where conflict with humans would be minimal. Rewilding would restore landscape linkages—employing devices from vegetated overpasses over highways to broad habitat corridors—allowing the free movement of fauna and flora and accommodation to climate change. The cooperation of government agencies and willing landowners would eventually create four continental scale wildways (formerly called MegaLinkages):

*Pacific Wildway:* From southern Alaska through the Coast Range of British Columbia, the Cascades, and the Sierra Nevada to the high mountains of northern Baja California.

*Spine of the Continent Wildway:* From the Brooks Range of Alaska through the Rocky Mountains to the uplands of Western Mexico.

*Atlantic Wildway:* From the Canadian Maritime south, mostly through the Appalachians to Okefenokee and the Everglades.

*Arctic-Boreal Wildway:* Northern North America from Alaska through the Canadian arctic/subarctic to Labrador with an extension into the Upper Great Lakes.

Many critical ecological processes are mediated by larger animals and plants, and the recovery, dispersal, and migration of these keystone and foundation species (species that are critical in maintaining the structure of communities disproportionately more than their relative abundance) is essential if nature is to adapt to stresses such a climate change and habitat loss caused by energy development, sprawl, and the proliferation of roads. Rewilding will help restore ecosystems in the Wildways to structural and functional states more like those that prevailed before industrial society accelerated the transformation of the continent. Similar rewilding projects on other continents are now in the implementation stage—as in the "Gondwana Link" in Australia (see Box 5.2). The possible downsides to rewilding include the spread of some diseases, invasive species, and fires and the social and economic consequences of increased livestock depredation caused by large, keystone predators (as have accompanied wolf reintroduction programs) (Maehr *et al.* 2001). Careful thought also is needed about the size of these Wildways; to be sure they are large enough for these species to again persist in their "old homes". Nonetheless, it seems clear that such potential costs of rewilding would be overwhelmed by the ecological and economic-cultural benefits that well designed and monitored reintroductions could provide.

### REFERENCES AND SUGGESTED READING

Donlan, J. C., Berger, J., Bock, C. E., *et al.* (2006). Pleistocene rewilding: an optimistic agenda for twenty-first century conservation. *American Naturalist*, **168**, 660–681.

Foreman, D. (2004). *Rewilding North America: a vision for conservation in the 21st Century*. Island Press, Washington, DC.

Maehr, D. S., Noss, R. F., and Larkin, J. L., eds (2001). *Large mammal restoration: ecological and sociological challenges in the 21st centuary*. Island Press, Washington, DC.

Soulé, M. E. and Terborgh, J. (1999). *Continental conservation: scientific foundations of regional reserve networks*. Island Press, Washington, DC.

Soulé, M. E., Estes, J. A., Miller, B., and Honnold, D. L. (2005). Highly interactive species: conservation policy, management, and ethics. *BioScience*, **55**, 168–176.

### 5.6.5 Plan for the long term

Landscape change is ongoing. Over the long-term, incremental destruction and fragmentation of habitats have profound consequences for conservation. Long-term planning is required to sustain present conservation values and prevent foreclosure of future options. Actions include:

• Using current knowledge to forecast the likely consequences if ongoing landscape change occurs.
• Developing scenarios as a means to consider alternative future options.
• Developing a long-term vision, shared by the wider community, of land use and conservation goals for a particular region.

### 5.6.6 Learn from conservation actions

Effective conservation in fragmented landscapes demands that we learn from current management in order to improve future actions. Several issues include:

• Integrating management and research to more effectively evaluate and refine conservation measures.
• Monitoring the status of selected species and ecological processes to evaluate the longer-term outcomes and effectiveness of conservation actions.

## Summary

• Destruction and fragmentation of habitats are major factors in the global decline of species, the modification of native plant and animal communities and the alteration of ecosystem processes.
• Habitat destruction, habitat fragmentation (or subdivision) and new forms of land use are closely intertwined in an overall process of landscape change.
• Landscape change is not random: disproportionate change typically occurs in flatter areas, at lower elevations and on more-productive soils.
• Altered physical processes (e.g. wind and water flows) and the impacts of human land-use have a profound influence on fragments and their biota, particularly at fragment edges.
• Different species have different ecological attributes (such as scale of movement, life-history stages, what constitutes habitat) which influence how a species perceives a landscape and its ability to survive in modified landscapes.
• Differences in the vulnerability of species to landscape change alter the structure of communities and modify interactions between species (e.g. pollination, parasitism).
• Changes within fragments, and between fragments and their surroundings, involve time-lags before the full consequences of landscape change are experienced.
• Conservation in fragmented landscapes can be enhanced by: protecting and increasing the amount of habitat, improving habitat quality, increasing connectivity, managing disturbance processes in the wider landscape, planning for the long term, and learning from conservation actions undertaken.

## Suggested reading

Forman, R. T. T. (1995). *Land mosaics. The ecology of landscapes and regions.* Cambridge University Press, Cambridge, UK.
Hobbs, R. J. and Yates, C. J. (2003). Turner Review No. 7. Impacts of ecosystem fragmentation on plant populations: generalising the idiosyncratic. *Australian Journal of Botany,* **51**, 471–488.
Laurance, W. F. and Bierregard, R. O., eds (1997). *Tropical forest remnants: ecology, management, and conservation of fragmented communities.* University of Chicago Press, Chicago, Illinois.
Lindenmayer, D. B. and Fischer, J. (2006). *Habitat fragmentation and landscape change. An ecological and conservation synthesis.* CSIRO Publishing, Melbourne, Australia.

## Relevant websites

• Sustainable forest partnerships: http://sfp.cas.psu.edu/fragmentation/fragmentation.html.
• Smithsonian National Zoological Park, Migratory Bird Center: http://nationalzoo.si.edu/ConservationAndScience/MigratoryBirds/Research/Forest_Fragmentation/default.cfm.

- United States Department of Agriculture, Forest Service: http://nationalzoo.si.edu/Conservation AndScience/MigratoryBirds/Research/Forest_ Fragmentation/default.cfm.
- Mongabay: http://www.mongabay.com.

## REFERENCES

Aguilar, R., Ashworth, L., Galetto, L., and Aizen, M. A. (2006). Plant reproductive susceptibility to habitat fragmentation: review and synthesis through a meta-analysis. *Ecology Letters*, **9**, 968–980.

Andrén, H. (1992). Corvid density and nest predation in relation to forest fragmentation: a landscape perspective. *Ecology*, **73**, 794–804.

Andrén, H. and Angelstam, P. (1988). Elevated predation rates as an edge effect in habitat islands: experimental evidence. *Ecology*, **69**, 544–547.

Atauri, J. A. and De Lucio, J. V. (2001). The role of landscape structure in species richness distribution of birds, amphibians, reptiles and lepidopterans in Mediterranean landscapes. *Landscape Ecology*, **16**, 147–159.

Benedick, S., Hill, J. K., Mustaffa, N., *et al.* (2006). Impacts of rainforest fragmentation on butterflies in northern Borneo: species richness, turnover and the value of small fragments. *Journal of Applied Ecology*, **43**, 967–977.

Bennett, A. F. (1999). *Linkages in the landscape. The role of corridors and connectivity in wildlife conservation*. IUCN-The World Conservation Union, Gland, Switzerland.

Bright, P. W., Mitchell, P., and Morris, P. A. (1994). Dormouse distribution: survey techniques, insular ecology and selection of sites for conservation. *Journal of Applied Ecology*, **31**, 329–339.

Brittingham, M. and Temple, S. (1983). Have cowbirds caused forest songbirds to decline? *BioScience*, **33**, 31–35.

Brooker, L. and Brooker, M. (2002). Dispersal and population dynamics of the blue-breasted fairy-wren, *Malurus pulcherrimus*, in fragmented habitat in the Western Australian wheatbelt. *Wildlife Research*, **29**, 225–233.

Chapman, C. A., Balcomb, S. R., Gillespie, T. R., Skorupa, J. P., and Struhsaker, T. T. (2000). Long-term effects of logging on African primate communities: a 28 year comparison from Kibale National Park, Uganda. *Conservation Biology*, **14**, 207–217.

Connor, E. F. and Mccoy, E. D. (1979). The statistics and biology of the species-area relationship. *American Naturalist*, **113**, 791–833.

Daily, G.C., Ceballos, G., Pacheco, J., Suzán, G., and Sánchez-Azofeifa, A. (2003). Countryside biogeography of neotropical mammals: conservation opportunities in agricultural landscapes of Costa Rica. *Conservation Biology*, **17**, 1814–1826.

Ehrlich, P. and Hanski, I., eds (2004). *On the wings of checkerspots: A model system for population biology*. Oxford University Press, Oxford, UK.

Fahrig, L. (2003). Effects of habitat fragmentation on biodiversity. *Annual Review of Ecology and Systematics*, **34**, 487–515.

Ferraz, G., Russell, G. J., Stouffer, P. C., Bierregaard, R. O., Pimm, S. L., and Lovejoy, T. E. (2003). Rates of species loss from Amazonian forest fragments. *Proceedings of the National Academy of Sciences, United States of America*, **100**, 14069–14073.

Ferris-Kaan, R. (1995). Management of linear habitats for wildlife in British forests. In D. A. Saunders, J. L. Craig, and E. M. Mattiske, eds *Nature conservation 4: the role of networks*, pp. 67–77. Surrey Beatty and Sons, Chipping Norton, Australia.

Fischer, J., Lindenmayer, D. B., and Fazey, I. (2004). Appreciating ecological complexity: habitat contours as a conceptual landscape model. *Conservation Biology*, **18**, 1245–1253.

Forman, R. T. T. and Godron, M. (1986). *Landscape ecology*. John Wiley and Sons, New York.

Gascon, C., Lovejoy, T. E., Bierregaard, R. O., Malcolm, J. R., Stouffer, P. C., Vasconcelos, H. L., Laurance, W. F., Zimmerman, B., Tocher, M., and Borges, S. (1999). Matrix habitat and species richness in tropical forest remnants. *Biological Conservation*, **91**, 223–229.

Grey, M. J., Clarke, M. F., and Loyn, R. H. (1997). Initial changes in the avian communities of remnant eucalypt woodlands following reduction in the abundance of noisy miners, *Manorina melanocephala*. *Wildlife Research*, **24**, 631–648.

Haddad, N. M., Bowne, D. R., Cunningham, A., Danielson, B. J., Levey, D. J., Sargent, S., and Spira, T. (2003). Corridors use by diverse taxa. *Ecology*, **84**, 609–615.

Hanski, I. (1999). *Metapopulation ecology*. Oxford University Press, Oxford, UK.

Harvey, C. A., Medina, A., Sanchez, D., *et al.* (2006). Patterns of animal diversity in different forms of tree cover in agricultural landscapes. *Ecological Applications*, **16**, 1986–1999.

Hill, J. K., Thomas, C. D., and Lewis, O. T. (1996). Effects of habitat patch size and isolation on dispersal by *Hesperia comma* butterflies: implications for metapopulation structure. *Journal of Animal Ecology*, **65**, 725–735.

Hobbs, R. J. (1993). Effects of landscape fragmentation on ecosystem processes in the Western Australian wheatbelt. *Biological Conservation*, **64**, 193–201.

Hobbs, R. J. and Yates, C. J. (2003). Turner Review No. 7. Impacts of ecosystem fragmentation on plant populations: generalising the idiosyncratic. *Australian Journal of Botany*, **51**, 471–488.

Laurance, W. F. (2008). Theory meets reality: how habitat fragmentation research has transcended island biogeographic theory. *Biological Conservation*, **141**, 1731–1744.

Lindenmayer, D. B. and Fischer, J. (2006). *Habitat fragmentation and landscape change. An ecological and conservation synthesis*. CSIRO Publishing, Melbourne, Australia.

Loyn, R. H. (1987). Effects of patch area and habitat on bird abundances, species numbers and tree health in fragmented Victorian forests. In D. A. Saunders, G. W. Arnold, A. A. Burbidge, and A. J. M. Hopkins, eds *Nature conservation: the role of remnants of native vegetation*, pp. 65–77. Surrey Beatty and Sons, Chipping Norton, Australia.

Lunt, I. D. and Spooner, P. G. (2005). Using historical ecology to understand patterns of biodiversity in fragmented agricultural landscapes. *Journal of Biogeography*, **32**, 1859–1873.

MacArthur, R. H. and Wilson, E. O. (1967). *The theory of island biogeography*. Princeton University Press, Princeton, New Jersey.

McGarigal, K. and Cushman, S. A. (2002). Comparative evaluation of experimental approaches to the study of habitat fragmentation effects. *Ecological Applications*, **12**, 335–345.

McIntyre, S. and Hobbs, R. (1999). A framework for conceptualizing human effects on landscapes and its relevance to management and research models. *Conservation Biology*, **13**, 1282–1292.

Newmark, W. D. (1991). Tropical forest fragmentation and the local extinction of understory birds in the Eastern Usambara Mountains, Tanzania. *Conservation Biology*, **5**, 67–78.

Patterson, B. D. and Atmar, W. (1986). Nested subsets and the structure of insular mammalian faunas and archipelagoes. *Biological Journal of the Linnean Society*, **28**, 65–82.

Peres, C. A. (2001). Synergistic effects of subsistence hunting and habitat fragmentation on Amazonian forest vertebrates. *Conservation Biology*, **15**, 1490–1505.

Peters, M. K., Likare, S., and Kraemar, M. (2008). Effects of habitat fragmentation and degradation on flocks of African ant-following birds. *Ecological Applications*, **18**, 847–858.

Prober, S. M. and Thiele, K. R. (1995). Conservation of the grassy white box woodlands: relative contributions of size and disturbance to floristic composition and diversity of remnants. *Australian Journal of Botany*, **43**, 349–366.

Radford, J. Q. and Bennett, A. F. (2007). The relative importance of landscape properties for woodland birds in agricultural environments. *Journal of Applied Ecology*, **44**, 737–747.

Radford, J. Q., Bennett, A. F., and Cheers, G. J. (2005). Landscape-level thresholds of habitat cover for woodland-dependent birds. *Biological Conservation*, **124**, 317–337.

Redpath, S. M. (1995). Habitat fragmentation and the individual: tawny owls *Strix aluco* in woodland patches. *Journal of Animal Ecology*, **64**, 652–661.

Ricketts, T. H. (2001). The matrix matters: effective isolation in fragmented landscapes. *The American Naturalist*, **158**, 87–99.

Saunders, D. A. and Hobbs, R. J., eds (1991). *Nature conservation 2: The role of corridors*. Surrey Beatty & Sons, Chipping Norton, New South Wales.

Saunders, D. A., Hobbs, R. J., and Arnold, G. W. (1993). The Kellerberrin project on fragmented landscapes: a review of current information. *Biological Conservation*, **64**, 185–192.

Saunders, D. A., Hobbs, R. J., and Margules, C. R. (1991). Biological consequences of ecosystem fragmentation: a review. *Conservation Biology*, **5**, 18–32.

Sekercioglu, C. H., Ehrlich, P. R., Daily, G. C., Aygen, D., Goehring, D., and Sandi, R. F. (2002). Disappearance of insectivorous birds from tropical forest fragments. *Proceedings of the National Academy of Sciences of the United States of America*, **99**, 263–267.

Soulé, M. E. (1986). Conservation biology and the "real world". In M. E. Soule, ed. *Conservation biology. The science of scarcity and diversity*, pp. 1–12. Sinauer Associates, Sunderland, Massachusetts.

Terborgh, J., Lopez, L., Nunez V. P., et al. (2001). Ecological meltdown in predator-free forest fragments. *Science*, **294**, 1923–1926.

Trzcinski, M. K., Fahrig, L., and Merriam, G. (1999). Independent effects of forest cover and fragmentation on the distribution of forest breeding birds. *Ecological Applications*, **9**, 586–593.

Virgos, E. (2001). Role of isolation and habitat quality in shaping species abundance: a test with badgers (*Meles meles* L.) in a gradient of forest fragmentation. *Journal of Biogeography*, **28**, 381–389.

Wilcove, D. S. (1985). Nest predation in forest tracts and the decline of migratory songbirds. *Ecology*, **66**, 1211–1214.

# CHAPTER 6

# Overexploitation

## Carlos A. Peres

In an increasingly human-dominated world, where most of us seem oblivious to the liquidation of Earth's natural resource capital (Chapters 3 and 4), exploitation of biological populations has become one of the most important threats to the persistence of global biodiversity. Many regional economies, if not entire civilizations, have been built on free-for-all extractive industries, and history is littered with examples of boom-and-bust economic cycles following the emergence, escalation and rapid collapse of unsustainable industries fuelled by raw renewable resources. The economies of many modern nation-states still depend heavily on primary extractive industries, such as fisheries and logging, and this includes countries spanning nearly the entire spectrum of per capita Gross National Product (GNP), such as Iceland and Cameroon.

Human exploitation of biological commodities involves resource extraction from the land, freshwater bodies or oceans, so that wild animals, plants or their products are used for a wide variety of purposes ranging from food to fuel, shelter, fiber, construction materials, household and garden items, pets, medicines, and cosmetics. Overexploitation occurs when the harvest rate of any given population exceeds its natural replacement rate, either through reproduction alone in closed populations or through both reproduction and immigration from other populations. Many species are relatively insensitive to harvesting, remaining abundant under relatively high rates of offtake, whereas others can be driven to local extinction by even the lightest levels of offtake. Fishing, hunting, grazing, and logging are classic consumer-resource interactions and in natural systems such interactions tend to come into equilibrium with the intrinsic productivity of a given habitat and the rates at which resources are harvested. Furthermore, efficiency of exploitation by consumers and the highly variable intrinsic resilience to exploitation by resource populations may have often evolved over long periods. Central to these differences are species traits such as the population density (or stock size), the per capita growth rate of the population, spatial diffusion from other less harvested populations, and the direction and degree to which this growth responds to harvesting through either positive or negative density dependence. For example, many long-lived and slow-growing organisms are particularly vulnerable to the additive mortality resulting from even the lightest offtake, especially if these traits are combined with low dispersal rates that can inhibit population diffusion from adjacent unharvested source areas, should these be available. These species are often threatened by overhunting in many terrestrial ecosystems, unsustainable logging in tropical forest regions, cactus "rustling" in deserts, overfishing in marine and freshwater ecosystems, or many other forms of unsustainable extraction. For example, overhunting is the most serious threat to large vertebrates in tropical forests (Cunningham et al. 2009), and overexploitation, accidental mortality and persecution caused by humans threatens approximately one-fifth (19%) of all tropical forest vertebrate species for which the cause of decline has been documented [Figure 6.1; IUCN (International Union for Conservation of Nature) 2007].

Overexploitation is the most important cause of freshwater turtle extinctions (IUCN 2007) and the third-most important for freshwater fish extinctions, behind the effects of habitat loss and introduced species (Harrison and Stiassny 1999). Thus, while population declines driven by habitat

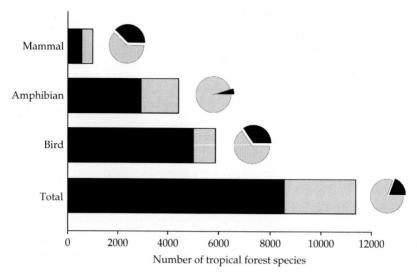

**Figure 6.1** Importance of threats to tropical forest terrestrial vertebrate species other than reptiles, which have not yet been assessed. Horizontal bars indicate the total number of species occurring in tropical forests; dark grey bars represent the fraction of those species classified as vulnerable, endangered, critically endangered or extinct in the wild according to the IUCN (2007) Red List of Threatened Species (www.iucnredlist.org). Dark slices in pie charts indicate the proportion of species for which harvesting, accidental mortality or persecution by humans is the primary cause of population declines.

loss and degradation quite rightly receive a great deal of attention from conservation biologists (MEA 2006), we must also contend with the specter of the 'empty' or 'half-empty' forests, savannahs, wetlands, rivers, and seas, even if the physical habitat structure of a given ecosystem remains otherwise unaltered by other anthropogenic processes that degrade habitat quality (see Chapter 4). Overexploitation also threatens frogs: with Indonesia the main exporter of frog legs for markets in France and the US (Warkentin et al. 2009). Up to one billion wild frogs are estimated to be harvested every year for human consumption (Warkentin et al. 2009).

I begin this chapter with a consideration of why people exploit natural populations, including the historical impacts of exploitation on wild plants and animals. This is followed by a review of effects of exploitation in terrestrial and aquatic biomes. Throughout the chapter, I focus on tropical forests and marine ecosystems because many plant and animal species in these realms have succumbed to some of the most severe and least understood overexploitation-related threats to population viability of contemporary times. I then explore impacts of exploitation on both target and non-target species, as well as cascading effects on the ecosystem. This leads to a reflection at the end of this chapter of resource management considerations in the real-world, and the clashes of culture between those concerned with either the theoretical underpinnings or effective policy solutions addressing the predicament of species imperiled by overexploitation.

## 6.1 A brief history of exploitation

Our rapacious appetite for both renewable and non-renewable resources has grown exponentially from our humble beginnings—when early humans exerted an ecological footprint no larger than that of other large omnivorous mammals—to currently one of the main driving forces in reorganizing the structure of many ecosystems. Humans have subsisted on wild plants and animals since the earliest primordial times, and most contemporary aboriginal societies remain primarily extractive in their daily quest for food, medicines, fiber and other biotic sources of raw materials to produce a wide range of utilitarian and ornamental artifacts. Modern hunter-gatherers and semi-subsistence farmers in tropical ecosystems, at varying stages of transition to an agricultural economy, still exploit a large number of plant and animal populations.

By definition, exploited species extant today have been able to co-exist with some background level of exploitation. However, paleontological evidence suggests that prehistoric peoples have been driving prey populations to extinction long before the emergence of recorded history. The late Paleolithic archaeology of big-game hunters in several parts of the world shows the sequential collapse of their majestic lifestyle. Flint spearheads manufactured by western European Cro-Magnons became gradually smaller as they shifted down to ever smaller kills, ranging in size from mammoths to rabbits (Martin 1984). Human colonization into previously people-free islands and continents has often coincided with a rapid wave of extinction events resulting from the sudden arrival of novel consumers. Mass extinction events of large-bodied vertebrates in Europe, parts of Asia, North and South America, Madagascar, and several archipelagos have all been attributed to post-Pleistocene human overkill (Martin and Wright 1967; Steadman 1995; McKinney 1997; Alroy 2001). These are relatively well corroborated in the (sub)fossil record but many more obscure target species extirpated by archaic hunters will remain undetected.

In more recent times, exploitation-induced extinction events have also been common as European settlers wielding superior technology greatly expanded their territorial frontiers and introduced market and sport hunting. One example is the decimation of the vast North American buffalo (bison; *Bison bison*) herds. In the 1850s, tens of millions of these ungulates roamed the Great Plains in herds exceeding those ever known for any other megaherbivore, but by the century's close, the bison was all but extinct. Another example is the extirpation of monodominant stands of Pau-Brasil legume trees (*Caesalpinia echinata*, Leguminosae-Mimosoidae) from eastern Brazil, a source of red dye and hardwood that gave Brazil its name. These were once extremely abundant and formed dense clusters along 3000 km of coastal Atlantic forest. This species sustained the first trade cycle between the new Portuguese colony and European markets and was relentlessly exploited from 1500 to 1875 when it finally became economically extinct (Dean 1996). Today, specimens of Pau-Brasil trees are largely confined to herbaria, arboreta and a few private collections. The aftershock of modern human arrival is still being felt in many previously inaccessible tropical forest frontiers, such as those in parts of Amazonia, where greater numbers of hunters wielding firearms are emptying vast areas of its harvest-sensitive megafauna (Peres and Lake 2003).

In many modern societies, the exploitative value of wildlife populations for either subsistence or commercial purposes has been gradually replaced by recreational values including both consumptive and non-consumptive uses. In 1990, over 20 million hunters in the United States spent over half a billion days afield in pursuit of wild game, and hunting licenses finance vast conservation areas in North America. In 2006, ~87.5 million US residents spent ~US$122.3 billion in wildlife-related recreational activities, including ~US$76.6 billion spent on fishing and/or hunting by 33.9 million people (US Census Bureau 2006). Some 10% of this total is spent hunting white-tailed deer alone (Conover 1997). Consumptive uses of wildlife habitat are therefore instrumental in either financing or justifying much of the conservation acreage available in the 21$^{st}$ century from game reserves in Africa, Australia and North America to extractive reserves in Amazonia, to the reindeer rangelands of Scandinavia and the saiga steppes of Mongolia.

Strong cultural or social factors regulating resource choice often affect which species are taken. For example, while people prefer to hunt large-bodied mammals in tropical forests, feeding taboos and restrictions can switch "on or off" depending on levels of game depletion (Ross 1978) as predicted by foraging theory. This is consistent with the process of de-tabooing species that were once tabooed, as the case of brocket deer among the Siona-Secoya (Hames and Vickers 1982). However, several studies suggest that cultural factors breakdown and play a lesser role when large-bodied game species become scarce, thereby forcing discriminate harvesters to become less selective (Jerozolimski and Peres 2003).

## 6.2 Overexploitation in tropical forests

### 6.2.1 Timber extraction

Tropical deforestation is driven primarily by frontier expansion of subsistence agriculture and large development programs involving resettlement, agriculture, and infrastructure (Chapter 4). However, animal and plant population declines are typically pre-empted by hunting and logging activity well before the *coup de grâce* of deforestation is delivered. It is estimated that between 5 and 7 million hectares of tropical forests are logged annually, approximately 68-79% of the area that was completely deforested each year between 1990 and 2005 [FAO (Food and Agriculture Organization of the United Nations) 2007]. Tropical forests account for ~25% of the global industrial wood production worth US$400 billion or ~2% of the global gross domestic product [WCFSD (World Commission on Forests and Sustainable Development) 1998]. Much of this logging activity opens up new frontiers to wildlife and non-timber resource exploitation, and catalyses the transition into a landscape dominated by slash-and-burn and large-scale agriculture.

Few studies have examined the impacts of selective logging on commercially valuable timber species and comparisons among studies are limited because they often fail to employ comparable methods that are adequately reported. The best case studies come from the most valuable timber species that have already declined in much of their natural ranges. For instance, the highly selective, but low intensity logging of broadleaf mahogany (*Swietenia macrophylla*), the most valuable widely traded Neotropical timber tree, is driven by the extraordinarily high prices in international markets, which makes it lucrative for loggers to open-up even remote wilderness areas at high transportation costs. Mechanized extraction of mahogany and other prime timber species impacts the forest by creating canopy gaps and imparting much collateral damage due to logging roads and skid trails (Grogan *et al.* 2008). Mahogany and other high-value tropical timber species worldwide share several traits that predispose them to commercial extirpation: excellent pliable wood of exceptional beauty; natural distributions in forests experiencing rapid conversion rates; low-density populations (often <1 tree/ha); and life histories generally characterized as non-pioneer late secondary, with fast growth rates, abiotic seed dispersal, and low-density seedlings requiring canopy disturbance for optimal seedling regeneration in the understory (Swaine and Whitmore 1988; Sodhi *et al.* 2008).

One of the major obstacles to implementing a sustainable forestry sector in tropical countries is the lack of financial incentives for producers to limit offtakes to sustainable levels and invest in regeneration. Economic logic often dictates that trees should be felled whenever their rate of volume increment drops below the prevailing interest rate (Pearce 1990). Postponing harvest beyond this point would incur an opportunity cost because profits from logging could be invested at a higher rate elsewhere. This partly explains why many slow-growing timber species from tropical forests and savannahs are harvested unsustainably (e.g. East African Blackwood (*Dalbergia melanoxylon*) in the Miombo woodlands of Tanzania; Ball 2004). This is particularly the case where land tenure systems are unstable, and where there are no disincentives against 'hit-and-run' operations that mine the resource capital at one site and move on to undepleted areas elsewhere. This is clearly shown in a mahogany study in Bolivia where the smallest trees felled are ~40 cm in diameter, well below the legal minimum size (Gullison 1998). At this size, trees are increasing in volume at about 4% per year, whereas real mahogany price increases have averaged at only 1%, so that a 40-cm mahogany tree increases in value at about 5% annually, slowing down as the tree becomes larger. In contrast, real interest rates in Bolivia and other tropical countries are often >10%, creating a strong economic incentive to liquidate all trees of any value regardless of resource ownership.

## 6.2.2 Tropical forest vertebrates

Humans have been hunting wildlife in tropical forests for over 100 000 years, but the extent of consumption has greatly increased over the last few decades. Tropical forest species are hunted for local consumption or sales in distant markets as food, trophies, medicines and pets. Exploitation of wild meat by forest dwellers has increased due to changes in hunting technology, scarcity of alternative protein, larger numbers of consumers, and greater access infrastructure. Recent estimates of the annual wild meat harvest are 23 500 tons in Sarawak (Bennett 2002), up to 164 692 tons in the Brazilian Amazon (Peres 2000), and up to 3.4 million tons in Central Africa (Fa and Peres 2001). Hunting rates are already unsustainably high across vast tracts of tropical forests, averaging sixfold the maximum sustainable harvest in Central Africa (Fa et al. 2001). Consumption is both by rural and urban communities, who are often at the end of long supply chains that extend into many remote areas (Milner-Gulland et al. 2003). The rapid acceleration in tropical forest defaunation due to unsustainable hunting initially occurred in Asia (Corlett 2007), is now sweeping through Africa, and is likely to move into the remotest parts of the neotropics (Peres and Lake 2003), reflecting human demographics in different continents.

Hunting for either subsistence or commerce can profoundly affect the structure of tropical forest vertebrate assemblages, as revealed by both village-based kill-profiles (Jerozolimski and Peres 2003; Fa et al. 2005) and wildlife surveys in hunted and unhunted forests. This can be seen in the residual game abundance at forest sites subjected to varying degrees of hunting pressure, where overhunting often results in faunal biomass collapses, mainly through declines and local extinctions of large-bodied species (Bodmer 1995; Peres 2000). Peres and Palacios (2007) provide the first systematic estimates of the impact of hunting on the abundances of a comprehensive set of 30 reptile, bird, and mammal species across 101 forest sites scattered widely throughout the Amazon Basin and Guianan Shield. Considering the 12 most harvest-sensitive species, mean aggregate population biomass was reduced almost eleven-fold from 979.8 kg/km$^2$ in unhunted sites to only 89.2 kg/km$^2$ in heavily hunted sites (see Figure 6.2). In Kilum Ijim, Cameroon, most large mammals, including elephants, buffalo, bushbuck, chimpanzees, leopards, and lions, have been lost as a result of hunting (Maisels et al. 2001). In Vietnam, 12 large vertebrate species have become virtually extinct over the last five decades primarily due to hunting (Bennett and Rao 2002). Pangolins and several other forest vertebrate species are facing regional-scale extinction throughout their range across southern Asia [Corlett 2007, TRAFFIC (The Wildlife Trade Monitoring Network) 2008], largely as a result of trade, and over half of all Asian freshwater turtle species are considered Endangered due to over-harvesting (IUCN 2007).

In sum, game harvest studies throughout the tropics have shown that most unregulated, commercial hunting for wild meat is unsustainable (Robinson and Bennett 2000; Nasi et al. 2008), and that even subsistence hunting driven by local demand can severely threaten many medium to large-bodied vertebrate populations, with potentially far-reaching consequences to other species. However, persistent harvesting of multi-species prey assemblages can often lead to post-depletion equilibrium conditions in which slow-breeding, vulnerable taxa are eliminated and gradually replaced by fast-breeding robust taxa that are resilient to typical offtakes. For example, hunting in West African forests could now be defined as sustainable from the viewpoint of urban bushmeat markets in which primarily rodents and small antelopes are currently traded, following a series of historical extinctions of vulnerable prey such as primates and large ungulates (Cowlishaw et al. 2005).

## 6.2.3 Non-timber forest products

Non-timber forest products (NTFPs) are biological resources other than timber which are extracted from either natural or managed forests (Peters 1994). Examples of exploited plant products include fruits, nuts, oil seeds, latex, resins, gums, medicinal plants, spices, dyes, ornamental plants, and raw materials such as firewood,

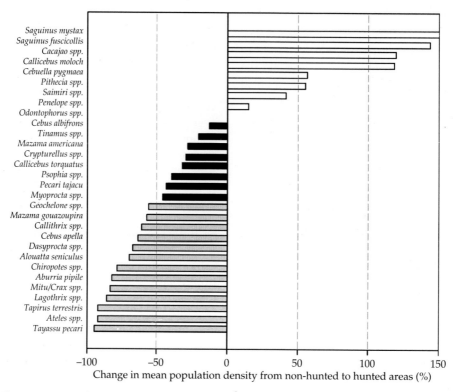

**Figure 6.2** Changes in mean vertebrate population density (individuals/km$^2$) between non-hunted and hunted neotropical forest sites (n = 101), including 30 mammal, bird, and reptile species. Forest sites retained in the analysis had been exposed to different levels of hunting pressure but otherwise were of comparable productivity and habitat structure. Species exhibiting higher density in hunted sites (open bars) are either small-bodied or ignored by hunters; species exhibiting the most severe population declines (shaded bars) were at least halved in abundance or driven to local extinction in hunted sites (data from Peres and Palacios 2007).

*Desmoncus* climbing palms, bamboo and rattan. The socio-economic importance of NTFP harvest to indigenous peoples cannot be underestimated. Many ethnobotanical studies have catalogued the wide variety of useful plants (or plant parts) harvested by different aboriginal groups throughout the tropics. For example, the Waimiri-Atroari Indians of central Amazonia make use of 79% of the tree species occurring in a single 1 ha terra firme forest plot (Milliken *et al.* 1992), and 1748 of the ~8000 angiosperm species in the Himalayan region spanning eight Asian countries are used medicinally and many more for other purposes (Samant *et al.* 1998).

Exploitation of NTFPs often involves partial or entire removal of individuals from the population, but the extraction method and whether vital parts are removed usually determine the mortality level in the exploited population. Traditional NTFP extractive practices are often hailed as desirable, low-impact economic activities in tropical forests compared to alternative forms of land use involving structural disturbance such as selective logging and shifting agriculture (Peters *et al.* 1989). As such, NTFP exploitation is usually assumed to be sustainable and a promising compromise between biodiversity conservation and economic development under varying degrees of market integration. The implicit assumption is that traditional methods of NTFP exploitation have little or no impact on forest ecosystems and tend to be sustainable because they have been practiced over many generations. However, virtually any form of NTFP exploitation in tropical forests has an ecological impact. The spatial extent and magnitude of this impact depends on the accessibility of the resource stock, the floristic composition of the forest, the nature and intensity of harvesting, and the particular species or plant part under exploitation.

Yet few studies have quantitatively assessed the demographic viability of plant populations sourcing NTFPs. One exception are Brazil nuts (*Bertholletia excelsa*, Lecythidaceae) which comprise the most important wild seed extractive industry supporting millions of Amazonian forest dwellers for either subsistence or income. This wild seed crop is firmly established in export markets, has a history of ~200 years of commercial exploitation, and comprises one of the most valuable non-timber extractive industries in tropical forests anywhere. Yet the persistent collection of *B. excelsa* seeds has severely undermined the patterns of seedling recruitment of Brazil nut trees. This has drastically affected the age structure of many natural populations to the point where persistently overexploited stands have succumbed to a process of senescence and demographic collapse, threatening this cornerstone of the Amazonian extractive economy (Peres *et al.* 2003).

A boom in the use of homeopathic remedies sustained by over-collecting therapeutic and aromatic plants is threatening at least 150 species of European wild flowers and plants and driving many populations to extinction (Traffic 1998). Commercial exploitation of the Pau-Rosa or rosewood tree (*Aniba rosaeodora*, Lauraceae), which contains linalol, a key ingredient in luxury perfumes, involves a one-off destructive harvesting technique that almost invariably kills the tree. This species has consequently been extirpated from virtually its entire range in Brazilian Amazonia (Mitja and Lescure 2000). Channel 5® and other perfumes made with Pau-Rosa fragrance gained wide market demand decades ago, but the number of processing plants in Brazil fell from 103 in 1966 to fewer than 20 in 1986, due to the dwindling resource base. Yet French perfume connoisseurs have been reluctant to accept replacing the natural Pau-Rosa fragrance with synthetic substitutes, and the last remaining populations of Pau-Rosa remain threatened. The same could be argued for a number of NTFPs for which the harvest by destructive practices involves a lethal injury to whole reproductive individuals. What then is the impact of NTFP extraction on the dynamics of natural populations? How does the impact vary with the life history of plants and animals harvested? Are current extraction rates truly sustainable? These are key questions in terms of the demographic sustainability of different NTFP offtakes, which will ultimately depend on the ability of the resource population to recruit new seedlings either continuously or in sporadic pulses while being subjected to a repeated history of exploitation.

Unguarded enthusiasm for the role of NTFP exploitation in rural development partly stems from unrealistic economic studies reporting high market values. For example, Peters *et al.* (1989) reported that the net-value of fruit and latex extraction in the Peruvian Amazon was US$6330/ha. This is in sharp contrast with a Mesoamerican study that quantified the local value of foods, construction materials, and medicines extracted from the forest by 32 indigenous Indian households (Godoy *et al.* 2000). The combined value of consumption and sale of forest goods ranged from US$18 to US$24 ha$^{-1}$ yr$^{-1}$, at the lower end of previous estimates (US$49 - US$1 089 ha$^{-1}$ yr$^{-1}$). NTFP extraction thus cannot be seen as a panacea for rural development and in many studies the potential value of NTFPs is exaggerated by unrealistic assumptions of high discount rates, unlimited market demands, availability of transportation facilities and absence of product substitution.

## 6.3 Overexploitation in aquatic ecosystems

Marine biodiversity loss, largely through overfishing, is increasingly impairing the capacity of the world's oceans to provide food, maintain water quality, and recover from perturbations (Worm *et al.* 2006). Yet marine fisheries provide employment and income for 0.2 billion people around the world, and fishing is the mainstay of the economy of many coastal regions; 41 million people worked as fishers or fish farmers in 2004, operating 1.3 million decked vessels and 2.7 million open boats (FAO 2007). An estimated 14 million metric tons of fuel was consumed by the fish-catching sector at a cost equivalent to US$22 billion, or ~25% of the total revenue of the sector. In 2004, reported catches from marine and inland capture fisheries were 85.8 million and 9.2 million tons, respectively, which was worth US$84.9 billion at first sale. Freshwater catches taken every year for food have declined recently but on average 500 000 tons are taken from the Mekong river in South-East Asia; 210 000 tons are taken from the Zaire river in Africa; and 210 000 tons of fish are taken from the Amazon river in South America. Seafood consumption is still high and rising in the First World and has doubled in China within the last decade. Fish contributes to, or exceeds 50% of the total animal protein consumption in many countries and regions, such as Bangladesh, Cambodia, Congo, Indonesia, Japan or the Brazilian Amazon. Overall, fish provides more than 2.8 billion people with ~20% or more of their average per capita intake of animal protein. The oscillation of good and bad years in marine fisheries can also modulate the protein demand from terrestrial wildlife populations (Brashares et al. 2004). The share of fish in total world animal protein supply amounted to 16% in 2001 (FAO 2004). These 'official' landing statistics tend to severely underestimate catches and total values due to the enormous unrecorded contribution of subsistence fisheries consumed locally.

Although the world's oceans are vast (see Box 4.3), most seascapes are relatively low-productivity, and 80% of the global catch comes from only ~20% of the area. Approximately 68% of the world's catch comes from the Pacific and northeast Atlantic. At current harvest rates, most of the economically important marine fisheries worldwide have either collapsed or are expected to collapse. Current impacts of overexploitation and its consequences are no longer locally nested, since 52% of marine stocks monitored by the FAO in 2005 were fully exploited at their maximum sustainable level and 24% were overexploited or depleted, such that their current biomass is much lower than the level that would maximize their sustained yield (FAO 2007). The remaining one-

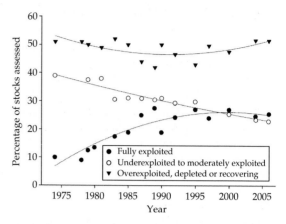

**Figure 6.3** Global trends in the status of world marine fish stocks monitored by FAO from 1974 to 2006 (data from FAO 2007).

quarter of the stocks were either underexploited or moderately exploited and could perhaps produce more (Figure 6.3).

The Brazilian sardine (*Sardinella brasiliensis*) is a classic case of an overexploited marine fishery. In the 1970s hey-day of this industry, 200 000 tons were captured in southeast Brazil alone every year, but landings suddenly plummeted to <20 000 tons by 2001. Despite new fishing regulations introduced following its collapse, it is unclear whether southern Atlantic sardine stocks have shown any sign of recovery. With the possible exception of herring and related species that mature early in life and are fished with highly selective equipment, many gadids (e.g. cod, haddock) and other non-clupeids (e.g. flatfishes) have experienced little, if any, recovery in as much as 15 years after 45–99% reductions in reproductive biomass (Hutchings 2000). Worse still, an analysis of 147 populations of 39 wild fish species concluded that historically overexploited species, such as North Sea herring, became more prone to extreme year-on-year variation in numbers, rendering them vulnerable to economic or demographic extinction (Minto et al. 2008).

Marine fisheries are an underperforming global asset—yields could be much greater if they were properly managed. The difference between the potential and actual net economic benefits from marine fisheries is in the order of US$50 billion per year—equivalent to over half the value of the global seafood trade (World Bank 2008). The cumulative economic loss to the global economy over the last three decades is estimated to be approximately US$2 trillion, and in many countries fishing operations are buoyed up by subsidies, so that the global fishery economy to the point of landing is already in deficit.

Commercial fishing activities disproportionately threaten large-bodied marine and freshwater species (Olden *et al.* 2007). This results in fishermen fishing down the food chain, targeting ever-smaller pelagic fish as they can no longer capture top predatory fish. This is symptomatic of the now widely known process of 'fishing down marine food webs' (see Box 6.1). Such sequential size-graded exploitation systems also take place in multi-species assemblages hunted in tropical forests (Jerozolimski and Peres et al 2003). In the seas, overexploitation threatens the persistence of ecologically significant populations of many large marine vertebrates, including sharks, tunas and sea turtles. Regional scale populations of large sharks worldwide have declined by 90% or more, and rapid declines of >75% of the coastal and oceanic Northwest Atlantic populations of scalloped hammerhead, white, and thresher sharks have occurred in the past 15 years (Baum *et al.* 2003; Myers and Worm 2003; Myers *et al.* 2007). Much of this activity is profligate and often driven by the surging global demand for shark fins. For example, in 1997 line-fishermen captured 186 000 sharks in southern Brazil alone, of which 83% were killed and discarded in open waters following the removal of the most lucrative body parts (C.M. Vooren, pers. comm.). Of the large-bodied coastal species affected by this trade, several have virtually disappeared from shallow waters (e.g. greynurse sharks, *Carcharias taurus*). Official figures show that 131 tons of shark fins, corresponding to US $2.4 million, were exported from Brazil to Asia in 2007.

Finally, we know rather little about ongoing extinction processes caused by harvesting. For example, from a compilation of 133 local, regional and global extinctions of marine fish populations, Dulvy *et al.* (2003) uncovered that exploitation was the main cause of extinctions (55% of all populations), but these were only reported after a median 53-year lag following their real-time disappearance. Some 80% of all extinctions were only discovered through historical comparisons; e.g. the near-extinction of large skates on both sides of the Atlantic was only brought to the world's attention several decades after the declines have occurred.

## 6.4 Cascading effects of overexploitation on ecosystems

All extractive systems in which the overharvested resource is one or more biological populations, can lead to pervasive trophic cascades and other unintended ecosystem-level consequences to non-target species. Most hunting, fishing, and collecting activities affect not only the primary target species, but also species that are taken accidentally or opportunistically. Furthermore, exploitation often causes physical damage to the environment, and has ramifications for other species through cascading interactions and changes in food webs.

In addition, overexploitation may severely erode the ecological role of resource populations in natural communities. In other words, overexploited populations need not be entirely extirpated before they become ecologically extinct. In communities that are "half-empty" (Redford and Feinsinger 2001), populations may be reduced to sufficiently low numbers so that, although still present in the community, they no longer interact significantly with other species (Estes *et al.* 1989). Communities with reduced levels of species interactions may become pale shadows of their former selves. Although difficult to measure, severe declines in large vertebrate populations may result in multi-trophic cascades that may profoundly alter the structure of marine ecosystems such as kelp forests, coral reefs and estuaries (Jackson *et al.* 2001), and analogous processes may occur in many terrestrial ecosystems. Plant reproduction in endemic island floras can be severely affected by population declines in flying foxes (pteropodid fruit bats) that serve as strong mutualists as pollinators and seed dispersers (Cox *et al.* 1991).

In some Pacific archipelagos, several species may become functionally extinct, ceasing to effectively disperse large seeds long before becoming rare (McConkey and Drake 2006). A key agenda for future research will involve understanding the non-linearities between functional responses to the numeric abundance of strong interactors reduced by exploitation pressure and the quality of ecological services that depleted populations can perform. For example, what is the critical density of any given exploited population below which it can no longer fulfill its community-wide ecological role?

In this section I concentrate on poorly known interaction cascades in tropical forest and marine environments, and discuss a few examples of how apparently innocuous extractive activities targeted to one or a few species can drastically affect the structure and functioning of these terrestrial and aquatic ecosystems.

### 6.4.1 Tropical forest disturbance

Timber extraction in tropical forests is widely variable in terms of species selectivity, but even highly selective logging can trigger major ecological changes in the understory light environment, forest microclimate, and dynamics of plant regeneration. Even reduced-impact logging (RIL) operations can generate enough forest disturbance, through elevated canopy gap fracture, to greatly augment forest understory desiccation, dry fuel loads, and fuel continuity, thereby breaching the forest flammability threshold in seasonally-dry forests (Holdsworth and Uhl 1997; Nepstad et al. 1999; Chapter 9). During severe dry seasons, often aggravated by increasingly frequent continental-scale climatic events, extensive ground fires initiated by either natural or anthropogenic sources of ignition can result in a dramatically reduced biomass and biodiversity value of previously unburnt tropical forests (Barlow and Peres 2004, 2008). Despite these undesirable effects, large-scale commercial logging that is unsustainable at either the population or ecosystem level continues unchecked in many tropical forest frontiers (Curran et al. 2004; Asner et al. 2005). Yet surface fires aggravated by logging disturbance represent one of the most powerful mechanisms of functional and compositional impoverishment of remaining areas of tropical forests (Cochrane 2003), and arguably the most important climate-mediated phase shift in the structure of tropical ecosystems (see also Chapters 8 and 9).

### 6.4.2 Hunting and plant community dynamics

Although the direct impacts of defaunation driven by overhunting can be predicted to some degree, higher-order indirect effects on community structure remain poorly understood since Redford's (1992) seminal paper and may have profound, long-term consequences for the persistence of other taxa, and the structure, productivity and resilience of terrestrial ecosystems (Cunningham et al. 2009). Severe population declines or extirpation of the world's megafauna may result in dramatic changes to ecosystems, some of which have already been empirically demonstrated, while others have yet to be documented or remain inexact.

Large vertebrates often have a profound impact on food webs and community dynamics through mobile-linkage mutualisms, seed predation, and seedling and sapling herbivory. Plant communities in tropical forests depleted of their megafauna may experience pollination bottlenecks, reduced seed dispersal, monodominance of seedling cohorts, altered patterns of seedling recruitment, other shifts in the relative abundance of species, and various forms of functional compensation (Cordeiro and Howe 2003; Peres and Roosmalen 2003; Wang et al. 2007; Terborgh et al. 2008; Chapter 3). On the other hand, the net effects of large mammal defaunation depends on how the balance of interactions are affected by population declines in both mutualists (e.g. high-quality seed dispersers) and herbivores (e.g. seed predators) (Wright 2003). For example, significant changes in population densities in wild pigs (*Suidae*) and several other ungulates and rodents, which are active seed predators, may have a major effect on seed and seedling survival and forest regeneration (Curran and Webb 2000).

Tropical forest floras are most dependent on large-vertebrate dispersers, with as many as

97% of all tree, woody liana and epiphyte species bearing fruits and seeds that are morphologically adapted to endozoochorous (passing through the gut of an animal) dispersal (Peres and Roosmalen 2003). Successful seedling recruitment in many flowering plants depends on seed dispersal services provided by large-bodied frugivores (Howe and Smallwood 1982), while virtually all seeds falling underneath the parent's canopy succumb to density-dependent mortality—caused by fungal attack, other pathogens, and vertebrate and invertebrate seed predators (see review in Carson *et al.* 2008).

A growing number of phytodemographic studies have examined the effects of large-vertebrate removal. Studies examining seedling recruitment under different levels of hunting pressure (or disperser abundance) reveal very different outcomes. At the community level, seedling density in overhunted forests can be indistinguishable, greater, or less than that in the undisturbed forests (Dirzo and Miranda 1991; Chapman and Onderdonk 1998; Wright *et al.* 2000), but the consequences of increased hunting pressure to plant regeneration depends on the patterns of depletion across different prey species. In persistently hunted Amazonian forests, where large-bodied primates are driven to local extinction or severely reduced in numbers (Peres and Palacios 2007), the probability of effective dispersal of large-seeded endozoochorous plants can decline by over 60% compared to non-hunted forests (Peres and Roosmalen 2003). Consequently, plant species with seeds dispersed by vulnerable game species are less abundant where hunters are active, whereas species with seeds dispersed by abiotic means or by small frugivores ignored by hunters are more abundant in the seedling and sapling layers (Nuñez-Iturri and Howe 2007; Wright *et al.* 2007; Terborgh *et al.* 2008). However, the importance of dispersal-limitation in the absence of large frugivores depends on the degree to which their seed dispersal services are redundant to any given plant species (Peres and Roosmalen 2003). Furthermore, local extinction events in large-bodied species are rarely compensated by smaller species in terms of their population density, biomass, diet, and seed handling outcomes (Peres and Dolman 2000).

Large vertebrates targeted by hunters often have a disproportionate impact on community structure and operate as "ecosystem engineers" (Jones *et al.* 1994; Wright and Jones 2006), either performing a key landscaping role in terms of structural habitat disturbance, or as mega-herbivores that maintain the structure and relative abundance of plant communities. For example, elephants exert a major role in modifying vegetation structure and composition as herbivores, seed dispersers, and agents of mortality for many small trees (Cristoffer and Peres 2003). Two similar forests with or without elephants show different succession and regeneration pathways, as shown by long-term studies in Uganda (Sheil and Salim 2004). Overharvesting of several other species holding a keystone landscaping role can lead to pervasive changes in the structure and function of ecosystems. For example, the decimation of North American beaver populations by pelt hunters following the arrival of Europeans profoundly altered the hydrology, channel geomorphology, biogeochemical pathways and community productivity of riparian habitats (Naiman *et al.* 1986).

Mammal overhunting triggers at least two additional potential cascades: the secondary extirpation of dependent taxa and the subsequent decline of ecological processes mediated by associated species. For instance, overhunting can severely disrupt key ecosystem processes including nutrient recycling and secondary seed dispersal exerted by relatively intact assemblages of dung beetles (Coleoptera: Scarabaeinae) and other coprophagous invertebrates that depend on large mammals for adult and larval food resources (Nichols *et al.* 2009).

### 6.4.3 Marine cascades

Apart from short-term demographic effects such as the direct depletion of target species, there is growing evidence that fishing also contributes to important genetic changes in exploited populations. If part of the phenotypic variation of target species is due to genetic differences among individuals then selective fishing will cause genetic changes in life-history traits such as ages and sizes at maturity (Law 2000). The genetic effects of fishing are increasingly seen as a long-term management issue, but this is not yet managed proactively as short-term regulations tend to merely focus on controlling mortality. However, the damage caused by overfishing extends well beyond the main target species with profound effects on: (i) low-productivity species in mixed fisheries; (ii) non-target species; (iii) food webs; and (iv) the structure of oceanic habitats.

*Low-productivity species in mixed fisheries*
Many multi-species fisheries are relatively unselective and take a wide range of species that vary in their capacity to withstand elevated mortality. This is particularly true in mixed trawl fisheries where sustainable mortality rates for a productive primary target species are often unsustainable for species that are less productive, such as skates and rays, thereby leading to widespread depletion and, in some cases, regional extinction processes. Conservation measures to protect unproductive species in mixed fisheries are always controversial since fishers targeting more productive species will rarely wish to sacrifice yield in order to spare less productive species.

*Bycatches*
Most seafood is captured by indiscriminate methods (e.g. gillnetting, trawl netting) that haul in large numbers of incidental captures (termed bycatches) of undesirable species, which numerically may correspond to 25–65% of the total catch. These non-target pelagic species can become entangled or hooked by the same fishing gear, resulting in significant bycatch mortality of many vulnerable fish, reptile, bird and mammal populations, thereby comprising a key management issue for most fishing fleets (Hall *et al.* 2000). For example, over 200 000 loggerhead (*Caretta caretta*) and 50 000 leatherback turtles (*Dermochelys coriacea*) were taken as pelagic longline bycatch in 2000, likely contributing to the 80–95% declines for Pacific loggerhead and leatherback populations over two decades (Lewison *et al.* 2004). While fishing pressure on target species relates to target abundance, fishing pressure on bycatch species is likely insensitive to bycatch abundance (Crowder and Murawski 1998), and may therefore result in "piggyback" extinctions. Bycatches have been the focus of considerable societal concern, often expressed in relation to the welfare of individual animals and the status of their populations. Public concerns over unacceptable levels of mortality of large marine vertebrates (e.g. sea turtles, seabirds, marine mammals, sharks) have therefore led to regional bans on a number of fishing methods and gears, including long drift-nets.

*Food webs*
Overfishing can create trophic cascades in marine communities that can cause significant declines in species richness, and wholesale changes in coastal food webs resulting from significant reductions in consumer populations due to overfishing (Jackson *et al.* 2001). Predators have a fundamental top-down role in the structure and function of biological communities, and many large marine predators have declined by >90% of their baseline population levels (Pauly *et al.* 1998; Myers and Worm 2003; see Box 6.1). Fishing affects

---

**Box 6.1 The state of fisheries**
**Daniel Pauly**

Industrial, or large-scale and artisanal, or small-scale marine fisheries, generate, at the onset of the 21$^{st}$ century, combined annual catches of 120–140 million tons, with an ex-vessel value of about US$100 billion. This is much higher than officially reported landings (80–90 million tons), which do not account for illegal, unreported and undocumented (IUU) catches (Pauly *et al.* 2002). IUU catches include, for example, the fish discarded by shrimp trawlers (usually 90% of their actual catch), the catch of high sea industrial fleets operating under flags of convenience, and the individually small catch by millions of artisanal fishers (including women and children) in developing countries, which turns out to be very high in the

*continues*

### Box 6.1 (Continued)

aggregate, but still goes unreported by national governments and international agencies.

This global catch, which, depending on the source, is either stagnating or slowly declining, is the culmination of the three-pronged expansion of fisheries which occurred following the Second World War: (i) an offshore/depth expansion, resulting from the depletion of shallow-water, inshore stocks (Morato et al. 2006); (ii) a geographic expansion, as the fleets of industrialized countries around the North Atlantic and in East Asia, faced with depleted stocks in their home waters, shifted their operations toward lower latitudes, and thence to the southern hemisphere (Pauly et al. 2002); and (iii) a taxonomic expansion, i.e. capturing and marketing previously spurned species of smaller fish and invertebrates to replace the diminishing supply of traditionally targeted, larger fish species (Pauly et al. 1998; see Box 6.1 Figure).

In the course of these expansions, fishing effort grew enormously, especially that of industrial fleets, which are, overall, 3–4 times larger than required. This is, among other things, a result of the US$30–34 billion they receive annually as government subsidies, which now act to keep fleets afloat that have no fish to exploit (Sumaila et al. 2008). In addition to representing a giant waste of economic resources, these overcapitalized fishing fleets have a huge, but long-neglected impact on their target species, on non-targeted species caught as by-catch, and on the marine ecosystems in which all species are embedded. Also, these fleets emit large amounts of carbon dioxide; for example trawlers nowadays often burn several tons of diesel fuel for every ton of fish landed (and of which 80% is water), and their efficiency declines over time because of declining fish stocks (Tyedmers et al. 2005).

Besides threatening the food security of numerous developing countries, for example in West Africa, these trends endanger marine biodiversity, and especially the continued existence of the large, long-lived species that have sustained fisheries for centuries (Worm et al. 2006).

The good news is that we know in principle how to avoid the overcapitalization of fisheries and the collapse of their underlying stocks. This would involve, besides an abolition of capacity-enhancing subsidies (e.g. tax-free fuel, loan guarantees for boat purchases (Sumaila et al. 2008), the creation of networks of large marine protected areas, and the reduction of fishing effort in the remaining exploited areas, mainly through the creation of dedicated access privilege (e.g. for adjacent small scale fisher communities), such as to reduce the "race for fish".

Also, the measures that will have to be taken to mitigate climate change offer the prospect of a reduction of global fleet capacity (via a reduction of their greenhouse gas emissions). This may lead to more attention being paid to small-scale fisheries, so far neglected, but whose adjacency to the resources they exploit, and use of fuel-efficient, mostly passive gear, offers a real prospect for sustainability.

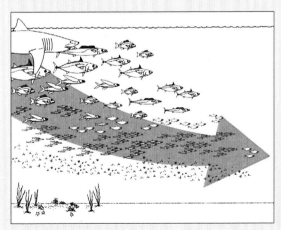

**Box 6.1 Figure** Schematic representation of the process, now widely known as 'fishing down marine food webs', by which fisheries first target the large fish, then, as these disappear, move on to smaller species of fish and invertebrates, lower in the food web. In the process, the functioning of marine ecosystems is profoundly disrupted, a process aggravated by the destruction of the bottom fauna by trawling and dredging.

### REFERENCES

Morato, T., Watson, R., Pitcher, T. J., and Pauly, D. (2006). Fishing down the deep. *Fish and Fisheries*, **7**, 24–34.

*continues*

> **Box 6.1 (Continued)**
>
> Pauly, D., Christensen, V., Dalsgaard, J., Froese, R., and Torres, F. C. Jr. (1998). Fishing down marine food webs. *Science*, **279**, 860–863.
>
> Pauly, D., Christensen, V., Guénette, S., *et al.* (2002). Towards sustainability in world fisheries. *Nature*, **418**, 689–695.
>
> Sumaila, U. R., Teh, L., Watson, R., Tyedmers, P., and Pauly, D. (2008). Fuel price increase, subsidies, overcapacity, and resource sustainability. *ICES Journal of Marine Science*, **65**, 832–840.
>
> Tyedmers, P., Watson, R., and Pauly, D. (2005). Fueling global fishing fleets. *AMBIO: a Journal of the Human Environment*, **34**, 635–638.
>
> Worm, B., Barbier, E. B., Beaumont, N., *et al.* (2006). Impacts of biodiversity loss on ocean ecosystem services. *Science*, **314**, 787–790.

predator-prey interactions in the fished community and interactions between fish and other species, including predators of conservation interest such as seabirds and mammals. For example, fisheries can compete for the prey base of seabirds and mammals. Fisheries also produce discards that can provide significant energy subsidies especially for scavenging seabirds, in some cases sustaining hyper-abundant populations. Current understanding of food web effects of overfishing is often too poor to provide consistent and reliable scientific advice.

*Habitat structure*

Overfishing is a major source of structural disturbance in marine ecosystems. The very act of fishing, particularly with mobile bottom gear, destroys substrates, degrades habitat complexity, and ultimately results in the loss of biodiversity (see Box 4.3). These structural effects are compounded by indirect effects on habitat that occur through removal of ecological or ecosystem engineers (Coleman and Williams 2002). Many fishing gears contact benthic habitats during fishing and habitats such as coral reefs are also affected by changes in food webs. The patchiness of impacts and the interactions between types of gears and habitats are critical to understanding the significance of fishing effects on habitats; different gears have different impacts on the same habitat and different habitats respond differently to the same gear. For some highly-structured habitats such as deep water corals, recovery time is so slow that only no fishing would be realistically sustainable (Roberts *et al.* 2006).

## 6.5 Managing overexploitation

This chapter has repeatedly illustrated examples of population declines induced by overexploitation even in the face of the laudable goals of implementing conservation measures in the real-world. This section will conclude with some comments about contrasts between theory and practice, and briefly explore some of the most severe problems and management solutions that can minimize the impact of harvesting on the integrity of terrestrial and marine ecosystems.

Unlike many temperate countries where regulatory protocols preventing overexploitation have been developed through a long and repeated history of trial and error based on ecological principles and hard-won field biology, population management prescriptions in the tropics are typically non-existent, unenforceable, and lack the personnel and scientific foundation on which they can be built. The concepts of game wardens, bag limits, no-take areas, hunting or fishing licenses, and duck stamps are completely unfamiliar to the vast majority of tropical subsistence hunters or fishers (see Box 6.2). Yet these resource users are typically among the poorest rungs in society and often rely heavily on wild animal populations as a critical protein component of their diet. In contrast, countries with a strong tradition in fish and wildlife management and carefully regulated harvesting policy in private and public areas, may include sophisticated legislation encompassing bag limits on the age and sex of different target species, as well as restrictions on hunting and fishing seasons and

## Box 6.2 Managing the exploitation of wildlife in tropical forests
### Douglas W. Yu

Hunting threatens the persistence of tropical wildlife, their ecological functions, such as seed dispersal, and the political will to maintain forests in the face of alternative land-use options. However, game species are important sources of protein and income for millions of forest dwellers and traders of wildlife (Peres 2000; Bulte and van Kooten 2001; Milner-Gulland et al. 2003; Bennett et al. 2007; this chapter).

Policy responses to the overexploitation of wildlife can be placed into two classes: (i) demand-side restrictions on offtake, to increase the cost of hunting, and (ii) the supply-side provisioning of substitutes, to decrease the benefit of hunting (Bulte and Damania 2005; Crookes and Milner-Gulland 2006). Restrictions on offtake vary from no-take areas, such as parks, to various partial limits, such as reducing the density of hunters via private property rights, and establishing quotas and bans on specific species, seasons, or hunting gear, like shotguns (Bennett et al. 2007). Where there are commercial markets for wildlife, restrictions can also be applied down the supply chain in the form of market fines or taxes (Clayton et al. 1997; Damania et al. 2005). Finally, some wildlife products are exported for use as medicines or decoration and can be subjected to trade bans under the aegis of the Convention on International Trade in Endangered Species (CITES) (Stiles 2004; Bulte et al. 2007; Van Kooten 2008).

Bioeconomic modeling (Ling and Milner-Gulland 2006) of a game market in Ghana has suggested that imposing large fines on the commercial sale of wild meat should be sufficient to recover wildlife populations, even in the absence of forest patrols (Damania et al. 2005). Fines reduce expected profits from sales, so hunters should shift from firearms to cheaper but less effective snares and consume more wildlife at home. The resulting loss of cash income should encourage households to reallocate labor toward other sources of cash, such as agriculture.

Offtake restrictions are, however, less useful in settings where governance is poor, such that fines are rarely expected and incursions into no-take areas go unpunished, or where subsistence hunting is the norm, such as over much of the Amazon Basin (Peres 2000). In the latter case, markets for wild meat are small or nonexistent, and human populations are widely distributed, exacerbating the already-difficult problem of monitoring hunting effort in tropical forests (Peres and Terborgh 1995; Peres and Lake 2003; Ling and Milner-Gulland 2006). Moreover, the largest classes of Amazonian protected areas are indigenous and sustainable development reserves (Nepstad et al. 2006; Peres and Nascimento 2006), within which inhabitants hunt legally.

Such considerations are part of the motivation for introducing demand-side remedies, such as alternative sources of protein. The logic is that local substitutes (e.g. fish from aquaculture) should decrease demand for wild meat and allow the now-excess labor devoted to hunting to be reallocated to competing activities, such as agriculture or leisure.

However, the nature of the substitute and the structure of the market matter greatly. If the demand-side remedy instead takes the form of increasing the opportunity cost of hunting by, for example, raising the profitability of agriculture, it is possible that total hunting effort will ultimately increase, since income is fungible and can be spent on wild meat (Damania et al. 2005). Higher consumer demand also raises market prices and can trigger shifts to more effective but more expensive hunting techniques, like guns (Bulte and Horan 2002; Damania et al. 2005). More generally, efforts to provide alternative economic activities are likely to be inefficient and amount to little more than 'conservation by distraction' (Ferraro 2001; Ferraro and Simpson 2002).

In many settings, the ultimate consumers are not the hunters, and demand-side remedies could take the form of educational programs aimed at changing consumer preferences or,

*continues*

### Box 6.2 (Continued)

alternatively, of wildlife farms (e.g. crocodilian ranches) that are meant to compete with and depress the price of wild-caught terrestrial vertebrates. The latter strategy could, however, lead to perverse outcomes if the relevant market is dominated by only a few suppliers, who have the power to maintain high prices by restricting supply to market (Wilkie et al. 2005; Bulte and Damania 2005; Damania and Bulte 2007). Then, the introduction of a farmed substitute can, in principle, induce intense price-cutting competition, which would increase consumer demand and lead to more hunting and lower wildlife stocks. Also, farmed substitutes can undermine efforts to stigmatize the consumption of wildlife products, increasing overall demand. Given these caveats, the strategy of providing substitutes for wildlife might best be focused on cases where the substitute is different from and clearly superior to the wildlife product, as is the case for *Viagra* versus aphrodisiacs derived from animal parts (von Hippel and von Hippel 2002).

Ultimately, given the large numbers of rural dwellers, the likely persistence of wildlife markets of all kinds, and the great uncertainties that remain embedded in our understanding of the ecology and economics of wildlife exploitation, any comprehensive strategy to prevent hunting from driving wildlife populations extinct must include no-take areas (Bennett et al. 2007)—the bigger the better. The success of no-take areas will in turn depend on designing appropriate enforcement measures for different contexts, from national parks to indigenous reserves and working forests to community-based management (Keane et al. 2008).

A potential approach is to use the economic theory of contracts and asymmetric information (Ferraro 2001, 2008; Damania and Hatch 2005) to design a menu of incentives and punishments that deters hunting in designated no-take areas, given that hunting is a hidden action. In the above bioeconomic model in Ghana (Damania et al. 2005), hidden hunting effort is revealed in part by sales in markets, which can be monitored, and the imposition of a punishing fine causes changes in the behavior of households that result ultimately in higher game populations.

It should also be possible to employ positive incentives in the form of payments for ecological services (Ferraro 2001; Ferraro and Simpson 2002; Ferraro and Kiss 2002). For example, in principle, the state might pay local communities in return for abundant wildlife as measured in regular censuses. In practice, however, the high stochasticity of such a monitoring mechanism, and the problem of free riders within communities, might make this mechanism unworkable. Alternatively, in the case of landscapes that still contain vast areas of high animal abundance, such as in many parks that host small human populations, a strategy that takes advantage of the fact that central-place subsistence hunters are distance limited is appropriate (Ling and Milner-Gulland 2008; Levi et al. 2009). The geographic distribution of settlements is then an easily monitored proxy for the spatial distribution of hunting effort. As a result, economic incentives to promote settlement sedentarism, which can range from direct payments to the provision of public services such as schools, would also limit the spread of hunting across a landscape.

### REFERENCES

Bennett, E., Blencowe, E., Brandon, K., et al. (2007). Hunting for consensus: Reconciling bushmeat harvest, conservation, and development policy in west and central Africa. *Conservation Biology*, 21, 884–887.

Bulte, E. H. and Damania, R. (2005). An economic assessment of wildlife farming and conservation. *Conservation Biology*, 19, 1222–1233.

Bulte, E. H. and Horan, R. D. (2002). Does human population growth increase wildlife harvesting? An economic assessment. *Journal of Wildlife Management*, 66, 574–580.

Bulte, E. H. and van Kooten, G. C. (2001). State intervention to protect endangered species: why history and bad luck matter. *Conservation Biology*, 15, 1799–1803.

Bulte, E. H., Damania, R., and Van Kooten, G. C. (2007). The effects of one-off ivory sales on elephant mortality. *Journal of Wildlife Management*, 71, 613–618.

Clayton, L., Keeling, M., and Milner-Gulland, E. J. (1997). Bringing home the bacon: a spatial model of wild pig hunting in Sulawesi, Indonesia. *Ecological Application*, 7, 642–652.

Crookes, D. J. and Milner-Gulland, E. J. (2006). Wildlife and economic policies affecting the bushmeat trade: a

*continues*

> **Box 6.2 (Continued)**
>
> framework for analysis. *South African Journal of Wildlife Research*, **36**, 159–165.
> Damania, R. and Bulte, E. H. (2007). The economics of wildlife farming and endangered species conservation. *Ecological Economics*, **62**, 461–472.
> Damania, R. and Hatch, J. (2005). Protecting Eden: markets or government? *Ecological Economics*, **53**, 339–351.
> Damania, R., Milner-Gulland, E. J., and Crookes, D.J. (2005). A bioeconomic analysis of bushmeat hunting. *Proceedings of Royal Society of London B*, **272**, 259–266.
> Ferraro, P. J. (2001). Global habitat protection: limitation of development interventions and a role for conservation performance payments. *Conservation Biology*, **15**, 990–1000.
> Ferraro, P. J. (2008). Asymmetric information and contract design for payments for environmental services. *Ecological Economics*, **65**, 810–821.
> Ferraro, P. J. and Kiss, A. (2002). Direct payments to conserve biodiversity. *Science*, **298**, 1718–1719.
> Ferraro, P. J. and Simpson, R. D. (2002). The cost-effectiveness of conservation payments. *Land Economics*, **78**, 339–353.
> Keane, A., Jones, J. P. G., Edwards-Jones, G., and Milner-Gulland, E. (2008). The sleeping policeman: understanding issues of enforcement and compliance in conservation. *Animal Conservation*, **11**, 75–82.
> Levi, T., Shepard, G. H., Jr., Ohl-Schacherer, J., Peres, C. A., and Yu, D.W. (2009). Modeling the long-term sustainability of indigenous hunting in Manu National Park, Peru: Landscape-scale management implications for Amazonia. *Journal of Applied Ecology*, **46**, 804–814.
> Ling, S. and Milner-Gulland, E. J. (2006). Assessment of the sustainability of bushmeat hunting based on dynamic bioeconomic models. *Conservation Biology*, **20**, 1294–1299.
> Ling, S. and Milner-Gulland, E. (2008). When does spatial structure matter in models of wildlife harvesting? *Journal of Applied Ecology*, **45**, 63–71.
> Milner-Gulland, E., Bennett, E. & and the SCB 2002 Annual Meeting Wild Meat Group (2003). Wild meat: the bigger picture. *Trends in Ecology and Evolution*, **18**, 351–357.
> Nepstad, D., Schwartzman, S., Bamberger, B., *et al.* (2006). Inhibition of Amazon deforestation and fire by parks and indigenous lands. *Conservation Biology*, **20**, 65–73.
> Peres, C. A. (2000). Effects of subsistence hunting on vertebrate community structure in Amazonian forests. *Conservation Biology*, **14**, 240–253.
> Peres, C. A. and Lake, I. R. (2003). Extent of nontimber resource extraction in tropical forests: accessibility to game vertebrates by hunters in the Amazon basin. *Conservation Biology*, **17**, 521–535.
> Peres, C. A. and Nascimento, H. S. (2006). Impact of game hunting by the Kayapo of south-eastern Amazonia: implications for wildlife conservation in tropical forest indigenous reserves. *Biodiversity and Conservation*, **15**, 2627–2653.
> Peres, C. A. and Terborgh, J. W. (1995). Amazonian nature reserves: an analysis of the defensibility status of existing conservation units and design criteria for the future. *Conservation Biology*, **9**, 34–46.
> Stiles, D. (2004). The ivory trade and elephant conservation. *Environmental Conservation*, **31**, 309–321.
> von Hippel, F. and von Hippel, W. (2002). Sex drugs and animal parts: will Viagra save threatened species? *Environmental Conservation*, **29**, 277–281.
> Van Kooten, G. C. (2008). Protecting the African elephant: A dynamic bioeconomic model of ivory trade. *Biological Conservation*, **141**, 2012–2022.
> Wilkie, D. S., Starkey, M., Abernethy, K. *et al.* (2005). Role of prices and wealth in consumer demand for bushmeat in Gabon, Central Africa. *Conservation Biology*, **19**, 1–7.

capture technology. Despite the economic value of wildlife (Peres 2000; Chardonnet *et al.* 2002; Table 6.1), terrestrial and aquatic wildlife in many tropical countries comprise an 'invisible' commodity and local offtakes often proceed unrestrained until the sudden perception that the resource stock is fully depleted. This is reflected in the contrast between carefully regulated and unregulated systems where large numbers of hunters may operate. For example, Minnesota hunters sustainably harvest over 700 000 wild white-tailed deer (*Odocoileus virginianus*) every year, whereas Costa Rica can hardly sustain an annual harvest of a few thousand without pushing the same cervid species, albeit in a different food environment, to local extinction (D. Janzen, pers. comm.).

An additional widespread challenge in managing any diffuse set of resources is presented when resources (or the landscape or seascape which they occupy) have no clear ownership. This is widely referred to as the 'tragedy of the commons' (Hardin 1968) in which open-access exploitation systems lead to much greater rates of exploitation than are safe for the long-term survival of the population. This is dreadful for both the resource and the consumers, because each user is capturing fewer units of the resource than they could if they had fewer competitors. Governments often respond by providing perverse subsidies that deceptively reduce costs, hence catalyzing a negative spiral leading to further overexploitation (Repetto and Gillis 1988). The capital invested in many extractive industries such as commercial fisheries and logging operations cannot be easily reinvested, so that exploiters have few options but to continue harvesting the depleted resource base. Understandably, this leads to resistance against restrictions on exploitation rates, thereby further exacerbating the problems of declining populations. In fact, exploitation can have a one-way ratchet effect, with governments propping up overexploitation when populations are low, and supporting investment in the activity when yields are high.

Laws against the international wildlife and timber trade have often failed to prevent supplies sourced from natural populations from reaching their destination, accounting for an estimated US $292.73 billion global market, most of it accounted for by native timber and wild fisheries (see Table 6.1). Global movement of animals for the pet trade alone has been estimated at ~350 million live animals, worth ~US$20 billion per year (Roe 2008; Traffic 2008). At least 4561 extant bird species are used by humans, mainly as pets and for food, including >3337 species traded internationally (Butchart 2008). Some 15 to 20 million wild-caught ornamental fish are exported alive every year through Manaus alone, a large city in the central Amazon (Andrews 1990).

Regulating illegal overharvesting of exorbitant-priced resource populations—such as elephant ivory, rhino horn, tiger bone or mahogany trees—presents an additional, and often insurmountable, challenge because the rewards accrued to violators often easily outweigh the enforceable penalties or the risks of being caught.

**Table 6.1** Total estimated value of the legal wildlife trade worldwide in 2005 (data from Roe 2008).

| Commodity | Estimated value (US$ million) |
|---|---|
| **Live animals** | |
| Primates | 94 |
| Cage birds | 47 |
| Birds of prey | 6 |
| Reptiles (incl. snakes and turtles) | 38 |
| Ornamental fish | 319 |
| **Animal products for clothing or ornaments** | |
| Mammal furs and fur products | 5000 |
| Reptile skins | 338 |
| Ornamental corals and shells | 112 |
| Natural pearls | 80 |
| **Animal products for food (excl. fish)** | |
| Game meat | 773 |
| Frog legs | 50 |
| Edible snails | 75 |
| **Plant products** | |
| Medicinal plants | 1300 |
| Ornamental plants | 13 000 |
| Fisheries food products (excl. aquaculture) | 81 500 |
| Timber | 190 000 |
| Total | $292.73 bill |

For example, giant bluefin tuna (*Thunnus thynnus*), which are captured illegally by commercial and recreational fishers assisted by high-tech gear, may be the most valuable animal on the planet, with a single 444-pound bluefin tuna sold wholesale in Japan a few years ago for US $173 600! In fact, a ban on harvesting of some highly valuable species has merely spawned a thriving illegal trade. After trade in all five species of rhino was banned, the black rhino became extinct in at least 18 African countries [CITES (Convention on International Trade in Endangered Species) 2008]. The long-term success of often controversial bans on wildlife trade depends on three factors. First, prohibition on trade must be accompanied by a reduction in demand for the banned products. Trade in cat and seal skins was crushed largely because ethical consumer campaigns destroyed demand at the same time as trade bans cut the legal supply.

Second, bans may curb legal trade, which often provides an economic incentive to maintain wildlife or their habitat. Some would therefore argue they undermine conservation efforts and may even create incentives to eliminate them. The American bison was doomed partly because its rangelands became more valuable for rearing cattle (Anderson and Hill 2004). Third, international trade agreements must be supported by governments and citizens in habitat-countries, rather than only conscious consumers in wealthy nations. But even well-meaning management prescriptions involving wildlife trade can be completely misguided bringing once highly abundant target species to the brink of extinction. The 97% decline of saiga antelopes (from >1 million to <30 000) in the steppes of Russia and Kazakhstan over a 10-year period has been partly attributed to conservationists actively promoting exports of saiga (*Saiga tatarica*) horn to the Chinese traditional medicine market as a substitute for the horn of endangered rhinos. In October 2002, saiga antelopes were finally placed on the Red List of critically endangered species following this population crash (Milner-Gulland *et al.* 2001). In sum, rather few happy stories can be told of illegal wildlife commerce resulting in the successful recovery of harvested wild populations. However, these tend to operate through a 'stick-and-carrot' approach at more than one linkage of the chain, controlling offtakes at the source, the distribution and transport by intermediate traders, and/or finally the consumer demand at the end-point of trade networks. In fact, successful management of any exploitation system will include enforceable measures ranging from demand-side disincentives to supply-side incentives (see Box 6.2), with the optimal balance between penalties on bad behavior or rewards on good behavior being highly context-specific.

Faced with difficulties of managing many semi-subsistence exploitation systems, such as small-scale fisheries and bushmeat hunting, conservation biologists are increasingly calling for more realistic control measures that manipulate the large-scale spatial structure of the harvest. One such method includes no-take areas, such as wildlife sanctuaries and marine protected areas (MPAs) that can be permanently or temporarily closed-off to maximize game and fish yields. Protection afforded by these spatial restrictions allows populations to increase through longer lifespans and higher reproductive success. Recovery of animal biomass inside no-take areas increases harvest levels in surrounding landscapes (or seascapes), and as stocks build up, juveniles and adults can eventually spill over into adjacent areas (e.g. Roberts *et al.* 2001). However, the theoretical and empirical underpinnings of marine reserves have advanced well beyond their terrestrial counterparts. Several typical life history traits of marine species such as planktonic larval dispersal are lacking in terrestrial game species, which differ widely in the degree to which surplus animals can colonize adjacent unharvested areas. However, many wild meat hunters may rely heavily on spillovers from no-take areas. A theoretical analysis of tapir hunting in Peruvian Amazonia showed that a source area of 9300 km$^2$ could sustain typical levels of hunting in a 1700 km$^2$ sink, if dispersal was directed towards that sink (Bodmer 2000). The degree to which source-sink population dynamics can inform real-world management problems remains at best an inexact science. In tropical forests, for example, we still lack basic data on the dispersal rate of most gamebird and large mammal species. Key management questions thus include the potential and realized dispersal rate of target species mediated by changes in density, the magnitude of the spillover effect outside no-take areas, how large these areas must be and still maintain accessible hunted areas, and what landscape configuration of no-take and hunted areas would work best. It is also critical to ensure that no-take areas are sufficiently large to maintain viable populations in the face of overharvesting and habitat loss or degradation in surrounding areas (Peres 2001; Claudet *et al.* 2008). In addition to obvious differences in life-history between organisms in marine and terrestrial systems, applying marine management concepts to forest reserves may be problematic due to differences in the local sociopolitical context in which no-take areas need to be accepted, demarcated and implemented (see Chapter 11). In particular, we need a better understanding of the opportunity costs in terms of income and livelihoods lost from community activities, such as bushmeat hunting and timber extraction, from designating no-take areas.

Finally, conservation biologists and policy-makers who bemoan our general state of data scarcity are akin to fiddlers while Rome burns. Although more fine-tuning data are still needed on the life-history characteristics and population dynamics of exploited populations, we already have a reasonably good idea of what control measures need to be implemented in many exploitation systems. Whether qualitative or quantitative restrictions are designed by resource managers seeking yield quotas based on economic optima or more preservationist views supporting more radical reductions in biomass extraction, control measures will usually involve reductions in harvest capacity and mortality in exploited areas, or more and larger no-take areas (Pauly *et al.* 2002). Eradication of perverse subsidies to unsustainable extractive industries would often be a win-win option leading to stock recovery and happier days for resource users. Co-management agreements with local communities based on sensible principles can also work provided we have the manpower and rural extension capacity to reach out to many source areas (Chapters 14 and 15). Ultimately, however, uncontrolled exploitation activities worldwide cannot be regulated unless we can count on political will and enforcement of national legislation prescribing sustainable management of natural resources, which are so often undermined by weak, absent, or corrupt regulatory institutions.

## Summary

- Human exploitation of biological commodities involves resource extraction from the land, freshwater bodies or oceans, so that wild animals, plants or their products are used for a wide variety of purposes.
- Overexploitation occurs when the harvest rate of any given population exceeds its natural replacement rate.
- Many species are relatively insensitive to harvesting, remaining abundant under relatively high rates of offtake, whereas others can be driven to local extinction by even the lightest levels of offtake.
- This chapter reviews the effects of overexploitation in terrestrial as well as aquatic biomes. Options to manage resource exploitation are also discussed.

## Relevant websites

- Bushmeat Crisis Task Force: http://www.bushmeat.org/portal/server.pt.
- Bioko Biodiversity Protection Program: http://www.bioko.org/conservation/hunting.asp.
- Wildlife Conservation Society: http://www.wcs.org/globalconservation/Africa/bushmeat.

## REFERENCES

Alroy, J. (2001). A multispecies overkill simulation of the late Pleistocene megafaunal mass extinction. *Science*, **292**, 1893-1896.

Anderson, T. L. and Hill, P. J. (2004). *The Not So Wild, Wild West: Property Rights on the Frontier*. Stanford University Press, Stanford, CA.

Andrews, C. (1990). The ornamental fish conservation. *Journal of Fish Biology*, **37**, 53–59.

Asner, G. P., Knapp, D. E., Broadbent, E. N. *et al.* (2005). Selective Logging in the Brazilian Amazon. *Science*, **310**, 480–482.

Ball, S. M. J. (2004). Stocks and exploitation of East African blackwood: a flagship species for Tanzania's Miombo woodlands. *Oryx*, **38**, 1–7.

Barlow, J. and Peres, C. A. (2004). Ecological responses to El Niño-induced surface fires in central Amazonia: management implications for flammable tropical forests. *Philosophical Transactions of the Royal Society of London B*, **359**, 367–380.

Barlow, J. and Peres, C. A. (2008). Fire-mediated dieback and compositional cascade in an Amazonian forest. *Philosophical Transactions of the Royal Society of London B*, **363**, 1787–1794.

Baum, J. K., Myers, R. A., Kehler, D. G. *et al.* (2003). Collapse and conservation of shark populations in the Northwest Atlantic. *Science*, **299**, 389–392.

Bennett, E. L. (2002). Is there a link between wild meat and food security? *Conservation Biology*, **16**, 590–592

Bennett, E. L. and Rao, M. (2002). *Hunting and wildlife trade in tropical and subtropical Asia: identifying gaps and developing strategies.* Unpublished report of the Wildlife Conservation Society, Bangkok, Thailand.

Bodmer, R. E. (1995). Managing Amazonian wildlife: biological correlates of game choice by detribalized hunters. *Ecological Applications,* **5,** 872–877.

Bodmer, R. (2000). Integrating hunting and protected areas in the Amazon. In N. Dunstone and A. Entwistle, eds *Future priorities for the conservation of mammals: has the Panda had its day?* pp. 277–290, Cambridge University Press, Cambridge, UK.

Brashares, J., Arcese, P., Sam, M. K., *et al.* (2004). Bushmeat hunting, wildlife declines, and fish supply in West Africa. *Science,* **306,** 1180–1183.

Butchart, S. M. (2008). Red List Indices to measure the sustainability of species use and impacts of invasive alien species. *Bird Conservation International,* **18,** 245–262

Carson, W. P., Anderson, J. T., Leigh, E. G., and Schnitzer, S. A. (2008). Challenges associated with testing and falsifying the Janzen–Connell hypothesis: A review and critique. In S Schnitzer and W Carson, eds *Tropical forest community ecology,* pp. 210–241. Blackwell Scientific, Oxford, UK.

Chapman, C. A. and Onderdonk, D. A. (1998). Forests without primates: primate/plant codependency. *American Journal of Primatology,* **45,** 127–141.

Chardonnet, P., des Clers, B., Fischer, J., *et al.* (2002). The value of wildlife. *Revue Scientifique et Technique Office Intational Des Épizooties,* **21,** 15–51.

CITES (2008). Convention on International Trade in Endangered Species of Wild Fauna and Flora. *UNEP-WCMC Species Database: CITES-Listed Species.* http://www.cites.org/eng/resources/species.html. Accessed 7 January, 2009.

Claudet, J., Osenberg, C. W., Benedetti-Cecchi, L., *et al.* (2008) Marine reserves: size and age do matter. *Ecology Letters,* **11,** 481–489

Cochrane, M. A. (2003). Fire science for rainforests. *Nature,* **421,** 913–919.

Coleman, F.c. and Williams, S. L. (2002). Overexploiting marine ecosystem engineers: potential consequences for biodiversity. *Trends in Ecology and Evolution,* **17,** 40–44.

Conover, M. R. (1997). Monetary and intangible valuation of deer in the United States. *Wildlife Society Bulletin,* **25,** 298–305.

Cordeiro, N. J. and Howe, H. F. (2003). Forest fragmentation severs mutualism between seed dispersers and an endemic African tree. *Proceedings of the National Academy of Sciences of the United States,* **100,** 14052–14056.

Corlett, R. T. (2007). The impact of hunting on the mammalian fauna of tropical Asian forests. *Biotropica,* **39,** 292–303.

Cowlishaw, G., Mendelson, S., and Rowcliffe, J. M. (2005). Evidence for post-depletion sustainability in a mature bushmeat market. *Journal of Applied Ecology,* **42,** 460–468.

Cox, P. A., Elmqvist, T., Pierson, E. D., and Rainey, W. E. (1991). Flying foxes as strong interactors in South Pacific Island ecosystems: a conservation hypothesis. *Conservation Biology,* **5,** 448–454.

Cristoffer, C. and Peres, C. A. (2003). Elephants vs. butterflies: the ecological role of large herbivores in the evolutionary history of two tropical worlds. *Journal of Biogeography,* **30,** 1357–1380.

Crowder, L. B. and Murawski, S. A. (1998). Fisheries bycatch: implications for management. *Fisheries,* **23,** 8–15.

Cunningham, A., Bennett, E., Peres, C. A., and Wilkie, D. (2009). The empty forest revisited. *Conservation Biology,* in review.

Curran, L. M. and Webb, C. O. (2000). Experimental tests of the spatiotemporal scale of seed predation in mast-fruiting Dipterocarpaceae. *Ecological Monographs,* **70,** 129–148.

Curran, L. M., Trigg, S. N., Mcdonald, A. K., *et al.* (2004). Lowland forest loss in protected areas of Indonesian Borneo. *Science,* **303,** 1000–1003.

Dean, W. (1996). *A Ferro e Fogo,* 2nd edn. Companhia das Letras, Rio de Janeiro, Brazil.

Dirzo, R. and Miranda A. (1991). Altered patterns of herbivory and diversity in the forest understory: a case study of the possible consequences of contemporary defaunation. In P. W. Price, T. M. Lewinsohn, G. W. Fernandes, and W. W. Benson WW, eds *Plant-animal interactions: evolutionary ecology in tropical and temperate regions,* pp. 273–287. New York: John Wiley & Sons, New York, NY.

Dulvy, N. K., Sadovy, Y., and Reynolds, J. D. (2003). Extinction vulnerability in marine populations. *Fish and Fisheries,* **4,** 25–64.

Estes, J. A., Duggins, D. O., and Rathbun, G. B. (1989). The ecology of extinctions in kelp forest communities. *Conservation Biology,* **3,** 252–264.

Fa, J. E. and Peres, C. A. (2001). Game vertebrate extraction in African and Neotropical forests: an intercontinental comparison. In: J. D. Reynolds, G. M. Mace, K. H. Redford and J.G. Robinson, eds *Conservation of exploited species,* pp. 203–241. Cambridge University Press, Cambridge, UK.

Fa, J. E, Peres, C. A., and Meeuwig, J. (2001). Bushmeat exploitation in tropical forests: an intercontinental comparison. *Conservation Biology,* **16,** 232–237.

Fa, J. E., Ryan, S. F., and Bell, D. J. (2005). Hunting vulnerability, ecological characteristics and harvest rates of bushmeat species in afrotropical forests. *Biological Conservation,* **121,** 167–176.

FAO. (2004). *The state of world fisheries and aquaculture 2004*. Food and Agriculture Organization of the United Nations, Rome.

FAO. (2007). *State of the World's Forests*. Food and Agriculture Organization of the United Nations, Italy, Rome.

Godoy, R., Wilkie, D., Overman, H., et al. (2000). Valuation of consumption and sale of forest goods from a Central American rain forest. *Nature*, **406**, 62–63.

Grogan, J., Jennings, S. B., Landis, R. M., et al. (2008). What loggers leave behind: impacts on big-leaf mahogany (*Swietenia macrophylla*) commercial populations and potential for post-logging recovery in the Brazilian Amazon. *Forest Ecology and Management*, **255**, 269–281

Gullison R. E. (1998). Will bigleaf mahogany be conserved through sustainable use? In E. J. Milner-Gulland and R. Mace, eds *Conservation of biological resources*, pp. 193–205. Blackwell Publishing, Oxford, UK.

Hall, M. A., Alverson, D. L., and Metuzals, K. I. (2000). Bycatch: problems and solutions. *Marine Pollution Bulletin*, **41**, 204–219.

Hames, R. B. and Vickers, W.t. (1982). Optimal diet breadth theory as a model to explain variability in Amazonian hunting. *American Ethnologist*, **9**, 358–378.

Hardin, G. (1968). The tragedy of the commons. *Science*, **162**, 1243–1248.

Harrison, I. J. and Stiassny, M. L. J. (1999). The quiet crisis. A preliminary listing of the freshwater fishes of the world that are extinct or 'missing in action'. In R. D. E. MacPhee, ed. *Extinctions in near time*, pp. 271–331. Kluwer Academic/Plenum Publishers, New York, USA.

Holdsworth, A. R. and Uhl, C. (1997). Fire in Amazonian selectively logged rain forest and the potential for fire reduction. *Ecological Applications*, **7**, 713–725.

Howe, H. F. and Smallwood, J. (1982). Ecology of seed dispersal. *Annual Review of Ecology and Systematics*, **13**, 201–218.

Hutchings, J. A. (2000). Collapse and recovery of marine fishes. *Nature*, **406**, 882–885.

IUCN. (2007). IUCN Red List of Threatened Species [www.iucnredlist.org]. International Union for Conservation of Nature and Natural Resources, Cambridge, UK.

Jackson, J. B. C., Kirby, M. X., Berger, W. H., et al. (2001). Historical overfishing and the recent collapse of coastal ecosystems. *Science*, **293**, 629–638.

Jerozolimski, A. and Peres, C. A. (2003). Bringing home the biggest bacon: a cross-site analysis of the structure of hunter-kill profiles in Neotropical forests. *Biological Conservation*, **111**, 415–425.

Jones, C. G., Lawton, J. H., and Shachak, M. (1994). Organisms as ecosystem engineers. *Oikos*, **69**, 373–386.

Law, R. (2000). Fishing, selection, and phenotypic evolution. *Journal of Marine Science*, **57**, 659–668.

Lewison, R. L., Freeman, S. A., and Crowder, L. B. (2004). Quantifying the effects of fisheries on threatened species: the impact of pelagic longlines on loggerhead and leatherback sea turtles. *Ecology Letters*, **7**, 221–231.

Maisels, F., Keming, E., Kemei, M., and Toh, C. (2001). The extirpation of large mammals and implications for montane forest conservation: the case of the Kilum-Ijim Forest, North-west Province, Cameroon. *Oryx*, **35**, 322–334.

Martin, P. S. (1984). Prehistoric overkill: the global model. In P. S. Martin and R. G. Klein, eds *Quaternary extinctions: a prehistoric revolution*, pp. 354–403. University of Arizona Press, Tucson, AZ.

Martin, P. S. and Wright, H. E., Jr., eds (1967). *Pleistocene extinctions: the search for a cause*. Yale University Press, New Haven, CN.

McConkey, K. R. and Drake, D. R. (2006). Flying foxes cease to function as seed dispersers long before they become rare. *Ecology*, **87**, 271–276.

McKinney, M. L. (1997). Extinction vulnerability and selectivity: combining ecological and paleontological views. *Annual Review of Ecology and Systematics*, **28**, 495–516.

MEA. (2006). Millennium Ecosystem Assessment. http://www.millenniumassessment.org/en/. Accessed November 10, 2008.

Milliken, W., Miller, R. P., Pollard, S. R., and Wandelli, E. V. (1992). *Ethnobotany of the Waimiri-Atroari Indians of Brazil*. Royal Botanic Gardens, Kew, UK.

Milner-Gulland, E. J., Kholodova, M. V., Bekenov, A., et al. (2001). Dramatic declines in saiga antelope populations. *Oryx*, **35**, 340–345.

Milner-Gulland, E. J., Bennett, E. L., and The SCB 2002 Annual Meeting Wild Meat Group. (2003). Wild meat – the bigger picture. *Trends in Ecology and Evolution*, **18**, 351–357.

Minto, C., Myers, R. A., and Blanchard, W. (2008). Survival variability and population density in fish populations. *Nature*, **452**, 344–347.

Mitja, D. and Lescure, J.-P. (2000). *Madeira para perfume: qual será o destino do pau-rosa? A Floresta em Jogo: o Extrativismo na Amazônia Central*. Editora UNESP, Imprensa Oficial do Estado, São Paulo, Brazil.

Myers, R. A. and Worm, B. (2003). Rapid worldwide depletion of predatory fish communities. *Nature*, **423**, 280–283.

Myers, R. A., Baum, J. K., Shepherd, T. D., et al. (2007). Cascading effects of the loss of apex predatory sharks from a coastal ocean. *Science*, **315**, 1846–1850.

Naiman, R. J., Melillo, J. M., and Hobbie, J. E. (1986). Ecosystem alteration of boreal forest streams by beaver (*Castor canadensis*). *Ecology*, **67**, 1254–1269.

Nasi, R., Brown, D., Wilkie, D., et al. (2008). *Conservation and use of wildlife-based resources: the bushmeat crisis*. Secretariat of the Convention on Biological Diversity,

Montreal, and Center for International Forestry Research (CIFOR), Bogor. Technical Series no. 33.

Nepstad, D. C., Verıssimo, A., Alencar, A., et al. (1999). Large-scale impoverishment of Amazonian forests by logging and fire. *Nature*, **398**, 505–508.

Nichols, E., Gardner, T. A., Peres, C. A., and Spector, S. (2009). Co-declining mammals and dung beetles: an impending ecological cascade. *Oikos*, **118**, 481–487.

Nuñez-Iturri, G., and Howe, H. F. (2007). Bushmeat and the fate of trees with seeds dispersed by large primates in a lowland rainforest in western Amazonia. *Biotropica*, **39**, 348–354.

Olden, J. D., Hogan, Z. S., and Zanden, M. J. V. (2007). Small fish, big fish, red fish, blue fish: size-biased extinction risk of the world's freshwater and marine fishes. *Global Ecology and Biogeography*, **16**, 694–701.

Pauly, D., Christensen, V., Dalsgaard, J., and Froese, R. (1998). Fishing down marine food webs. *Science*, **279**, 860–863.

Pauly, D., Christensen, V., Guenette, S., et al. (2002). Towards sustainability in world fisheries. *Nature*, **418**, 689–695.

Pearce, P. (1990). *Introduction to forestry economics*. University of British Columbia Press, Vancouver, Canada.

Peres, C. A. (2000). Effects of subsistence hunting on vertebrate community structure in Amazonian forests. *Conservation Biology*, **14**, 240–253.

Peres, C. A. (2001). Synergistic effects of subsistence hunting and habitat fragmentation on Amazonian forest vertebrates. *Conservation Biology*, **15**, 1490–1505.

Peres, C. A. and Dolman, P. (2000). Density compensation in neotropical primate communities: evidence from 56 hunted and non-hunted Amazonian forests of varying productivity. *Oecologia*, **122**, 175–189.

Peres, C. A. and Lake, I. R. (2003). Extent of nontimber resource extraction in tropical forests: accessibility to game vertebrates by hunters in the Amazon basin. *Conservation Biology*, **17**, 521–535.

Peres, C. A. and van Roosmalen, M. (2003). Patterns of primate frugivory in Amazonia and the Guianan shield: implications to the demography of large-seeded plants in overhunted tropical forests. In D. Levey, W. Silva and M. Galetti, eds *Seed dispersal and frugivory: ecology, evolution and conservation*, pp. 407–423. CABI International, Oxford, UK.

Peres C. A. and Palacios, E. (2007). Basin-wide effects of game harvest on vertebrate population densities in Amazonian forests: implications for animal-mediated seed dispersal. *Biotropica*, **39**, 304–315.

Peres, C. A., Baider, C., Zuidema, P. A., et al. (2003). Demographic threats to the sustainability of Brazil nut exploitation. *Science*, **302**, 2112–2114.

Peters, C. M. (1994). *Sustainable harvest of non-timber plant resources in tropical moist forest:an ecological primer*. Biodiversity Support Program, Washington, DC.

Peters, C. M., Gentry, A. H., and Mendelsohn, R. (1989). Valuation of an Amazonian rainforest. *Nature*, **339**, 655–656.

Redford, K. H. (1992). The empty forest. *BioScience*, **42**, 412–422.

Redford, K. H. and P. Feinsinger. (2001). The half-empty forest: sustainable use and the ecology of interactions. In J.D. Reynolds, G.M. Mace, K.H. Redford and J.G. Robinson, eds *Conservation of exploited species*, pp. 370–399. Cambridge University Press, Cambridge, UK.

Repetto, R. and Gillis, M., eds (1988). *Public policies and the misuse of forest resources*. Cambridge University Press, Cambridge, UK.

Roberts, C. M., Bohnsack, J. A., Gell, F., et al. (2001). Effects of marine reserves on adjacent fisheries. *Science*, **294**, 1920–1923.

Roberts, J. M, Wheeler, A. J, and Freiwald, A. (2006). Reefs of the deep: the biology and geology of cold-water coral ecosystems. *Science*, **312**, 543–547.

Robinson, J. G. and Bennett, E. L., eds (2000). *Hunting for sustainability in tropical forests*. Columbia University Press, New York.

Roe, D. (2008). *Trading Nature. A report, with case studies, on the contribution of wildlife trade management to sustainable livelihoods and the Millennium Development Goals*. TRAFFIC International and WWF International.

Ross, E. B. (1978). Food taboos, diet, and hunting strategy: the adaptation to animals in Amazon cultural ecology. *Current Anthropology*, **19**, 1–36.

Samant, S. S., Dhar, U., and Palni, L. M. S. (1998). *Medicinal plants of Indian Himalaya: diversity distribution potential values*. G. B. Pant Institute of Himalayan Environment and Development, Almora, India.

Sheil, D. and Salim, A. (2004). Forest trees, elephants, stem scars and persistence. *Biotropica*, **36**, 505–521.

Sodhi, N. S., Koh, L. P., Peh, K. S.-H., et al. (2008). Correlates of extinction proneness in tropical angiosperms. *Diversity and Distributions*, **14**, 1–10.

Steadman, D. A. (1995). Prehistoric extinctions of Pacific islands birds: biodiversity meets zooarcheology. *Science*, **267**, 1123–1131.

Swaine, M. D. and Whitmore, T. C. (1988). On the definition of ecological species groups in tropical rain forests. *Vegetatio*, **75**, 81–86.

Terborgh, J., Nunez-Iturri, G., Pitman, N. C. A., et al. (2008). Tree recruitment in an empty forest. *Ecology*, **89**, 1757–1768.

TRAFFIC. (1998). *Europe's medicinal and aromatic plants: their use, trade and conservation*. TRAFFIC International, Cambridge, UK.

TRAFFIC. (2008). *What's driving the wildlife trade? A review of expert opinion on economic and social drivers of the wildlife trade and trade control efforts in Cambodia, Indonesia, Lao PDR and Vietnam*. World Bank, Washington, DC.

US Census Bureau. (2006). *2006 National survey of fishing, hunting, and wildlife-associated recreation*. U.S. Department of the Interior, Fish and Wildlife Service, and U.S. Department of Commerce, US Census Bureau, Shepherdston, WV.

Wang, B. C., Leong, M. T., Smith, T. B., and Sork, V. L. (2007). Hunting of mammals reduces seed removal and dispersal from the Afrotropical tree, *Antrocaryon klaineanum* (Anacardiaceae). *Biotropica*, **39**, 340–347.

Warkentin, I. G., Bickford, D., Sodhi, N. S., and Bradshaw, C. J. A. (2009). Eating frogs to extinction. *Conservation Biology*, **23**, 1056–1059.

WCFSD. (1998). *Final Report on Forest Capital*. World Commission of Forests and Sustainable Development., Cambridge University Press, Cambridge, UK.

World Bank. (2008). *The sunken billions: the economic justification for fisheries reform*. Agriculture and Rural Development Department. The World Bank and Food and Agriculture Organization, Washington, DC.

Worm, B., Barbier, E. B., Beaumont, N., *et al.* (2006). Impacts of biodiversity loss on ocean ecosystem services. *Science*, **314**, 787–790.

Wright, J. P. and Jones, C. G. (2006). The concept of organisms as ecosystem engineers ten years on: progress, limitations and challenges. *BioScience*, **56**, 203–209.

Wright, S. J. (2003). The myriad effects of hunting for vertebrates and plants in tropical forests. *Perspectives in Plant Ecology, Evolution and Systematics*, **6**, 73–86.

Wright, S. J., Zeballos, H., Dominguez, I., *et al.* (2000). Poachers alter mammal abundance, seed dispersal and seed predation in a Neotropical forest. *Conservation Biology*, **14**, 227–239.

Wright, S. J., Hernandez, A., and Condit, R. (2007). The bushmeat harvest alters seedling banks by favoring lianas, large seeds and seeds dispersed by bats, birds and wind. *Biotropica*, **39**, 363–371.

# CHAPTER 7

# Invasive species

## Daniel Simberloff

An invasive species is one that arrives (often with human assistance) in a habitat it had not previously occupied, then establishes a population and spreads autonomously. Species invasions are one of the main conservation threats today and have caused many species extinctions. The great majority of such invasions are by species introduced from elsewhere, although some native species have become invasive in newly occupied habitats (see Box 7.1). In some areas of the world—especially islands (see Box 7.2)—introduced species comprise a large proportion of all species. For instance, for the Hawaiian islands, almost half the plant species, 25% of insects, 40% of birds, and most freshwater fishes are introduced, while the analogous figures for Florida are 27% of plant species, 8% of insects, 5% of birds, and 24% of freshwater fishes. Not all introduced species become invasive, however. Many plant species imported as ornamentals persist in

---

**Box 7.1 Native invasives**
**Daniel Simberloff**

Although the great majority of invasive species are introduced, occasionally native plant species have become invasive, spreading rapidly into previously unoccupied habitats. These invasions fall into two categories, both involving human activities. In the first, a native species that is rather restricted in range and habitat is supplemented with introductions from afar that have new genotypes, and the new genotypes, or recombinants involving the new genotypes, become invasive. An example in North America is common reed (*Phragmites australis*), which was present for at least thousands of years and is probably native, but which spread widely, became much more common, and began occupying more habitats beginning in the mid- nineteenth century. This invasion is wholly due to the introduction of Old World genotypes at that time, probably in soil ballast (Saltonstall 2002). Similarly, reed canarygrass (*Phalaris arundinacea*), native to North America but previously uncommon, became highly invasive in wetland habitats with the introduction of European genotypes as a forage crop in the 19th century (Lavergne and Molofsky 2007).

The second category of native invasives arises from human modification of the environment. For instance, in western Europe, the grass *Elymus athericus*, previously a minor component of high intertidal vegetation, began spreading seaward because of increased nitrogen in both aerial deposition and runoff, and it now occupies most of the intertidal in many areas (Valéry *et al.* 2004). The plant apparently uses the nitrogen to increase its tolerance or regulation of salt. In various regions of the western United States, Douglas fir (*Pseudotsuga menziesii*) and several other tree species have invaded grasslands and shrublands as a result of fire suppression, increased grazing by livestock, or both. Natural fire had precluded them, and when fire was suppressed, livestock served the same role (Simberloff 2008). By contrast, Virginia pine (*Pinus virginiana*) in the eastern United States

*continues*

### Box 7.1 (Continued)

invaded serpentine grasslands when fires were suppressed and long-time grazing practices were restricted (Thiet and Boerner 2007).

#### REFERENCES

Lavergne, S. and Molofsky, J. (2007). Increased genetic variation and evolutionary potential drive the success of an invasive grass. *Proceedings of the National Academy of Sciences of the United States of America*, **104**, 3883–3888.

Saltonstall, K. (2002). Cryptic invasion by a non-native genotype of the common reed, *Phragmites australis*, into North America. *Proceedings of the National Academy of Sciences of the United States of America*, **99**, 2445–2449.

Simberloff, D. (2008). Invasion biologists and the biofuels boom: Cassandras or colleagues? *Weed Science*, **56**, 867–872.

Thiet, R. K and Boerner, R. E. J. (2007). Spatial patterns of ectomycorrhizal fungal inoculum in arbuscular mycorrhizal barrens communities: implications for controlling invasion by *Pinus virginiana*. *Mycorrhiza*, **17**, 507–517.

Valéry, L., Bouchard, V., and Lefeuvre, J.-C. (2004). Impact of the invasive native species *Elymus athericus* on carbon pools in a salt marsh. *Wetlands*, **24**, 268–276.

### Box 7.2 Invasive species in New Zealand
Daniel Simberloff

Many islands have been particularly afflicted by introduced species, even large islands such as those comprising New Zealand (Allen and Lee 2006). New Zealand had no native mammals, except for three bat species but now has 30 introduced mammals. Among these, several are highly detrimental to local fauna and/or flora. The Australian brushtail possum (*Trichosurus vulpecula*; Box 7.2 Figure) now numbers in the millions and destroys broadleaved native trees, eating bird eggs and chicks as well. Pacific and Norway rats are also devastating omnivores that particularly plague native birds. Introduced carnivores—the stoat (*Mustela erminea*), weasel (*M. nivalis*), ferret (*M. furo*), and hedgehog (*Erinaceus europaeus*) —are all widespread and prey on various combinations of native birds, insects, skinks, geckos, and an endemic reptile (*Sphenodon punctatus*). Many ungulates have been introduced, of which European red deer (*Cervus elaphus*) is most numerous. Trampling and grazing by ungulates has greatly damaged native vegetation in some areas. Feral pigs (*Sus scrofa*) are now widespread in forest and scrub habitats, and their rooting causes erosion, reduces populations of some plant species, and changes nutrient cycling by mixing organic and mineral layers of the soil. Of 120 introduced bird species, 34 are established. To some extent they probably compete with native birds and prey on native invertebrates, but their impact is poorly studied and certainly not nearly as severe as that of introduced mammals. European brown trout (*Salmo trutta*) are widely established and have caused the local extirpation of a number of fish species.

Among the estimated 2200 established introduced invertebrate species in

**Box 7.2 Figure** Brushtail possum. Photograph by Rod Morris.

*continues*

> **Box 7.2 (Continued)**
>
> New Zealand, German wasps (*Vespula germanica*) and common wasps (*V. vulgaris*) have probably had the most impact, especially by monopolizing the honeydew produced by native scale insects that had supported several native bird species, including the kaka (*Nestor meridionalis*), the tui (*Prosthemadera novaeseelandiae*), and the bellbird (*Anthornis melanura*).
>
> About 2100 species of introduced plants are now established in New Zealand, outnumbering native species. Several tree species introduced about a century ago are now beginning to spread widely, the lag caused by the fact that trees have long life cycles. Most of the introduced plants in New Zealand, including trees, invade largely or wholly when there is some sort of disturbance, such as land-clearing or forestry. However, once established, introduced plants have in some instances prevented a return to the original state after disturbance stopped. New Zealand also has relatively few nitrogen-fixing plant species, and even these have been outcompeted by introduced nitrogen-fixers such as gorse (*Ulex europaeus*), Scotch broom (*Cytisus scoparius*), and tree lupine (*Lupinus arboreus*). As in other areas (see above), in parts of New Zealand these nitrogen-fixers have, by fertilizing the soil, favored certain native species over others and have induced an invasional meltdown by allowing other introduced plant species to establish.
>
> Given the enormous number of introduced species invading New Zealand and the many sorts of impacts these have generated, it is not surprising that New Zealand enacted the first comprehensive national strategy to address the entire issue of biological invasions, the Biosecurity Act of 1993.
>
> **REFERENCE**
>
> Allen, R. B. and Lee, W. G., eds (2006). *Biological invasions in New Zealand*. Springer, Berlin, Germany.

gardens with human assistance but cannot establish in less modified habitats. The fraction of introduced species that establish and spread is a matter under active research, but for some organisms it can be high. For example, half of the freshwater fish, mammal, and bird species introduced from Europe to North America or vice-versa have established populations, and of these, more than half became invasive (Jeschke and Strayer 2005).

Invasive species can produce a bewildering array of impacts, and impacts often depend on context; the same introduced species can have minimal effects on native species and ecosystems in one region but can be devastating somewhere else. Further, the same species can affect natives in several different ways simultaneously. However, a good way to begin to understand the scope of the threat posed by biological invasions is to classify the main types of impacts.

## 7.1 Invasive species impacts

### 7.1.1 Ecosystem modification

The greatest impacts of invasive species entail modifying entire ecosystems, because such modifications are likely to affect most of the originally resident species. Most obviously, the physical structure of the habitat can be changed. For instance, in Tierra del Fuego, introduction of a few North American beavers (*Castor canadensis*) in 1946 has led to a population now over 50 000, and in many areas they have converted forests of southern beech (*Nothofagus* spp.) to grass- and sedge-dominated meadows (Lizarralde et al. 2004). In the Florida Everglades, introduced Australian paperbark (*Melaleuca quinquenervia*) trees have effected the opposite change, from grass- and sedge-dominated prairies to nearly monospecific paperbark forests (Schmitz et al. 1997). In parts of Hawaii, Asian and American mangrove species have replaced beach communities of herbs and small shrubs with tall mangrove forests (Allen 1998).

Introduced plant species can modify an entire ecosystem by overgrowing and shading out native species. South American water hyacinth (*Eichhornia crassipes*) now covers parts of Lake Victoria in Africa (Matthews and Brand 2004a), many lakes and rivers in the southeastern United States (Schardt 1997), and various waterbodies in Asia and Australia (Matthews and Brand 2004b), often smothering native submersed vegetation. Vast quantities of rotting water hyacinth, and consequent drops in dissolved oxygen, can also affect many aquatic animal species. Similar overgrowth occurs in the Mediterranean Sea, where *Caulerpa taxifolia* (Figure 7.1), an alga from the tropical southwest Pacific Ocean, replaces seagrass meadows over thousands of hectares, greatly changing the animal community (Meinesz 1999).

A new species of cordgrass (*Spartina anglica*) arose in England in the late nineteenth century by hybridization between a native cordgrass and an introduced North American species. The new species invaded tidal mudflats and, trapping much more sediment, increased elevation and converted mudflats to badly drained, dense salt marshes with different animal species (Thompson 1991). The hybrid species was later introduced to New Zealand and the state of Washington with similar impacts.

Introduced species can change entire ecosystems by changing the fire regime (see Chapter 9). The invasion of the Florida Everglades by Australian paperbark trees, noted above, is largely due to the fact that paperbark catches fire easily and produces hotter fires than the grasses and sedges it replaces. The opposite transformation, from forest to grassland, can also be effected by a changed fire regime. In Hawaii, African molassesgrass (*Melinis minutiflora*) and tropical American tufted beardgrass (*Schizachyrium condensatum*) have replaced native-dominated woodland by virtue of increased fire frequency and extent (D'Antonio and Vitousek 1992).

Introduced plants can change entire ecosystems by modifying water or nutrient regimes. At Eagle Borax Spring in California, Mediterranean salt cedars (*Tamarix* spp.) dried up a large marsh (McDaniel *et al.* 2005), while in Israel, Australian eucalyptus trees were deliberately introduced to drain swamps (Calder 2002). By fertilizing nitrogen-poor sites, introduced nitrogen-fixing plants can favor other exotic species over natives. On the geologically young, nitrogen-poor volcanic island of Hawaii, firetree (*Morella faya*), a nitrogen-fixing shrub from the Azores, creates conditions that favor other introduced species that previously could not thrive in the low-nutrient

**Figure 7.1** *Caulerpa taxifolia*. Photograph by Alex Meinesz.

soil and disfavor native plants that had evolved to tolerate such soil (Vitousek 1986).

Pathogens that eliminate a previously dominant plant can impact an entire ecosystem. In the first half of the twentieth century, Asian chestnut blight (*Cryphonectria parasitica*) ripped through eastern North America, effectively eliminating American chestnut (*Castanea dentata*), a tree that had been common from Georgia through parts of Canada and comprised at least 30% of the canopy trees in many forests (Williamson 1996). This loss in turn led to substantial structural changes in the forest, and it probably greatly affected nutrient cycling, because chestnut wood, high in tannin, decomposes slowly, while the leaves decompose very rapidly (Ellison *et al.* 2005). Chestnut was largely replaced by oaks (*Quercus* spp.), which produce a recalcitrant litter. Because this invasion occurred so long ago, few of its effects were studied at the time, but it is known that at least seven moth species host-specific to chestnut went extinct (Opler 1978). Such pathogens are also threats to forest industries founded on introduced species as well as natives, as witness the vast plantations in Chile of North American Monterrey pine (*Pinus radiata*) now threatened by recently arrived *Phytophthora pinifolia* (Durán *et al.* 2008).

### 7.1.2 Resource competition

In Great Britain, the introduced North American gray squirrel (*Sciurus carolinensis*) forages for nuts more efficiently than the native red squirrel (*Sciurus vulgaris*), leading to the decline of the latter species (Williamson 1996). The same North American gray squirrel species has recently invaded the Piedmont in Italy and is spreading, leading to concern that the red squirrel will also decline on the mainland of Europe as it has in Britain (Bertolino *et al.* 2008). The house gecko (*Hemidactylus frenatus*) from Southeast Asia and parts of Africa has invaded many Pacific islands, lowering insect populations that serve as food for native lizards, whose populations have declined in some areas (Petren and Case 1996).

### 7.1.3 Aggression and its analogs

The red imported fire ant (*Solenopsis invicta*) from southern South America has spread through the southeastern United States and more recently has invaded California. It attacks other ant species it encounters, and in disturbed habitats (which comprise much of the Southeast) this aggression has caused great declines in populations of native ant species (Tschinkel 2006). The Argentine ant (*Linepithema humile*), also native to South America, similarly depresses populations of native ant species in the United States by attacking them (Holway and Suarez 2004). The Old World zebra mussel (*Dreissena polymorpha*; Figure 7.2), spreading throughout much of North America, threatens the very existence of a number of native freshwater bivalve species, primarily by settling on them in great number and suturing their valves together with byssal threads, so that they suffocate or starve (Ricciardi *et al.* 1998). Although plants do not attack, they have an analogous ability to inhibit other species, by producing or sequestering chemicals. For example, the African crystalline ice plant (*Mesembryanthemum crystallinum*) sequesters salt, and when leaves fall and decompose, the salt remains in the soil, rendering it inhospitable to native plants in California that cannot tolerate such high salt concentrations (Vivrette and Muller 1977). Diffuse knapweed (*Centaurea diffusa*) from Eurasia and spotted knapweed (*C. stoebe*) from Europe are both major invaders of rangelands in the American West. One reason they dominate native range plants in the United States is that they produce

**Figure 7.2** Zebra mussel. Photograph by Tony Ricciardi.

root exudates that are toxic to native plants (Callaway and Ridenour 2004). An invasive introduced plant can also dominate a native species by interfering with a necessary symbiont of the native. For instance, many plants have established mutualistic relationships with arbuscular mycorrhizal fungi, in which the fungal hyphae penetrate the cells of the plants' roots and aid the plants to capture soil nutrients. Garlic mustard (*Alliaria petiolata*) from Europe, Asia, and North Africa is a highly invasive species in the ground cover of many North American woodlands and floodplains. Root exudates of garlic mustard, which does not have mycorrhizal associates, are toxic to arbuscular mycorrhizal fungi found in North American soils (Callaway *et al.* 2008).

## 7.1.4 Predation

One of the most dramatic and frequently seen impacts of introduced species is predation on native species. Probably the most famous cases are of mammalian predators such as the ship rat (*Rattus rattus*), Norway rat (*R. norvegicus*), Pacific rat (*R. exulans*), small Indian mongoose (*Herpestes auropunctatus*), and stoat (*Mustela erminea*) introduced to islands that formerly lacked such species. In many instances, native bird species, not having evolved adaptations to such predators, nested on the ground and were highly susceptible to the invaders. Introduced rats, for example, have caused the extinction of at least 37 species and subspecies of island birds throughout the world (Atkinson 1985). The brown tree snake (*Boiga irregularis*; Figure 7.3), introduced to Guam from New Guinea in cargo after World War II, has caused the extinction or local extirpation of nine of the twelve native forest bird species on Guam and two of the eleven native lizard species (Lockwood *et al.* 2007). For these native species, an arboreal habitat was no defense against a tree-climbing predator. Another famous introduced predator that has wreaked havoc with native species is the Nile perch (*Lates niloticus*), deliberately introduced to Lake Victoria in the 1950s in the hope that a fishery would be established to provide food and jobs to local communities (Pringle 2005). Lake Victoria is home of one of the great evolutionary species radiations, the hundreds of species of cichlid fishes. About half of them are now extinct because of predation by the perch, and several others are maintained only by captive rearing (Lockwood *et al.* 2007).

Many predators have been deliberately introduced for "biological control" of previously introduced species (see below), and a number of these have succeeded in keeping populations of the target species at greatly reduced levels. For instance, introduction of the Australian vedalia

**Figure 7.3** Brown tree snake. Photograph by Gad Perry.

ladybeetle (*Rodolia cardinalis*) in 1889 controlled Australian cottony-cushion scale (*Icerya purchasi*) on citrus in California (Caltagirone and Doutt 1989). However, some predators introduced for biological control have attacked non-target species to the extent of causing extinctions. One of the worst such disasters was the introduction of the rosy wolf snail (*Euglandina rosea*), native to Central America and Florida, to many Pacific islands to control the previously introduced giant African snail (*Achatina fulica*). The predator not only failed to control the targeted prey (which grows to be too large for the rosy wolf snail to attack it) but caused the extinction of over 50 species of native land snails (Cowie 2002). The small Indian mongoose, implicated as the sole cause or a contributing cause in the extinction of several island species of birds, mammals, and frogs, was deliberately introduced to all these islands as a biological control agent for introduced rats (Hays and Conant 2006). The mosquitofish (*Gambusia affinis*) from Mexico and Central America has been introduced to Europe, Asia, Africa, Australia, and many islands for mosquito control. Its record on this score is mixed, and there is often evidence that it is no better than native predators at controlling mosquitoes. However, it preys on native invertebrates and small fishes and in Australia is implicated in extinction of several fish species (Pyke 2008).

### 7.1.5 Herbivory

Introduced herbivores can devastate the flora of areas lacking similar native species, especially on islands. Goats (*Capra aegagrus hircus*) introduced to the island of St. Helena in 1513 are believed to have eliminated at least half of ~100 endemic plant species before botanists had a chance to record them (Cronk 1989). European rabbits (*Oryctolagus cuniculus*) introduced to islands worldwide have devastated many plant populations, often by bark-stripping and thus killing shrubs and seedling and sapling trees. Rabbits also often cause extensive erosion once vegetation has been destroyed (Thompson and King 1994). Damage to forests and crop plants by introduced herbivores is often staggering. For instance, the South American cassava mealybug (*Phenacoccus manihoti*), invading extensive cassava-growing parts of Africa, often destroys more than half the crop yield (Norgaard 1988), while in the United States, the Russian wheat aphid (*Diuraphis noxia*) caused US$600 million damage in just three years (Office of Technology Assessment 1993). In forests of the eastern United States, the European gypsy moth (*Lymantria dispar*) caused a similar amount of damage in only one year (Office of Technology Assessment 1993). In high elevation forests of the southern Appalachian Mountains, the Asian balsam woolly adelgid (*Adelges piceae*) has effectively eliminated the previously dominant Fraser fir tree (Rabenold *et al.* 1998), while throughout the eastern United States the hemlock woolly adelgid (*A. tsugae*) is killing most hemlock trees, which often formed distinct moist, cool habitats amidst other tree species (Ellison *et al.* 2005).

Plant-eating insects have been successful in many biological control projects for terrestrial and aquatic weeds. For instance, in Africa's Lake Victoria, a massive invasion of water hyacinth was brought under control by introduction of two South American weevils, *Neochetina eichhorniae* and *N. bruchi* (Matthews and Brand 2004a); these have also been introduced to attack water hyacinth in tropical Asia (Matthews and Brand 2004b). The South American alligator-weed flea beetle (*Agasicles hygrophila*) has minimized the invasion of its South American host plant (*Alternanthera philoxeroides*) in Florida (Center *et al.* 1997) and contributed greatly to its control in slow-moving water bodies in Asia (Matthews and Brand 2004b). A particularly famous case was the introduction of the South American cactus moth (*Cactoblastis cactorum*) to Australia, where it brought a massive invasion of prickly pear cactus (*Opuntia* spp.) under control (Zimmermann *et al.* 2001). In probably the first successful weed biological control project, a Brazilian cochineal bug (*Dactylopius ceylonicus*) virtually eliminated the smooth prickly pear (*Opuntia vulgaris*) from India (Doutt 1964). In 1913, the same insect was introduced to South Africa and effectively eliminated the same plant (Doutt 1964).

However, occasionally, biological control introductions of herbivorous insects have devastated non-target native species. The same cactus moth introduced to Australia was introduced to control pest prickly pear on the island of Nevis in the West Indies. From there, it island-hopped through the West Indies and reached Florida, then spread further north and west. In Florida, it already threatens the very existence of the native semaphore cactus (*O. corallicola*), and there is great concern that this invasion, should it reach the American Southwest and Mexico, would not only threaten other native *Opuntia* species but also affect economically important markets for ornamental and edible *Opuntia* (Zimmermann *et al.* 2001). The Eurasian weevil (*Rhinocyllus conicus*), introduced to Canada and the United States to control introduced pest thistles, attacks several native thistles as well (Louda *et al.* 1997), and this herbivory has led to the listing of the native Suisun thistle (*Cirsium hygrophilum* var. *hygrophilum*) on the U.S. Endangered Species List (US Department of the Interior 1997). In each of these cases of herbivorous biological control agents threatening natives, the introduced herbivore was able to maintain high numbers on alternative host plants (such as the targeted hosts), so decline of the native did not cause herbivore populations to decline.

### 7.1.6 Pathogens and parasites

Many introduced plant pathogens have modified entire ecosystems by virtually eliminating dominant plants. The chestnut blight was discussed above. A viral disease of ungulates, rinderpest, introduced to southern Africa from Arabia or India in cattle in the 1890s, attacked many native ungulates, with mortality in some species reaching 90%. The geographic range of some ungulate species in Africa is still affected by rinderpest. Because ungulates often play key roles in vegetation structure and dynamics, rinderpest impacts affected entire ecosystems (Plowright 1982).

Of course, many introduced diseases have affected particular native species or groups of them without modifying an entire ecosystem. For instance, avian malaria, caused by *Plasmodium relictum capristranoae*, introduced with Asian birds and vectored by previously introduced mosquitoes, contributed to the extinction of several native Hawaiian birds and helps restrict many of the remaining species to upper elevations, where mosquitoes are absent or infrequent (Woodworth *et al.* 2005). In Europe, crayfish plague (*Aphanomyces astaci*), introduced with the North American red signal crayfish (*Pacifastacus lenusculus*; Figure 7.4 and Plate 7) and also vectored by the subsequently introduced Lousiana crayfish

**Figure 7.4** North American red signal crayfish (right) and a native European crayfish (*Astacus astacus*). Photograph by David Holdich.

(*Procambarus clarkii*), has devastated native European crayfish populations (Goodell *et al.* 2000). The European fish parasite *Myxosoma cerebralis*, which causes whirling disease in salmonid fishes, infected North American rainbow trout (*Oncorhynchus mykiss*) that had been previously introduced to Europe and were moved freely among European sites after World War II. Subsequently, infected frozen rainbow trout were shipped to North America, and the parasite somehow got into a trout hatchery in Pennsylvania, from which infected rainbow trout were shipped to many western states. In large areas of the West, most rainbow trout contracted the disease and sport fisheries utterly collapsed (Bergersen and Anderson 1997). Introduced plant parasites can greatly damage agriculture. For example, parasitic witchweed (*Striga asiatica*) from Africa reached the southeastern United States after World War II, probably arriving on military equipment. It inflicts great losses on crops that are grasses (including corn) and has been the target of a lengthy, expensive eradication campaign (Eplee 2001).

Introduction of vectors can also spread not only introduced pathogens (e.g. the mosquitoes vectoring avian malaria in Hawaii) but also native ones. For example, the native trematode *Cyathocotyle bushiensis*, an often deadly parasite of ducks, has reached new regions along the St. Lawrence River recently as its introduced intermediate host, the Eurasian faucet snail (*Bithynia tentaculata*), has invaded (Sauer *et al.* 2007). Introduced parasites or pathogens and vectors can interact in complicated ways to devastate a native host species. Chinese grass carp (*Ctenopharyngodon idella*) infected with the Asian tapeworm *Bothriocephalus acheilognathi* were introduced to Arkansas in 1968 to control introduced aquatic plants and spread to the Mississippi River. There the tapeworm infected native fishes, including a popular bait fish, the red shiner (*Notropis lutrensis*). Fishermen or bait dealers then carried infected red shiners to the Colorado River, from which by 1984 they had reached a Utah tributary, the Virgin River. In the Virgin River, the tapeworm infected and killed many woundfin (*Plagopterus argentissimus*), a native minnow already threatened by dams and water diversion projects (Moyle 1993).

Parasites and pathogens have also been used successfully in biological control projects against introduced target hosts. For instance, the South American cassava mealybug in Africa, discussed above, has been partly controlled by an introduced South American parasitic wasp, *Epidinocarsis lopezi* (Norgaard 1988), while the European yellow clover aphid (*Therioaphis trifolii*), a pest of both clover and alfalfa, is controlled in California by three introduced parasitic wasps, *Praon palitans*, *Trioxys utilis*, and *Aphelinus semiflavus* (Van Den Bosch *et al.* 1964). The New World myxoma virus, introduced to mainland Europe (where the European rabbit is native) and Great Britain and Australia (where the rabbit is introduced), initially caused devastating mortality (over 90%). However, the initially virulent viral strains evolved to be more benign, while in Great Britain and Australia, rabbits evolved to be more resistant to the virus. Mortality has thus decreased in each successive epidemic (Bartrip 2008).

### 7.1.7 Hybridization

If introduced species are sufficiently closely related to native species, they may be able to mate and exchange genes with them, and a sufficient amount of genetic exchange (introgression) can so change the genetic constitution of the native population that we consider the original species to have disappeared—a sort of genetic extinction. This process is especially to be feared when the invading species so outnumbers the native that a native individual is far more likely to encounter the introduced species than a native as a prospective mate. The last gasp of a fish native to Texas, *Gambusia amistadensis*, entailed the species being hybridized to extinction through interbreeding with introduced mosquito fish *G. amistadensis* (Hubbs and Jensen 1984), while several fishes currently on the United States Endangered Species List are threatened at least partly by hybridization with introduced rainbow trout. The North American mallard (*Anas platyrhynchos*), widely introduced as a game bird, interbreeds extensively with many congeneric species and threatens the very existence of the endemic New Zealand grey duck (*A. superciliosa superciliosa*) and the Hawaiian duck (*A. wyvilliana*), as well as, perhaps, the yellowbilled duck (*A. undulata*) and the Cape shoveller (*A. smithii*) in Africa (Rhymer and Simberloff 1996, Matthews and Brand 2004a). European populations of the white-headed duck (*Oxyura leucocephala*) restricted to Spain, are threatened by hybridization and introgression with North American ruddy ducks (*O. jamaicensis*) (Muñoz-Fuentes *et al.* 2007). The latter had been introduced years earlier to Great Britain simply as an ornamental; they subsequently crossed the Channel, spread through France, and reached Spain.

Availability and increasing sophistication of molecular genetic techniques has led to the recognition that hybridization and introgression between introduced and native species is far more common than had been realized. Such hybridization can even lead to a new species. In the cordgrass (*Spartina*) case discussed above, occasional hybrids were initially sterile, until a chromosomal mutation (doubling of chromosome number) in one of them produced a fertile new polyploid species, which became highly invasive (Thompson 1991). A similar case involves Oxford ragwort (*Senecio squalidus*), a hybrid of two species from Italy, introduced to the Oxford Botanical Garden ca. 1690. *S. squalidus* escaped, first spread through Oxford, and then during the Industrial Revolution through much of Great Britain along railroad lines, producing sterile hybrids with several native British species of *Senecio*. A chromosomal mutation (doubling of chromosome number) of a hybrid between *S. squalidus* and *S. vulgaris* (groundsel) produced the new polyploid species *S. cambrensis* (Welsh groundsel) (Ashton and Abbott 1992).

It is possible for hybridization to threaten a species even when no genetic exchange occurs. Many populations of the European mink (*Mustela lutreola*) are gravely threatened by habitat destruction. North America mink (*M. vison*), widely introduced in Europe to foster a potential fur-bearing industry, have escaped and established many populations. In some sites, many female European mink hybridize with male American mink, which become sexually mature and active before the European mink males. The European mink females subsequently abort the hybrid embryos, so no genes can be exchanged between the species, but these females cannot breed again during the same season, a severe handicap to a small, threatened population (Maran and Henttonen 1995).

### 7.1.8 Chain reactions

Some impacts of introduced species on natives entail concatenated chains of various interactions: species A affecting species B, then species B affecting species C, species C affecting species D, and so forth. The spread of the Asian parasitic tapeworm from Arkansas ultimately to infect the woundfin minnow (*Plagopterus argentissimus*) in Utah is an example. However, chains can be even more complex, almost certainly unforeseeable. An example involves the devastation of European rabbit populations in Britain by New World myxoma virus, described above. Caterpillars of the native large blue butterfly (*Maculina arion*) in Great Britain required development in underground nests of the native ant *Myrmica sabuleti*. The ant avoids nesting in overgrown areas, which for centuries had not been problematic because of grazing and cultivation. However, changing land use patterns and decreased grazing led to a situation in which rabbits were the main species maintaining suitable habitat for the ant. When the virus devastated rabbit populations, ant populations declined to the extent that the large blue butterfly was extirpated from Great Britain (Ratcliffe 1979). In another striking chain reaction, landlocked kokanee salmon (*Oncorhynchus nerka*), were introduced to Flathead Lake, Montana in 1916, replacing most native cutthroat trout (*O. clarki*) and becoming the main sport fish. The kokanee were so successful that they spread far from the lake, and their spawning populations became so large that they attracted large populations of bald eagles (*Haliaeetus leucocephalus*), grizzly bears (*Ursus arctos horribilis*), and other predators. Between 1968 and 1975, opossum shrimp (*Mysis relicta*), native to large deep lakes elsewhere in North America and in Sweden, were introduced to three lakes in the upper portion of the Flathead catchment in order to increase production of kokanee; the shrimp drifted downstream into Flathead Lake by 1981 and caused a sharp, drastic decline in populations of cladocerans and copepods they preyed on. However, the kokanee also fed on these prey, and kokanee populations fell rapidly, in turn causing a precipitous decline in local bald eagle and grizzly bear numbers (Spencer et al. 1991; Figure 7.5).

## 7.1.9 Invasional meltdown

An increasing number of studies of invasion effects have pointed to a phenomenon called "invasional meltdown" in which two or more introduced species interact in such a way that the probability of survival and/or the impact of at least one of them is enhanced (Simberloff and Von Holle 1999). In the above example of an introduced faucet snail (*Bithynia tentaculata*), vectoring a native trematode parasite of ducks and thereby expanding the trematode's range, a recent twist is the arrival of a European trematode (*Leyogonimus polyoon*). *Bithynia* also vectors this species, which has turned out also to be lethal to ducks (Cole and Friend 1999). So in this instance, the introduced snail and the introduced trematode combine to produce more mortality in ducks than either would likely have accomplished alone. This is but one of myriad instances of meltdown.

Sometimes introduced animals either pollinate introduced plants or disperse their seeds. For instance, figs (*Ficus* spp.) introduced to Florida had until ca. 20 years ago remained where they were planted, the species unable to spread because the host-specific fig wasps that pollinate the figs in their native ranges were absent, so the figs could not produce seeds. That situation changed abruptly upon the arrival of the fig-wasps of three of the fig species, which now produce seeds. One of them, *F. microcarpa*, has become an invasive weed, its seeds dispersed by birds and ants (Kauffman et al. 1991). On the island of La Réunion, the red-whiskered bulbul (*Pycnonotus jocosus*), introduced from Asia via Mauritius, disperses seeds of several invasive introduced plants, including *Rubus alceifolius*, *Cordia interruptus*, and *Ligustrum robustum*, which have become far more problematic since the arrival of the bulbul (Baret et al. 2006). The Asian common myna (*Acridotheres tristis*) was introduced to the Hawaiian islands as a biological

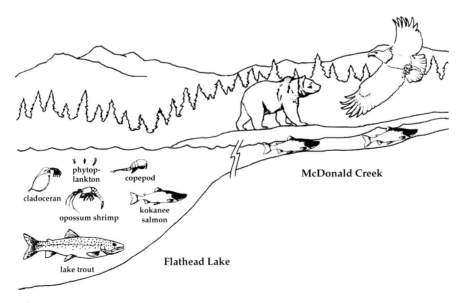

**Figure 7.5** Shrimp stocking, salmon collapse, and eagle displacement. Reprinted from Spencer et al. (1991) © American Institute of Biological Sciences.

control for pasture insects but has ended up dispersing one of the worst weeds, New World *Lantana camara*, throughout the lowlands and even into some native forests (Davis *et al.* 1993). Also in Hawaii, introduced pigs selectively eat and thereby disperse several invasive introduced plant species, and by rooting and defecating they also spread populations of several introduced invertebrates, while themselves fattening up on introduced, protein-rich European earthworms (Stone 1985).

Habitat modification by introduced plants can lead to a meltdown process with expanded and/or accelerated impacts. As noted above, the nitrogen-fixing *Morella faya* (firetree) from the Azores has invaded nitrogen-deficient volcanic regions of the Hawaiian Islands. Because there are no native nitrogen-fixing plants, firetree is essentially fertilizing large areas. Many introduced plants established elsewhere in Hawaii had been unable to colonize these previously nutrient-deficient areas, but their invasion is now facilitated by the activities of firetree (Vitousek 1986). In addition, firetree fosters increased populations of introduced earthworms, and the worms increase the rate of nitrogen burial from firetree litter, thus enhancing the effect of firetree on the nitrogen cycle (Aplet 1990). Finally, introduced pigs and an introduced songbird (the Japanese white-eye, *Zosterops japonicus*) disperse the seeds of the firetree (Stone and Taylor 1984, Woodward *et al.* 1990). In short, all these introduced species create a complex juggernaut of species whose joint interactions are leading to the replacement of native vegetation.

Large, congregating ungulates can interact with introduced plants, pathogens, and even other animals in dramatic cases of invasional meltdown. For instance, Eurasian hooved livestock devastated native tussock grasses in North American prairie regions but favored Eurasian turfgrasses that had coevolved with such animals and that now dominate large areas (Crosby 1986). In northeastern Australia, the Asian water buffalo (*Bubalus bubalis*), introduced as a beast of burden and for meat, damaged native plant communities and eroded stream banks. The Central American shrub *Mimosa pigra* had been an innocuous minor component of the vegetation in the vicinity of the town of Darwin, but the water buffalo, opening up the flood plains, created perfect germination sites of *Mimosa* seedlings, and in many areas native sedgelands became virtual monocultures of *M. pigra*. The mimosa in turn aided the water buffalo by protecting them from aerial hunters (Simberloff and Von Holle 1999).

Aquatic plants and animals can also facilitate one another. In North America, the introduced zebra mussel filters prodigious amounts of water, and the resulting increase in water clarity favors certain plants, including the highly invasive Eurasian watermilfoil (*Myriophyllum spicatum*). The milfoil then aids the mussel by providing a settling surface and facilitates the movement of the mussel to new water bodies when fragments of the plant are inadvertently transported on boat propellers or in water (Simberloff and Von Holle 1999).

Some instances of invasional meltdown arise when one introduced species is later reunited with a coevolved species through the subsequent introduction of the latter. The fig species and their pollinating fig wasps in Florida are an example; the coevolved mutualism between the wasps and the figs is critical to the impact of the fig invasion. However, meltdown need not be between coevolved species. The water buffalo from Asia and *Mimosa pigra* from Central America could not have coevolved, nor could the Asian myna and the New World *Lantana camara* in Hawaii.

### 7.1.10 Multiple effects

Many introduced species have multiple direct and indirect effects on native species, harming some and favoring others at the same time. For example, the round goby (*Neogobius melanostomus*), an Old World fish that arrived in ballast water, is widely recognized in the North American Great Lakes as a harmful invader, feeding on native invertebrates and eggs and larvae of several native fishes. It also competes for food and space with other native fish species. However, the round goby also feeds on the harmful zebra mussel and related quagga mussel (*Dreissena bugensis*), although the impact on their populations is not known. It also now is by far the main food source for the threatened endemic Lake Erie water snake (*Nerodia sipedon insularum*), constituting over 92% of all prey consumed. Further, snakes that feed on the goby grow faster and achieve large size, which may well decrease predation on the snake and increase population size (King *et al.* 2006). On balance, almost all observers would rather not have the round goby in this region, but it is well to bear in mind the complexity of its impacts.

## 7.2 Lag times

Introduced species may be innocuous in their new homes for decades or even centuries before abruptly increasing in numbers and range to generate major impacts. The case of the hybrid cordgrass *Spartina anglica*, discussed above, is an excellent example. The introduced progenitor, North American *S. alterniflora*, had been present in Great Britain at least since the early nineteenth century and had even hybridized with the native *S. maritima* occasionally, but the hybrids were all sterile until one underwent a chromosomal mutation ca. 1891, producing a highly invasive weed (Thompson 1991). Brazilian pepper (*Schinus terebinthifolius*) had been present in Florida since the mid-nineteenth century as isolated individual trees, but it became invasive only when it began to spread rapidly ca. 1940 (Ewel 1986). Giant reed (*Arundo donax*) was first introduced from the Mediterranean region to southern California in the early nineteenth century as a roofing material and for erosion control, and it remained restricted in range and unproblematic until the mid-twentieth century, when it spread widely, becoming a fire hazard, damaging wetlands, and changing entire ecosystems (Dudley 2000). The Caribbean brown anole lizard (*Anolis sagrei*) first appeared in Florida in the nineteenth century, but it was restricted to extreme south Florida until the 1940s, when its range began an expansion that accelerated in the 1970s, ultimately to cover most of Florida (Kolbe *et al.* 2004).

Many such invasion lags remain mysterious. For instance, the delay for giant reed in California has yet to be explained. In other instances, a change in the physical or biotic environment can account for a sudden explosion of a formerly restricted introduced species. The spread of Brazilian pepper in Florida after a century of harmless presence was caused by hydrological changes—draining farmland, various flood control projects, and lowering of the water table for agricultural and human use. As described earlier, the sudden invasion by long-present figs in south Florida was spurred by the arrival of pollinating fig wasps. In some instances, demography of a species dictates that it cannot build up population sizes rapidly even if the environment is suitable; trees, for example, have long life cycles and many do not begin reproducing for a decade or more.

As genetic analysis has recently rapidly expanded with the advent of various molecular tools, it appears that some, and perhaps many, sudden expansions after a lag phase occur because of the introduction of new genotypes to a previously established but restricted population. The brown anole population in Florida was augmented in the twentieth century by the arrival of individuals from different parts of the native range, so that the population in Florida now has far more genetic diversity than is found in any native population. It is possible that the rapid range expansion of this introduction results from introductions to new sites combined with the advent of new genotypes better adapted to the array of environmental conditions found in Florida (Kolbe *et al.* 2004). The northward range expansion of European green crab (*Carcinus maenas*) along the Atlantic coast of North America was produced by the introduction of new, cold-tolerant genotypes into the established population (Roman 2006).

An improved understanding of lag times is important in understanding how best to manage biological invasions (Boggs *et al.* 2006). It is not feasible to attempt active management (see next section) of all introduced species—there are simply too many. Typically in each site we focus on those that are already invasive or that we suspect will become invasive from observations elsewhere. However, if some currently innocuous established introduced species are simply biological time bombs waiting to explode when the right conditions prevail in the future, the existing approach clearly will not suffice.

## 7.3 What to do about invasive species

By far the best thing to do about invasive introduced species is to keep them out in the first place. If we fail to keep them out and they establish populations, the next possibility is to attempt to find them quickly and perhaps to eradicate them. If they have already established and begun to spread widely, we may still try to eradicate them, or we can instead try to keep their populations at sufficiently low levels that they do not become problems.

### 7.3.1 Keeping them out

Introductions can be either planned (deliberate) or inadvertent, and preventing these two classes involves somewhat different procedures. In each instance, prevention involves laws, risk analyses, and border control. For planned introductions, such as of ornamental plants or new sport fish or game species, the law would be either a "white list," a "black list," or some combination of the two. A white list is a list of species approved for introduction, presumably after some risk analysis in which consideration is given to the features of the species intended for introduction and the outcome in other regions where it has been introduced. The most widely used risk analyses currently include versions of the Australian Weed Risk Assessment, which consists of a series of questions about species proposed for introduction and an algorithm for combining the answers to those questions to give a score, for which there is a threshold above which a species cannot be admitted (Pheloung *et al.* 1999). A black list is a list of species that cannot be admitted under any circumstances, and for which no further risk analysis is needed. Examples of black lists include the United States Federal Noxious Weed list and a short list of animals forbidden for entry to the US under the Lacey Act.

For such lists to be effective, the risk analyses have to be accurate enough, and the lists sufficiently large, that the great majority of species that would become invasive are actually identified as such and placed on black lists or kept off white lists. There are grave concerns that neither criterion is met. For instance, the black list of the Lacey Act is very short, and many animal species that have a high probability of becoming invasive if introduced are not on the list. The risk assessment tools, on the other hand, all yield some percentage of false negatives—that is, species assessed as unlikely to cause harm, therefore eligible for a white list, when in fact they will become harmful. Much active research (e.g. Kolar and Lodge 2002) is aimed at improving the accuracy of risk analyses—especially lowering the rate of false negatives while not inflating the rate of false positives (species judged likely to become invasive when, in fact, they would not).

For inadvertent introductions, one must first identify pathways by which they occur (Ruiz and Carlton 2003). For instance, many marine organisms are inadvertently carried in ballast water (this is probably how the zebra mussel entered North America). Insects stow away on ornamental plants or agricultural products. The Asian longhorned beetle (*Anoplophora glabripennis*), a dangerous forest pest, hitchhiked to North America in untreated wooden packing material from Asia, while snails have been transported worldwide on paving stones and ceramics. The Asian tiger mosquito (*Aedes albopictus*) arrived in the United States in water transported in used tires. Once these pathways have been identified, their use as conduits of introduction must be restricted. For ballast water, for example, water picked up as ballast in a port can be exchanged with water from the open ocean to lower the number of potential invaders being transported. For insects and pathogens carried in wood, heat and chemical treatment may be effective. For agricultural products, refrigeration, and/or fumigation are often used. The general problem is that each of these procedures entails a cost, and there has historically been opposition to imposing such costs on the grounds that they interfere with free trade and make goods more expensive. Thus it remains an uphill battle to devise and to implement regulations sufficiently stringent that they constrict these pathways.

Whatever the regulations in place for both deliberate and unplanned introductions, inspections at ports of entry are where they come into play, and here a variety of detection technologies are available and improvements are expected. Trained sniffer dogs are commonplace in ports in many countries, and various sorts of machinery, including increasingly accurate X-ray equipment, are widely in use (Baskin 2002). Although technologies have improved to aid a port inspector to identify a potential invader once it has been detected, in many nations these are not employed because of expense or dearth of qualified staff. Also, improved detection and identification capabilities are only half of the solution to barring the introduction of new species either deliberately or by accident (as for example, in dirt on shoes, or in untreated food). The other half consists of penalties sufficiently severe that people fear the consequences if they are caught introducing species. Many nations nowadays have extensive publicity at ports of entry, on planes and ships, and sometimes even in popular media, that combine educational material about the many harmful activities of invasive species and warnings about penalties for importing them.

### 7.3.2 Monitoring and eradication

The key to eradicating an introduced species before it can spread widely is an early warning-rapid response system, and early warning requires an ongoing monitoring program. Because of the great expense of trained staff, few if any nations adequately monitor consistently for all sorts of invasions, although for specific habitats (e.g. waters in ports) or specific groups of species (e.g. fruit fly pests of agriculture) intensive ongoing monitoring exists in some areas. Probably the most cost-effective way to improve monitoring is to enlist the citizenry to be on the lookout for unusual plants or animals and to know what agency to contact should they see something (see Figure 7.6 and Plate 8). Such efforts entail public education and wide dissemination in popular media and on the web, but they can yield enormous benefits. For instance, the invasion of the Asian longhorned beetle to the Chicago region was discovered by a citizen gathering firewood who recognized the beetle from news reports and checked his identification on a state agency website. This early warning and a quick, aggressive response by authorities led to successful regional extirpation of this insect after a five-year campaign. Similarly, the invasion in California of the alga *Caulerpa taxifolia* was discovered probably within a year of its occurrence by a diver who had seen publicity about the impact of this species in the Mediterranean. This discovery led to successful eradication after a four-year effort, and citizens have been alerted to watch for this and other non-native algal species in both Mediterranean nations and California.

Many introduced species have been successfully eradicated, usually when they are found early but occasionally when they have already established widespread populations. The keys to successful eradication have been as follows; (i) Sufficient resources must be available to see the project through to completion; the expense of finding and removing the last few individuals may exceed that of quickly ridding a site of the majority of the population; (ii) Clear lines of authority must exist so that an individual or agency can compel cooperation. Eradication is, by its nature, an all-or-none operation that can be subverted if a few individuals decide not to cooperate (for instance, by forbidding access to private property, or forbidding the use of a pesticide or herbicide); (iii) The biology of the target organism must be studied well enough that a weak point in its life cycle is identified; and (iv) Should the eradication succeed, there must be a reasonable prospect that reinvasion will not occur fairly quickly.

In cases where these criteria have been met, successful eradications are numerous. Many are on islands, because they are often small and because reinvasion is less likely, at least for isolated islands. Rats have been eradicated from many islands worldwide; the largest to date is 113 km$^2$. Recently, large, longstanding populations of feral goats and pigs have been

Figure 7.6 Maryland's aquatic invasive species. Poster courtesy of the Maryland Department of Natural Resources.

eradicated from Santiago Island (585 km²) in the Galapagos (Cruz *et al.* 2005). The giant African snail has been successfully eradicated from sites in both Queensland and Florida (Simberloff 2003). Even plants with soil seed banks have been eradicated, such as sand bur (*Cenchrus echinatus*) from 400 ha Laysan Island (Flint and Rehkemper 2002). When agriculture or public health are issues, extensive and expensive eradication campaigns have been undertaken and have often been successful, crowned by the global eradication of smallpox. The African mosquito (*Anopheles gambiae*), vector of malaria, was eradicated from a large area in northeastern Brazil (Davis and Garcia 1989), and various species of flies have been eradicated from many large regions, especially in the tropics (Klassen 2005). The pasture weed *Kochia scoparia* was eradicated from a large area of Western Australia (Randall 2001), and the witchweed eradication campaign in the southeastern United States mentioned above is nearing success. These successes suggest that, if conservation is made a high enough priority, large-scale eradications purely for conservation purposes may be very feasible.

A variety of methods have been used in these campaigns: males sterilized by X-rays for fruitflies, chemicals for *Anopheles gambiae* and for rats, hunters and dogs for goats. Some campaigns that probably would have succeeded were stopped short of their goals not for want of technological means but because of public objections to using chemicals or to killing vertebrates. A notable example is the cessation, because of pressure from animal-rights groups, of the well-planned campaign to eradicate the gray squirrel before it spreads in Italy (Bertolino and Genovese 2003).

### 7.3.3 Maintenance management

If eradication is not an option, many available technologies may limit populations of invasive species so that damage is minimized. There are three main methods—mechanical or physical control, chemical control, and biological control. Sometimes these methods can be combined, especially mechanical and chemical control. In South Africa, the invasive Australian rooikrans tree (*Acacia cyclops*) can be effectively controlled by mechanical means alone—cutting and pulling roots—so long as sufficient labor is available (Matthews and Brand 2004a). Sometimes chemical control alone can keep a pest at low numbers. The Indian house crow (*Corvus splendens*), is an aggressive pest in Africa, attacking native birds, competing with them for food, preying on local wildlife, stripping fruit trees, and even dive-bombing people and sometimes stealing food from young children. It can be controlled by a poison, Starlicide, so long as the public does not object (Matthews and Brand 2004a). Many invasive plants have been kept at acceptable levels by herbicides. For instance, in Florida, water hyacinth was drastically reduced and subsequently managed by use of the herbicide 2,4-D, combined with some mechanical removal (Schardt 1997). For lantana in South Africa, a combination of mechanical and chemical control keeps populations minimized in some areas (Matthews and Brand 2004a). A South African public works program, Working for Water, has had great success using physical, mechanical, and chemical methods to clear thousands of hectares of land of introduced plants that use prodigious amounts of water, such as mesquite (*Prosopis* spp.) and several species of *Acacia* (Matthews and Brand 2004a). Similarly, in the Canadian province of Alberta, Norway rats have been kept at very low levels for many years by a combination of poisons and hunting by the provincial Alberta Rat Patrol (Bourne 2000).

However, long-term use of herbicides and pesticides often leads to one or more problems. First is the evolution of resistance in the target species, so that increasing amounts of the chemical have to be used even on a controlled population. This has happened recently with the use of the herbicide used to control Asian *Hydrilla verticillata* in Florida (Puri *et al.* 2007), and it is a common phenomenon in insect pests of agriculture. A second, related problem is that chemicals are often costly, and they can be prohibitively expensive if used over large areas. Whereas the market value of an agricultural product may be perceived as large enough to warrant such great expense, it may be difficult to convince a government agency that it is worth controlling an introduced species affecting conservation values that are not easily quantified. Finally, chemicals often have non-target impacts, including human health impacts. The decline of raptor populations as DDT residues caused thin eggshells is a famous example (Lundholm 1997). Many later-generation herbicides and pesticides have few if any non-target impacts when used properly, but expense may still be a major issue.

These problems with pesticides have led to great interest in the use of classical biological control—deliberate introduction of a natural enemy (predator, parasite, or disease) of an introduced pest. This is the philosophy of fighting fire with fire. Although only a minority of well-planned biological control projects actually end up controlling the target pest, those that have succeeded are often dramatically effective and conferred low-cost control in perpetuity. For instance, massive infestations of water hyacinth in the Sepik River catchment of New Guinea were well controlled by introduction of the two South American weevils that had been used for this purpose in Lake Victoria, *Neochetina eichhorniae* and *N. bruchi* (Matthews and Brand 2004b). A recent success on the island of St. Helena is the control of a tropical American scale insect (*Orthezia insignis*) that had threatened the existence of the endemic gumwood tree (*Commidendrum robustum*). A predatory South American lady beetle (*Hyperaspis pantherina*) now keeps the scale insect population at low densities (Booth *et al.* 2001). Even when a biological control agent successfully controls a target pest at one site, it may fail to do so elsewhere. The same two weevils that control water hyacinth in New Guinea and Lake Victoria had minimal effects on the hyacinth in Florida, even though they did manage to establish populations (Schardt 1997).

However, in addition to the fact that most biological control projects have not panned out, several biological control agents have attacked non-target species and even caused extinctions—the cases involving the cactus moth, rosy wolf snail, small Indian mongoose, mosquitofish, and thistle-eating weevil have been mentioned earlier. In general, problems of this sort have been associated with introduced biological control agents such as generalized predators that are not specialized to use the specific target host. However, even species that are restricted to a single genus of host, such as the cactus moth, can create problems.

## Summary

- Invasive species cause myriad sorts of conservation problems, many of which are complicated, some of which are subtle, and some of which are not manifested until long after a species is introduced.
- The best way to avoid such problems is to prevent introductions in the first place or, failing that, to find them quickly and eradicate them.
- However, many established introduced species can be managed by a variety of technologies so that their populations remain restricted and their impacts are minimized.

## Suggested reading

Baskin, Y. (2002). *A plague of rats and rubbervines*. Island Press, Washington, DC.
Davis, M. A. (2009). *Invasion biology*. Oxford University Press, Oxford.
Elton, C. E. (1958). *The ecology of invasions by animals and plants*. Methuen, London (reprinted by University of Chicago Press, 2000).
Lockwood, J. L., Hoopes, M. F., and Marchetti, M. P. (2007). *Invasion ecology*. Blackwell, Malden, Massachusetts.
Van der Weijden, W., Leewis, R., and Bol, P. (2007). *Biological globalisation*. KNNV Publishing, Utrecht, the Netherlands.

## Relevant websites

- World Conservation Union Invasive Species Specialist Group: http://www.issg.org/index.html.
- National Invasive Species Council of the United States: http://www.invasivespecies.gov.
- National Agriculture Library of the United States: http://www.invasivespeciesinfo.gov.
- European Commission: http://www.europealiens.org.

## REFERENCES

Allen, J. A. (1998). Mangroves as alien species: the case of Hawaii. *Global Ecology and Biogeography Letters*, **7**, 61–71.

Aplet, G. H. (1990). Alteration of earthworm community biomass by the alien *Myrica faya* in Hawaii. *Oecologia*, **82**, 411–416.

Ashton, P. A. and Abbott, R. J. (1992). Multiple origins and genetic diversity in the newly arisen allopolyploid species, *Senecio cambrensis* Rosser (Compositae). *Heredity*, **68**, 25–32.

Atkinson I. A. E. (1985). The spread of commensal species of *Rattus* to oceanic islands and their effects on island avifaunas. In P.J. Moors, ed. *Conservation of island birds*, pp. 35–81. International Council of Bird Conservation Technical Publication No.3.

Baret, S., Rouget, M., Richardson, D.M., *et al.* (2006). Current distribution and potential extent of the most invasive alien plant species on La Réunion (Indian Ocean, Mascarene islands). *Austral Ecology*, **31**, 747–758.

Bartrip, P. W. J. (2008). *Myxomatosis: A history of pest control and the rabbit*. Macmillan, London.

Baskin, Y. (2002). *A plague of rats and rubbervines*. Island Press, Washington, DC.

Bergersen, E. P. and Anderson, D. E. (1997). The distribution and spread of *Myxobolus cerebralis* in the United States. *Fisheries*, **22**, 6–7.

Bertolino, S. and Genovese, P. (2003). Spread and attempted eradication of the grey squirrel (*Sciurus carolinensis*) in Italy, and consequences for the red squirrel (*Sciurus vulgaris*) in Eurasia. *Biological Conservation*, **109**, 351–358.

Bertolino, S., Lurz, P. W. W., Sanderson, R., and Rushton, S. P. (2008). Predicting the spread of the American grey squirrel (*Sciurus carolinensis*) in Europe: A call for a co-ordinated European approach. *Biological Conservation*, **141**, 2564–2575.

Boggs, C., Holdren, C. E., Kulahci, I. G., *et al.* (2006). Delayed population explosion of an introduced butterfly. *Journal of Animal Ecology*, **75**, 466–475.

Booth, R. G., Cross, A. E., Fowler, S. V., and Shaw, R.H. (2001). Case study 5.24. Biological control of an insect to save an endemic tree on St. Helena. In R. Wittenberg and M. J. W. Cock, eds *Invasive alien species: A toolkit of best prevention and management practices*, p. 192. CAB International, Wallingford, UK.

Bourne, J. (2000). *A history of rat control in Alberta*. Alberta Agriculture, Food and Rural Development, Edmonton.

Calder, I. R. (2002). Eucalyptus, water and the environment. In J. J. W. Coppen, ed. *Eucalyptus. The genus Eucalyptus*, pp. 36–51. Taylor and Francis, New York.

Callaway, R. M. and Ridenour, W. M. (2004). Novel weapons: Invasive success and the evolution of increased competitive ability. *Frontiers in Ecology and the Environment*, **2**, 436–443.

Callaway, R. M., Cipollini, D., Barto, K., *et al.* (2008). Novel weapons: invasive plant suppresses fungal mutualisms in America but not in its native Europe. *Ecology*, **89**, 1043–1055.

Caltagirone L. E. and Doutt, R. L. (1989). The history of the *Vedalia* beetle importation to California and its impact on the development of biological control. *Annual Review of Entomology*, **34**, 1–16.

Center, T. D., Frank, J. H., and Dray, F. A. Jr. (1997). Biological control. In D. Simberloff, D. C. Schmitz, and T. C. Brown, eds *Strangers in paradise. Impact and management of nonindigenous species in Florida*, pp. 245–263. Island Press, Washington, DC.

Cole, R. A. and Friend, M. (1999). Miscellaneous parasitic diseases. In M. Friend and J. C. Franson, eds *Field manual of wildlife diseases*, pp. 249–262. U.S. Geological Survey, Biological Resources Division, National Wildlife Health Center, Madison, Wisconsin.

Cowie, R. H. (2002). Invertebrate invasions on Pacific islands and the replacement of unique native faunas: a synthesis of the land and freshwater snails. *Biological Invasions*, **3**, 119–136.

Cronk, Q. C. B. (1989). The past and present vegetation of St Helena. *Journal of Biogeography*, **16**, 47–64.

Crosby, A. W. (1986). *Ecological imperialism. The biological expansion of Europe, 900–1900*. Cambridge University Press, Cambridge.

Cruz, F., Donlan, C. J., Campbell, K., and Carrion, V. (2005). Conservation action in the Galápagos: feral pig (*Sus scrofa*) eradication from Santiago Island. *Biological Conservation*, **121**, 473–478.

D'Antonio, C. M. and Vitousek, P. M. (1992). Biological invasions by exotic grasses, the grass/fire cycle, and global change. *Annual Review of Ecology and Systematics*, **23**, 63–87.

Davis, C. J., Yoshioka, E., and Kageler, D. (1993). Biological control of lantana, prickly pear, and hamakua pamakane in Hawai'i: a review and update. In C.P. Stone, C. W. Smith, and J. T. Tunison, eds *Alien plant invasions in native ecosystems of Hawaii*, pp. 411–431. University of Hawaii Press, Honolulu.

Davis, J. R. and Garcia, R. (1989). Malaria mosquito in Brazil. In D. L. Dahlsten and R. Garcia, eds *Eradication of exotic pests*, pp. 274–283. Yale University Press, New Haven, Connecticut.

Doutt, R. L. (1964). The historical development of biological control. In P. DeBach, ed. *Biological control of insect pests and weeds*, pp. 21–42. Chapman and Hall, London.

Dudley, T. L. (2000). *Arundo donax* L. In C. C. Bossard, J. M. Randall, and M. C. Hoshovsky, eds *Invasive plants of*

*California's wildlands*, pp. 53–58. University of California Press, Berkeley, California.

Durán, A., Gryzenhout, M., Slippers, B., et al. (2008). *Phytophthora pinifolia* sp. nov. associated with a serious needle disease of *Pinus radiata* in Chile. *Plant Pathology*, **57**, 715–727.

Ellison, A. M., Bank, M. S., Clinton, B. D., et al. (2005). Loss of foundation species: consequences for the structure and dynamics of forested ecosystems. *Frontiers in Ecology and the Environment*, **9**, 479–486.

Eplee, R. E. (2001). Case study 2.10. Co-ordination of witchweed eradication in the U.S.A. In R. Wittenberg and M. J. W. Cock, eds *Invasive alien species: A toolkit of best prevention and management practices*, p. 36. CAB International, Wallingford, UK.

Ewel, J. J. (1986). Invasibility: Lessons from south Florida. In H. A. Mooney and J. A. Drake, eds *Ecology of biological invasions of North America and Hawaii*, pp. 214–230. Springer-Verlag, New York.

Flint, E. and Rehkemper, C. (2002). Control and eradication of the introduced grass, *Cenchrus echinatus*, at Laysan Island, Central Pacific Ocean. In C. R. Veitch and M. N. Clout, eds *Turning the tide: the eradication of invasive species*, pp. 110–115. IUCN SSC Invasive Species Specialist Group. IUCN, Gland, Switzerland.

Goodell, K., Parker, I. M., and Gilbert, G. S. (2000). Biological impacts of species invasions: Implications for policy makers. In National Research Council (US), *Incorporating science, economics, and sociology in developing sanitary and phytosanitary standards in international trade*, pp. 87–117. National Academy Press, Washington, DC.

Hays, W. T. and Conant, S. (2006). Biology and impacts of Pacific Islands invasive species. 1. A worldwide review of effects of the small Indian mongoose, *Herpestes javanicus* (Carnivora: Herpestidae). *Pacific Science*, **61**, 3–16.

Holway, D.A. and Suarez, A.V. (2004). Colony structure variation and interspecific competitive ability in the invasive Argentine ant. *Oecologia*, **138**, 216–222.

Hubbs, C. and Jensen, B. L. (1984). Extinction of *Gambusia amistadensis*, an endangered fish. *Copeia*, **1984**, 529–530.

Jeschke, J. M. and Strayer, D. L. (2005). Invasion success of vertebrates in Europe and North America. *Proceedings of the National Academy of Sciences of the United States of America*, **102**, 7198–7202.

Kauffman, S., McKey, D. B., Hossaert-McKey, M., and Horvitz, C. C. (1991). Adaptations for a two-phase seed dispersal system involving vertebrates and ants in a hemiepiphytic fig (*Ficus microcarpa*: Moraceae). *American Journal of Botany*, **78**, 971–977.

King, R. B., Ray, J. M., and Stanford, K. M. (2006). Gorging on gobies: beneficial effects of alien prey on a threatened vertebrate. *Canadian Journal of Zoology*, **84**, 108–115.

Klassen, W. (2005). Area-wide integrated pest management and the sterile insect technique. In V. A. Dyck, J. Hendrichs and A. S. Robinson, eds *Sterile insect technique. Principles and practice in area-wide integrated pest management*, pp. 39–68. Springer, Dordrecht, the Netherlands.

Kolar, C. S. and Lodge, D. M. (2002). Ecological predictions and risk assessment for alien fishes in North America. *Science*, **298**, 1233–1236.

Kolbe, J. J., Glor, R. E., Schettino, L. R., Lara, A. C., Larson, A., and Losos, J.B. (2004). Genetic variation increases during biological invasion by a Cuban lizard. *Nature*, **431**, 177–181.

Lizarralde. M., Escobar, J., and Deferrari, G. (2004). Invader species in Argentina: a review about the beaver (*Castor canadensis*) population situation on Tierra del Fuego ecosystem. *Interciencia*, **29**, 352–356.

Lockwood, J. L., Hoopes, M. F., and Marchetti, M. P. (2007). *Invasion ecology*. Blackwell, Malden, Massachusetts.

Louda, S. M., Kendall, D., Connor, J., and Simberloff, D. (1997). Ecological effects of an insect introduced for the biological control of weeds. *Science*, **277**, 1088–1090.

Lundholm, C. E. (1997). DDE-Induced eggshell thinning in birds. *Comparative Biochemistry and Physiology Part C: Pharmacology, Toxicology and Endocrinology*, **118**, 113–128.

Maran, T. and Henttonen, H. (1995). Why is the European mink (*Mustela lutreola*) disappearing?—A review of the process and hypotheses. *Annales Zoologici Fennici*, **32**, 47–54.

Matthews, S. and Brand, K. (2004a). *Africa invaded*. Global Invasive Species Programme, Cape Town, South Africa.

Matthews, S. and Brand, K. (2004b). *Tropical Asia invaded*. Global Invasive Species Programme, Cape Town, South Africa.

McDaniel, K. C., DiTomaso, J. M., and Duncan, C. A. (2005). Tamarisk or saltcedar, *Tamarix* spp. In J. K. Clark and C. L. Duncan, eds *Assessing the economic, environmental and societal losses from invasive plants on rangeland and wildlands*, pp. 198–222. Weed Science Society of America, Champaign, Illinois.

Meinesz, A. (1999). *Killer algae*. University of Chicago Press, Chicago.

Moyle, P. B. (1993). *Fish: An enthusiast's guide*. University of California Press, Berkeley.

Muñoz-Fuentes, V., Vilà, C., Green, A. J., Negro, J., and Sorenson, M. D. (2007). Hybridization between white-headed ducks and introduced ruddy ducks in Spain. *Molecular Ecology*, **16**, 629–638.

Norgaard, R. B. (1988). The biological control of cassava mealybug in Africa. *American Journal of Agricultural Economics*, **70**, 366–371.

Office of Technology Assessment (US Congress). (1993). *Harmful non-indigenous species in the United States*. OTA-F-565. US Government Printing Office, Washington, DC.

Opler, P. A. (1978). Insects of American chestnut: possible importance and conservation concern. In J. McDonald, ed. *The American chestnut symposium*, pp. 83–85. West Virginia University Press, Morgantown, West Virginia.

Petren, K. and Case, T. J. (1996). An experimental demonstration of exploitation competition in an ongoing invasion. *Ecology*, **77**, 118–132.

Pheloung, P. C., Williams, P. A., and Halloy, S. R. (1999). A weed risk assessment model for use as a biosecurity tool evaluating plant introductions. *Journal of Environmental Management*, **57**, 239–251.

Plowright, W. (1982). The effects of rinderpest and rinderpest controlon wildlife in Africa. *Symposia of the Zoological Society of London*, **50**, 1–28.

Pringle, R. M. (2005). The origins of the Nile perch in Lake Victoria. *BioScience*, **55**, 780–787.

Puri, A., MacDonald, G. E., and Haller, W. T. (2007). Stability of fluridone-resistant hydrilla (*Hydrilla verticillata*) biotypes over time. *Weed Science*, **55**, 12–15.

Pyke, G. H. (2008). Plague minnow or mosquito fish? A review of the biology and impacts of introduced *Gambusia* species. *Annual Review of Ecology, Evolution, and Systematics*, **39**, 171–191.

Rabenold, K. N., Fauth, P. T., Goodner, B. W., Sadowski, J. A., and Parker, P. G. (1998). Response of avian communities to disturbance by an exotic insect in spruce-fir forests of the southern Appalachians. *Conservation Biology*, **12**, 177–189.

Randall, R. (2001). Case study 5.5. Eradication of a deliberately introduced plant found to be invasive. In R. Wittenberg and M. J. W. Cock, eds *Invasive alien species: A toolkit of best prevention and management practices*, p. 174. CAB International, Wallingford, UK.

Ratcliffe, D. (1979). The end of the large blue butterfly. *New Scientist*, **8**, 457–458.

Rhymer, J. and Simberloff, D. (1996). Extinction by hybridization and introgression *Annual Review of Ecology and Systematics*, **27**, 83–109.

Ricciardi, A., Neves, R. J., and Rasmussen, J. B. (1998). Impending extinctions of North American freshwater mussels (Unionoida) following the zebra mussel (*Dreissena polymorpha*) invasion. *Journal of Animal Ecology*, **67**, 613–619.

Roman, J. (2006). Diluting the founder effect: cryptic invasions expand a marine invader's range. *Proceedings of the Royal Society of London B*, **273**, 2453–2459.

Ruiz, G. M. and Carlton, J. T., eds (2003). *Invasive species. Vectors and management strategies*. Island Press, Washington, DC.

Sauer, J. S., Cole, R. A., and Nissen, J. M. (2007). *Finding the exotic faucet snail (Bithynia tentaculata): Investigation of waterbird die-offs on the upper Mississippi River National Wildlife and Fish Refuge*. US Geological Survey Open-File Report 2007–1065, US Geological Survey, Washington, DC.

Schardt, J. D. (1997). Maintenance control. In D. Simberloff, D. C. Schmitz, and T. C. Brown, eds *Strangers in paradise. Impact and management of nonindigenous species in Florida*, pp. 229–243. Island Press, Washington, DC.

Schmitz, D. C., Simberloff, D., Hofstetter, R. H., Haller, W., and Sutton, D. (1997). The ecological impact of nonindigenous plants. In D. Simberloff, D. C. Schmitz, and T. C. Brown, eds *Strangers in paradise. Impact and management of nonindigenous species in Florida*, pp. 39–61. Island Press, Washington, DC.

Simberloff, D. (2003). How much information on population biology is needed to manage introduced species? *Conservation Biology*, **17**, 83–92.

Simberloff, D. and Von Holle, B. (1999). Positive interactions of nonindigenous species: invasional meltdown? *Biological Invasions*, **1**, 21–32.

Spencer C. N., McClelland, B. R., and Stanford, J. A. (1991). Shrimp stocking, salmon collapse, and eagle displacement. *BioScience*, **41**, 14–21.

Stone, C. P. (1985). Alien animals in Hawai'i's native ecosystems: toward controlling the adverse effects of introduced vertebrates. In C. P. Stone and J. M. Scott, eds *Hawai'i's terrestrial ecosystems: Preservation and management*, pp. 251–297. University of Hawaii, Honolulu, Hawaii.

Stone, C. P. and Taylor, D. D. (1984). Status of feral pig management and research in Hawaii Volcanoes National Park. *Proceedings of the Hawaii Volcanoes National Park Science Conference*, **5**, 106–117.

Thompson, H. V. and King, C. M., eds (1994). *The European rabbit. The history and ecology of a successful colonizer*. Oxford University Press, Oxford.

Thompson, J. D. (1991). The biology of an invasive plant: What makes *Spartina anglica* so successful? *BioScience*, **41**, 393–401.

Tschinkel, W. R. (2006). *The fire ants*. Harvard University Press, Cambridge, Massachusetts.

US Department of the Interior (Fish and Wildlife Service). (1997). Endangered and threatened wildlife and plants; Determination of Endangered status for two tidal marsh plants—*Cirsium hydrophilum* var. *hydrophilum* (Suisun Thistle) and *Cordylanthus mollis* ssp. *mollis* (Soft Bird's-Beak) from the San Francisco Bay area of California. 50 CFR Part 17. *Federal Register*, **62**, 61916–61921.

Van Den Bosch, R., Schlinger, E. I., Dietrick, E. J., Hall, J. C., and Puttler, B. (1964). Studies on succession, distribution, and phenology of imported parasites of *Therioaphis trifolii* (Monell) in southern California. *Ecology*, **45**, 602–621.

Vitousek, P. (1986). Biological invasions and ecosystem properties: can species make a difference? In H. A. Mooney and J. A. Drake, eds *Ecology of biological invasions of North America and Hawaii*, pp. 163–176. Springer, New York.

Vivrette, N. J. and Muller, C. H. (1977). Mechanism of Invasion and Dominance of Coastal Grassland by Mesembryanthemum crystallinum *Ecological Monographs*, **47**, 301–318.

Williamson, M. (1996). *Biological invasions*. Chapman and Hall, London, UK.

Woodward, S. A., Vitousek, P. M., Matson, K., Hughes, F., Benvenuto, K., and Matson, P. (1990). Use of the exotic tree *Myrica faya* by native and exotic birds in Hawai'i Volcanoes National Park. *Pacific Science*, **44**, 88–93.

Woodworth, B. L., Atkinson, C. T., LaPointe, D. A., *et al.* (2005). Host population persistence in the face of introduced vector-borne disease: Hawaii amakihi and avian malaria. *Proceedings of the National Academy of Sciences of the United States of America*, **102**, 1531–1536.

Zimmermann, H. G., Moran, V. C., and Hoffmann, J. H. (2001). The renowned cactus moth, *Cactoblastis cactorum* (Lepidoptera: Pyralidae): Its natural history and threat to native *Opuntia* floras in Mexico and the United States of America. *Florida Entomologist*, **84**, 543–551.

# CHAPTER 8

# Climate change

Thomas E. Lovejoy

In 1896 Swedish physicist Arrhenius asked a new and important question, namely why is the temperature of the Earth so suitable for humans and other forms of life? From that emerged the concept of the greenhouse effect, namely that the concentrations of various atmospheric gases [e.g. carbon dioxide ($CO_2$), methane, nitrous oxide, chlorofluorocarbons; also called greenhouse gasses] was such that some of the radiant heat received from the sun is trapped, rendering the earth a considerably warmer planet than it otherwise would be. Arrhenius even did a manual calculation of the effect of doubling the pre-industrial level of $CO_2$. His results are precisely what the supercomputer models of Earth's climate predict. We are well on the way toward that $CO_2$ concentration, having started at pre-industrial levels of 280 ppm (parts per million). Current atmospheric levels are 390 ppm of $CO_2$, and are increasing at a rate above the worst case scenario of the Intergovernmental Panel for Climate Change (IPCC) (Canadell *et al.* 2007).

Modern science is able to study past climate, so we now know that the last 10 000 years were a period of unusual stability in the global climate. This probably has been extremely beneficial to the human species for that period includes all our recorded history as well as the origins of agriculture and of human settlements. It is easy to conclude that the entire human enterprise is based on a freak stretch of relatively unchanging climatic conditions.

A bit less obvious is the realization that ecosystems have adjusted to that stable climate also so they – as well as the benefits society receives in ecosystem goods and services (see Chapter 3) – are vulnerable to climate change as well. Indeed, it is rapidly becoming clear that the natural world is as – or more – sensitive to climate than anything else society is concerned about.

The current levels of greenhouse gas concentration have already led to an overall rise in global temperature of 0.75 degree Celsius (see Figure 8.1). In addition, because there is a lag between attaining a concentration level and the consequent trapping of heat energy, the planet is slated for an additional 0.5 degree (for a total of 1.25 degrees Celsius) even if greenhouse gas concentrations were to cease to increase immediately.

This chapter highlights the effects of human-induced climate change on Earth's physical environments and biodiversity. Possible mitigation options of this predicament are also briefly discussed.

## 8.1 Effects on the physical environment

Already there are widespread changes in the physical environment, primarily involving the solid and liquid phases of water. Northern hemisphere lakes are freezing later in the autumn and the ice is breaking up earlier in the spring. Glaciers are in retreat in most parts of the world, and those on high peaks in the tropics like Mount Kilimanjaro (Tanzania) are receding at a rate that they will likely cease to exist in 15 years (UNEP 2007). The melt rate of Greenland glaciers is increasing and the seismic activity they generate is accelerating.

Arctic sea ice is retreating at unprecedented rates, as would be predicted by the increased heat absorption capacity of dark open water as compared to reflective ice. This represents a positive feedback, namely the more dark water replaces what had been reflecting ice the more heat is absorbed and the more the Earth warms.

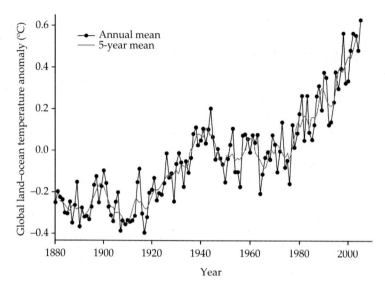

**Figure 8.1** Global annual mean temperature anomaly relative to 1951–80. Reprinted from Hansen et al. (2006) © National Academy of Sciences, USA.

The danger of positive feedbacks is that they accelerate climate change and can lead to a "runaway greenhouse effect". The first summer with an ice free Arctic Ocean once predicted for 2100 is now possible in 2030, with some predictions suggesting as soon as in next five years.

In addition, there is a statistically significant increase in wildfires in the American West because longer summers and earlier melt of the snow pack have led to dryer environments and higher fire vulnerability (Flannigan et al. 2000). Argentina, the American southwest, and Australia in 2009 were experiencing unusual drought, and parts of southern Australia had extraordinarily high temperatures and devastating fires in the summer of 2008–2009. In addition there is the possible increase in the number of intense tropical cyclones like Katrina, although there is still some uncertainty on the matter. Another additional system change was previewed in 2005 when Atlantic circulation changes triggered the greatest drought in recorded history in the Amazon. The Hadley Center global climate model and other work predict similar but relatively permanent change at 2.5 degrees Centigrade with consequent Amazon dieback (mostly in the eastern half of the basin) (Malhi et al. 2009).

Other possible examples of system change would be methane release from thawing permafrost in the tundra – another dangerous positive feedback loop. The first signs of this have been observed in Siberia and Alaska. These are all part of how the Earth system functions. Although understanding of the Earth system is only preliminary it clearly includes thresholds and teleconnections (changes in one part of the globe can trigger changes in some far distant part). Increasing climate change is taking the planet in that dangerous direction.

Oceans are also threatened by acidification caused by elevated $CO_2$ levels in the atmosphere. A significant part of that $CO_2$ is absorbed by the oceans but some of it becomes carbonic acid. As a consequence the acidity of the oceans has increased 0.1 pH unit since pre-industrial times – a number that sounds trivial but being on a logarithmic scale is equivalent to 30% more acid.

All these changes to the physical environment have consequences for biodiversity.

## 8.2 Effects on biodiversity

Populations, species and ecosystems are responding to these physical changes all over the planet.

Many species are changing the timing of their life histories (phenology) (Root *et al.* 2003; Parmesan 2006). Wherever there are good records in the northern hemisphere many plant species are flowering earlier in the spring as in central England (Miller-rushing and Primack 2008). Similarly, animal species are changing the timing in their life cycles, such as tree swallows (*Tachycineta bicolor*) nesting and laying their eggs earlier (Dunn and Winkler 1999). Some species are changing their migration times and in North America, one hummingbird species has ceased to migrate (Parmesan 2006).

In addition, the geographical distribution of some species is changing. In western North America, the change both northward and upward in altitude of Edith's checkerspot butterfly (*Euphydryas editha*) is well documented (Parmesan 2006). In Europe, many butterfly species have moved northward as well, including the sooty copper (*Heodes tityrus*), which now occurs and breeds in Estonia (Parmesan *et al.* 1999).

There is considerable change among Arctic species because so many life histories are tied to the ice which decreased dramatically both in area and thickness in 2007 and 2008. The polar bear (*Ursus maritimus*) is the best known by far with stress/decline being observed in a number of the populations (Stirling *et al.* 1999). Many bird species feed on the Arctic cod (*Arctogadus glacialis*), a species that occurs near the edge and just under the ice. Nesting seabirds like the black guillemot (*Cepphus grylle*) fly from their nests on land to the edge of the ice to feed and return to feed their young. So as the distance to the edge of the ice increases, there is a point at which the trip is too great and first the individual nest, then eventually the seabird colony fails.

Species that occur at high altitudes will, as a class, be very vulnerable to climate change simply because as they move upslope to track their required conditions, they ultimately will have no further up to go. The American pika (*Ochotona princeps*), a lagomorph species with a fascinating harvesting aspect to its natural history, is a prime example. It is comprised of roughly a dozen populations in different parts of the Rocky Mountains that we can anticipate will wink out one by one.

Temperature increase also will be greater in high latitudes and particularly in the northern hemisphere where there is more terrestrial surface. Climate change of course is not only about temperature it is also about precipitation. On land the two most important physical parameters for organisms are temperature and precipitation. In aquatic ecosystems the two most important are temperature and pH. Drying trends are already affecting Australia, the Argentine pampas, the American southwest and the prairie pothole region of the upper Midwest northward into Canada. Prairie potholes are a critical landscape feature supporting the great central flyway of migratory birds in North America.

For well known species such as the sugar maple (*Acer saccharum*), the environmental requirements are fairly well known so it is possible to model how the geography of those requirements is likely to change along with climate. In this case all the major climate models show that at double pre-industrial levels of greenhouse gases, the distribution of this species – so characteristic of the northeastern United States that its contribution to fall foliage is the basis of a significant tourism industry – will move north to Canada. While the tourism and the appeal of maple sugar and syrup are not significant elements of the northeast US economy, they are significant with respect to a sense of place, and are partly why these states have taken a leadership role on climate change. In the mid-Atlantic states, the Baltimore oriole (*Icterus galbula*) will no longer occur in Baltimore due to climate-driven range shift.

In the northern oceans there are changes in plankton (small organisms drifting along the ocean currently) and fish distributions. The eel grass (*Zostera marina*) communities of the great North American estuary, the Chesapeake Bay, have a sensitive upper temperature limit. Accordingly, the southern boundary has been moving steadily northward year after year (http://www.chesapeakebay.net/climatechange.aspx). Similarly, plankton populations have been moving northwards in response to water temperature increase (Dybas 2006). This trend, for example, has resulted in low plankton densities around

Scotland, likely reducing the densities of plankton-eating fish and bird species there (Dybas 2006).

Changes have been observed not only in the Arctic and temperate regions but also in the tropics (see Box 8.1). There are more than 60 vertebrate species endemic to Australia's rainforests including the grey-headed robin (*Heteromyias albispecularis*) and the ringtail possum (*Pseudocheirus peregrinus*). With climate change the amount of suitable habitat available for them shrinks dramatically such that at 5 degrees Centigrade increase most are doomed to extinction (Shoo *et al.* 2005). The Monteverde cloud forest in Costa Rica, an ecosystem type almost entirely dependent on condensation from clouds for moisture has been encountering more frequent dry days as the elevation at which clouds form has risen. Nest predators like toucans are moving up into the cloud forest from the dry tropical forest below (Pounds *et al.* 1999). The charismatic golden toad (*Bufo periglenes*) of Monteverde could

---

**Box 8.1 Lowland tropical biodiversity under global warming**
Navjot S. Sodhi

Global warming may drive species poleward or towards higher elevations. However, how tropical species, particularly those occupying lowlands, will respond to global warming remains poorly understood. Because the latitudinal gradient in temperature levels off to a plateau between the Tropic of Cancer and the Tropic of Capricorn, latitudinal range shifts are not likely for species confined to the tropics. This leaves upslope range shifts as the primary escape route for tropical species already living near their thermal limit. One scenario is that tropical lowland biodiversity may decline with global warming, because there is no "species pool" to replace lowland species that migrate to higher elevations. Colwell *et al.* (2008) speculated on the effects of projected global warming on lowland biotas by using relatively large datasets of plants and insects from Costa Rica. Data on the distribution of 1902 species of epiphytes, understory rubiaceous plants, geometrid moths, and ants were collected from a transect that traversed from sea level to 2900 m elevation. Colwell *et al.* (2008) developed a graphic model of elevational range shifts in these species under climatic warming. Assuming 600-m upslope shifts with 3.2°C temperate increase over the next century, they estimated that 53% of species will be candidates for *lowland biotic attrition* (decline or disappearance in the lowlands) and 51% will encounter the spatial gaps between their current and projected ranges (Box 8.1 Figure). A number of these species will likely face both challenges. Authors cautioned that their local-level data may have underestimated regional elevation ranges and must, in this regard, be considered as a worst case scenario. However, it is also plausible that their results represent a best case scenario, considering that other drivers such as habitat loss, fire, overharvesting and invasive species can synergistically drive species to decline and extinction (Brook *et al.* 2008).

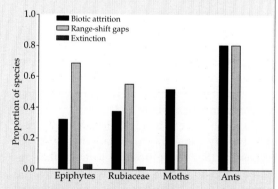

**Box 8.1 Figure** Proportion of species projected to be affected by global warming. Data for the analysis were collected from a lowland elevational transect in Costa Rica. Proportion sums are greater than one because a species may have more than one response. Reprinted from Colwell *et al.* (2008).

*continues*

> **Box 8.1 (Continued)**
>
> Most previous studies determining the effects of global warming on tropical species have focused on montane species, reporting their elevation shifts or disappearances (e.g. Pounds *et al.* 1999). Colwell *et al.*'s (2008) findings remind us that lowland tropical biodiversity remains equally vulnerable to the changing climate. Their study is yet another reminder that we need to urgently mitigate the effects of human generated climate changes.
>
> **REFERENCES**
>
> Brook, B. W., Sodhi, N. S., and Bradshaw, C. J. A. (2008). Synergies among extinction drivers under global change. *Trends in Ecology and Evolution*, **23**, 453–460.
>
> Colwell, R. K., Brehm, G., Cardelús, C. L., Gilman, A. C., and Longino, J. T. (2008). Global warming, elevational range shifts, and lowland biotic attrition in the wet tropics. *Science*, **322**, 258–261.
>
> Pounds, J. A., Fogden, M. P. L., and Campbell, J. H. (1999). Biological response to climate change on a tropical mountain. *Nature*, **398**, 611–615.

well be the first documented terrestrial extinction caused by climate change (Figure 8.2; Pounds *et al.* 1999). The rapid extinction of large numbers of amphibian species in which a chytrid fungus plays a major role may well be in synergy with climate change (Crump *et al.* 1992; Collins and Storfer 2003).

In tropical oceans, coral reefs are quite temperature sensitive. Only a slight increase in temperature causes the basic partnership between a coral animal and an alga to break down. The coral animal expels the alga triggering what are called bleaching events in which most of the color of the communities is lost and productivity, biodiversity and the ecosystems services of the reefs crash. Such occurrences were virtually unknown 40 years ago and become more frequent every year, likely due to the elevation of sea temperature (Hoegh-guldberg 1999). Coral reefs around the globe are threatened (Pandolfi *et al.* 2003). It is hard to envision a reasonable future for tropical coral reefs and the diversity of marine life they support.

Species of coastal regions will encounter problems with sea level rise. Some will succeed in adapting and others probably will not. The rate of sea level rise will be of significance: generally speaking the more rapid the rise the more species will encounter difficulty in adapting. Low lying island species constitute another class highly vulnerable to climate change, principally because of sea level rise. Islands of course have major numbers of endemic species such as the key deer (*Odocoileus virginianus clavium*) in the Florida Keys. Island species have been particularly vulnerable to extinction because of limited populations. Sea level rise caused by climate change will be the coup de grace for species of low lying islands.

To change the basic chemistry of two thirds of the planet, i.e. ocean, is staggering to contemplate in itself. In addition, the implications for the tens of thousands of marine species that build shells and skeletons of calcium carbonate are very grave. They depend on the calcium carbonate equilibrium to mobilize the basic molecules of their shells and skeletons. This includes obvious

**Figure 8.2** The golden toad. Photograph from the US Fish and Wildlife Service.

organisms like mollusks and vertebrates but also tiny plankton like pteropods (tiny snails with their "foot" modified to flap like a wing to maintain them in the water column). At a certain point of increasing acidity the shells of such organisms will go into solution while they are still alive. Effects have been seen already at the base of the food chain in the North Atlantic and off of Alaska.

Freshwater species will be affected as well. They all have characteristic temperature ranges that will be affected by climate change. Cold-water species like trout and the many species of the food chains on which they depend will no longer be able to survive in many places where they occur today (Allan et al. 2005).

These kinds of changes are relatively minor ripples in the living world but are occurring virtually everywhere. Nature is on the move and this no longer is a matter of individual examples but is statistically robust. And this is with only 0.75 degrees Celsius increase in global temperature with at least that much and probably more in store by century's end. The first projection of what double pre-industrial levels of $CO_2$ might portend for the biota estimated extinction of 18–35% of all species (Thomas et al. 2004) – a range confirmed by the 2007 report of the Intergovernmental Panel on Climate Change (IPCC 2007).

With more climate change, the impacts upon and response of biological diversity will change qualitatively and become more complex and harder to manage. Climate change of course is nothing new in the history of life on Earth. Glaciers came and went on a major scale in the northern and temperate latitudes in the last hundreds of thousands of years. Species were able to move and track their required climatic conditions without much loss of biological diversity. The difference today is that the landscapes within which species would move in response to climate change have been highly modified by human activity through deforestation, agricultural conversion, wetland drainage and the like. Landscapes have been converted into obstacle courses for dispersing organisms. Former National Zoo Director Michael Robinson stated that species would move but "Philadelphia will be in the way". Basically these landscapes will result in substantial extinction if they remain in their current condition.

A second difference is that we know from studies of past response to climate change that biological communities do not move as a unit, but rather it is the individual species that move each at its own rate and in its own direction. The consequence is that ecosystems as we know them will disassemble and the surviving species will assemble into new ecosystem configurations that largely defy the ability to foresee. Certainly that was the case as species moved in Europe after the last retreat of the glaciers (Hewitt and Nichols 2005). The management challenge to respond to this is therefore hard to understand let alone plan to address.

We also know that in contrast to the climate change models run on super computers that change will be neither linear nor gradual. We know there have been discontinuities in the physical climate system in the past. For example the global conveyor belt – the gigantic ocean current that distributes heat around the globe – has shut down in the past. Equally disturbing, abrupt threshold change is already occurring in ecosystems. Bleaching in coral reef systems is clearly an example in the oceans (see above).

## 8.3 Effects on biotic interactions

Relationships between two species can depend on relatively precise timing. Sometimes the timing mechanism of one is based on day length and the other on temperature and has worked well because of the relative climate stability. The seabird nesting-Arctic Cod coupling is just such an example and under climate change leads to "decoupling" (see above). The Arctic hare (*Lepus arcticus*), for example, changes from a white winter pelage that camouflages it in wintry white landscapes to a brownish pelage that blends into the vegetation after the snow and ice disappear. As spring thaw advances earlier with climate change, Arctic hares become vulnerable to predators as they are conspicuously white in no longer wintry landscapes.

Similarly, in terrestrial ecosystems threshold change is occurring in coniferous forests in

North America and Europe as climate change tips the balance in favor of native pine bark beetles. Milder winters allow more to overwinter and longer summers permit an additional generation of beetles. The consequence is vast stretches of forest in which 70% of the trees have been killed. It is an enormous forest management and fire management problem, and being without known precedent it is not clear how these ecosystems will respond. Yet more, there are the first signs of system change, i.e., change on yet a greater scale.

## 8.4 Synergies with other biodiversity change drivers

Climate change will also have synergistic effects with other kinds of environmental problems such as invasive species (Chapter 7). The emerald ash borer (*Agrilus planipennis*), an Asian species, is causing major mortality of American ash trees (*Fraxinus americana*) – from which baseball bats are manufactured – from the mid-west to Mid-Atlantic States (http://www.emeraldashborer.info/). The borer is over wintering in greater numbers because of milder winters and has a longer active boring season because of longer summers. Another example will be the impact of the introduced bird malaria vector mosquito which causes mortality in most species of the endemic Hawaiian honeycreepers (see Figure 12.4). Of the surviving honeycreeper species most of the vulnerable ones persist only above an altitude – the mosquito line – above which the temperature is too low for the mosquitoes. With climate change the mosquito line will move up and the area safe for honeycreepers diminishes (Pratt 2005).

## 8.5 Mitigation

All of this bears on probably the most critical environmental question of all time, namely at what point is climate change "dangerous", i.e., where should it be limited. For a long time conservationists asserted that 450 ppm of $CO_2$ (roughly equivalent to 2 degrees Centigrade warming) should be the limit beyond which it is dangerous. This means limiting peak concentration levels to as low a figure as possible and seeking ways to draw $CO_2$ out of the atmosphere to return to a lower ppm as soon as possible. It is clear that the grave risk and urgency of climate change has not been recognized (Sterman 2008; Solomon *et al.* 2009). The IPCC (2007) synthesis report suggests 450 ppm of $CO_2$ itself is dangerous. Remember, the earth is 0.75 degree warmer than pre-industrial times with another 0.5 already in the pipeline. Yet at 0.75 ecosystem threshold change is already occurring.

The last time the Earth was two degrees Centigrade warmer, sea level was four to six meters higher. The current changes in Arctic sea ice, the accelerating melting of the Greenland ice sheet, together with major ecosystem disruption all suggest that 350 ppm of $CO_2$ is the level above which it is not "safe". That is James Hansen's conclusion as a climate scientist. The insights emerging about biological diversity and ecosystems are convergent with 350 ppm. Yet atmospheric $CO_2$ is at 390 ppm and climbing at rates beyond the worst case projections.

This means the agenda for "adaptation" – to use the climate convention's terminology – is indeed urgent. Conservation strategies need revision and amplification and the conservation biology of adaptation is a rapidly developing field. Restoring natural connections in the environment will facilitate the movement of organisms as they respond to changing climate (Box 8.2). Reducing other stresses on ecosystems reduces the probability of negative synergies with climate change. Downscaled climate projections to one square kilometer, for example, or similar will provide managers with useful data for making needed decisions.

While existing protected areas will no longer be fulfilling their original purpose, e.g., Joshua trees (*Yucca brevifolia*) will no longer exist inside of the Joshua Tree National Park, they will have the new value of being the safe havens from which species can move and create the new biogeographic pattern. That together with the need for new protected areas for the new locations of important biodiversity plus the need for natural

> **Box 8.2 Derivative threats to biodiversity from climate change**
> Paul R. Ehrlich
>
> Besides the obvious direct impacts on biodiversity, climate disruption will have many other effects. For instance, if climatologists are correct, humanity is likely to be faced with a millennium or more of continuously changing patterns of precipitation that likely in itself will be devastating for biodiversity (Solomon et al. 2009). But those changes will also require humanity to continually reconstruct water-handling and food-producing infrastructure around the globe. New dams, canals, and pipelines will need to be built, often with devastating impacts on stream and river ecosystems. Lakes behind new dams will flood terrestrial habitats, and changing river flows will have impacts on estuaries and coral reefs, among the most productive of marine environments. Reefs are especially sensitive to the siltation that often accompanies major upstream construction projects.
>
> Changing water flows means that new areas will be cleared for crop agriculture and subjected to grazing, as old areas become unproductive. Roads and pipelines will doubtless need to be built to service new agricultural areas. What the net effects of these shifts will mean is almost impossible to estimate, especially where old areas may be available for rewilding (Box 5.3). It is also likely that warming will open much of the Arctic to commerce, with an accompanying increase in the construction of infrastructure – ports, roads, towns, and so on.
>
> Human society in response to growing climatic problems will also begin to revise energy-mobilizing infrastructure across the planet. Large areas of desert may be claimed by solar-energy capturing devices. Wind turbines are likely to dot landscapes and some near-shore seascapes. New high-speed rail lines may be constructed, natural ecosystems may be plowed under to plant crops for conversion to biofuels (Box 13.3). This is already happening with deforestation in the Amazon now accelerating in response to demand for biofuel crops. Expanding farming operations are also destroying the prairie pothole ecosystem of the northern plains of North America (http://www.abcbirds.org/newsandreports/stories/080226_biofuels.html). That is critical habitat for many bird populations, among other fauna, including ducks much in demand by duck hunters who have in the past proven to be allies of conservationists.
>
> All of these changes will cause multitudes of populations, and likely many species, to disappear, so that conservation biologists should be consulted on each project, and society should be made very aware as soon as possible of the potential conflicts between human and natural capital inherent in revision of water, energy, and transport infrastructure.
>
> **REFERENCES**
>
> Solomon, S., Plattner, G-K., Knuttic R., and Friedlingsteind, P. (2009). Irreversible climate change due to carbon dioxide emissions. *Proceedings of the National Academy of Sciences of the United States of America*, **106**, 1704–1709.

connections between natural areas, clearly mean that more conservation is needed not less.

Simultaneously, the "mitigation" agenda – to use the convention's term for limiting the growth of greenhouse gas concentrations in the atmosphere – becomes a matter of huge global urgency because the greater the climate change the more difficult is adaptation. Transforming the energy base for human society is the dominant center of mitigation, but biology and conservation play a significant role as well (Box 8.2).

Tropical deforestation (see Chapter 4) plays an important role in greenhouse gas emissions: literally 20% of annual emissions come from the destruction of biomass, principally tropical deforestation and burning (IPCC 2007). In the current rank order of emitting nations after China and the United States are Indonesia and

Brazil because of their deforestation. There is now gathering effort to include "Reductions in Emissions from Deforestation and Degradation" (= REDD) as part of the negotiations. Obviously there are multiple benefits in doing so in reduction of emissions (and thus atmospheric concentration levels), biodiversity benefits and ecosystem services (Chapter 3). There are technical problems in monitoring and measuring as well as issues about "leakage" – when protection of one forest simply deflects the deforestation to another – but none of it seems intractable.

All greenhouse gas emissions involve the release of solar energy trapped by photosynthesis whether ancient (fossil fuels) or present deforestation and other ecosystem degradation. That raises the important question of what role biology and biodiversity might play in removing some of the $CO_2$ accumulated in the atmosphere. Twice in the history of life on earth high levels of $CO_2$ concentrations had been reduced to levels on the order of pre-industrial. The first was associated with the origin of land plants and the second with the expansion of angiosperms (Beerling 2007). This suggests substantial potential if the biosphere is managed properly.

In the past three centuries, terrestrial ecosystems have lost 200 billion tons of carbon and perhaps more depending on hard to estimate losses of soil carbon. What is clear is that to the extent that terrestrial ecosystems can be restored, a substantial amount of carbon could be withdrawn from the atmosphere rather than lingering for a hundred to a thousand years. If that number is 160 billion tons of carbon, it probably equates to reducing atmospheric concentrations of it by 40 ppm.

This would be tantamount to planetary engineering with ecosystems – essentially a regreening of what Beerling (2007) terms the Emerald Planet. All other planetary or geo-engineering schemes have potential negative consequences, and only deal with temperature to the total neglect of ocean acidification (Lovelock and Rapley 2007; Shepherd et al. 2007). This takes the agenda beyond forests to all terrestrial ecosystems, grasslands, wetlands, and even agro-ecosystems. Essentially it is conservation on a planetary scale: managing the living planet to make the planet more habitable for humans and all forms of life.

## Summary

- Massive releases of greenhouse gasses by humans have altered the climate.
- Rapid global warming is responsible for abiotic changes such as receding of glaciers and increase in wildfires.
- Increased $CO_2$ concentrations in the atmosphere have acidified the oceans.
- Populations, species, and ecosystems are responding to these climatic conditions.
- Urgent actions are needed to reverse the climatic changes.

## Suggested reading

Lovejoy, T. E. and Hannah, L., eds (2005). *Climate change and biodiversity*. Yale University Press, New Haven, CT.

## Relevant websites

- Intergovernmental Panel on Climate Change: http://www.ipcc.ch/.
- Nature reports on climate change: http://www.nature.com/climate/index.html.
- United States Environmental Protection Agency: http://www.epa.gov/climatechange/.

## REFERENCES

Allan, J. D., Palmer, M. E., and Poff, N. L. (2005). Climate change and freshwater ecosystem. In T. E. Lovejoy and L. Hannah, eds *Climate change and biodiversity*, pp. 274–290. Yale University Press, New Haven, CT.

Beerling, D. (2007). *The emerald planet: how plant's changed Earth's history*. Oxford University Press, Oxford, UK.

Canadell, J. G., Quéré, C. L., Raupach, M. R., *et al.* (2007). Contributions to accelerating atmospheric $CO_2$ growth from economic activity, carbon intensity, and efficiency of natural sinks. *Proceedings of the National Academy of Sciences of the United States of America*, **104**, 18866–18870.

Collins, J. P. and Storfer, A. (2003). Global amphibian decline: sorting the hypotheses. *Diversity and Distributions*, **9**, 89–98.

Crump, M. L., Hensley, F. R., and Clark, K. L. (1992). Apparent decline of the golden toad: underground or extinct? *Copia*, **1992**, 413–420.

Dunn, P. O. and Winkler, D. W. (1999). Climatic change has affected breeding date of tree swallows throughout North America. *Proceedings of the Royal Society of London B*, **266**, 2487–2490.

Dybas, C. L. (2006). On collision course: ocean plankton and climate change. *BioScience*, **56**, 642–646.

Flannigan, M. D., Stocks, B. J., and Wotton, B. M. (2000). Climate change and forest fires. *The Science of the Total Environment*, **262**, 221–229.

Hansen, J., Sato, M., and Ruedy, R. (2006). Global temperature change. *Proceedings of the National Academy of Sciences of the United States of America*, **103**, 14288–14293.

Hewitt, G. M. and Nichols, R. A. (2005). Genetic and evolutionary impacts of climate change. In T. E. Lovejoy and L. Hannah, eds *Climate change and biodiversity*, pp. 176–192. Yale University Press, New Haven, CT.

Hoegh-Guldberg, O. (1999). Climate change, coral bleaching and the future of world's coral reefs. *Marine Freshwater Research*, **50**, 839–866.

IPCC. (2007). *Climate Change 2007 – impacts, adaptation and vulnerability*. Cambridge University Press, Cambridge, UK.

Lovelock, J. E. and Rapley, C. G. (2007). Ocean pipes could help the Earth to cure itself. *Nature*, **449**, 403.

Malhi, Y., Aragão, L. E. O. C., Galbraith, D., *et al.* (2009). Exploring the likelihood and mechanism of a climate-change-induced dieback of the Amazon rainforest. *Proceedings of the National Academy of Sciences of the United States of America*, in press.

Miller-Rushing, A. and Primack, R. B. (2008). Global warming and flowing times in Thoreau's Concord: a community perspective. *Ecology*, **89**, 332–341.

Pandolfi, J. M., Bradbury, R. H., Sala, E., *et al.* (2003). Global trajectory of the long-term decline in coral reef ecosystems. *Science*, **301**, 955–958.

Parmesan, C. (2006). Ecological and evolutionary responses to recent climate change. *Annual Review of Ecology, Evolution and Systematics*, **37**, 637–669.

Parmesan, C., Ryrholm, N., Stefanescu, C., *et al.* (1999). Poleward shifts in geographic ranges of butterfly species associated with regional warming. *Nature*, **399**, 579–583.

Pounds, J. A., Fodgen, M. P. L., and Campbell, J. H. (1999). Biological response to climate change on a tropical mountain. *Nature*, **398**, 611–615.

Pratt, D. H. (2005). *Hawaiian honeycreepers*. Oxford University Press, Oxford, UK.

Root, T. L, Price, J. T., Hall, K. R., *et al.* (2003). Fingerprints of global warming on wild animals and plants. *Nature*, **421**, 57–60.

Shepherd, J., Iglesias-rodriguez, D., and Yool, A. (2007). Geo-engineering might cause, not cure, problems. *Nature*, **449**, 781.

Shoo, L. P., Williams, S. E., and Hero, J.-M. (2005). Climate warming and the rainforest birds of the Australian wet tropics: using abundance data as a sensitivity predictor of change in total population. *Biological Conservation*, **125**, 335–343.

Solomon, S., Plattner, G.-S., Knutti, R., and Friedlingstein, P. (2009). Irrevesible climate change due to carbon dioxide emissions. *Proceedings of the National Academy of Sciences of the United States of America*, **106**, 1704–1709.

Sterman, J. D. (2008). Risk communication on climate: mental models and mass balance. *Science*, **322**, 532–533.

Stirling, I., Lunn, N. J., and Iacozza, J. (1999). Long-term population trends in population ecology of polar bears in western Hudson Bay in relation to climate change. *Arctic*, **52**, 294–306.

Thomas, C. D., Cameron, A., Green, R. E., *et al.* (2004). Extinction risk from climate change. *Nature*, **427**, 145–148.

United Nations Environment Programme (UNEP). (2007). *Global outlook for ice & snow*. UNEP, Nairobi, Kenya.

# CHAPTER 9

# Fire and biodiversity

David M.J.S Bowman and Brett P. Murphy

In a famous passage in the concluding chapter of *The Origin of Species,* Darwin (1859, 1964) invites the reader to "contemplate an entangled bank, clothed with many plants of many kinds, with birds singing on the bushes, with various insects flitting about, and with worms crawling through the damp earth" and "reflect that these elaborately constructed forms, so different from each other, and dependent on each other in so complex a manner, have all been produced by laws acting around us." Likewise, let us consider a tropical savanna ablaze with hovering raptors catching insects fleeing the fire-front, where flames sweep past tree trunks arising from dry crackling grass. Within weeks the blackened savanna trees are covered in green shoots emerging from thick bark, woody juveniles are resprouting from root stocks, and herbivores are drawn to grass shooting from growing tips buried beneath the surface soil (Figure 9.1). In this chapter we will show that the very same evolutionary and ecological principles that Darwin espoused in that brilliant passage relate to landscape fire. This is so because fire is enmeshed in the evolution and ecology of terrestrial life, including our own species. This perspective is deeply challenging to the classical view of the "Balance of Nature" that is still held by a broad cross-section of ecologists, naturalists and conservationists, most of who have trained or live in environments where landscape fire is a rare event, and typically catastrophic (Bond and Van Wilgen 1996; Bond and Archibald 2003). Only in the past decade have books been published outlining the general principles of fire ecology (Whelan 1995; Bond and Van Wilgen 1996) and journals established to communicate the latest findings in fire ecology and

**Figure 9.1** A eucalypt savanna recovering from a fire that has occurred in the early dry season in Kakadu National Park, northern Australia. Note the flowering *Livistona* palm and the strong resprouting response of juvenile woody plants on the still bare ground surface. Photograph by David Bowman.

wildfire management (see http://www.firecology.net and http://www.iawfonline.org).

## 9.1 What is fire?

At the most basic level fire can be considered a physiochemical process that rapidly releases energy via the oxidation of organic compounds, and can be loosely considered as "anti-photosynthesis". This physiochemical process if often summarized in a classic "fire triangle" made up of the three key factors to cause combustion: oxygen, fuel and ignitions (Whelan 1995; Pyne 2007). Atmospheric oxygen levels create a "window" that controls fire activity because ignitions are constrained by atmospheric oxygen (Scott and Glasspool 2006). Fire cannot occur when levels fall below 13% of the atmosphere at sea level, and under the current oxygen levels (21%) fire activity is limited by fuel moisture, yet at 35% even moist fuels will burn. Because of substantial fluctuations in atmospheric oxygen, fire risk has changed significantly through geological time. In the Permian Period (between 290 and 250 million years ago) for example, oxygen levels were substantially higher than at present and even moist giant moss (lycopod) forests would have been periodically burnt (Scott and Glasspool 2006). However, fire in the biosphere should not be considered merely a physicochemical process but rather a fundamental biogeochemical process. Fires instantaneously link biomass with the atmosphere by releasing heat, gases (notably water vapor), and the geosphere by releasing nutrients and making soils more erodible and thus changing the nutrient content of streams and rivers (hydrosphere). Fire is therefore quite unlike other natural disturbances, such as floods and cyclones, given the complex web of interactions and numerous short and long-range feedbacks. Some ecologists have suggested that landscape fires should be considered as being "biologically constructed", and have drawn parallels with herbivory (Bond and Keeley 2005) or decomposition (Pyne 2007). Such tight coupling between fire and life bedevils simple attribution of cause and effect, and raises fascinating questions about the potential coevolution of fire and life.

## 9.2 Evolution and fire in geological time

There is evidence from the fossil record that wildfires started to occur soon after vegetation established on the land surface (about 420 million years ago) (Scott and Glasspool 2006). The long history of exposure of terrestrial life to fire leads to the idea that fire is an important evolutionary factor, and more controversially, that fire and life have coevolved (Mutch 1970). While gaining some support from modeling (Bond and Midgley 1995), this is difficult to prove because adaptations to fire cannot be unambiguously identified in the fossil record. For example, in many fire-prone environments, seeds are often contained in woody fruits that only open after a fire event, a feature known as serotiny. However, woody fruits may also be a defense against seed predators such as parrots, and seeds are released once mature, irrespective of fire (Bowman 2000). In most cases it is impossible to know if fossilized woody fruits are truly serotinous, thus woody fruit occurrence is not clear evidence of an adaptation to fire. Much care is required in the attribution of fire adaptations. For example, microevolution can result in switching from possible fire-adaptations, such as the serotinous state. More problematic for understanding the evolution of flammability, Schwilk and Kerr (2002) have proposed a hypothesis they call "genetic niche-hiking" that flammable traits may spread without any "direct fitness benefit of the flammable trait".

Insights into the evolution of flammability have been gained by tracking the emergence of highly fire-adapted lineages such as *Eucalyptus*. Eucalypts are renowned for their extraordinarily prolific vegetative recovery of burnt trunks via epicormic buds (Figure 9.2). Recently, Burrows (2002) has shown that eucalypt epicormics are anatomically unique. Unlike other plant lineages, which have fully developed dormant buds on the trunks, eucalypts, have strips of "precursor" cells that span the cambium layer that, given the right cues, develop rapidly into epicormic buds. The advantage of this system is that should the trunk be severely burnt the tree retains the capacity to develop epicormic buds from cells protected in the cambium. The molecular phylogeny of

**Figure 9.2** Prolific epicormic sprouts on a recently burnt tall eucalypt forest in eastern Tasmania. Photograph by David Bowman.

eucalypts, dated using the fossil record, suggests that this trait existed before the "bloodwood" eucalypt clade split off from other eucalypts some 30 million years ago, given that it occurs in both these lineages. Such an ancient feature to the lineage suggests that eucalypts had developed a vegetative response to landscape fire, which appears to have become more common in the Australian environment associated with a dry climate and nutrient impoverished soils. This interpretation is concordant with the fresh insights about the evolution of the Australian biota derived from numerous molecular phylogenies of quintessentially Australian plants and animals (Bowman and Yeates 2006).

## 9.3 Pyrogeography

Satellite sensors have revolutionized our understanding of fire activity from landscape to global scales. Global compilations of satellite data have demonstrated the occurrence of landscape fire on every vegetated continent, yet the incidence of fire is not random across the globe (Justice *et al.* 2003). Fire has predictable features regarding how it spreads across landscapes and the frequency and season of occurrence. Such predictability has lead to the idea of the "fire regime".

Key aspects of the fire regime include types of fuels consumed (e.g. grass vs. canopies), spatial pattern (area burnt and shape), and consequences (severity relative to impacts on the vegetation and/or soils) (Gill 1975; Bond and Keeley 2005). For example, savanna fires are often of low intensity and high frequency (often annual), while forest fires are often of low frequency (once every few centuries) and very high intensity. Fire regimes are part of the habitat template that organizes the geographic distribution of biodiversity, and, in turn, species distributions influence the spread of fire. Some authors have even applied "habitat suitability modeling" to predict where fire is most likely to occur at the global to local level.

Fire activity is strongly influenced by climate variability. Fire managers have developed empirical relationships that combine climate data, such as the intensity of antecedent moisture deficit, wind speed, relative humidity, and air temperature, to calculate fire danger (see http://www.firenorth.org.au). Mathematical models combining such climate data with fuel loads and topography have been developed to predict how a fire may behave as it spreads across a landscape (Cary *et al.* 2006). The spread of fire is also strongly influenced by vegetation type (Figure 9.3). For example, grassy environments carry fire frequently because of the rapid accumulation of fuel while rainforests burn

**Figure 9.3** Landscape scale patterns of fire spread in southwestern Tasmania. Fire spread is controlled by topography, vegetation, and the meteorological conditions that prevailed at the time of the fire creating strongly non-random patterns of burnt and unburnt areas. Photograph by David Bowman.

infrequently because of microclimates that keep fuels moist under all but drought conditions. Climate cycles such as the El Niño Southern Oscillation (ENSO) also strongly influence fire activity. For example, fire activity typically increases in arid environments after a wet period because of the build-up of fine fuels. Conversely, fire activity increases after a long drought period in moist forests.

The satellite record has been extraordinarily useful in understanding fire activity in highly fire prone environments (see http://www.cfa4wd.org/information/Forest_FDI.htm). Yet the limited time-depth of this record may mask the occurrence of infrequent fire events that occur in long-lived fire-prone vegetation such as the boreal forests of Canada and Siberia. Understanding the "fire regimes" of long-lived forests like those of the boreal zone demands historical reconstruction such as dendrochronology (tree-ring analysis) to determine the timing of "stand replacing fires" which initiate a cohort of regeneration to replace the burnt forest. Statistical analysis of forest stand-age structures can be used to determine the inter-fire intervals (Johnson and Gutsell 1994). Dendrochronology has also been used to date precisely past fire events by identifying injuries to growth rings (fire scars) on the trunks on long-lived trees (Swetnam 1993).

The study by Sibold *et al.* (2006) captures many of the above complexities in understanding fire extent and occurrence. They combined tree ring analyses and geographic information systems (GIS) techniques to identify the influence of vegetation type and structure, elevation and aspect, and regional climate influences on fire activity in the Rocky Mountains National Park, Colorado, USA. Their analysis identified the primary importance of ENSO for fire activity, yet this climatic effect was modulated by landscape setting and vegetation type. Over the 400-year record, fire activity was common in the dry, low elevation slopes that support fire-prone lodgepole pine (*Pinus contorta*) forests but at higher elevation there were large areas of long unburnt mesic spruce-fir (*Picea engelmannii*) forest. On the moist western side of the mountain range were fewer, but larger, fires compared to the drier eastern sides of the mountain range. This example shows that while climate is a driver of fire risk, the linkage between fire, climate and vegetation is complex, frustrating simple attribution of "cause and effect". Finally, human fire usage has a profound effect on fire activity, disturbing "natural" fire regimes. For example, tropical rainforests are currently being transformed to pasture by burning (see Box 9.1) yet in some environments, like the forests of the western USA, fire managers have effectively eliminated fire from some fire prone landscapes.

## 9.4 Vegetation–climate patterns decoupled by fire

A classic view of plant geography is that vegetation and climate are closely coupled. Recently Bond *et al.* (2005) challenged this view by asking the question of whether the vegetation of the Earth is significantly influenced by landscape fire. The approach they took was via dynamic global vegetation models (DGVMs), which are computer simulations of vegetation based on physiological principles. The effect of landscape fire on global vegetation patterns is implicit in DGVMs because they include "fire modules" that introduce frequent disturbances to modeled vegetation patterns and processes. Such modules are necessary in order to recreate actual

---

### Box 9.1 Fire and the destruction of tropical forests
### David M. J. S. Bowman and Brett P. Murphy

Each year, extensive areas of tropical forest are unintentionally burnt by anthropogenic fires, and are severely degraded or destroyed as a result (see Chapter 4). Enormous conflagrations can occur in response to drought events associated with ENSO, most notably the Indonesian fires of 1997–1998, which burnt around 8 million hectares of forest (Cochrane 2003). Until recent decades, most tropical forests experienced fires very infrequently, with fire return intervals in the order of centuries, although it is now clear that fire frequency has increased dramatically in the past few decades. Current human land-use activities promote forest fires by fragmenting (see Chapter 5) and degrading forests and providing ignition sources, which would otherwise be rare. These three factors can act synergistically to initiate a series of positive feedbacks that promote the massive tropical forest fires that have become common in recent decades (see Box 9.1 Figure). Forest edges tend to be much more susceptible to fire than forest cores, because they tend to be more desiccated by wind and sun, have higher rates of tree mortality and hence, woody fuel accumulation and grassy fuel loads tend to be higher. The result is that fire frequency tends to increase with proximity to a forest edge, such that highly fragmented forests have high fire frequencies. Forests degraded by selective logging are also at risk of fire due to their reduced canopy cover, which allows the forest to become desiccated and light to penetrate

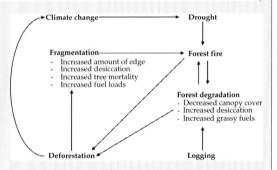

**Box 9.1 Figure** The synergistic effects of habitat fragmentation and degradation on the occurrence of tropical forest fires. Adapted from Cochrane (2003).

and encourage grass growth. The waste biomass from logging operations can also dramatically elevate fuel loads. Similarly, forests degraded by an initial fire tend to be more susceptible to repeat fires, further enhancing the feedback loop.

The negative impacts of frequent, intense fires on tropical forest biodiversity are likely to be enormous, given the existing threats posed by the direct effects of deforestation (Chapter 4) and overharvesting (Chapter 6). Intense fires easily kill a large proportion of tropical forest tree species, and repeated fires can be especially detrimental to species regenerating vegetatively or from seed. Generally, repeated fires lead to a loss of primary forest tree species, with these replaced by an impoverished set of pioneer species (Barlow and Peres 2008). The effects of fire on forest animals are less well

*continues*

> **Box 9.1 (Continued)**
>
> understood, although studies following the 1997–1998 Indonesian fires suggest severe impacts on many groups, especially those reliant on fruit-trees and arthropod communities in leaf litter (Kinnaird and O'Brien 1998). On Borneo, endangered orangutan (*Pongo pygmaeus*) populations suffered declines of around 33% following the 1997–1998 fires (Rijksen and Meijaard 1999).
>
> In many tropical regions, climate change is expected to exacerbate forest fires. There is evidence that extreme weather events, such as the ENSO droughts that triggered the 1997–1998 Indonesian fires, and tropical storms, may become more frequent (Timmermann *et al.* 1999; Mann and Emanuel 2006). Additionally, we can expect strong positive feedbacks between forest fire occurrence and climate change, because tropical forest fires result in enormous additions of greenhouse gases to the atmosphere, leading to even more rapid climate change. For example, the 1997–1998 Indonesian fires released 0.8–2.6 Gt of carbon to the atmosphere, equivalent to 13–40% of global emissions due to burning fossil fuels, making a large contribution to the largest recorded annual increase in atmospheric $CO_2$ concentration (Page *et al.* 2002).
>
> **REFERENCES**
>
> Barlow, J. and Peres, C. A. (2008). Fire-mediated dieback and compositional cascade in an Amazonian forest. *Philosophical Transactions of the Royal Society of London B*, **363**, 1787–1794.
>
> Cochrane, M. A. (2003). Fire science for rainforests. *Nature*, **421**, 913–919.
>
> Kinnaird, M. F. and O'Brien, T. G. (1998). Ecological effects of wildfire on lowland rainforest in Sumatra. *Conservation Biology*, **12**, 954–956.
>
> Rijksen, H. D. and Meijaard, E. (1999). *Our vanishing relative: the status of wild orang-utans at the close of the twentieth century*. Tropenbos Publications, Wageningen, the Netherlands.
>
> Mann, M. E. and Emanuel, K. A. (2006). Atlantic hurricane trends linked to climate change. *Eos, Transactions of the American Geophysical Union*, **87**, doi:10.1029/2006EO240001.
>
> Page, S. E., Siegert, F., Rieley, J. O., Boehm, H. D. V., Jaya, A., and Limin, S. (2002). The amount of carbon released from peat and forest fires in Indonesia during 1997. *Nature*, **420**, 61–65.
>
> Timmermann, A., Oberhuber, J., Bacher, A., Esch, M., Latif, M., and Roeckner, E. (1999). Increased El Niño frequency in a climate model forced by future greenhouse warming. *Nature*, **398**, 694–697.

vegetation patterns. Bond *et al.* (2005) found that a world without fire had very different vegetation zones compared with the actual vegetation geography. For example, when fire was "switched off", dense (>80%) tree cover increased from 27% to 56% of the vegetated Earth surface and more than half (52%) of the current global distribution of tropical savannas were transformed to angiosperm-dominated forests. The core message of this analysis is that fire causes the "decoupling" of vegetation patterns from climate.

Arguably the most well known decoupling of vegetation and climate concerns the geographic distribution of forest and savanna. Savannas are among the most fire-prone biomes on Earth, and are characterized by varying mixtures of both tree and grass biomass. The question of how both trees and grasses can coexist in the long-term has long puzzled savanna ecologists. Conventional ecological theory of plant succession suggests that highly productive savannas are unstable and should gradually progress toward closed canopy forest. While it seems that in less productive savannas, such as in low rainfall areas, tree biomass is indeed constrained by the limitation of resources, such as water, recent research suggests that in more productive savannas, recurrent disturbance plays an important role in maintaining a tree–grass balance (Sankaran *et al.* 2005). Given the high flammability of savannas, it seems that disturbance due to fire is of particular importance.

The most widely accepted explanation of how frequent fires limit tree biomass in savannas assumes that a "tree demographic-bottleneck" occurs. It is accepted that fire frequency controls the recruitment of savanna trees, particularly the growth of saplings into the tree layer. Unlike mature trees, saplings are too short in stature to avoid fire-damage and unlike juveniles, if they are damaged they cannot rapidly return to their previous size from root stocks (Hoffman and Solbrig 2003). Thus saplings must have the ability to tolerate recurrent disturbance until they have sufficient reserves to escape through a disturbance-free "recruitment window" into the canopy layer where they suffer less fire damage. Recurrent disturbance by fire can stop savanna tree populations from attaining maximal tree biomass by creating bottlenecks in the transition of the relatively fire-sensitive sapling stage to the fire tolerant tree stage (Sankaran et al. 2004). In the extreme case, a sufficient frequency of burning can result in the loss of all trees and the complete dominance of grass. Conversely, fire protection can ultimately result in the recruitment of sufficient saplings to result in a closed canopy forest.

Large herbivores may also interact with fire activity because high levels of grazing typically reduce fire frequency, and this can enable woody plants to escape the "fire trap", and increase in dominance (Sankaran et al. 2004; Werner 2005). For example, extensive woody plant encroachment has occurred in mesic grassland and savanna in Queensland, Australia, and has been attributed to cattle grazing and changed fire regimes (Crowley and Garnett 1998). This trend can be reversed by reduced herbivory coupled with sustained burning—a methodology used by pastoralists to eliminate so called "woody weeds" from overgrazed savannas. Bond and Archibald (2003) suggest that in southern African savannas there is a complex interplay between fire frequency and herbivory. Heavily grazed savannas support short grass "lawns", dominated by species in the sub-family Chloridioideae, which do not burn. These lawns support a diversity of large grazers including white rhino (*Ceratotherium simum*), wildebeest (*Connochaetes* spp.), impala (*Aepyceros melampus*), warthog (*Phacochoerus africanus*), and zebra (*Equus* spp.) (Figure 9.4). Under less intense grazing, these lawns can switch to supporting bunch grass, in the sub-family Andropogoneae, which support a less diverse mammal assemblage adapted to gazing tall grasses, such as African buffalo (*Syncerus caffer*).

**Figure 9.4** Zebra and wildebeest grazing on a 'lawn' in a humid savanna in Hluhluwe-Umfolozi Park, South Africa. Bond and Archibald (2003) suggest that intense grazing by African mammals may render savannas less flammable by creating mosaics of lawns that increase the diversity of the large mammal assemblage. Large frequent fires are thought to switch the savannas to more flammable, tall grasses with a lower diversity of large mammals. Photograph by David Bowman. See similar Figure 4.6.

The high biomass of the bunch grasslands render these systems highly flammable. Bond and Archibald (2003) propose a model where frequent large fires can result in a loss of lawns from a landscape with corresponding declines in mammal diversity. The mechanism for this is that resprouting by grasses following fire causes a lowering in overall grazing pressure across the landscape. Fully understanding the drivers of the expansion of woody vegetation into rangelands, including the role of fire and herbivory, remains a major ecological challenge (see http://ag.arizonal.edu/research/archer/research/biblio1.html).

How savanna vegetation evolved is unclear. Some authors suggest that falling atmospheric carbon dioxide ($CO_2$) concentrations may have stimulated the development of grasses that now dominate tropical savannas (Bond *et al.* 2003). Tropical savanna grasses have the C4 photosynthetic pathway that is highly productive in hot, wet climates, and under low $CO_2$ concentrations these grasses have a physiological advantage over woody vegetation that has the C3 photosynthetic pathway. The production of large quantities of fine and well-aerated fuels may have greatly increased the frequency of landscape fire disadvantaging woody plants and promoting further grassland expansion. The development of monsoon climates might have also been as important a driver as low atmospheric concentrations of $CO_2$ (Keeley and Rundel 2003). The monsoon climate is particularly fire-prone because of the characteristic alternation of wet and dry seasons. The wet season allows rapid accumulation of grass fuels, while the dry season allows these fuels to dry out and become highly flammable. Furthermore, the dry season tends to be concluded by intense convective storm activity that produces high densities of lightning strikes (Bowman 2005).

## 9.5 Humans and their use of fire

Our ancestors evolved in tropical savannas and this probably contributed to our own species' mastery of fire. Indeed, humans can be truly described as a fire keystone species given our dependence on fire; there is no known culture that does not routinely use fire. For example, the Tasmanian Aborigines always carried fire with them, as it was an indispensable tool to survive the cold wet environment (Bowman 1998). The expansion of humans throughout the world must have significantly changed the pattern of landscape burning by either intentionally setting fire to forests to clear them or accidentally starting fires. How prehistoric human fire usage changed landscape fire activity and ecosystem processes remains controversial and this issue has become entangled in a larger debate about the relative importance of humans vs. climate change in driving the late Pleistocene megafaunal extinctions (Barnosky *et al.* 2004; Burney and Flannery 2005). Central to this debate is the Aboriginal colonization of Australia that occurred some 40 000 years ago. Some researchers believe that human colonization caused such substantial changes to fire regimes and vegetation distribution patterns that the marsupial megafauna were driven to extinction. This idea has recently been supported by the analysis of stable carbon isotopes ($\delta^{13}C$) in fossil eggshells of emus and the extinct giant flightless bird *Genyornis newtoni* in the Lake Eyre Basin of central Australia. Miller *et al.* (2005a) interpreted these results as indicating that sustained Aboriginal landscape burning during colonization in the late Pleistocene caused the transformation of the central Australian landscape from a drought-adapted mosaic of trees, shrubs, and nutritious grasslands to the modern fire-adapted desert scrub. Further, climate modeling suggests that the switch from high to low leaf-area-index vegetation may explain the weak penetration of the Australian summer monsoon in the present, relative to previous periods with similar climates (known as "interglacials") (Miller *et al.* 2005b).

Yet despite the above evidence for catastrophic impacts following human colonization of Australia, it is widely accepted that at the time of European colonization Aboriginal fire management was skilful and maintained stable vegetation patterns (Bowman 1998). For example, recent studies in the savannas of Arnhem Land, northern Australia, show that areas under Aboriginal fire management are burnt in patches to increase kangaroo densities (Figure 9.5; Murphy and Bowman

2007). Further, there is evidence that the cessation of Aboriginal fire management in the savannas has resulted in an increase in flammable grass biomass and associated high levels of fire activity consistent with a "grass–fire cycle" (see Box 9.2). It is unrealistic to assume that there should only be one uniform ecological impact from indigenous fire usage. Clearly working out how indigenous people have influenced landscapes demands numerous studies, in order to detect local-scale effects and understand the underlying "logic" of their landscape burning practices (e.g. Murphy and Bowman 2007). Also of prime importance is study of the consequences of prehistoric human colonization of islands such as New Zealand. In this case, there is clear evidence of dramatic loss of forest cover and replacement with grasslands (McGlone 2001).

---

### Box 9.2 The grass–fire cycle
### David M. J. S. Bowman and Brett P. Murphy

D'Antonio and Vitousek (1992) described a feedback between fire and invasive grasses that has the capacity to radically transform woodland ecosystems, a process they described as the "grass–fire cycle". The cycle begins with invasive grasses establishing in native vegetation, increasing the abundance of quick-drying and well-aerated fine fuels that promote frequent, intense fires. While the invasive grasses recover rapidly from these fires via regeneration from underground buds or seeds, woody plants tend to decrease in abundance. In turn, this increases the abundance of the invasive grasses, further increasing fire frequency and intensity. The loss of woody biomass can also result in drier microclimates, further adding momentum to the grass–fire cycle. Eventually the grass–fire cycle can convert a diverse habitat with many different species to grassland dominated by a few exotics.

The consequences of a grass–fire cycle for ecosystem function can be enormous. The increase in fire frequency and intensity can result in massive losses of carbon, both directly, via combustion of live and dead biomass, and indirectly, via the death of woody plants and their subsequent decomposition or combustion. For example, invasion of cheatgrass (*Bromus tectorum*) in the Great Basin of the United States and the establishment of a grass–fire cycle has led to a loss of 8 Mt of carbon to the atmosphere and is likely to result in a further 50 Mt loss in coming decades (Bradley *et al.* 2006). During fires, nitrogen is also volatilized and lost in smoke, while other nutrients, such as phosphorus, are made more chemically mobile and thus susceptible to leaching. Thus, nutrient cycles are disrupted, with a consequent decline in overall stored nutrients for plants. This change can further reinforce the grass–fire cycle because the fire-loving grasses thrive on the temporary increase in the availability of nutrients.

An example of an emerging grass–fire cycle is provided by the tropical savannas of northern Australia, where a number of African grasses continue to be deliberately spread as improved pasture for cattle. Most notably, gamba grass (*Andropogon gayanus*) rapidly invades savanna vegetation, resulting in fuel loads more than four times that observed in non-invaded savannas (Rossiter *et al.* 2003). Such fuel loads allow extremely intense savanna fires, resulting in rapid reductions in tree biomass (see Box 9.2 Figure). The conversion of a savanna woodland, with a diverse assemblage of native grasses, to a grassland monoculture is likely to have enormous impacts on savanna biodiversity as gamba grass becomes established over large tracts of northern Australia. Despite the widely acknowledged threat posed by gamba grass, it is still actively planted as a pasture species in many areas. Preventing further spread of gamba grass must be a management priority, given that, once established, reversing a grass fire–cycle is extraordinarily difficult. This is because woody juveniles have little chance of reaching maturity given the high frequency of intense fires and intense competition from grasses.

*continues*

### Box 9.2 (Continued)

**Box 9.2 Figure** An example of a grass-fire cycle becoming established in northern Australian savannas. African gamba grass is highly invasive and promotes enormously elevated fuel loads and high intensity fires, resulting in a rapid decline in woody species. Photograph by Samantha Setterfield.

### REFERENCES

Bradley, B. A., Houghtonw, R. A., Mustard, J. F., and Hamburg, S. P. (2006). Invasive grass reduces aboveground carbon stocks in shrublands of the Western US. *Global Change Biology*, **12**, 1815–1822.

D'Antonio, C. M. and Vitousek, P. M. (1992). Biological invasions by exotic grasses, the grass/fire cycle, and global change. *Annual Review of Ecology and Systematics*, **23**, 63–87.

Rossiter, N. A., Setterfield, S. A., Douglas, M. M., and Hutley, L. B. (2003). Testing the grass-fire cycle: alien grass invasion in the tropical savannas of northern Australia. *Diversity and Distributions*, **9**, 169–176.

---

Agricultural expansion is often enabled by using fire as a tool to clear forests, a pattern that has occurred since the rise of civilization. Currently, this process is occurring most in the tropics. The fire-driven destruction of forests has been studied in close detail in the Amazon Basin, and is characterized by an ensemble of positive feedbacks greatly increasing the risk of fires above the extremely low background rate (Cochrane *et al.* 1999; Cochrane 2003; see Box 9.1). Recurrent burning can therefore trigger a landscape-level transformation of tropical rainforests into flammable

**Figure 9.5** Traditional land management using fire is still practiced by indigenous people in many parts of northern and central Australia. Recent work in Arnhem Land suggests that skilful fire management results in a fine-scale mosaic of burnt patches of varying age, which is thought to be critically important for maintaining populations of many small mammals and granivorous birds. Photograph by Brett Murphy.

scrub and savanna, exacerbated by the establishment of a "grass–fire cycle" (see Box 9.2).

## 9.6 Fire and the maintenance of biodiversity

### 9.6.1 Fire-reliant and fire-sensitive species

Many species in fire-prone landscapes are not only fire tolerant, but depend on fire to complete their life-cycles and to retain a competitive edge in their environment. Such species typically benefit from the conditions that prevail following a fire, such as increased resource availability associated with the destruction of both living and dead biomass, nutrient-rich ash, and high light conditions (see Box 9.3). For example, fire is critically important for the regeneration of many plant species of the fire-prone heath communities typical of the world's Mediterranean climates (e.g. South African fynbos, southwestern Australian kwongan, Californian chaparral). Many species in these communities have deeply dormant seeds that only germinate following fire, when normally limited resources, such as light and nutrients, are abundant. Many hard-seeded heath species, especially *Acacia* species and other legumes, are stimulated to germinate by heat, while many others are stimulated by chemicals in smoke (Bell *et al.* 1993; Brown 1993). Other species in these communities typically only flower following a fire (e.g. Denham and Whelan 2000).

---

**Box 9.3  Australia's giant fireweeds**
**David M.J.S Bowman and Brett P. Murphy**

Australian botanists have been remarkably unsuccessful in reaching agreement as to what constitutes an Australian rainforest (Bowman 2000). The root of this definitional problem lies with the refusal to use the term "rainforest" in the literal sense, which would involve including the tall eucalypt forests that occur in Australia's high rainfall zones (see Box 9.3 Figure). This is despite the fact that the originator of the term, German botanist Schimper, explicitly included eucalypts in his conception of rainforest. The reason why eucalypt forests are excluded from the term "rainforest" by Australians is that these forests require fire disturbance to regenerate, in contrast to true rainforests that are comparatively fire-sensitive. Typically, infrequent very intense fires kill all individual eucalypts, allowing prolific regeneration from seed to occur, facilitated by the removal of the canopy and creation of a nutrient-rich bed of ash. Without fire, regeneration from seed does not occur, resulting in very even-aged stands of mature eucalypts.

The gigantic (50–90 m tall) karri (*Eucalyptus diversicolor*) forests of southwestern Australia underscore the complexity of the term "rainforest" in Australia. These forests grow in a relatively high rainfall environment (>1100 mm per annum) with a limited summer drought of less than three months duration. Elsewhere in Australia, such a climate would support rainforest if protected from fire. However, in southwestern Australia there are no continuously regenerating and fire intolerant rainforest species to compete with karri, although geological and biogeographic evidence point to the existence of rainforest in the distant past. The cause of this disappearance appears to be Tertiary aridification and the accompanying increased occurrence of landscape fire. For example, a pollen core from 200 km north of Perth shows that by 2.5 million years ago the modern character of the vegetation, including charcoal evidence of recurrent landscape fires, had established in this region, although some rainforest pollen (such as *Nothofagus* and *Phyllocladus*) indicates that rainforest pockets persisted in the landscape at this time (Dodson and Ramrath 2001).

*continues*

### Box 9.3 (Continued)

The gigantic size of karri and a regeneration strategy dependent upon fire disturbance, including mass shedding of tiny seeds with limited reserves onto ashbeds, suggests convergent evolution with other, distantly related, eucalypts such as mountain ash (*E. regnans*) in southeastern Australian and *E. grandis* in northeastern and eastern Australia. Such convergence suggests that all have been exposed to similar natural selection pressures and have evolved to compete with rainforest species by using fire as an agent of inter-specific competition (e.g. Bond and Midgley 1995). The extraordinary diversity of the genus *Eucalyptus* and convergent evolution of traits such as gigantism in different lineages in this clade, and similar patterns of diversification in numerous other taxonomic groups, leads to the inescapable conclusion that fire had been an integral part of the Australian environment for millions of years before human colonization. Aborigines, therefore, learnt to live with an inherently flammable environment.

### REFERENCES

Bowman, D. M. J. S. (2000). *Australian rainforests: islands of green in a sea of fire*. Cambridge University Press, Cambridge, UK.

Bond, W. J. and Midgley, J. J. (1995). Kill thy neighbor—an individualistic argument for the evolution of flammability. *Oikos*, **73**, 79–85.

Dodson, J. R. and Ramrath, A. (2001). An Upper Pliocene lacustrine environmental record from south-Western Australia—preliminary results. *Palaeogeography Palaeoclimatology Palaeoecology*, **167**, 309–320.

**Box 9.3 Figure** Giant *Eucalyptus regnans* tree in southern Tasmania. The life-cycle of these trees depends upon infrequent fire to enable seedling establishment. Without fire a dense temperate *Nothofagus* rainforest develops because of the higher tolerance of rainforest seedlings to low light conditions. Photograph by David Bowman.

Even within fire-prone landscapes, there may be species and indeed whole communities that are fire-sensitive. Typically these occur in parts of the landscape where fire frequency or severity is low, possibly due to topographic protection. For example, when fire sensitive rainforest communities occur within a flammable matrix of grassland and savanna, as throughout much of the tropics, they are often associated with rocky gorges, incised gullies (often called "gallery forests"), and slopes on the lee-side of "fire-bearing" winds (Bowman 2000). Several factors lead to this association: fires burn more intensely up hill, especially if driven by wind; rocks tend to limit the amount of grassy fuel that can accumulate; deep gorges are more humid, reducing the flammability of fuels; and high soil moisture may lead to higher growth rates of the canopy trees, increasing their chances of reaching maturity, or a fire-resistant size, between fires.

Somewhat counter-intuitively, many fire-sensitive species in fire-prone landscapes are favored by moderate frequencies of low intensity fires, especially if they are patchy. Such fires greatly reduce fuel loads and thus the likelihood of large, intense fires. In addition, because low intensity fires are typically more patchy than high intensity fires, they tend to leave populations of fire-sensitive species undamaged providing a seed source for regeneration. Such an example is provided by the decline of the fire-sensitive endemic Tasmanian conifer King Billy pine (*Athrotaxis selaginoides*) following the cessation of Aboriginal landscape burning (Brown 1988; http://www.anbg.gov.au/fire ecology/fire-and-biodiversity.html). The relatively high frequency of low-intensity fires under the Aboriginal regime appears to have limited the occurrence of spatially extensive, high intensity fires. Under the European regime, no deliberate burning took place, so that when wildfires inevitably occurred, often started by lightning, they were large, intense, and rapidly destroyed vast tracts of King Billy pine. Over the last century, about 30% of the total coverage of King Billy pine has been lost.

A similar situation has resulted in the decline of the cypress pine (*Callitris intratropica*) in northern Australian savannas (Bowman and Panton 1993). Cypress pine is a fire-sensitive conifer found across much of tropical Australia. Mature trees have thick bark and can survive mild but not intense fires, and if stems are killed it has very limited vegetative recovery. Seedlings cannot survive even the coolest fires. Thus, it is aptly described as an "obligate seeder". Populations of cypress pine can survive mild fires occurring every 2–8 years, but not frequent or more intense fires because of the delay in seedlings reaching maturity and the cumulative damage of fires to adults. Cessation of Aboriginal land management has led to a decline of cypress pine in much of its former range, and it currently persists only in rainforest margins and savanna micro-sites such as in rocky crevasses or among boulders or drainage lines that protect seedlings from fire (Figure 9.6). Fire sensitive species such as King Billy pine and cypress pine are powerful bio-indicators of altered fire regimes because changes in their distribution,

**Figure 9.6** Recently killed individuals of cypress pine (*Callitris intratopica*), a conifer that is an obligate seeder. Changes in fire regime following the breakdown of traditional Aboriginal fire management have seen a population crash of this species throughout its range in northern Australia. Photograph by David Bowman.

density, and stand structure signal departure from historical fire regimes.

## 9.6.2 Fire and habitat complexity

A complex fire regime can create habitat complexity for wildlife by establishing mosaics of different patch size of regenerating vegetation following fires. Such habitat complexity provides a diversity of microclimates, resources, and shelter from predators. It is widely believed that the catastrophic decline of mammal species in central Australia, where clearing of native vegetation for agriculture has not occurred, is a direct consequence of the homogenization of fine-scale habitat mosaics created by Aboriginal landscape burning. This interpretation has been supported by analysis of "fire scars" from historical aerial photography and satellite imagery. For example, Burrows and Christensen (1991) compared fire scars present in Australia's Western Desert in 1953, when traditional Aboriginal people still occupied the region, with those present in 1986, when the area had become depopulated of its original inhabitants. In 1953, the study area contained 372 fire scars with a mean area of 34 ha, while in 1986, the same area contained a single fire scar, covering an area of 32 000 ha. Clearly, the present regime of large, intense and infrequent fires associated with lightning strikes has obliterated the fine-grained mosaic of burnt patches of varying ages that Aboriginal people had once maintained (Burrows et al. 2006). The cessation of Aboriginal landscape burning in central Australia has been linked to the range contraction of some mammals such as the rufous hare-wallaby (*Lagorchestes hirsutus*) (Lundie-jenkins 1993). Recent research in northern Australia's tropical savannas, where small mammals and granivorous birds are in decline, also points to the importance of unfavorable fire regimes that followed European colonization (Woinarski et al. 2001). A prime example is the decline of the partridge pigeon (*Geophaps smithii*). This bird is particularly vulnerable to changes in fire regime because it is feeds and nests on the ground and has territories of less than 10 ha. Their preferred habitat is a fine-grained mosaic of burnt and unburnt savanna, where it feeds on seeds on burnt ground but nests and roosts in unburnt areas (Fraser et al. 2003). Aboriginal landscape burning has been shown to produce such a fine-grained mosaic (Bowman et al. 2004).

### 9.6.3 Managing fire regimes for biodiversity

The contrasting requirements of different species and communities within fire-prone landscapes highlights the difficulties faced by those managing fire regimes for biodiversity conservation. How does one manage for fire-reliant and fire-sensitive species at the same time? Lessons can clearly be learnt from traditional hunter-gatherer societies that extensively used, and in some cases still use, fire as a land management tool. While it is unlikely that the enormous complexity of traditional fire use can ever be fully encapsulated in fire regimes imposed by conservation managers, it is clear that spatial and temporal complexity of the regime must be maximized to ensure the maximum benefits to biodiversity. Clearly, in the case of fire regimes designed for biodiversity conservation, one size can't fit all. The quest for sustainable fire regimes demands trialing approaches and monitoring outcomes while balancing biodiversity outcomes against other priorities such as protection of life and property. This quest for continuous improvement in land management has been formalized in a process known as "adaptive management". This iterative process is most applicable when faced with high levels of uncertainty, and involves continually monitoring and evaluating the outcomes of management actions, and modifying subsequent actions accordingly.

### 9.7 Climate change and fire regimes

There is mounting concern that the frequency and intensity of wildfires may increase in response to global climate change (see Chapter 8), due to the greater incidence of extreme fire weather. While the effect is likely to vary substantially on a global scale, regions that are likely to experience substantial increases in temperature and reductions in rainfall are also likely to experience more extreme fire weather. Indeed, such a trend is already apparent in southeastern Australia (Lucas et al. 2007) and the western United States (Westerling et al. 2006).

In addition to the effects of climate change, an increase in atmospheric $CO_2$ concentration is likely to affect the abundance and composition of fuel loads, and hence the frequency and intensity of fires. Elevated $CO_2$ concentration is likely to increase plant productivity, especially that of species utilizing the C3 photosynthetic pathway (mainly woody plants and temperate grasses), such that there have been suggestions that fuel production will increase in the future (Ziska et al. 2005). Further, elevated $CO_2$ concentration may lower the nitrogen content of foliage, slowing decomposition and resulting in heavier fuel build up (Walker 1991). However, to state that an increase in $CO_2$ concentration will increase fuel loads, and hence fire frequency and intensity, is likely to be a gross over-generalization; the effects of elevated $CO_2$ are in fact likely to vary substantially between biomes. For example, in tropical savannas, it is likely that increases in $CO_2$

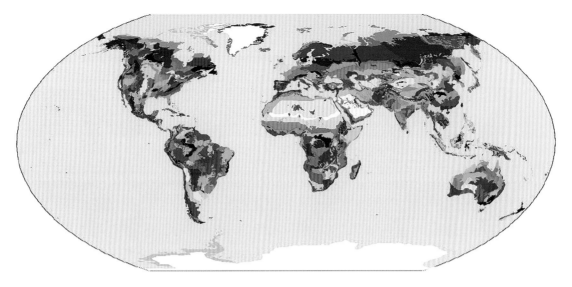

**Plate 1** The terrestrial ecoregions. Reprinted from Olson *et al.* (2001). (see page 32).

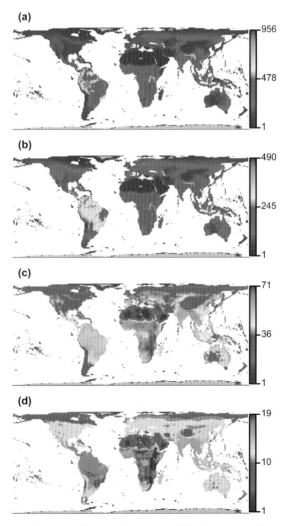

**Plate 2** Global richness patterns for birds of (a) species, (b) genera, (c) families, and (d) orders. Reprinted from Thomas *et al.* (2008). (see page 36).

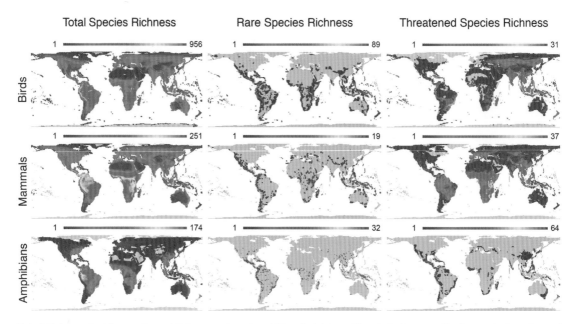

**Plate 3** Global species richness patterns of birds, mammals, and amphibians, for total, rare (those in the lower quartile of range size for each group) and threatened (according to the IUCN criteria) species. Reprinted from Grenyer *et al*. (2006). (see page 37).

**Plate 4** A few of the many human threats to marine ecosystems around the world. (A) The seafloor before and after bottom trawl fishing occurred [courtesy CSIRO (Australian Commonwealth Scientific and Research Organization) Marine Research], (B) coastal development in Long Beach, California (courtesy California Coastal Records Project), (C) shrimp farms in coastal Ecuador remove coastal habitat (courtesy Google Earth), and (D) commercial shipping and ports produce pollution and introduce non-native species (courtesy public commons). (see page 83).

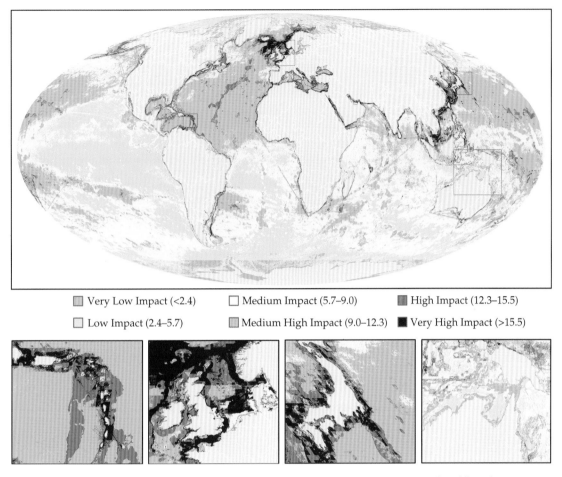

**Plate 5** Global map of the cumulative human impact on marine ecosystems, based on 20 ecosystem types and 17 different human activities. Grayscale colors correspond to overall condition of the ocean as indicated in the legend, with cumulative impact score cutoff values for each category of ocean condition indicated. (see page 84).

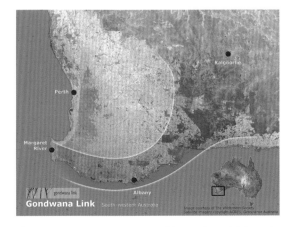

**Plate 6** Diagrammatic representation of the Gondwana link in southwest Western Australia. Shaded areas indicate remnant native vegetation. (see page 101).

**Plate 7** North American red signal crayfish (right) and a native European crayfish (*Astacus astacus*). Photograph by David Holdich. (see page 138).

**Plate 8** Maryland's aquatic invasive species. Poster courtesy of the Maryland Department of Natural Resources. (see page 146).

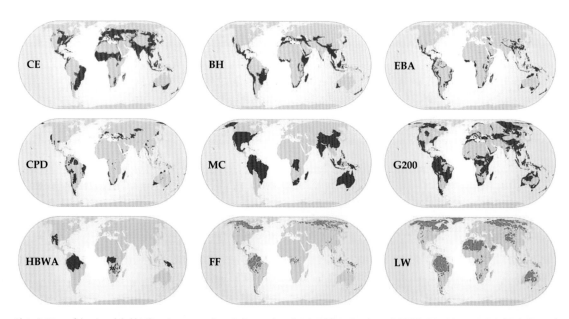

**Plate 9** Maps of the nine global biodiversity conservation priority templates (reprinted from Brooks et al. 2006): CE, crisis ecoregions (Hoekstra et al. 2005); BH, biodiversity hotspots (Mittermeier et al. 2004); EBA, endemic bird areas (Stattersfield et al. 1998); CPD, centers of plant diversity (WWF and IUCN 1994–7); MC, megadiversity countries (Mittermeier et al. 1997); G200, global 200 ecoregions (Olson and Dinerstein 1998); HBWA, high-biodiversity wilderness areas (Mittermeier et al. 2003); FF, frontier forests (Bryant et al. 1997); and LW, last of the wild (Sanderson et al. 2002a). With permission from AAAS (American Association for the Advancement of Science). (see page 200).

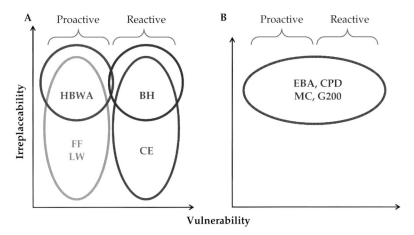

**Plate 10** Global biodiversity conservation priority templates placed within the conceptual framework of irreplaceability and vulnerability (reprinted from Brooks et al. 2006). Template names follow the Figure 11.1 legend. (A) Purely reactive (prioritizing high vulnerability) and purely proactive (prioritizing low vulnerability) approaches. (B) Approaches that do not incorporate vulnerability as a criterion (all prioritize high irreplaceability). With permission from AAAS (American Association for the Advancement of Science). (see page 201).

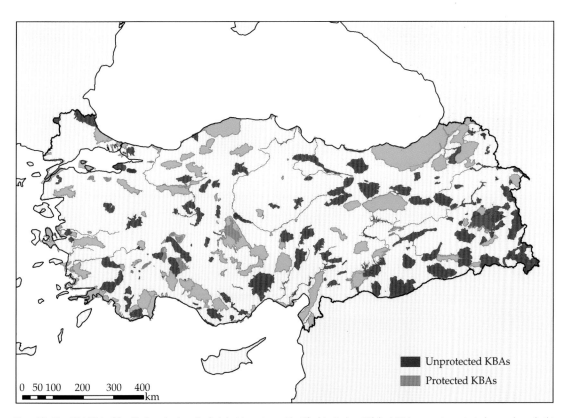

**Plate 11** The 294 KBAs (Key Biodiversity Areas) of global importance identified in Turkey. While 146 incorporate protected areas (green), this protection still covers <5% of Turkey's land area. The remaining 148 sites (red) are wholly unprotected. (see page 209).

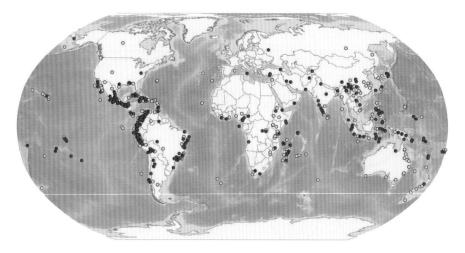

**Plate 12** Map of 595 sites of imminent species extinction (reprinted from Ricketts et al. 2005). Yellow sites are either fully protected or partially contained within declared protected areas (*n* = 203 and 87, respectively), and red sites are completely unprotected or have unknown protection status (*n* = 257 and 48, respectively). In areas of overlap, unprotected (red) sites are mapped above protected (yellow) sites to highlight the more urgent conservation priorities. (see page 211).

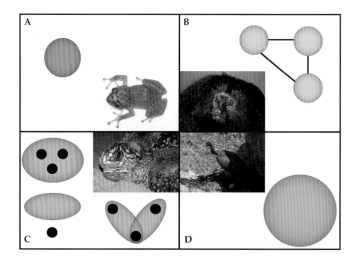

**Plate 13** Scale requirements for the conservation of globally threatened species in the short- to medium term (reprinted from Boyd *et al.* 2008). (A, dark green) Species best conserved at a single site (e.g. *Eleutherodactylus corona*); (B, pale green) Species best conserved at a network of sites (e.g. black lion-tamarin *Leontopithecus chrysopygus*); (C, dark blue) Species best conserved at a network of sites complemented by broad-scale conservation action (e.g. leatherback turtle *Dermochelys cariacea*); (D, pale blue) Species best conserved through broad-scale conservation action (e.g. Indian vulture *Gyps indicus*). Photographs by S. B. Hedges (A), R.A. Mittermeier (B), O. Langrand (C), and A. Rahmani (D). (see page 212).

**Plate 14** Akohekohe (Palmeria dolei), an endangered Hawaiian honeycreeper. Like many Hawaiian honeycreepers, it is endangered by a combination of habitat destruction and diseases transmitted by introduced mosquitoes. Photograph by Jaan Lepson. (see page 229).

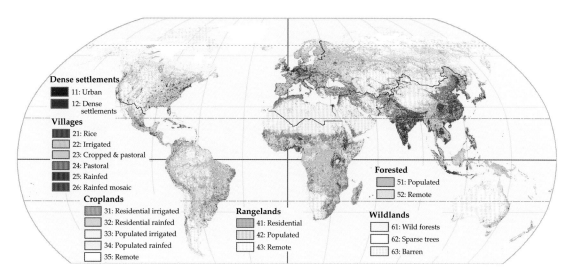

**Plate 15** Anthropogenic biomes. Global land-cover analysis reveals that that less than a quarter of the Earth's ice-free land can still be considered as wild. Biomes displayed on the map are organized into groups and are ranked according to human population density. Reprinted from Ellis and Ramankutty (2008). (see page 239).

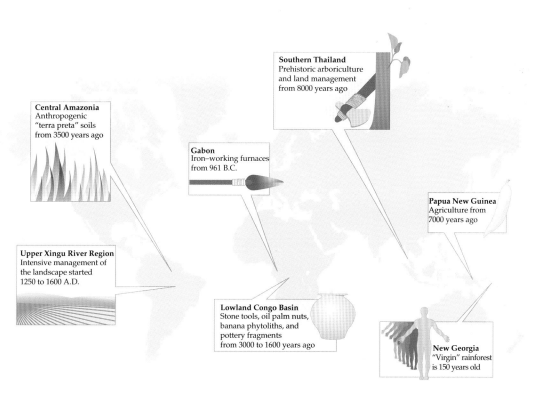

**Plate 16** Evidence of human modification of "pristine" tropical rainforest. Archaeological and paleoecological studies suggest that rainforests in the Amazon basin, the Congo basin, and Southeast Asia have regenerated from disturbance by prehistoric human settlements. Reprinted from Willis et al. (2004) with permission from AAAS (American Association for the Advancement of Science). (see page 240).

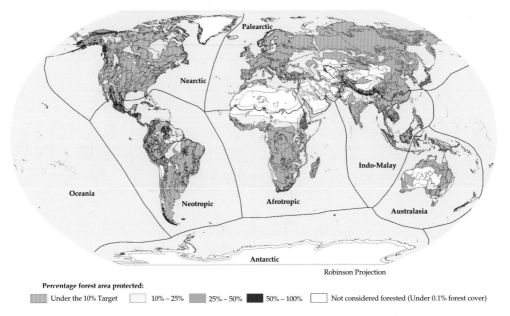

**Plate 17** Distribution of the percentage of protected forest area within WWF ecoregions. The highest levels of protection can be seen in parts of Australia, the Amazon, Southeast Asia, and Alaska. Notable areas of low protection include the Congo Basin in Central Africa and Northern Boreal forests. Black lines indicate biogeographic realms. White areas indicate no forest cover. Reprinted from Schmitt et al. (2009). (see page 241).

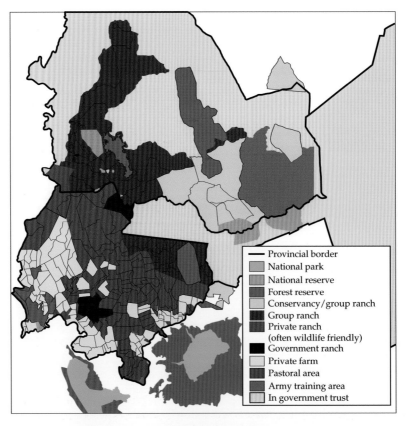

**Plate 18** The biodiversity of the Ewaso ecosystem in central Kenya is relatively intact due in large part to a strong set of protected areas. However, even these are not sufficient to preserve the patterns and processes of biodiversity and to reach conservation objectives. To do so, conservationists are working in the complex matrix of land uses beyond the protected areas, with a vast array of stakeholders, and using actions that benefit both people and biodiversity. (see page 294).

concentration will strongly favor woody plants, especially trees, at the expense of grasses and other herbaceous plants (Bond and Midgley 2000). A shift from highly flammable grassy fuels to fuels based on woody plants is likely to reduce fire frequency and intensity in savannas. Indeed, Bond and Archibald (2003) have argued that managers should consider increasing fire frequencies to counteract the increase in growth rates of savanna trees that would result in higher tree densities due to a weakening of the "tree demographic bottle-neck". In contrast, in more arid biomes where fire occurrence is strongly limited by antecedent rainfall (Allan and Southgate 2002), an increase in productivity is indeed likely to increase the frequency with which fires can occur with a corresponding decrease in woody cover.

Climate change is set to make fire management even more complicated, given that climate change simultaneously changes fire risk, ecosystem function, and the habitat template for most organisms, including invasive species. A recent report by Dunlop and Brown (2008) discussing the impact of climate change on nature reserves in Australia succinctly summarizes the problem conservation biologists now face. They write:

> "The question is how should we respond to the changing fire regimes? Efforts to maintain 'historic' fire regimes through hazard reduction burning and vigorous fire suppression may be resource intensive, of limited success, and have a greater impact on biodiversity than natural changes in regimes. It might therefore be more effective to allow change and manage the consequences. The challenge is to find a way to do this while ensuring some suitable habitat is available for sensitive species, and simultaneously managing the threat to urban areas, infrastructure, and public safety."

Again this demands an adaptive management approach, the key ingredients of which include: (i) clear stated objectives; (ii) comprehensive fire mapping programs to track fire activity across the landscape; (iii) monitoring the population of biodiversity indicator species and/or condition and extent of habitats; and (iv) rigorous evaluation of the costs and benefits of management interventions.

An important concept is "thresholds of potential concern" which predefines acceptable changes in the landscape in response to different fire regimes (Bond and Archibald 2003). Bradstock and Kenny (2003) provided an example of this approach for assessing the effect of inter-fire interval on species diverse scherophyll vegetation in the Sydney region of southeastern Australia. This vegetation supports a suite of species that are obligate seeders whose survival is held in a delicate balance by fire-frequency. Fire intervals that are shorter than the time required for maturation of plant species result in local extinction because of the absence of seeds while longer fire intervals also ultimately result in regeneration failure because adults die and seed-banks become exhausted. Bradstock and Kenny (2003) found that to sustain the biodiversity of sclerophyll vegetation, fire intervals between 7 and 30 years are required. Monitoring is required to ensure that the majority of the landscape does not move outside these "thresholds of potential concern".

Fire management is set to remain a thorny issue for conservation biologists given the need to devise fire regimes to achieve multiple outcomes that on the one hand protect life and property and on the other maintain biodiversity and ecosystem services. The accelerating pace of global environmental change, of which climate change is but one component, makes the quest for sustainable fire management both more critical and more complex. The current quest for ecologically sustainable fire management can draw inspiration from indigenous societies that learnt to coexist with fire to create ecologically sustainable and biodiverse landscapes (also see Box 1.1). Modern solutions will undoubtedly be science based and use space-age technologies such as satellites, global positioning systems, computer models and the web.

## Summary

- The Earth has a long history of landscape fire given: (i) the evolution of terrestrial carbon based vegetation; (ii) levels of atmospheric oxygen that are sufficient to

support the combustion of both living and dead organic material; and (iii) abundant and widespread ignitions from lightning, volcanoes and humans.
- There is a clear geographic pattern of fire activity across the planet reflecting the combined effects of climate, vegetation type and human activities. Most fire activity is concentrated in the tropical savanna biome.
- Fire activity shows distinct spatial and temporal patterns that collectively can be grouped into "fire regimes". Species show preferences for different fire regimes and an abrupt switch in fire regime can have a deleterious effect on species and in extreme situations, entire ecosystems. A classic example of this is the establishment of invasive grasses, which dramatically increase fire frequency and intensity with a cascade of negative ecological consequences.
- Climate change presents a new level of complexity for fire management and biodiversity conservation because of abrupt changes in fire risk due to climate change and simultaneous stress on species. Further, elevated atmospheric $CO_2$ concentration may result in changes in growth and fuel production due to changes in growth patterns, water use efficiency and allocation of nutrients.
- Numerous research challenges remain in understanding the ecology and evolution of fire including: (i) whether flammability changes in response to natural selection; (ii) how life-history traits of both plants and animals are shaped by fire regimes; and (iii) how to manage landscape fire in order to conserve biodiversity.

## Suggested reading

- Bond, W. J. and Van Wilgen, B. W. (1996). *Fire and plants*. Chapman and Hall, London, UK.
- Bowman, D. M. J. S. (2000). *Australian rainforests: islands of green in a land of fire*. Cambridge University Press, Cambridge, UK.
- Flannery, T. F. (1994). *The future eaters: an ecological history of the Australasian lands and people*. Reed Books, Chatswood, New South Wales, Australia.
- Whelan, R. J. (1995). *The ecology of fire*. Cambridge University Press, Melbourne, Australia.
- Gill, A. M., Bradstock, R. A., and Williams, J. E. (2002). *Flammable Australia: the fire regimes and biodiversity of a continent*. Cambridge University Press, Cambridge, UK.

- Pyne, S. J. (2001). *Fire: a brief history*. University of Washington Press, Seattle.

## Relevant websites

- Online journal of the Association for Fire Ecology: http://www.firecology.net.
- International Journal of Wildland Fire, journal of the International Association of Wildland Fire: http://www.iawfonline.org.
- North Australian Fire Information: http://www.fire-north.org.au.
- Forest Fire Danger Meter: http://www.cfa4wd.org/information/Forest_FDI.htm.
- Proliferation of woody plants in grasslands and savannas – a bibliography: http://ag.arizona1.edu/research/archer/research/biblio1.html.
- How fires affect biodiversity: http://www.anbg.gov.au/fire_ecology/fire-and-biodiversity.html.
- Kavli Institute of Theoretical Physics Miniconference: Pyrogeography and Climate Change (May 27–30, 2008): http://online.itp.ucsb.edu/online/pyrogeo_c08.

## Acknowledgements

The Kavli Institute for Theoretical Physics Miniconference: Pyrogeography and Climate Change meeting (http://online.itp.ucsb.edu/online/pyrogeo_c08) helped us organize our thinking. We thank the co-convener of that meeting, Jennifer Balch, for commenting on this chapter. An Australian Research Council grant (DP0878177) supported this work.

## REFERENCES

Allan, G. E. and Southgate, R. I. (2002). Fire regimes in the spinifex landscapes of Australia. In R. A. Bradstock, J. E. Williams, and M. A. Gill, eds *Flammable Australia: the Fire Regimes and Biodiversity of a Continent*, pp. 145–176. Cambridge University Press, Cambridge, UK.

Barnosky, A. D., Koch, P. L., Feranec, R. S., Wing, S. L., and Shabel, A. B. (2004). Assessing the causes of Late Pleistocene extinctions on the continents. *Science*, **306**, 70–75.

Bell, D. T., Plummer, J. A., and Taylor, S. K. (1993). Seed germination ecology in southwestern Western Australia. *Botanical Review*, **59**, 24–73.

Bond, W. J. and Archibald, S. (2003). Confronting complexity: fire policy choices in South African savanna parks. *International Journal of Wildland Fire*, **12**, 381–389.

Bond, W. J. and Keeley, J. E. (2005). Fire as a global 'herbivore': the ecology and evolution of flammable ecosystems. *Trends in Ecology and Evolution*, **20**, 387–394.

Bond, W. J. and Midgley, J. J. (1995). Kill thy neighbor—an individualistic argument for the evolution of flammability. *Oikos*, **73**, 79–85.

Bond, W. J. and Midgley, G. F. (2000). A proposed $CO_2$-controlled mechanism of woody plant invasion in grasslands and savannas. *Global Change Biology*, **6**, 865–869.

Bond, W. J, and Van Wilgen, B. W. (1996). *Fire and plants*. Chapman and Hall, London, UK.

Bond, W. J., Midgley, G. F., and Woodward, F. I. (2003). The importance of low atmospheric $CO_2$ and fire in promoting the spread of grasslands and savannas. *Global Change Biology*, **9**, 973–982.

Bond, W. J., Woodward, F. I., and Midgley, G. F. (2005). The global distribution of ecosystems in a world without fire. *New Phytologist*, **165**, 525–538.

Bowman, D. M. J. S. (1998). The impact of Aboriginal landscape burning on the Australian biota. *New Phytologist*, **140**, 385–410.

Bowman, D. M. J. S. (2000). *Australian rainforests: islands of green in a sea of fire*. Cambridge University Press, Cambridge, UK.

Bowman, D. M. J. S. (2005). Understanding a flammable planet—climate, fire and global vegetation patterns. *New Phytologist*, **165**, 341–345.

Bowman, D. M. J. S. and Panton, W. J. (1993). Decline of *Callitris intratropica* Baker, R.T. and Smith, H.G. in the Northern Territory—implications for pre-European and post-European colonization fire regimes. *Journal of Biogeography*, **20**, 373–381.

Bowman, D. M. J. S. and Yeates, D. (2006). A remarkable moment in Australian biogeography. *New Phytologist*, **170**, 208–212.

Bowman, D. M. J. S., Walsh, A., and Prior, L. D. (2004). Landscape analysis of Aboriginal fire management in Central Arnhem Land, north Australia. *Journal of Biogeography*, **31**, 207–223.

Bradstock, R. A. and Kenny, B. J. (2003). An application of plant functional types to fire management in a conservation reserve in southeastern Australia. *Journal of Vegetation Science*, **14**, 345–354.

Brown, M. J. (1988). *The distribution and conservation of King Billy pine*. Forestry Commission of Tasmania, Hobart, Tasmania, Australia.

Brown, N. A. C. (1993). Promotion of germination of fynbos seeds by plant-derived smoke. *New Phytologist*, **123**, 575–583.

Burney, D. A. and Flannery, T. F. (2005). Fifty millennia of catastrophic extinctions after human contact. *Trends in Ecology and Evolution*, **20**, 395–401.

Burrows, G. E. (2002). Epicormic strand structure in Angophora, Eucalyptus and Lophostemon (Myrtaceae)—implications for fire resistance and recovery. *New Phytologist*, **153**, 111–131.

Burrows, N. D. and Christensen, P. E. S. (1991). A survey of Aboriginal fire patterns in the Western Desert of Australia. In S.C. Nodvin, and T.A. Waldrop, eds 1991. *Fire and The Environment: ecological and cultural perspectives*, pp. 297–305. Proceedings of an International Symposium; 1990 March 20–24; Gen. Tech. rep. SE-69. Asheville, NC: U.S. Dept. of Agriculture, Forest Service, Southeastern Forest Experiment Station. Knoxville, TN.

Burrows, N. D., Burbidge, A. A., Fuller, P. J., and Behn, G. (2006). Evidence of altered fire regimes in the Western Desert regime of Australia. *Conservation Science Western Australia*, **5**, 272–284.

Cary, G. J., Keane, R. E., Gardner, R. H., *et al.* (2006). Comparison of the sensitivity of landscape-fire-successional models to variation in terrain, fuel pattern, climate and weather. *Landscape Ecology*, **21**, 121–137.

Cochrane, M. A. (2003). Fire science for rainforests. *Nature*, **421**, 913–919.

Cochrane, M. A., Alencar, A., Schulze, M. D. *et al.* (1999). Positive feedbacks in the fire dynamic of closed canopy tropical forests. *Science*, **284**, 1832–1835.

Crowley, G. M. and Garnett, S. T. (1998). Vegetation changes in the grasslands and grassy woodlands of east-central Cape York Peninsula, Australia. *Pacific Conservation Biology*, **4**, 132–148.

Darwin, C. (1859, 1964). *On the origin of species*. Harvard University Press, Cambridge, Massachusetts.

Denham, A. J. and Whelan, R. J. (2000). Reproductive ecology and breeding system of *Lomatia silaifolia* (Proteaceae) following a fire. *Australian Journal of Botany*, **48**, 261–269.

Dunlop, M. and Brown, P. R. (2008). *Implications of climate change for Australia's national reserve system: a preliminary assessment*. Australian Government, Department of Climate Change, Canberra, Australia.

Fraser, F., Lawson, V., Morrison, S., Christopherson, P., Mcgreggor, S., and Rawlinson, M. (2003). Fire management experiment for the declining partridge pigeon, Kakadu National Park. *Environmental Management and Restoration*, **4**, 94–102.

Gill, A. M. (1975). Fire and the Australia flora: a review. *Australian Forestry*, **38**, 4–25.

Hoffmann, W. A. and Solbrig, O. T. (2003). The role of topkill in the differential response of savanna woody species to fire. *Forest Ecology and Management*, **180**, 273–286.

Johnson, E. A. and Gutsell, S. L. (1994). Fire frequency models, methods and interpretations. *Advances in Ecological Research*, **25**, 239–287.

Justice, C. O., Smith, R., Gill, A. M., and Csiszar, I. (2003). A review of current space-based fire monitoring in Australia and the GOFC/GOLD program for international coordination. *International Journal of Wildland Fire*, **12**, 247–258.

Keeley, J. E. and Rundel, P. W. (2003). Evolution of CAM and $C_4$ carbon-concentrating mechanisms. *International Journal of Plant Sciences*, **164**, S55–S77.

Lucas, C., Hennessey, K., Mills, G., and Bathols, J. (2007). *Bushfire weather in Southeast Australia: recent trends and projected climate change impacts*. Consultancy report prepared for the Climate Institute of Australia by the Bushfire CRC and Australian Bureau of Meteorology.

Lundie-Jenkins, G. (1993). Ecology of the rufous hare-wallaby, *Lagorchestes hirsutus* Gould (Marsupialia: Macropodidae), in the Tanami Desert, Northern Territory. I. Patterns of habitat use. *Wildlife Research*, **20**, 457–476.

McGlone, M. S. (2001). The origin of the indigenous grasslands of southeastern South Island in relation to pre-human woody ecosystems. *New Zealand Journal of Ecology*, **25**, 1–15.

Miller, G. H., Fogel, M. L., Magee, J. W., Gagan, M. K., Clarke, J. S., and Johnson, B. J. (2005a). Ecosystem collapse in Pleistocene Australia and a human role in megafaunal extinction. *Science*, **309**, 287–290.

Miller, G. H., Mangan, J., Pollard, D., Thompson, S., Felzer, B., and Magee, J. (2005b). Sensitivity of the Australian Monsoon to insolation and vegetation: implications for human impact on continental moisture balance. *Geology*, **33**, 65–68.

Murphy, B. P. and Bowman, D. M. J. S. (2007). The interdependence of fire, grass, kangaroos and Australian Aborigines: a case study from central Arnhem Land, northern Australia. *Journal of Biogeography*, **34**, 237–250.

Mutch, R. W. (1970). Wildland fires and ecosystems—a hypothesis. *Ecology*, **51**, 1046–1051.

Pyne, S. J. (2007). Problems, paradoxes, paradigms: triangulating fire research. *International Journal of Wildland Fire*, **16**, 271–276.

Sankaran, M., Ratnam, J., and Hanan, N. P. (2004). Tree-grass coexistence in savannas revisited—insights from an examination of assumptions and mechanisms invoked in existing models. *Ecology Letters*, **7**, 480–490.

Sankaran, M., Hanan, N. P., Scholes, R. J., *et al.* (2005). Determinants of woody cover in African savannas. *Nature*, **438**, 846–849.

Schwilk, D. W. and Kerr, B. (2002). Genetic niche-hiking: an alternative explanation for the evolution of flammability. *Oikos*, **99**, 431–442.

Scott, A. C. and Glasspool, I. J. (2006). The diversification of Paleozoic fire systems and fluctuations in atmospheric oxygen concentration. *Proceedings of the National Academy of Sciences United States of America*, **103**, 10861–10865.

Sibold, J. S., Veblen, T. T., and Gonzalez, M. E. (2006). Spatial and temporal variation in historic fire regimes in subalpine forests across the Colorado Front Range in Rocky Mountain National Park, Colorado, USA. *Journal of Biogeography*, **33**, 631–647.

Swetnam, T. W. (1993). Fire history and climate change in Giant Sequoia groves. *Science*, **262**, 885–889.

Walker, B. H. (1991). Ecological consequences of atmospheric and climate change. *Climatic Change*, **18**, 301–316.

Werner, P. A. (2005). Impact of feral water buffalo and fire on growth and survival of mature savanna trees: an experimental field study in Kakadu National Park, northern Australia. *Austral Ecology*, **30**, 625–647.

Westerling, A. L., Hidalgo, H. G., Cayan, D. R., and Swetnam, T. W. (2006). Warming and earlier spring increase western US forest wildfire activity. *Science*, **313**, 940–943.

Whelan, R. J. (1995). *The ecology of fire*. Cambridge University Press, Cambridge, UK.

Woinarski, J. C. Z., Milne, D. J., and Wanganeen, G. (2001). Changes in mammal populations in relatively intact landscapes of Kakadu National Park, Northern Territory, Australia. *Austral Ecology*, **26**, 360–370.

Ziska, L. H., Reeves, J. B., and Blank, B. (2005). The impact of recent increases in atmospheric $CO_2$ on biomass production and vegetative retention of Cheatgrass (*Bromus tectorum*): implications for fire disturbance. *Global Change Biology*, **11**, 1325–1332.

# CHAPTER 10

# Extinctions and the practice of preventing them

## Stuart L. Pimm and Clinton N. Jenkins

In this chapter, we will outline why we consider species extinction to be the most important problem conservation science must address. Species extinction is irreversible, is progressing at a high rate and is poised to accelerate. We outline the global features of extinctions — how fast and where they occur. Such considerations should guide global allocation of conservation efforts; they do to some extent, though the priorities of some global conservation organizations leave much to be desired.

We conclude by asking how to go from these insights to what tools might be used in a practical way. That requires a translation from scales of about 1 million km$^2$ to mere tens of km$^2$ at which most conservation actions take place. Brooks (Chapter 11) considers this topic in some detail, and we shall add only a few comments. Again, the match between what conservation demands and common practice is not good.

## 10.1 Why species extinctions have primacy

"Biodiversity" means three broad things (Norse and McManus 1980; Chapter 2): (i) there is diversity within a species — usually genetic-based, but within our own species, there is a large, but rapidly shrinking cultural diversity (Pimm 2000); (ii) the diversity of species themselves, and; (iii) the diversity of the different ecosystems they comprise.

The genetic diversity within a species is hugely important as an adaptation to local conditions. Nowhere is this more obvious than in the different varieties of crops, where those varieties are the source of genes to protect crops from disease. Genetic uniformity can be catastrophic — the famous example is the potato famine in Ireland in the 1840s.

We simply do not know the genetic diversity of enough species for it to provide a practical measure for mapping diversity at a large scale. There is, however, a rapidly increasing literature on studies of the genetic diversity of what were once thought to be single species and are now known to be several. These studies can significantly alter our actions, pointing as they sometimes do to previously unrecognized species that need our attention.

Martiny (Box 10.1) argues for the importance of distinct populations within species, where the diversity is measured simply geographically. She argues, *inter alia*, that the loss of local populations means the loss of the ecosystem services species provide locally. She does not mention that, in the USA at least, "it's the law." Population segments, such as the Florida panther (*Puma concolor coryi*) or grizzly bears (*Ursus arctos horribilis*) in the continental USA are protected under the Endangered Species Act (see Chapter 12) as if they were full species. Indeed, the distinction is likely not clear to the average citizen, but scientific committees (National Research Council 1995) affirm Martiny's point and the public perception. Yes, it's important to have panthers in Florida, and grizzly bears in the continental USA, not just somewhere else.

That said, species extinction is irreversible in a way that population extinction is not. Some species have been eliminated across much of their ranges and later restored. And some of these flourished — turkeys in the eastern USA, for example. Aldo Leopold's dictum applies: the first law of intelligent

## Box 10.1 Population conservation
Jennifer B. H. Martiny

Although much of the focus of biodiversity conservation concentrates on species extinctions, population diversity is a key component of biodiversity. Imagine, for instance, that no further species are allowed to go extinct but that every species is reduced to just a single population. The planet would be uninhabitable for human beings, because many of the benefits that biodiversity confers on humanity are delivered through populations rather than species. Furthermore, the focus on species extinctions obscures the extent of the biodiversity crisis, because population extinction rates are orders of magnitude higher than species extinction rates.

When comparing species versus population diversity, it is useful to define population diversity as the number of populations in an area. Delimiting the population units themselves is more difficult. Historically, populations can be defined both demographically (by abundance, distribution, and dynamics) and genetically (by the amount of genetic variation within versus between intraspecific groups). Luck et al. (2003) also propose that populations be defined for conservation purposes as "service-providing units" to link population diversity explicitly to the ecosystem services that they provide.

The benefits of population diversity include all the reasons for saving species diversity and more (Hughes et al. 1998). In general, the greater the number of populations within a species, the more likely that a species will persist; thus, population diversity is directly linked to species conservation. Natural ecosystems are composed of populations of various species; as such systems are disrupted or destroyed, the benefits that those ecosystems provide are diminished. These benefits include aesthetic values, such as the firsthand experience of observing a bird species in the wild or hiking in an old growth forest. Similarly, many of the genetic benefits that biodiversity confers to humanity, such as the discovery and improvement of pharmaceuticals and agricultural crops, are closely linked to population diversity. For instance, genetically uniform strains of the world's three major crops (rice, wheat, and maize) are widely planted; therefore, population diversity among wild crop relatives is a crucial source of genetic material to resist diseases and pests.

Perhaps the most valuable benefit of population diversity is the delivery of ecosystem services such as the purification of air and water, detoxification and decomposition of wastes, generation and maintenance of soil fertility, and the pollination of crops and natural vegetation (see Chapter 3). These services are typically provided by local biodiversity; for a region to receive these benefits, populations that carry out the ecosystem services need to exist nearby. For instance, native bee populations deliver valuable pollination services to agriculture but only to fields within a few kilometers of the populations' natural habitats (Kremen et al. 2002; Ricketts et al. 2004).

Estimates of population extinctions due to human activities, although uncertain, are much higher than species extinctions. Using a model of habitat loss that has previously been applied to species diversity, it is estimated that millions of populations are going extinct per year (Hughes et al. 1997). This rate is three orders of magnitude higher than that of species extinction. Studies on particular taxa confirm these trends; population extinctions are responsible for the range contractions of extant species of mammals and amphibians (Ceballos and Ehrlich 2002; Wake and Freedenberg 2008).

### REFERENCES

Ceballos, G. and Ehrlich, P. R. (2002). Mammal population losses and the extinction crisis. Science, 296, 904–907.

Hughes, J. B., Daily, G. C., and Ehrlich, P. R. (1997). Population diversity: Its extent and extinction. Science, 278, 689–692.

Hughes, J. B., Daily, G. C., and Ehrlich, P. R. (1998). Population diversity and why it matters. In P. H. Raven, ed. Nature and human society, pp. 71–83. National Academy Press, Washington, DC.

*continues*

> **Box 10.1 (Continued)**
>
> Kremen, C., Williams, N. M., and Thorp, R. W. (2002). Crop pollination from native bees at risk from agricultural intensification. *Proceedings of the National Academy of Sciences of the United States of America*, **99**, 16812–16816.
>
> Luck, G. W., Daily, G. C., and Ehrlich, P. R. (2003). Population diversity and ecosystem services. *Trends in Ecology and Evolution*, **18**, 331–336.
>
> Ricketts, T. H., Daily, G. C., Ehrlich, P. R., and Michener, C. D. (2004). Economic value of tropical forest to coffee production. *Proceedings of the National Academy of Sciences of the United States of America*, **101**, 12579–12582.
>
> Wake, D. B. and Greenburg, V. T. (2008). Are we in the midst of the sixth mass extinction? A view from the world of amphibians. *Proceedings of the National Academy of Sciences of the United States of America*, **105**, 11466–11473.

tinkering is to keep every cog and wheel (Leopold 1993). So long as there is one population left, however bleak the landscapes from which it is missing, there is hope. Species extinction really is forever — and, as we shall soon present, occurring at unprecedented rates.

There are also efforts to protect large-scale ecosystems for their intrinsic value. For example, in North America, the Wildlands Project has as one of its objectives connecting largely mountainous regions from Yellowstone National Park (roughly 42°N) to the northern Yukon territory (roughly 64°N)—areas almost 3000 km away (Soulé and Terborgh 1999). A comparably heroic program in Africa is organized by the Peace Parks Foundation (Hanks 2003). It has already succeeded in connecting some of the existing network of already large national parks in southern Africa particularly through transboundary agreements. These efforts proceed with little regard to whether they contain species at risk of extinction, but with the clear understanding that if one does maintain ecosystems at such scales then the species within them will do just fine. Indeed, for species that need very large areas to survive—wild dog and lion in Africa — such areas may hold the only hope for saving these species in the long-term.

## 10.2 How fast are species becoming extinct?

There are ~10 000 species of birds and we know their fate better than any other comparably sized group of species. So we ask first: at what rate are birds becoming extinct? Then we ask: how similar are other less well-known taxa?

To estimate the rate of extinctions, we calculate the extinction rate as the number of extinctions per year per species or, to make the numbers more reasonable, per million species-years — MSY (Pimm *et al.* 1995; Pimm and Brooks 2000). With the exception of the past five mass extinction events, estimates from the fossil record suggest that across many taxa, an approximate background rate is one extinction per million species-years, (1 E/MSY) (Pimm *et al.* 1995). This means we should observe one extinction in any sample where the sum of all the years over all the species under consideration is one million. If we consider a million species, we should expect one extinction per year. Follow the fates of 10 000 bird species and we should observe just one extinction per 100 years.

### 10.2.1 Pre-European extinctions

On continents, the first contact with modern humans likely occurred ~15 000 years ago in the Americas and earlier elsewhere — too far back to allow quantitative estimates of impacts on birds. The colonization of oceanic islands happened much more recently. Europeans were not the first trans-oceanic explorers. Many islands in the Pacific and Indian Oceans received their first human contact starting 5000 years ago and many only within the last two millennia (Steadman 1995; Gray *et al.* 2009).

Counting the species known to have and estimated to have succumbed to first contact suggests that between 70 and 90 endemic species were lost to human contact in the Hawaiian Islands alone, from an original terrestrial avifauna estimated to be 125 to 145 species (Pimm et al. 1994). Comparable numbers emerge from similar studies across the larger islands of the Polynesian expansion (Pimm et al. 1994). One can also recreate the likely species composition of Pacific islands given what we know about how large an island must be to support a species of (say) pigeon and the geographical span of islands that pigeons are known to have colonized. Curnutt and Pimm (2001) estimated that in addition to the ~200 terrestrial bird species taxonomists described from the Pacific islands from complete specimens, ~1000 species fell to first contact with the Polynesians.

Species on other oceanic islands are likely to have suffered similar fates within the last 1500 years. Madagascar lost 40% of its large mammals after first human contact, for example (Simons 1997). The Pacific extinctions alone suggest one extinction every few years and extinctions elsewhere would increase that rate. An extinction every year is a hundred times higher than background (100 E/MSY) and, as we will soon show, broadly comparable to rates in the last few centuries.

## 10.2.2 Counting historical extinctions

Birdlife International produces the consensus list of extinct birds (BirdLife International 2000) and a regularly updated website (Birdlife International 2006). The data we now present come from Pimm et al. (2006) and website downloads from that year. In 2006, there were 154 extinct or presumed extinct species and 9975 bird species in total. The implied extinction rate is ~31 E/MSY — one divides the 154 extinctions by 506 years times the 9975 species (~ 5 million species-years) on the assumption that these are the bird extinctions since the year 1500, when European exploration began in earnest. (They exclude species known from fossils, thought to have gone before 1500.)

As Pimm et al. (2006) emphasize, the count of extinctions over a little more than 500 years has an unstated assumption that science has followed the fates of *all* the presently known species of bird over *all* these years. Scientific description though only began in the 1700s, increased through the 1800s, and continues to the present. Linnaeus described many species that survive to the present and the Alagoas curassow (*Mitu mitu*) that became extinct in the wild ~220 years later. By contrast, the po'o uli (*Melamprosops phaeosoma*), described in 1974, survived a mere 31 years after its description. If one sums all the years that a species has been known across all species, the total is only about 1.6 million species-years and the corresponding extinction rate is ~ 85 E/MSY, that is, slightly less than one bird extinction *per year*. This still underestimates the true extinction rate for a variety of reasons (Pimm et al. 2006).

## 10.2.3 Extinction estimates for the 21$^{st}$ century

Birdlife International (2006) lists 1210 bird species in various classes of risk of extinction, that combined we call, "threatened," for simplicity. The most threatened class is "critically endangered." Birdlife International (2006) list 182 such species, including the 25 species thought likely to have gone extinct but for conservation actions. For many of these species there are doubts about their continued existence. For all of these species, expert opinion expects them to become extinct with a few decades without effective conservation to protect them. Were they to expire over the next 30 years, the extinction rate would be 5 species per year or 500 E/MSY. If the nearly 1300 threatened or data deficient species were to expire over the next century, the average extinction rate would exceed 1300 E/MSY. This is an order of magnitude increase over extinctions-to-date.

Such calculations suggest that species extinction rates will now increase rapidly. Does this make sense, especially given our suggestion that the major process up to now, the extinction on islands, might slow because those species sensitive to human impacts have already perished? Indeed, it does, precisely because of a rapid increase in extinction on continents where there have been few recorded extinctions to date. To fully justify that, we must examine what we know about the global extinction process. First, however, we consider whether these results for birds seem applicable to other taxa.

### 10.2.4 Other taxa: what we don't know may make a very large difference

Birds play an important part in this chapter because they are well-known and that allows a deeper understanding of the processes of extinction than is possible with other taxa (e.g. Pimm *et al.* 1993). That said, birds constitute only roughly one thousandth of all species. (Technically, of eukaryote species, that is excluding bacteria and viruses.) Almost certainly, what we know for birds greatly underestimates the numbers of extinctions of other taxa, both past and present, for a variety of reasons.

On a percentage basis, a smaller fraction of birds are presently deemed threatened than mammals, fish, and reptiles, according to IUCN's Redlist (www.iucnredlist.org), or amphibians (Stuart *et al.* 2004). For North America, birds are the second least threatened of 18 well-known groups (The Nature Conservancy 1996). Birds may also be intrinsically less vulnerable than other taxa because of their mobility, which often allows them to persist despite substantial habitat destruction. Other explanations are anthropogenic.

Because of the widespread and active interest in birds, the recent rates of bird extinctions are far lower than we might expect had they not received special protection (Pimm *et al.* 2006; Butchart *et al.* 2006). Millions are fond of birds, which are major ecotourism attractions (Chapter 3). Many presently endangered species survive entirely because of extraordinary and expensive measures to protect them.

The most serious concern is that while bird taxonomy is nearly complete, other taxa are far from being so well known. For flowering plants worldwide, 16% are deemed threatened among the ~300 000 already described taxonomically (Walter and Gillett 1998). Dirzo and Raven (2003) estimate that about 100 000 plant species remain to be described. First, the majority of these will likely already be rare, since a local distribution is one of the principal factors in their escaping detection so far. Second, they are also certainly likely to be deemed threatened with extinction since most new species, in addition to being rare, live in tropical forests that are rapidly shrinking. We justify these two assumptions shortly.

Suppose we take Dirzo and Raven's estimate at face value. Then one would add the roughly 48 000 threatened species to the 100 000 as-yet unknown, but likely also threatened species, for a total of 148 000 threatened species out of 400 000 plants — or 37% of all plants.

With Peter Raven, we have been exploring whether his and Dirzo's estimate is reasonable. It comes from what plant taxonomists think are the numbers as-yet unknown. It is a best guess — and it proves hard to confirm. If it were roughly correct, we ought to see a decline in the numbers of species described each year — because fewer and fewer species are left undiscovered.

Consider birds again: Figure 10.1 shows the "discovery curve" — the number of species described per year. It has an initial spike with Linnaeus, then a severe drop (until Napoleone di Buonaparte was finally eliminated as a threat to world peace) and then a rapid expansion to about 1850. As one might expect, the numbers of new species then declined consistently, indicating that the supply of unknown species was drying up. That decline was not obvious, however, until a good half of all the species had been described (as shown by the graph of the cumulative number of species described.)

Interestingly, since 1950 there have been almost 300 new bird species added and the numbers per year have been more or less constant (Figure 10.1) Of these, about 10% were extinct when described, some found as only remains, others reassessments of older taxonomy. Of the rest, 27% are not endangered, 16% are near-threatened, 9% have insufficient data to classify, but 48% are threatened or already extinct. Simply, even for well-studied birds, there is a steady trickle of new species each year and most are threatened. Of course, we may never describe some bird species if their habitats are destroyed before scientists find them.

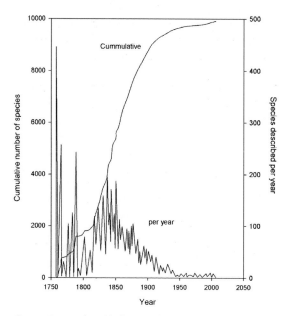

**Figure 10.1** Number of bird species described per year and the cumulative number of known bird species. Data from Pimm *et al.* 2006.

Now consider the implications for plants: plant taxonomy has rapidly increased the number of known species since about 1960, when modern genetic techniques became available. For example, there are ~30 000 species of orchids, but C. A. Luer (http://openlibrary.org/a/OL631100A) and other taxonomists have described nearly 800 species from Ecuador alone since 1995 — and there are likely similar numbers from other species-rich tropical countries! There is no decline in the numbers of new species — no peak in the discovery curve as there is for birds around 1850.

Might Dirzo and Raven have seriously underestimated the problem given that the half-way point for orchids might not yet have been reached? If orchids are typical, then there could be literally hundreds of thousands of species of as-yet unknown plants. By analogy to birds, most have tiny geographical ranges, live in places that are under immediate threat of habitat loss, and are in imminent danger of extinction. The final caveat for birds applies here, a fortiori. Many plants will never be described because human actions will destroy them (and their habitats) before taxonomists find them.

Well, Peter Raven (pers. comm., January 2009) argues that orchids might not be typical of other plants being under-collected. They are a group for which international laws make their export difficult, while their biology means they are often not in flower when found and so must be propagated. All this demands that we estimate numbers of missing taxa generally and, whenever possible, where they are likely to be.

Ceballos and Ehrlich (2009) have recently examined these issues for mammals, a group thought to be well-known. In fact, taxonomists described more than 400 mammal species since 1993 — ~10% of the total. Most of these new species live in areas where habitats are being destroyed and over half have small geographical ranges. As we show below, the combination of these two powerful factors predicts the numbers of species on the verge of extinction.

## 10.3 Which species become extinct?

Of the bird extinctions discussed, more than 90% have been on islands. Comparably large percentages of extinctions of mammals, reptiles, land snails, and flowering plants have been on islands too. So, will the practice of preventing extinction simply be a matter of protecting insular forms?

The answer is an emphatic "no" because the single most powerful predictor of past and likely future extinctions is the more general "rarity" — not island living itself. Island species are rare because island life restricts their range. Continental species of an equivalent level of rarity — very small geographical ranges — may not have suffered extinction yet, but they are disproportionately threatened with extinction. Quite against expectation, island species (and those that live in montane areas) are *less* likely to be threatened at range sizes smaller than 100 000 km$^2$ (Figure 10.2).

Certainly, species on islands may be susceptible to introduced predators and other enemies, but they (and montane species) have an offsetting advantage. They tend to be much more abundant locally than species with comparable range sizes living on continents.

Local rarity is a powerful predictor of threat in its own right. While species with large ranges tend to be locally common, there are obvious

EXTINCTIONS AND THE PRACTICE OF PREVENTING THEM 187

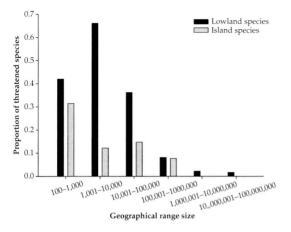

**Figure 10.2** The proportion of bird species in the Americas that are threatened declines as the size of a species' geographical range increases. While more than 90% of all extinctions have been on islands, for ranges less than 100 000 km², island species are presently *less* likely to be threatened with future extinction. Simplified from Manne et al. (1999).

exceptions—large carnivores, for example. Such species are at high risk. Manne and Pimm (2001) and Purvis et al. (2000) provide statistical analyses of birds and mammals, respectively, that expand on these issues. None of this is in any way surprising. Low total population size, whether because of small range, local rarity or both, exacerbated in fragmented populations and in those populations that fluctuate greatly from year-to-year (Pimm et al. 1988), likely brings populations to the very low numbers from which they cannot recover.

Given this importance of range size and local abundance, we now turn to the geography of species extinction.

## 10.4 Where are species becoming extinct?

### 10.4.1 The laws of biodiversity

There are at least seven "laws" to describe the geographical patterns of where species occur. By "law," we mean a general, widespread pattern, that is, one found across many groups of species and many regions of the world. Recall that Wallace (1855) described the general patterns of evolution in his famous "Sarawak Law" paper. (He would uncover natural selection, as the mechanism behind those laws, a few years later, independently of Darwin.) Wallace reviews the empirical patterns and then concludes:

*LAW 1. 'the following law may be deduced from these [preceding] facts: — Every species has come into existence coincident both in space and time with a pre-existing closely allied species'.*

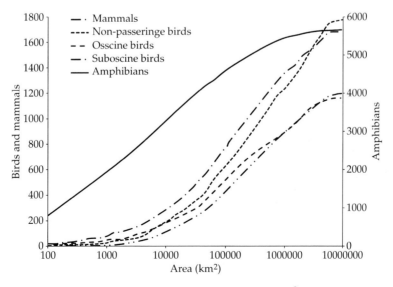

**Figure 10.3** Cumulative numbers of species with increasing size of geographical range size (in km²) for amphibians (worldwide; right hand scale), and mammals and three groups of bird species (for North and South America; left hand scale). Note that area is plotted on a log scale.

There are other generalities, too.

*LAW 2. Most species' ranges are very small; few are very large.*

Figure 10.3 shows cumulative distributions of range sizes for amphibians (worldwide) and for the mammals and three long-isolated lineages of birds in the Americas. The ranges are highly skewed. Certainly there are species with very large ranges — some greater than 10 million km$^2$, for example. Range size is so strongly skewed, however, that (for example) over half of all amphibian species have ranges smaller than ~6000 km$^2$. The comparable medians for the other taxa range from ~240 000 km$^2$ (mammals) to ~570 000 km$^2$ (non-passerine birds).

*LAW 3. Species with small ranges are locally scarce.*

There is a well-established relationship across many geographical scales and groups of species that links a species' range to its local abundance (Brown 1984). The largest-scale study is that of Manne and Pimm (2001) who used data on bird species across South America (Parker *et al.* 1996). The latter use an informal, if familiar method to estimate local abundances. A species is "common" if one is nearly guaranteed to see it in a day's fieldwork, then "fairly common," "uncommon"

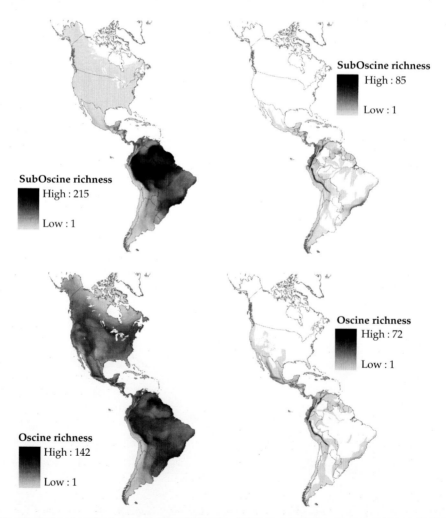

**Figure 10.4** Numbers of sub-oscine and oscine passerine birds, showing all species (at left) and those with geographical ranges smaller than the median.

down to "rare" — meaning it likely takes several days of fieldwork to find one even in the appropriate habitat. Almost all bird species with ranges greater than 10 million km$^2$ are "common," while nearly a third of species with ranges of less than 10 000 km$^2$ are "rare" and very few are "common."

LAW 4. *The number of species found in an area of given size varies greatly and according to some common factors.*

Figure 10.4 shows the numbers of all species (left hand side) and of those species with smaller than the median geographic range (right hand side) for sub-oscine passerine birds (which evolved in South America when it was geographically isolated) and oscine passerines (which evolved elsewhere.) Several broad factors are apparent, of which three seem essential (Pimm and Brown 2004).

*Geological history*
The long geographical isolation of South America that ended roughly 3 million years ago allowed suboscine passerines to move into North America across the newly formed Isthmus of Panama. The suboscines, nonetheless, have not extensively colonized North America and there are no small ranged suboscines north of Mexico.

*Ecosystem type*
Forests hold more species than do drier or colder habitats, even when other things (latitude, for example) are taken into consideration. Thus, eastern North American deciduous forests hold more species than the grasslands to their west, while the tropical forests of the Amazon and the southeast Atlantic coast of South America have more species than in the drier, cerrado habitats that separate them.

*Geographical constraints*
Extremes, such as high latitudes have fewer species, but interestingly — if less obvious — so too do peninsulas such as Baja California and Florida. Colwell *et al.* (2004) show there must be geographical constraints — by chance alone, there will be more species in the middle than at the extremes, given the observed distribution of geographical range sizes.

LAW 5. *Species with small ranges are often geographically concentrated and . . .*

LAW 6 . . . *those concentrations are generally not where the greatest numbers of species are found. They are, however, often in the same general places in taxa with different origins.*

Since the results on species extinction tell us that the most vulnerable species are those with small geographical ranges, we should explore where such species occur. The simplest expectation is that they will simply mirror the pattern of all species. That is, where there are more species, there will be more large-ranged, medium-ranged, and small-ranged species. Reality is strikingly different (Curnutt *et al.* 1994; Prendergast *et al.* 1994)!

Figure 10.4 shows that against the patterns for all species, small-ranged species are geographically concentrated, and not merely mirrored. Moreover, the concentrations of small-ranged species are, generally, not where the greatest numbers of species are. Even more intriguing, as Figure 10.4 also shows, is that the concentrations are in similar places for the two taxa *despite their very different evolutionary origins*. Maps of amphibians (Pimm and Jenkins 2005) and mammals (unpublished data) show these patterns to be general ones. At much coarser spatial resolution, they mirror the patterns for plants (Myers *et al.* 2000).

These similarities suggest common processes generate small-ranged species that are different from species as a whole.

*Island effects*
Likely it is that islands — real ones surrounded by water and "montane" islands of high elevation habitat surrounded by lowlands — provide the isolation needed for species formation. Figure 10.4 shows that it is just such places where small-ranged species are found.

*Glaciation history*
This is not a complete explanation, for some mountains — obviously those in the western USA and Canada — do not generate unusual numbers of small ranged species. Or perhaps they once did and those species were removed by intermittent glaciation.

Finally, there are simply anomalies: the Appalachian mountains of the eastern USA generate concentrations of small-ranged salamander species, but not birds or mammals. The mountains of western North America generate concentrations of small-ranged mammals but not birds.

### 10.4.2 Important consequences

Several interesting consequences emerge.

- The species at greatest risk of extinction are concentrated geographically and, broadly, such species in different taxa are concentrated into the same places. As argued previously, similar processes may create similar patterns across different taxa. This is of huge practical significance for it means that conservation efforts can be concentrated in these special places. Moreover, priorities set for one taxonomic group may be sensible for some others, at least at this geographical scale.
- A second consequence of these laws is far more problematical. Europe and North America have highly distorted selections of species. While most species have small ranges and are rare within them, these two continents have few species, very few species indeed with small ranges, and those ranges are not geographically concentrated. Any conservation priorities based on European and North American experiences are likely to be poor choices when it comes to preventing extinctions, a point to which we shall return.

### 10.4.3 Myers' Hotspots

By design, we have taken a mechanistic approach to draw a conclusion that extinctions will concentrate where there are many species with small ranges — *other things being equal*. Other things are not equal of course and the other important driver is human impact.

Figure 10.5 shows the distribution of threatened species of birds in The Americas. The concentration is in the eastern coast of South America, a place that certainly houses many species with small geographical ranges, but far from being the only place with such concentrations. What

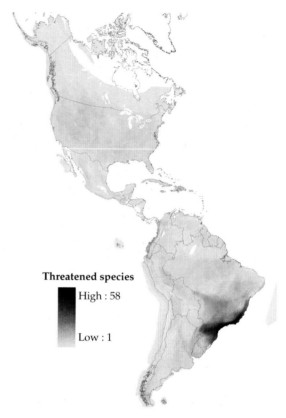

**Figure 10.5** The number of species of birds threatened with extinction in the Americas.

makes this region so unfortunately special is the exceptional high levels of habitat destruction.

Myers approached these topics from a "top down" perspective, identifying 10 and later 25 areas with more than 1000 endemic plants (Myers 1988, 1990; Myers *et al.* 2000). There are important similarities in the map of these areas (Figure 10.6) to the maps of Figure 10.4 (which only consider the Americas.) Central America, the Andes, the Caribbean, and the Atlantic Coast forests of South America stand out in both maps. California and the cerrado of Brazil (drier, inland forest) are important for plants, but not birds.

Myers added the second — and vital criterion — that these regions have less than 30% of their natural vegetation remaining. Myers' idea is a very powerful one. It creates the "number of

**Figure 10.6** The 25 hotspots as defined by Myers *et al.* 2000 (in black). The map projection is by Buckminster Fuller (who called it Dymaxion). It has no "right way" up and neither does the planet, of course.

small ranged species times habitat loss equals extinction" idea with another key and surprising insight. What surprises is that there are few examples of concentrations of small-ranged species that do not also meet the criterion of having lost 70% of more of their natural habitat. The island of New Guinea is an exception. Hotspots have disproportionate human impact measured in other ways besides their habitat loss. Cincotta *et al.* (2000) show that hotspots have generally higher human population densities and that almost all of them have annual population growth rates that are higher (average = 1.6% per annum) than the global average (1.3= per annum).

### 10.4.4 Oceanic biodiversity

Concerns about the oceans are usually expressed in terms of over-exploitation of relatively widespread, large-bodied and so relatively rare species (Chapter 6) — such as Steller's sea cow (*Hydrodamalis gigas*) and various whale populations. That said, given what we know about extinctions on the land, *where else* would we look for extinctions in the oceans?

As for the land, oceanic inventories are likely very incomplete. For example, there are more than 500 species of the lovely and medically important genus of marine snail, *Conus*. Of the 316 species of *Conus* from the Indo-Pacific region, Röckel *et al.* (1995) find that nearly 14% were described in the 20 years before their publication. There is no suggestion in the discovery curve that the rate of description is declining.

The first step would be to ask whether the laws we present apply to the oceans. We can do so using the data that Roberts *et al.* (2002) present geographically on species of lobster, fish, molluscs, and corals. Figure 10.7 shows the size of their geographical ranges, along with the comparable data for birds. Expressed as the cumulative percentages of species with given range sizes, (not total numbers of species as Figure 10.3), the scaling relationships are remarkably similar. For all but corals, the data show that a substantial fraction of marine species have very small geographical ranges. The spatial resolution of these data is coarse — about 1 degree latitude/longitude or ~10 000 km$^2$ — and likely overestimates actual ranges. Many of the species depend on

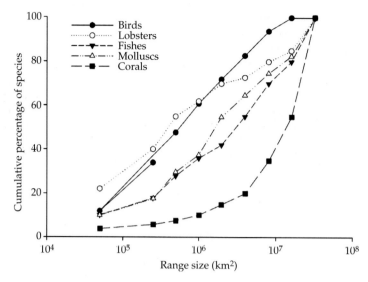

**Figure 10.7** The cumulative number of species of marine organisms (lobsters, fishes, molluscs, and corals) with birds for comparison (data from Roberts et al. 2002). Unlike Figure 10.3, these are scaled to 100% of the total number of species.

coral reefs, for example, that cover only a small fraction of the area within the 1-degree latitude/longitude cell where a species might occur.

The interesting generality here is that there are large fractions of marine species with very small geographical ranges — just as there are on land. The exception are the corals, most of which appear to occupy huge geographical ranges. Even here, this may be more a reflection of the state of coral taxonomy than of nature itself.

Roberts et al. (2002) also show that the other laws apply. Species-rich places are geographically concentrated in the oceans (Figure 10.8). They further show that as with the land, a small number of areas have high concentrations of species with small ranges and they are often not those places with the greatest number of species. Certainly, the islands between Asia and Australia have both many species and many species with small ranges. But concentrations of small range species also occur in the islands south of Japan, the Hawaiian Islands, and the Gulf of California — areas not particularly rich in total species. Finally, Bryant et al. (1998) do for reefs what Myers did for the land — and show that areas with concentrations of small-ranged species are often particularly heavily impacted by human actions.

Were we to look for marine extinctions, it would be where concentrations of small-ranged species collide with unusually high human impacts. Given that the catalogue of *Conus* species is incomplete, that many have small geographical ranges, and those occur in areas where reefs are being damaged, it seems highly unlikely to us that as few as four *Conus* species (<1%) are threatened with extinction as IUCN suggest (www.iucnredlist.org).

## 10.5 Future extinctions

### 10.5.1 Species threatened by habitat destruction

The predominant cause of bird species endangerment is habitat destruction (BirdLife International 2000). It is likely to be so for other taxa too. While large tracts of little changed habitat remain worldwide, most of the planet's natural ecosystems have been replaced or fragmented (Pimm 2001). Some species have benefited from those changes, but large numbers have not. The most important changes are to forests, particularly tropical forests for these ecosystems house most of the world's bird species (and likely other taxa as well). We now show that the numbers of extinctions predicted by a simple quantitative model match what we expect from the amount of forest lost. We then extend these ideas to more recently deforested areas to predict the numbers of species likely

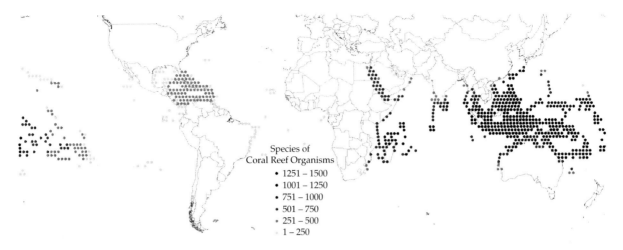

**Figure 10.8** Species richness of coral reef organisms (data from Roberts et al. 2002).

to become extinct eventually. The observed numbers of threatened species match those predictions, suggesting that we understand the mechanisms generating the predicted increase in extinction rate.

Rarity — either through small range size or local scarcity — does not itself cause extinction. Rather, it is how human impacts collide with such susceptibilities. As Myers reminds us, extinctions will concentrate where human actions impact concentrations of small ranged species. Without such concentrations, human impacts will have relatively little effect. The eastern USA provides a case history.

### 10.5.2 Eastern North America: high impact, few endemics, few extinctions

Europeans settled Eastern North America in the early 1600s and moved inland from the mid-1700s, settling the prairie states in the late 1800s. Along the way, they cleared most of the deciduous forest at one time or another. Despite this massive deforestation, only four species of land birds became extinct — the Carolina parakeet (*Conuropsis carolinensis*), passenger pigeon (*Ectopistes migratorius*), ivory-billed woodpecker (*Campephilus principalis*), and Bachman's warbler (*Vermivora bachmanii*) — out of a total of about 160 forest species.

Pimm and Askins (1995) considered why so few species were lost, despite such extensive damage. They considered a predictive model of how many species should be lost as a function of the fraction of habitat lost. This model follows from the familiar species-area law that describes the number of species found on islands in relation to island area. There is an obvious extension to that law that posits that as area is reduced (from $A_o$ to $A_n$) then the original number of species $S_o$ will shrink to $S_n$ in a characteristic way.

*LAW 7. The fraction of species ($S_n/S_o$) remaining when human actions reduce the area of original habitat $A_o$ to $A_n$ is $(A_n/A_o)^{0.25}$.*

We call this a law because we now show it to hold across a variety of circumstances.

First, Pimm and Askins noticed that while few forests were uncut, the deforestation was not simultaneous. European colonists cleared forests along the eastern seaboard, then moved across the Appalachians and then into the lake states. When settlers realized they could grow crops in the prairies, the eastern forests began to recover. At the low point, perhaps half of the forest remained. Applying the formula, the region should have retained 84% of its species and so lost 16%. Now 16% of 160 species is ~26 species and that is clearly not the right answer.

Second, Pimm and Askins posed the obvious thought-experiment: how many species should have been lost if all the forest was cleared? The answer is not 160, because most of those species have ranges outside of eastern North America — some across the forests of Canada, others in the western USA, some down into Mexico. They would survive elsewhere, even if all the forest were cut. Indeed only 30 species have sufficiently small ranges to be endemic to the region and so at risk if all the forest were lost. Applying the formulae to these one predicts that there would be 4.8 species at risk — surprisingly close to the right answer, given that another eastern species, the red-cockaded woodpecker (*Picoides borealis*), is threatened with extinction!

Simply, that there were so few extinctions — and so few species at risk — is largely a consequence of there being so few species with small ranges. So what happens when there are many species with small ranges?

### 10.5.3 Tropical areas with high impact, many endemics, and many species at risk

Case histories comparing how many species are threatened with extinction with how many are predicted to become extinct using Law 7 include birds in the Atlantic coast forest of Brazil (Brooks and Balmford 1996), birds and mammals in insular southeast Asia (Brooks *et al.* 1997; Brooks *et al.* 1999a), plants, invertebrates, and vertebrates of Singapore (Brook *et al.* 2003), and birds, mammals, amphibians, reptiles, and plants across the 25 biodiversity "hotspots" that we now introduce.

These studies, by choice, look at areas where there are many species with small geographical ranges, for the number of predicted extinctions depends linearly on the number of such species. But notice that Law 7 implies a highly non-linear relationship to the amount of habitat destruction. Losing the first half of eastern North America's forests resulted in a predicted loss of 16% of its species. Losing the remaining half would have exterminated the remaining 84%! The studies the previous paragraph cites looked at areas with far more extensive habitat destruction than eastern North America.

Pimm and Raven (2000) applied this recipe to each of the 25 hotspots using the statistics on endemic bird species, original area, and the present area of remaining natural vegetation. This provides a best-case scenario of what habitat might remain. They predicted that ~1700 species of birds should be lost eventually. Species can obviously linger in small habitat fragments for decades before they expire — as evidenced by the rediscovery of species thought extinct for up to a century. They suggest that bird extinctions among doomed species have a half-life of ~50 years (Brooks *et al.* 1999b; Ferraz *et al.* 2003). So perhaps three quarters of these species — 1250 — will likely go extinct this century — a number very similar to the number Birdlife considers to be at risk.

These estimates of extinction rates (~1000 E/MSY) come from human actions to date. Two extrapolations are possible. The worst-case scenario for the hotspots assumes that the only habitats that will remain intact will be the areas currently protected. This increases the prediction of number of extinctions to 2200 (Pimm and Raven 2000). The second adds in species from areas not already extensively deforested. If present trends continue, large remaining areas of tropical forest that house many species (such as the Amazon, the Congo, and Fly basin of New Guinea) will have extinction rates that exceed those in the hotspots by mid-century. For example, the Amazon basin is often ignored as a concentration of vulnerable species because its ~300 endemic bird species are found across ~5 million $km^2$. At current rates of deforestation, most of the Amazon will be gone by mid-century. There are plans for infrastructure development that would accelerate that rate of forest clearing (Laurance *et al.* 2001). If this were to happen, then many of the Amazon's species will become threatened or go extinct.

### 10.5.4 Unexpected causes of extinction

There are various unexpected causes of extinction and they will add to the totals suggested from habitat destruction. The accidental introduction of the brown tree snake (*Boiga irregularis*) to Guam eliminated the island's birds in a couple of decades (Savidge 1987; Wiles *et al.* 2003). In the oceans, increases in long-line fisheries (Tuck *et al.*

2003) are a relatively new and very serious threat to three-quarters of the 21 albatross species (Birdlife International 2006).

### 10.5.5 Global change and extinction

Finally, one of the most significant factors in the extinction of species will undoubtedly be climate change (see Chapter 8), a factor not included in any of the estimates presented above. Thomas *et al.* (2004) estimate that climate change threatens 15–37% of species within the next 50 years depending on which climate scenario unfolds. Even more species are at risk if one looks to climate changes beyond 50 years. More detailed, regional modeling exercises in Australia (Williams *et al.* 2003) and South Africa (Erasmus *et al.* 2002) have led to predictions of the extinction of many species with narrowly-restricted ranges during this or longer intervals.

The critical question is whether these extinctions, which are predominantly of small-ranged species, are the same as those predicted from habitat destruction or whether they are additional (Pimm 2008). In many cases, they are certainly the latter.

For example, the Atlantic coast humid forests of Brazil have the greatest numbers of bird species at risk of extinction within the Americas (Manne *et al.* 1999). The current threat comes from the extensive clearing of lowland forest. Upland forests have suffered less. Rio de Janeiro State has retained relatively more of its forests — 23% survives compared to <10% for the region as a whole. Less than 10% of the forest below 200 m remains though, whereas some 84% of the forest remains above 1300 m. It is precisely the species in these upper elevations that are at risk from global warming, for they have no higher elevations into which to move when the climate warms. These upland, restricted-range species will suffer the greatest risk from global warming, not the lowland species that are already at risk. Thus, the effects of direct habitat destruction and global warming are likely to be additive.

How large an additional threat is global warming? For New World passerine birds, a quarter live 1000 m above sea-level. Detailed modeling can certainly provide predictions of which species are at most risk (Sekercioglu *et al.* 2008), but the basic concerns are clear. If that fraction of species in mountains is typical of other taxa and other places, then a quarter of those species are at risk — a very substantial addition to species already threatened with extinction (Pimm 2008).

## 10.6 How does all this help prevent extinctions?

Thus far, we have guided the reader to areas of roughly one million km$^2$ — many orders of magnitude larger than the tens or at best hundreds of km$^2$ at which practical conservation actions unfold. Brooks (Chapter 11) considers formal tools for setting more local conservation priorities. We have rarely used such approaches in our work, though we understand the need for them.

This chapter establishes a recipe for conservation action that transcends scales. One can quite literally zoom in on Figure 10.5 to find out exactly where the greatest concentrations of threatened species are and, moreover, plot their ranges on maps of remaining forest. Our experiences are shaped by two places where our operational arm, www.savingspecies.org, has worked to date: the Atlantic Coastal Forest of Brazil and the island of Madagascar.

We have told this story in detail elsewhere (Harris *et al.* 2005; Jenkins 2003; Pimm and Jenkins 2005; Jenkins and Pimm 2006). For the Americas, we start with the species map of Figure 10.5 (but much enlarged). This shows the very highest concentration of threatened species to be in the State of Rio de Janeiro — an area of ∼40 000 km$^2$. At that point, what compels us most strongly is satellite imagery that shows what forest remains — not ever more detail about where species are found. There is not much forest — and very little indeed of the lowland forest remains. And that forest is in fragments.

Whatever conservation we do here is driven by these facts. We do not worry about the issues of capturing as many species in a given area (Pimm and Lawton 1998), and then write philosophical papers about weighting species because of their various "values" — taxonomic distinctiveness, for example. We do not fret about whether our priorities for birds match those for orchids for which we have only crude range information (Pimm 1996) or nematodes about which we know even less. What few remaining fragments remain will be the priorities for every taxon.

The practical solution is obvious too. The land between isolated forests needs to be brought into protection and reforested. That is exactly what we have helped our Brazilian colleagues achieve (www.micoleao.org.br). Connecting isolated forest fragments by reforesting them in areas rich in small-ranged species is an effective and cheap way of preventing extinctions. We commend this solution to others.

## Summary

- Extinctions are irreversible, unlike many other environmental threats that we can reverse.
- Current and recent rates of extinction are 100 times faster than the background rate, while future rates may be 1000 times faster.
- Species most likely to face extinction are rare; rare either because they have very small geographic ranges or have a low population density with a larger range.
- Small-ranged terrestrial vertebrate species tend to be concentrated in a few areas that often do not hold the greatest number of species. Similar patterns apply to plants and many marine groups.
- Extinctions occur most often when human impacts collide with the places having many rare species.
- While habitat loss is the leading cause of extinctions, global warming is expected to cause extinctions that are additive to those caused by habitat loss.

## Suggested reading

Brooks, T. M., Pimm, S. L., and Oyugi. J. O. (1999). Time Lag between deforestation and bird extinction in tropical forest fragments. *Conservation Biology*, **13**, 1140–1150.

Ferraz, G., Russell, G. J., Stouffer, P C., Bierregaard, R. O., Pimm, S. L., and Lovejoy, T. E. (2003). Rates of species loss from Amazonian forest fragments. *Proceedings of the National Academy of Sciences of the United States of America*, **100**, 14069–14073.

Manne, L. L, Brooks, T. M., and Pimm S. L. (1999). Relative risk of extinction of passerine birds on continents and islands. *Nature*, **399**, 258–261.

Myers, N., Mittermeier, R. A., Mittermeier, C. G., Fonseca, G. A. B., and Kent, J. (2000). Biodiversity hotspots for conservation priorities. *Nature*, **403**, 853–858.

Pimm S. L. (2001). *The World According to Pimm: A Scientist Audits the Earth*. McGraw–Hill, New York, NY.

Pimm, S. L. and Askins, R. (1995). Forest losses predict bird extinctions in eastern North America. *Proceedings of the National Academy of Sciences of the United States of America*, **92**, 9343–9347.

Pimm, S. L. and C. Jenkins. (2005). Sustaining the variety of Life. *Scientific American*, **September**, 66–73.

## Relevant web sites

- BirdLife International, Data Zone: http://www.birdlife.org/datazone.
- Threatened amphibians: http://www.globalamphibians.org.
- The IUCN Red List: http://www.iucnredlist.org.
- Saving species: http://savingspecies.org.

## REFERENCES

BirdLife International (2000). *Threatened Birds of the World*. Lynx Edicions and BirdLife International, Cambridge, UK.

Birdlife web site (2006). http://www.birdlife.org.

Brook, B. W., Sodhi, N. S., and Ng, P. K. L. (2003). Catastrophic extinctions follow deforestation in Singapore. *Nature*, **424**, 420–423.

Brooks, T. and Balmford, A. (1996). Atlantic forest extinctions, *Nature*, **380**, 115.

Brooks, T. M., Pimm, S. L., and Collar, N. J. (1997). Deforestation predicts the number of threatened birds in insular southeast Asia. *Conservation Biology*, **11**, 382–384.

Brooks, T. M., Pimm, S. L., Kapos V., and Ravilious C. (1999a). Threat from deforestation to montane and lowland birds and mammals in insular Southeast Asia. *Journal of Animal Ecology*, **68**, 1061–1078.

Brooks, T. M., Pimm, S. L., and Oyugi. J. O. (1999b). Time lag between deforestation and bird extinction in tropical forest fragments. *Conservation Biology*, **13**, 1140–1150.

Brown, J. H. (1984). On the relationship between abundance and distribution of species. *The American Naturalist*, **124**, 255–279.

Bryant, D., Burke, L., McManus, J., and Spalding, M. (1998). *Reefs at risk: a map-based indicator of the threats to the world's coral reefs*. Joint publication by World Resources Institute, International Center for Living Aquatic Resources Management, World Conservation Monitoring Centre, and United Nations Environment Programme, Washington, DC.

Butchart S. H. M, Stattersfield A, and Collar, N. (2006). How many bird extinctions have we prevented? *Oryx*, **40**, 266–278.

Ceballos, G and Ehrlich, P. R. (2009). Discoveries of new mammal species and their implications for conservation and ecosystem services. *Proceedings of the National Academy of Sciences of the United States of America*, **106**, 3841–3846.

Cincotta, R. P., Wisnewski, J., and Engelman, R. (2000). Human population in the biodiversity hotspots. *Nature* **404**, 990–992.

Colwell, R. K., Rahbek, C., and Gotelli, N. J. (2004). The mid-domain effect and species richness patterns: what have we learned so far? *The American Naturalist*, **163**, E1–E23.

Curnutt, J. and Pimm S. L. (2001). How many bird species in Hawai'i and the Central Pacific before first contact? In J. M. Scott, S. Conant and C. van Riper III, eds *Evolution, ecology, conservation, and management of Hawaiian birds: a vanishing avifauna*, pp. 15–30. Studies in Avian Biology 22, Allen Press Inc., Lawrence, KS.

Curnutt, J., Lockwood, J., Luh, H.-K., Nott, P., and Russell, G. (1994). Hotspots and species diversity. *Nature* **367**, 326.

Dirzo R. and Raven P. (2003). Global state of biodiversity loss. *Annual Reviews in Environment and Resources*, **28**, 137–167.

Erasmus, B. F. N., van Jaarsveld, A. S., Chown, S. L., Kshatriya, M., and Wessels, K. (2002). Vulnerability of South African animal taxa to climate change. *Global Change Biology*, **8**, 679–693.

Ferraz, G., Russell, G. J., Stouffer, P. C., Bierregaard, R. O., Pimm, S. L., and Lovejoy, T. E. (2003). Rates of species loss from Amazonian forest fragments. *Proceedings of the National Academy of Sciences of the United States of America*, **100**, 14069–14073.

Gray, R.D., Drummond, A. J., and Greenhill, S. J. (2009). Language phylogenies reveal expansion pulses and pauses in Pacific Settlment. *Science*, **323**, 479–483.

Hanks, J. (2003). Transfrontier Conservation Areas (TFCAs) in Southern Africa: their role in conserving biodiversity, socioeconomic development and promoting a culture of peace. *Journal of Sustainable Forestry*, **17**, 121–142.

Harris, G. M., Jenkins, C. N., and Pimm, S. L. (2005). Refining biodiversity conservation priorities. *Conservation Biology*, **19**, 1957–1968.

Jenkins, C. N. (2003). Importância dos remanescentes de Mata Atlântica e dos corredores ecológicos para a preservação e recuperação da avifauna do estado do Rio de Janeiro. Chapter in Índice de Qualidade dos Municípios – Verde II. Fundação CIDE, 2003, Rio de Janeiro, Brazil.

Jenkins, C. N. and Pimm S.L. (2006). Definindo Prioridades de Conservação em um Hotspot de Biodiversidade Global (Defining conservation priorities in a global biodiversity hotspot). Chapter in Biologia da Conservação: Essências. RiMa Editora, São Carlos, SP. Rocha, C.F.D.; H.G. Bergallo; M. Van Sluys & M.A.S. Alves. (Orgs.).

Laurance, W. F., Cochrane, M. A., Bergen, S., Fearnside, P. M., Delamonica, P., Barber, C., D'Angelo, S., and Fernandes, T. (2001). The future of the Brazilian Amazon. *Science*, **291**, 438–439.

Leopold, A. (1993). *Round River*. Oxford University Press, Oxford, UK.

Manne, L. L. and Pimm, S. L. (2001). Beyond eight forms of rarity: which species are threatened and which will be next? *Animal Conservation*, **4**, 221–230.

Manne, L. L, Brooks, T. M., and Pimm S. L. (1999). Relative risk of extinction of passerine birds on continents and islands. *Nature*, **399**, 258–261.

Myers, N. (1988). Threatened biotas: 'hotspots' in tropical forests. *The Environmentalist*, **8**, 1–20.

Myers, N. (1990). The biodiversity challenge: expanded hotspots analysis. *The Environmentalist*, **10**, 243.

Myers, N., Mittermeier, R. A., Mittermeier, C. G., Fonseca, G. A. B., and Kent, J. (2000). Biodiversity hotspots for conservation priorities. *Nature*, **403**, 853–858.

National Research Council (1995). *Science and the Endangered Species Act*. National Academy Press, Washington, DC.

Norse, E. A. and McManus, R. E. (1980). Ecology and living resources: biological diversity. In *Environmental quality 1980: the eleventh annual report of the Council on Environmental Quality*, pp. 31–80. Council on Environmental Quality, Washington, DC.

Parker, T. A., III., Stotz, D. F., and Fitzpatrick, J. W. (1996). Ecological and distributional databases for neotropical birds. In D. F. Stotz, T. A. Parker, J. W. Fitzpatrick and D. Moskovits, eds *Neotropical birds: ecology and conservation*, pp.113-436. University of Chicago Press, Chicago, IL.

Pimm, S. L. (1996). Lessons from a kill. *Biodiversity and Conservation*, **5**, 1059–1067.

Pimm, S. L. (2000). Biodiversity is us. *Oikos*, **90**, 3–6.

Pimm S. L. (2001). *The World According to Pimm: A Scientist Audits the Earth*. McGraw–Hill, New York, NY.

Pimm, S. L. (2008). Biodiversity: climate change or habitat loss — which will kill more species? *Current Biology*, **18**, 117–119.

Pimm, S. L. and Askins, R. (1995). Forest losses pr ict bird extinctions in eastern North America. *Proceedings of the National Academy of Sciences of the United States of America*, **92**, 9343–9347.

Pimm, S. L. and Brooks, T. M. (2000). The Sixth Extinction: How large, how soon, and where? In P. Raven, ed. *Nature and Human Society: the quest for a sustainable world*, pp. 46–62. National Academy Press, Washington, DC.

Pimm, S. L. and Brown, J. H. (2004). Domains of diversity. *Science*, **304**, 831–833.

Pimm, S. L. and C. Jenkins. (2005). Sustaining the variety of Life. *Scientific American*, **September**, 66–73.

Pimm, S. L. and Lawton. J. H. (1998). Planning for biodiversity. *Science*, **279**, 2068–2069.

Pimm, S. L. and Raven, P. (2000). Extinction by numbers. *Nature* **403**, 843–845.

Pimm, S. L., Jones, H. L., and J. M. Diamond. (1988). On the risk of extinction. *The American Naturalist*, **132**, 757–785.

Pimm, S. L., Diamond, J., Reed, T. R., Russell, G. J., and Verner, J. (1993). Times to extinction for small populations of large birds. *Proceedings of the National Academy of Sciences of the United States of America*, **90**, 10871–10875.

Pimm, S. L., Moulton M. P., and Justice J. (1994). Bird extinctions in the central Pacific. *Philosophical Transactions of the Royal Society B*, **344**, 27–33.

Pimm, S. L., Raven, P., Peterson, A., Sekercioglu, C. H., and Ehrlich P. R. (2006). Human impacts on the rates of recent, present, and future bird extinctions. *Proceedings of the National Academy of Sciences of the United States of America*, **103**, 10941–10946.

Pimm, S. L., Russell, G. J., Gittleman, J. L., and Brooks T. M. (1995). The future of biodiversity. *Science*, **269**, 347–350.

Prendergast, J. R., Quinn, R. M., Lawton, J. H., Eversham, B. C., and Gibbons, D. W. (1994). Rare species, the coincidence of diversity hotspots and conservation strategies. *Nature*, **365**, 335–337.

Purvis, A., Gittleman, J. L., Cowlishaw, G., and Mace, G. (2000). Predicting extinction risk in declining species. *Proceedings of the Royal Society of London B*, **267**, 1947–1952.

Roberts C. M., McClean, C. J., Veron, J. E., *et al.* (2002). Marine biodiversity hotspots and conservation priorities for tropical reefs. *Science* **295**, 1280–1284.

Röckel, D., Korn, W., and Kohn A. J. (1995). *Manual of the Living Conidae*, Vol 1. Springer-Verlage, New York, NY.

Savidge, J. A. (1987). Extinction of an island forest avifauna by an introduced snake. *Ecology*, **68**, 660–668.

Sekercioglu, C. H., Schneider, S. H., Fay, J. P., and Loarie, S. R. (2008). Climate change, elevational range shifts, and bird extinctions. *Conservation Biology*, **22**, 140–150.

Simons E. L. (1997). Lemurs: old and new. In S. M. Goodman and B. D. Patterson, eds *Natural change and human impact in Madagascar*, pp. 142–156. Smithsonian Institution Press, Washington, DC.

Soulé, M. E. and Terborgh, J., eds (1999). *Continental Conservation: Scientific Foundations of Regional Reserve Networks*. Island Press, Washington, DC.

Steadman D. W. (1995). Prehistoric extinctions of Pacific island birds: biodiversity meets zooarcheology. *Science*, **267**, 1123–1131.

Stuart S. N., Chanson J. S., Cox N. A., *et al.* (2004). Status and trends of amphibian declines and extinctions worldwide. *Science* **306**, 1783–1786.

The Nature Conservancy (1996). *TNC Priorities for Conservation: 1996 Annual Report Card for U.S. Plant and Animal Species*. The Nature Conservancy, Arlington, VA.

Thomas, C. D., Cameron, A., and Green, R. E., *et al.* (2004). Extinction risk from climate change. *Nature* **427**, 145–148.

Tuck G. N., Polacheck T., and Bulman C. M. (2003). Spatiotemporal trends of longline fishing effort in the Southern Ocean and implications for seabird bycatch. *Biological Conservation*, **114**, 1–27.

Wallace, A. R. (1855). On the law which has regulated the introduction of new species. *Annals and Magazine of Natural History*, September 1855.

Walter K. S. and Gillett H. J., eds (1998). *1997 IUCN Red List of Threatened Plants*. IUCN, Gland, Switzerland.

Wiles G. J., Bart J., Beck R. E., and Aguon C. F. (2003). Impacts of the brown tree snake: patterns of decline and species persistence in Guam's avifauna. *Conservation Biology*, **17**, 1350–1360.

Williams, S. E., Bolitho, E. E., and Fox, S. (2003). Climate change in Australian tropical rainforests: an impending environmental catastrophe *Proceedings of the Royal Society of London B.*, **270**, 1887–1892.

# CHAPTER 11

# Conservation planning and priorities

Thomas Brooks

Maybe the first law of conservation science should be that human population—which of course drives both threats to biodiversity and its conservation—is distributed unevenly around the world (Cincotta *et al.* 2000). This parallels a better-known first law of biodiversity science, that biodiversity itself is also distributed unevenly (Gaston 2000; Chapter 2). Were it not for these two patterns, conservation would not need to be planned or prioritized. A conservation investment in one place would have the same effects as that in another. As it is, though, the contribution of a given conservation investment towards reducing biodiversity loss varies enormously over space. This recognition has led to the emergence of the sub-discipline of systematic conservation planning within conservation biology.

Systematic conservation planning now dates back a quarter-century to its earliest contributions (Kirkpatrick 1983). A seminal review by Margules and Pressey (2000) established a firm conceptual framework for the sub-discipline, parameterized along axes derived from the two aforementioned laws. Variation in threats to biodiversity (and responses to these) can be measured as vulnerability (Pressey and Taffs 2001), or, put another way, the breadth of options available over time to conserve a given biodiversity feature before it is lost. Meanwhile, the uneven distribution of biodiversity can be measured as irreplaceability (Pressey *et al.* 1994), the extent of spatial options available for the conservation of a given biodiversity feature. An alternative measure of irreplaceability is complementarity—the degree to which the biodiversity value of a given area adds to the value of an overall network of areas.

This chapter charts the history, state, and prospects of conservation planning and prioritization, framed through the lens of vulnerability and irreplaceability. It does not attempt to be comprehensive, but rather focuses on the boundary between theory and practice, where successful conservation implementation has been explicitly planned from the discipline's conceptual framework of vulnerability and irreplaceability. In other words, the work covered here has successfully bridged the "research–implementation gap" (Knight *et al.* 2008). The chapter is structured by scale. Its first half addresses global scale planning, which has attracted a disproportionate share of the literature since Myers' (1988) pioneering "hotspots" treatise. The remainder of the chapter tackles conservation planning and prioritization on the ground (and in the water). This in turn is organized according to three levels of increasing ecological and geographic organization: from species, through sites, to seascapes and landscapes.

## 11.1 Global biodiversity conservation planning and priorities

Most conservation is parochial—many people care most about what is in their own backyard (Hunter and Hutchinson 1994). As a result, maybe 90% of the ~US$6 billion global conservation budget originates in, and is spent in, economically wealthy countries (James *et al.* 1999). Fortunately, this still leaves hundreds of millions of dollars of globally flexible conservation investment that can theoretically be channeled to wherever would deliver the greatest benefit. The bulk of these resources are invested through multilateral agencies [in particular, the Global Environment Facility (GEF) (www.gefweb.org)], bilateral donors, and non-governmental organizations. Where should they be targeted?

### 11.1.1 History and state of the field

Over the last two decades, nine major templates of global terrestrial conservation priorities have been developed by conservation organizations, to guide their own efforts and attract further attention (Figure 11.1 and Plate 9; Brooks et al. 2006). Brooks et al. (2006) showed that all nine templates fit into the vulnerability/irreplaceability framework, although in a variety of ways (Figure 11.2 and Plate 10). Specifically, two of the templates prioritize regions of high vulnerability, as "reactive approaches", while three prioritized regions of low vulnerability, as "proactive approaches". The remaining four are silent regarding vulnerability. Meanwhile, six of the templates prioritize regions of high irreplaceability; the remaining three do not incorporate irreplaceability. To understand these global priority-setting approaches, it is important to examine the metrics of vulnerability and irreplaceability that they use, and the spatial units among which they prioritize.

Wilson et al.'s (2005) classification recognizes four types of vulnerability measures: environmental and spatial variables, land tenure, threatened species, and expert opinion. All five of the global prioritization templates that incorporated vulnerability did so using the first of these measures, specifically habitat extent. Four of these utilized proportionate habitat loss, which is useful as a measure of vulnerability because of the consistent relationship between the number of species in an area and the size of that area (Brooks et al. 2002). However, it is an imperfect metric, because it is difficult to assess in xeric and aquatic systems, it ignores threats such as invasive species and hunting, and it is retrospective rather than predictive (Wilson et al. 2005). The "frontier forests" approach (Bryant et al. 1997) uses absolute forest cover as a measure, although this is only dubiously reflective of vulnerability (Innes and Er 2002). Beyond habitat loss, one template also incorporates land tenure, as protected area coverage (Hoekstra et al. 2005), and two incorporate human population

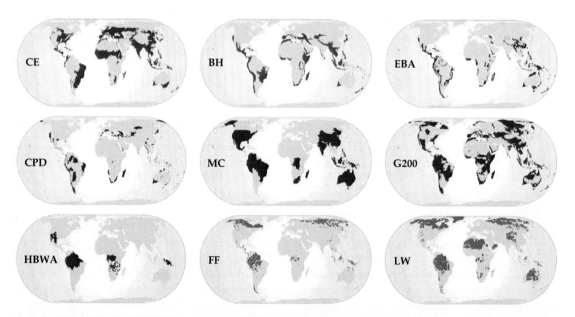

**Figure 11.1** Maps of the nine global biodiversity conservation priority templates (reprinted from Brooks et al. 2006): CE, crisis ecoregions (Hoekstra et al. 2005); BH, biodiversity hotspots (Mittermeier et al. 2004); EBA, endemic bird areas (Stattersfield et al. 1998); CPD, centers of plant diversity (WWF and IUCN 1994–7); MC, megadiversity countries (Mittermeier et al. 1997); G200, global 200 ecoregions (Olson and Dinerstein 1998); HBWA, high-biodiversity wilderness areas (Mittermeier et al. 2003); FF, frontier forests (Bryant et al. 1997); and LW, last of the wild (Sanderson et al. 2002a). With permission from AAAS (American Association for the Advancement of Science).

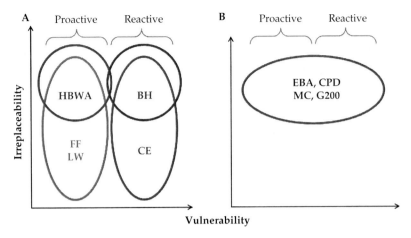

**Figure 11.2** Global biodiversity conservation priority templates placed within the conceptual framework of irreplaceability and vulnerability (reprinted from Brooks et al. 2006). Template names follow the Figure 11.1 legend. (A) Purely reactive (prioritizing high vulnerability) and purely proactive (prioritizing low vulnerability) approaches. (B) Approaches that do not incorporate vulnerability as a criterion (all prioritize high irreplaceability). With permission from AAAS (American Association for the Advancement of Science).

density (Mittermeier et al. 2003; Sanderson et al. 2002a).

The most common measure of irreplaceability is plant endemism, used by four of the templates, with a fifth (Stattersfield et al. 1998) using bird endemism. The logic behind this is that the more endemic species in a region, the more biodiversity lost if the region's habitat is lost (although, strictly, any location with even one endemic species is irreplaceable). Data limitations have restricted the plant endemism metrics to specialist opinion estimates, and while this precludes replication or formal calculation of irreplaceability (Brummitt and Lughadha 2003), subsequent tests have found these estimates accurate (Krupnick and Kress 2003). Olson and Dinerstein (1998) added taxonomic uniqueness, unusual phenomena, and global rarity of major habitat types as measures of irreplaceability, although with little quantification. Although species richness is popularly but erroneously assumed to be important in prioritization (Orme et al. 2005), none of the approaches relies on this. This is because species richness is driven by common, widespread species, and so misses exactly those species most in need of conservation (Jetz and Rahbek 2002).

One of the priority templates uses countries as its spatial unit (Mittermeier et al. 1997). The remaining eight utilize spatial units based on biogeography, one using regions defined *a posteriori* from the distributions of restricted-range bird species (Stattersfield et al. 1998), and the other seven using units like "ecoregions", defined *a priori* (Olson et al. 2001). This latter approach brings ecological relevance, but also raises problems because ecoregions vary in size, and because they themselves have no repeatable basis (Jepson and Whittaker 2002). The use of equal area grid cells would circumvent these problems, but limitations on biodiversity data compilation so far have prevented their general use. Encouragingly, some initial studies (Figure 11.3) for terrestrial vertebrates (Rodrigues et al. 2004b) and, regionally, for plants (Küper et al. 2004) show considerable correspondence with many of the templates (da Fonseca et al. 2000).

What have been the costs and benefits of global priority-setting? The costs can be estimated to lie in the low millions of dollars, mainly in the form of staff time. The benefits are hard to measure, but large. The most tractable metric, publication impact, reveals that Myers et al. (2000), the benchmark paper on hotspots, was the single most cited paper in the ISI Essential Science Indicators category "Environment/Ecology" for the decade preceding 2005. Much more important is the impact that these prioritization templates have had on resource allocation. Myers (2003) estimated that over the preceding 15 years, the hotspots concept had focused

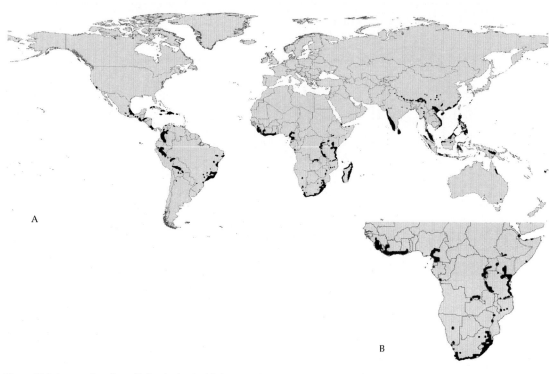

**Figure 11.3** Incorporating primary biodiversity data in global conservation priority-setting (reprinted from Brooks et al. 2006). Global conservation prioritization templates have been based almost exclusively on bioregional classification and specialist opinion, rather than primary biodiversity data. Such primary datasets have recently started to become available under the umbrella of the IUCN Species Survival Commission (IUCN 2007), and they allow progressive testing and refinement of templates. (A) Global gap analysis of coverage of 11 633 mammal, bird, turtle, and amphibian species (~40% of terrestrial vertebrates) in protected areas (Rodrigues et al. 2004a). It shows unprotected half-degree grid cells characterized simultaneously by irreplaceability values of at least 0.9 on a scale of 0–1, and of the top 5% of values of an extinction risk indicator based on the presence of globally threatened species (Rodrigues et al. 2004b). (B) Priorities for the conservation of 6269 African plant species (~2% of vascular plants) across a 1-degree grid (Küper et al. 2004). These are the 125 grid cells with the highest product of range-size rarity (a surrogate for irreplaceability) of plant species distributions and mean human footprint (Sanderson et al. 2002a). Comparison of these two maps, and between them and Fig. 11.2, reveals a striking similarity among conservation priorities for vertebrates and those for plants, in Africa.

US$750 million of globally flexible conservation resources. Entire funding mechanisms have been established to reflect global prioritization, such as the US$150 million Critical Ecosystem Partnership Fund (www.cepf.net) and the US$100 million Global Conservation Fund (web.conservation.org/xp/gef); and the ideas have been incorporated into the Resource Allocation Framework of the Global Environment Facility, the largest conservation donor.

### 11.1.2 Current challenges and future directions

Six major research fronts can be identified for the assessment of global biodiversity conservation priorities (Mace et al. 2000; Brooks et al. 2006).

First, it remains unclear the degree to which priorities set using data for one taxon reflect priorities for others, and, by extension, whether priorities for well-known taxa like vertebrates and plants reflect those for the poorly-known, megadiverse invertebrates, which comprise the bulk of life on earth. Lamoreux et al. (2006), for example, found high congruence between conservation priorities for terrestrial vertebrate species. In contrast, Grenyer et al. (2006) reported low congruence between conservation priorities for mammals, birds, and amphibians. However, this result was due to exclusion of unoccupied cells; when this systematic bias is corrected, the same data actually show remarkably high congruence

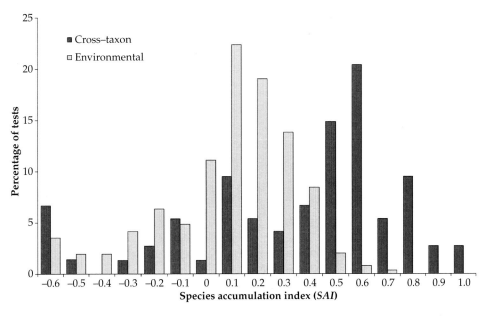

**Figure 11.4** Frequency distribution of values of a species accumulation index (SAI) of surrogate effectiveness for comparison between tests on terrestrial cross-taxon and on environmental surrogates; the SAI has a maximum value of 1 (perfect surrogacy), and indicates random surrogacy when it has a value of 0, and surrogacy worse than random when it is negative (reprinted from Rodrigues and Brooks 2007).

(Rodrigues 2007). More generally, a recent review found that positive (although rarely perfect) surrogacy is the norm for conservation priorities between different taxa; in contrast, environmental surrogates rarely function better than random (Figure 11.4 and Plate X; Rodrigues and Brooks 2007).

While surrogacy may be positive within biomes, none of the conservation prioritization templates to date have considered freshwater or marine biodiversity, and at face value one might expect that conservation priorities in aquatic systems would be very different from those on land (Reid 1998). Remarkably, two major studies from the marine environment suggest that there may in fact be some congruence between conservation priorities on land and those at sea. Roberts et al. (2002), found that 80% of their coral hotspots, although restricted to shallow tropical reef systems, were adjacent to Myers et al. (2000) terrestrial hotspots. More recently, Halpern et al. (2008) measured and mapped the intensity of pressures on the ocean (regardless of marine biodiversity); the pressure peaks on their combined map are surprisingly close to biodiversity conservation priorities on land (the main exception being the North Sea) (see also Box 4.3). While much work remains in marine conservation prioritization, and that for freshwater biodiversity has barely even begun, these early signs suggest that there may be some geographic similarity in conservation priorities even between biomes.

Another open question is the extent to which conservation priorities represent not just current diversity but also evolutionary history. For primates and carnivores globally, Sechrest et al. (2002) showed that biodiversity hotspots hold a disproportionate concentration of phylogenetic diversity, with the ancient lineages of Madagascar a key driver of this result (Spathelf and Waite 2007). By contrast, Forest et al. (2007) claimed to find that incorporating botanical evolutionary history for the plants of the Cape Floristic hotspot substantially altered the locations of conservation priorities. Using simulations, Rodrigues et al. (2005) argued that phylogeny will only make a difference to conservation prioritization under specific conditions: where very deep lineages endemics are endemic to species-poor regions. Addressing the question globally across entire classes remains an important research priority.

Even if existing conservation priorities capture evolutionary history well, this does not necessarily mean that they capture evolutionary process. Indeed, a heterodox view proposes that the young, rapidly speciating terminal twigs of phylogenetic trees should be the highest conservation priorities (Erwin 1991)—although some work suggests that existing conservation priority regions are actually priorities for both ancient and young lineages (Fjeldså and Lovett 1997). Others argue that much speciation is driven from ecotonal environments (Smith *et al.* 1997) and that these are poorly represented in conservation prioritization templates (Smith *et al.* 2001). The verdict is still out.

The remaining research priorities for global conservation prioritization concern intersection with human values. Since the groundbreaking assessment of Costanza *et al.* (1997), much work has been devoted to the measurement of ecosystem service value—although surprisingly little to prioritizing its conservation (but see Ceballos and Ehrlich 2002). Kareiva and Marvier (2003) suggested that existing global biodiversity conservation priorities are less important than other regions for ecosystem service provision. Turner *et al.* (2007), by contrast, showed considerable congruence between biodiversity conservation priority and potential ecosystem service value, at least for the terrestrial realm. Moreover, that there is correspondence of both conservation priorities and ecosystem service value with human population (Balmford *et al.* 2001) and poverty (Balmford *et al.* 2003) suggests that biodiversity conservation may be delivering ecosystem services where people need them most.

Maybe the final frontier of global priority-setting is the incorporation of cost of conservation. This is important, because conservation costs per unit area vary over seven orders of magnitude, but elusive, because they are hard to measure (Polasky 2008). Efforts over the last decade, however, have begun to develop methods for estimating conservation cost (James *et al.* 1999, 2001; Bruner *et al.* 2004). These have in turn allowed assessment of the impact of incorporating costs into conservation prioritization—with initial indications suggesting that this makes a substantial difference within regions (Ando *et al.* 1998; Wilson *et al.* 2006), across countries (Balmford *et al.* 2000), and globally (Carwardine *et al.* 2008). Further, and encouragingly, it appears that incorporation of costs may actually decrease the variation in conservation priorities caused by consideration of different biodiversity datasets, at least at the global scale (Bode *et al.* 2008). The development of a fine-scale, spatial, global estimation of conservation costs is therefore an important priority for global conservation prioritization.

## 11.2 Conservation planning and priorities on the ground

For all of the progress of global biodiversity conservation priority-setting, planning at much finer scales is necessary to allow implementation on the ground or in the water (Mace *et al.* 2000; Whittaker *et al.* 2005; Brooks *et al.* 2006). Madagascar can and should attract globally flexible conservation resources because it is a biodiversity hotspot, for example, but this does not inform the question of where within the island these resources should be invested (see Box 12.1). Addressing this question requires consideration of three levels of ecological organization—species, sites, and sea/landscapes—addressed in turn here.

### 11.2.1 Species level conservation planning and priorities

Many consider species the fundamental unit of biodiversity (Wilson 1992). Conversely, avoiding species extinction can be seen as the fundamental goal of biodiversity conservation, because while all of humanity's other impacts on the Earth can be repaired, species extinction, Jurassic Park fantasies notwithstanding, is irreversible. It is fitting, then, that maybe the oldest, best-known, and most widely used tool in the conservationist's toolbox informs conservation planning at the species level. This is the IUCN Red List of Threatened Species (www.iucnredlist.org).

*History and state of the field*
The IUCN Red List now dates back nearly 50 years, with its first volumes published in the

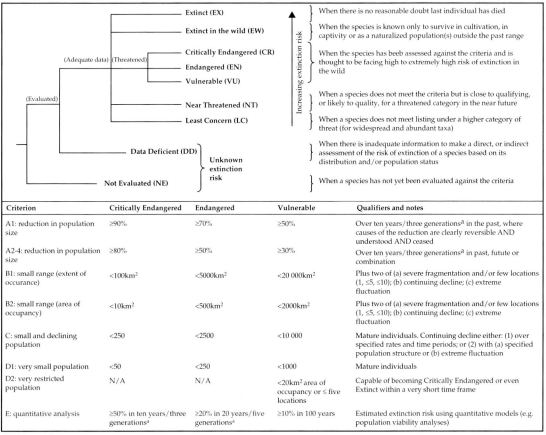

**Figure 11.5** The IUCN Red List categories and criteria (reprinted from Rodrigues et al. 2006 © Elsevier). For more details see Rodrigues et al. (2006).

1960s (Fitter and Fitter 1987). Over the last two decades it has undergone dramatic changes, moving from being a simple list of qualitative threat assessments for hand-picked species to its current form of quantitative assessments across entire taxa, supported by comprehensive ancillary documentation (Rodrigues et al. 2006). The heart of the IUCN Red List lies in assessment of vulnerability at the species level, specifically in estimation of extinction risk (Figure 11.5). Because the requirements for formal population viability analysis (Brook et al. 2000) are too severe to allow application for most species, the IUCN Red List is structured through assessment of species status against threshold values for five quantitative criteria (IUCN 2001). These place species into broad categories of threat which retrospective analyses have shown to be broadly equivalent between criteria (Brooke et al. 2008), and which are robust to the incorporation of uncertainty (Akçakaya et al. 2000).

As of 2007, 41 415 species had been assessed against the IUCN Red List categories and criteria, yielding the result that 16 306 of these are globally threatened with a high risk of extinction in the medium-term future (IUCN 2007). This includes comprehensive assessments of all mammals (Schipper et al. 2008), birds (BirdLife International 2004) and amphibians (Stuart et al. 2004), as well as partially complete datasets for many other taxa (Baillie et al. 2004). Global assessments are underway for reptiles, freshwater species (fish, mollusks, odonata, decapod crustaceans), marine species (fish, corals), and plants.

It is worth a short digression here concerning irreplaceability at the species level, where phylogenetic, rather than geographic, space provides the dimension over which irreplaceability can be measured. A recent study by Isaac et al. (2007) has pioneered the consideration of this concept of "phylogenetic irreplaceability" alongside the IUCN Red List to derive species-by-species conservation priorities. A particularly useful application of this approach may prove to be in prioritizing efforts in *ex situ* conservation.

The benefits of the IUCN Red List are numerous (Rodrigues et al. 2006), informing site conservation planning (Hoffmann et al. 2008), environmental impact assessment (Meynell 2005), national policy (De Grammont and Cuarón 2006), and intergovernmental conventions (Brooks and Kennedy 2004), as well as strengthening the conservation constituency through the workshops process. Data from the assessments for mammals, birds, amphibians, and freshwater species to date suggest that aggregate costs for the IUCN Red List process average around US$200 per species, including staff time, data management, and, in particular, travel and workshops. This cost is expected to decrease as the process moves into assessments of plant and invertebrate species, because these taxa have many fewer specialists per species than do vertebrates (Gaston and May 1992). However, it is expected that the benefits of the process will also decrease for invertebrate taxa, because the proportion of data deficiency will likely rise compared to the current levels for vertebrate groups (e.g. ~23% for amphibians: Stuart et al. 2004). However, a sampled Red List approach is being developed to allow inexpensive insight into the conservation status of even the megadiverse invertebrate taxa (Baillie et al. 2008).

*Current challenges and future directions*
The main challenge facing the IUCN Red List is one of scientific process: how to expand the Red List's coverage in the face of constraints of taxonomic uncertainty, data deficiency, lack of capacity, and demand for training (Rodrigues et al. 2006). Some of the answer to this must lie in coordination of the IUCN Red List with national red listing processes, which have generated data on thousands of species not yet assessed globally (Rodriguez et al. 2000). To this end, IUCN have developed guidelines for sub-global application of the Red List criteria (Gärdenfors et al. 2001), but much work is still needed to facilitate the data flow between national and global levels.

One specific scientific challenge worth highlighting here is the assessment of threats driven by climate change. Climate change is now widely recognized as a serious threat to biodiversity (Thomas et al. 2004). However, it hard to apply the Red List criteria against climate change threats, especially for species with short generation times (Akçakaya et al. 2006), because climate change is rather slow-acting (relative to the time scale of the Red List criteria). Research is underway to address this limitation.

## 11.2.2 Site level conservation planning and priorities

With 16 306 species known to be threatened with extinction, threat rates increasing by the year (Butchart et al. 2004), and undoubtedly many thousands of threatened plants and invertebrates yet to be assessed, the task of biodiversity conservation seems impossibly daunting. Fortunately, it is not necessary to conserve these thousands of species one at a time. Examination of those threatened species entries on the Red List for which threats are classified reveals that habitat destruction is the overwhelming driver, threatening 90% of threatened species (Baillie et al. 2004). The logical implication of this is that the cornerstone of conservation action must be conserving the habitats in which these species live—establishing protected areas (Bruner et al. 2001). This imperative for protecting areas is not new, of course—it dates back to the roots of conservation itself—but analyses of the World Database on Protected Areas show that there are now 104 791 protected areas worldwide covering 12% of the world's land area (Chape et al. 2005). Despite this, however, much biodiversity is still wholly unrepresented within protected areas (Rodrigues et al. 2004a). The Programme of Work on Protected Areas of the Convention on Biological Diversity (www.cbd.int/protected) therefore calls for gap analysis to allow planning of "comprehensive, effectively managed, and ecologically representative" protected area systems. How can such planning best take place?

*History and state of the field*
Broadly, approaches to planning protected area systems can be classified into four groups. The oldest is *ad hoc* establishment, which often increases protected area coverage with minimal value for biodiversity (Pressey and Tully 1994). The 1990s saw the advent of the rather more successful consensus workshop approach, which allowed for data sharing and stake-holder buy-in, and certainly represented a considerable advance over *ad hoc* approaches (Hannah *et al.* 1998). However, the lack of transparent data and criteria still limited the reliability of workshop-based site conservation planning. Meanwhile, developments in theory (Margules and Pressey 2000) and advances in supporting software (e.g. Marxan; www.uq.edu.au/marxan), led to large scale applications of wholly data-driven conservation planning, most notably in South Africa (Cowling *et al.* 2003). However, the black-box nature of these applications led to limited uptake in conservation practice, which some have called the "research–implementation gap" (Knight *et al.* 2008).

To overcome these limitations, the trend in conservation planning for implementation on the ground is now towards combining data-driven with stakeholder-driven techniques (Knight *et al.* 2007; Bennun *et al.* 2007). This approach actually has a long history in bird conservation, with the first application of "important bird areas" dating back to the work of Osieck and Mörzer Bruyns (1981). This "site-specific synthesis" (Collar 1993–4) of bird conservation data has gained momentum to the point where important bird area identification is now close to being complete worldwide (BirdLife International 2004). Over the last decade, the approach has been extended to numerous other taxa (e.g. plants: Plantlife International 2004), and thence generalized into the "key biodiversity areas" approach (Eken *et al.* 2004). Several dozen countries have now completed key biodiversity area identification as part of their commitment towards national gap analysis under the Convention on Biological Diversity's Programme of Work on Protected Areas (e.g. Madagascar: Figure 11.6; Turkey: Box 11.1), and a comprehensive guidance manual published to support this work (Langhammer *et al.* 2007). Furthermore, all of the world's international conservation organizations, and many national ones, have come together as the Alliance for Zero Extinction (AZE), to identify and implement action for the very highest priorities for site-level conservation (Ricketts *et al.* 2005, Figure 11.7 and Plate 12).

The key biodiversity areas approach, in alignment with the conceptual framework for conservation planning (Margules and Pressey 2000), is based on metrics of vulnerability and irreplaceability (Langhammer *et al.* 2007). Their vulnerability criterion is derived directly from the IUCN Red List, through the identification of sites regularly holding threshold populations of one or more threatened species. The irreplaceability

**Figure 11.6** Location and protection status of the Key Biodiversity Areas (KBAs) of Madagascar (reprinted from Langhammer *et al.* 2007).

criterion is based on regular occurrence at a site of a significant proportion of the global population of a species. This is divided into sub-criteria to recognize the various situations under which this could occur, namely for restricted range species, species with clumped distributions, congregatory populations (species that concentrate during a portion of their life cycle), source populations, and biome-restricted assemblages. The reliance on occurrence data undoubtedly causes omission errors (where species occur in unknown sites) and hence the approach overestimates irreplaceability. These omission errors could in theory be reduced by use of modeling or extrapolation techniques, but these instead yield dangerous commission errors, which could lead to extinction through a species wrongly considered to be safely represented (Rondinini et al. 2006). Where such techniques are of proven benefit is in identifying research priorities (as opposed to conservation priorities) for targeted field surveys (Raxworthy et al. 2003).

To facilitate implementation and gap analysis, key biodiversity areas are delineated based on existing land management units, such as protected areas, indigenous or community lands, private concessions or ranches, and military or other public holdings (Langhammer et al. 2007). Importantly, this contrasts with subdivision of the entire landscape into, for example, grid cells, habitat types, or watersheds. While grid cells have the advantage of analytical rigor, and habitats and watersheds deliver ecological coherence, these spatial units are of minimal relevance to the stakeholders on whom conservation on the ground fundamentally depends. Indeed, the entire key biodiversity areas process is designed to build the constituency for local conservation, while following global standards and criteria (Bennun et al. 2007). The costs and benefits of site conservation planning approaches have yet to be fully evaluated, but some early simulation work suggests that the benefits of incorporation of primary biodiversity data are large (Balmford and Gaston 1999).

### Current challenges and future directions

Three important challenges can be discerned as facing site level conservation planning. The first stems from the fact that most applications of these approaches to date come from fragmented habitats—it often proves difficult to identify sites of global biodiversity conservation significance in regions that retain a wilderness character, for instance, in the Amazon (Mittermeier et al. 2003). Under such circumstances, the omission errors attendant on use of occurrence data (because of very low sampling density) combine with difficulty in delineating sites (because of overlapping or non-existent land tenure). These problems can, and indeed must, be overcome by delineating very large key biodiversity areas, which is still a possibility in such environments (e.g. Peres 2005).

The second challenge facing site level conservation planning is its extension to aquatic environments. Human threats to both freshwater and marine biodiversity are intense, but species assessments in these biomes are in their infancy (see above), seriously hampering conservation planning. Difficulties of low sampling density and delineation are also challenging for conservation planning below the water, as in wilderness regions on land. Nevertheless, initial scoping suggests that the application of the key biodiversity areas approach will be desirable in both freshwater (Darwall and Vié 2005) and marine (Edgar et al. 2008a) environments, and proof-of-concept from the Eastern Tropical Pacific shows that it is feasible (Edgar et al. 2008b).

The third research front for the key biodiversity areas approach is prioritization (Langhammer et al. 2007)—once sites have been identified and delineated as having global biodiversity conservation significance, which should be assigned the most urgent conservation action? This requires the measurement not just of irreplaceability and species vulnerability, but also of site vulnerability (Bennun et al. 2005). This is because site vulnerability interacts with irreplaceability: where irreplaceability is high (e.g., in AZE sites), the most threatened sites are priorities, while where irreplaceability is lower, the least vulnerable sites should be prioritized. This is particularly important in considering resilience (i.e. low vulnerability) of sites in the face of climate change. As with global prioritization (see above), it is also important to strive towards incorporating

**Box 11.1 Conservation planning for Key Biodiversity Areas in Turkey**
Güven Eken, Murat Ataol, Murat Bozdoğan, Özge Balkız, Süreyya İsfendiyaroğlu, Dicle Tuba Kılıç, and Yıldıray Lise

An impressive set of projects has already been carried out to map priority areas for conservation in Turkey. These include three inventories of Important Bird Areas (Ertan et al. 1989; Magnin and Yarar 1997; Kılıç and Eken 2004), a marine turtle areas inventory (Yerli and Demirayak 1996), and an Important Plant Areas inventory (Özhatay et al. 2003). These projects, collectively, facilitated on-the-ground site conservation in Turkey and drew attention to gaps in the present protected areas network.

We used the results of these projects as inputs to identify the Key Biodiversity Areas (KBAs) of Turkey, using standard KBA criteria across eight taxonomic groups: plants, dragonflies, butterflies, freshwater fish, amphibians, reptiles, birds, and mammals. As a result of this study, an inventory of two volumes (1112 pages) was published in Turkish fully documenting the country's KBAs (Eken et al. 2006).

We used the framework KBA criteria developed by Eken et al. (2004) and assessed 10 214 species occurring in Turkey against these criteria. Two thousand two hundred and forty six species triggered one or more KBA criteria. These include 2036 plant species (out of 8897 in Turkey; 23%), 71 freshwater fish (of 200; 36%), 36 bird (of 364; 10%), 32 reptile (of 120; 27%), 28 mammal (of 160; 18%), 25 butterfly (of 345; 7%), 11 amphibian (of 30; 37%), and 7 dragonfly (of 98; 7%) species. Then, we assessed all available population data against each KBA criterion and its threshold to select KBAs.

We identified 294 KBAs qualifying on one or more criteria at the global scale (Box 11.1 Figure and Plate 11). Two KBAs met the criteria for seven taxon groups, while 11 sites met them for six and 18 for five taxon groups. The greatest number of sites, 94, met the KBA criteria for two taxon groups, while 86 sites (29%) triggered the criteria for one taxon group only.

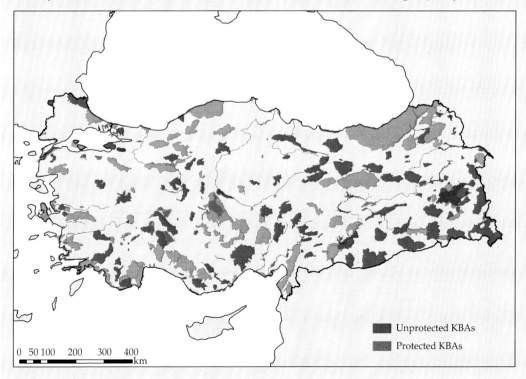

**Box 11.1 Figure** The 294 KBAs (Key Biodiversity Areas) of global importance identified in Turkey. While 146 incorporate protected areas (light), this protection still covers <5% of Turkey's land area. The remaining 148 sites (dark) are wholly unprotected.

*continues*

> **Box 11.1 (Continued)**
>
> The greatest number of sites, 223, was selected based on the criteria for plants, followed by reptiles and birds with 108 and 106 sites selected respectively. For other groups, smaller numbers of sites triggered the KBA criteria at the global scale: 95 KBAs were selected for mammals, 66 for butterflies, 61 for freshwater fish, and 29 each for amphibians and dragonflies. The number of sites selected for plants is actually rather low, given the high number of plant species in Turkey which trigger the KBA criteria. This can be explained by the overlapping distributions of restricted-range and threatened plants. The other taxon groups have relatively greater numbers of sites. For instance, the seven dragonfly species triggered the KBA criteria for 29 sites. One exception is the freshwater fish, which, like plants, have highly overlapping ranges.
>
> Large scale surface irrigation, drainage, and dam projects form the most significant threats to Turkey's nature. Irrigation and drainage projects affect 74% of the KBAs and dams have an effect on at least 49%. Inefficient use of water, especially in agriculture, is the root cause of these threats. A total of 40 billion m³ of water is channeled annually to agriculture (75%), industry (10%), and domestic use (15%), but 50–90% of water used for agriculture is lost during the transportation from dams to arable land. As a result of these threats, wetlands and associated grasslands are Turkey's most threatened habitat types. At least five wetland KBAs (Eşmekaya Marshes, Hotamış Marshes, Sultan Marshes, Ereğli Plain, and Seyfe Lake) have been lost entirely over the last decade, and other sites have lost at least 75% of their area during the same period.
>
> Less than 5% of the surface area of Turkey's KBAs is legally protected, and so this should be expanded rapidly and strategically. Steppe habitats, river valleys, and Mediterranean scrublands are particularly poorly covered by the current network of protected areas. Wildlife Development Reserves, Ramsar Sites and, in the future, Natura 2000 Sites, would likely be appropriate protected area categories for this expansion.
>
> **REFERENCES**
>
> Eken, G., Bennun, L., Brooks, T. M., *et al.* (2004). Key biodiversity areas as site conservation targets. *BioScience*, **54**, 1110–1118.
>
> Eken, G., Bozdoğan, M., İsfendiyaroğlu, S., Kılıç, D. T., and Lise, Y. (2006). *Türkiye'nin önemli doğa alanları*. Doğa Derneği, Ankara, Turkey.
>
> Ertan, A., Kılıç, A., and Kasparek, M. (1989). *Türkiye'nin önemli kuşalanları*. Doğal Hayatı Koruma Derneği and International Council for Bird Preservation, Istanbul, Turkey.
>
> Kılıç, D. T. and Eken, G. (2004). *Türkiye'nin önemli kuş alanları – 2004 güncellemesi*. Doğa Derneği, Ankara, Turkey.
>
> Magnin, G. and Yarar, M. (1997). *Important bird areas in Turkey*. Doğal Hayatı Koruma Derneği, Istanbul, Turkey.
>
> Özhatay, N., Byfield, A., and Atay, S. (2003). *Türkiye'nin önemli bitki alanları*. WWF-Türkiye, Istanbul, Turkey.
>
> Yerli, S. and Demirayak, F. (1996). *Türkiye'de denizkaplumbağaları ve üreme kumsalları üzerine bir değerlendirme*. Doğal Hayatı Koruma Derneği, Istanbul, Turkey.

cost of conservation. Given these complexities, considerable promise may lie in adapting conservation planning software to the purpose of prioritizing among conservation actions across key biodiversity areas.

### 11.2.3 Sea/landscape level conservation planning and priorities

The conservation community has more than 40 years experience with conservation planning at the species level, and more than 20 at the site level. However, the recent growth of the field of landscape ecology (Turner 2005) sounds a warning that while species and site planning are essential for effective biodiversity conservation, they are not sufficient. Why not, and how, then, can conservation plan beyond representation, for persistence?

*History and state of the field*
The first signs that conserving biodiversity in isolated protected areas might not ensure

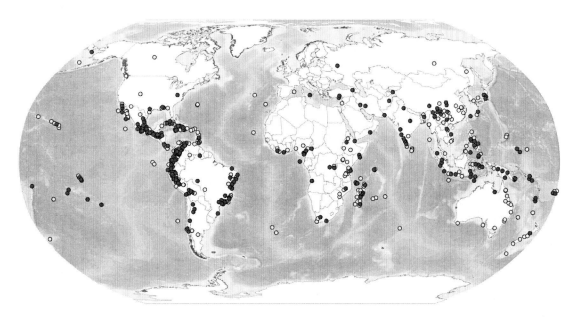

**Figure 11.7** Map of 595 sites of imminent species extinction (reprinted from Ricketts *et al.* 2005). Yellow sites are either fully protected or partially contained within declared protected areas ($n = 203$ and 87, respectively), and red sites are completely unprotected or have unknown protection status ($n = 257$ and 48, respectively). In areas of overlap, unprotected (red) sites are mapped above protected (yellow) sites to highlight the more urgent conservation priorities.

persistence came from evidence of long-term extinctions of mammal species from North American national parks (Newmark 1987). Over the following decade, similar patterns were uncovered across many taxa, unfolding over the timescale of decades-to-centuries, for megadiverse tropical ecosystems in Latin America (e.g. Robinson 1999), Africa (e.g. Brooks *et al.* 1999), and Asia (e.g. Brook *et al.* 2003). Large-scale experiments, most notably the Manaus Biological Dynamics of Forest Fragments project, provide increasingly refined evidence (Bierregaard *et al.* 2001). The mechanisms determining persistence—or extinction—in individual sites spans the full spectrum from the genetic scale (Saccheri *et al.* 1998; see Chapters 2 and 16) through populations (Lens *et al.* 2002) and communities (Terborgh *et al.* 2001), to the level of ecosystem processes across entire landscapes (Saunders *et al.* 1991; see Chapter 5).

The first recommendations of how conservation planning might address persistence at landscape scales were generic design criteria for the connectivity of protected areas (Diamond 1975). Conservation agencies were quick to pick up the concept, and over the last twenty years a number of large scale conservation corridors have been designed (Crooks and Sanjayan 2006), for example, the "Yellowstone to Yukon" (Raimer and Ford 2005) and Mesoamerican Biological Corridor (Kaiser 2001). There is no doubt that the implementation of corridors benefits biodiversity (Tewksbury *et al.* 2002). However, the establishment of generic corridors has also been criticized, in that they divert conservation resources from higher priorities in protected area establishment, and, even worse, have the potential to increase threats, such as facilitating the spread of disease, invasive, or commensal species (Simberloff *et al.* 1992).

Given these concerns, there has been a shift towards specification of the particular objectives for any given corridor (Hobbs 1992). A promising avenue of enquiry here has been to examine the needs of "landscape species" which require broad scale conservation (Sanderson *et al.* 2002b). Boyd *et al.* (2008) have generalized this approach, reviewing the scales of conservation required for all threatened terrestrial vertebrate species (Figure 11.8 and Plate 13). They found that 20% (793) of these threatened species required urgent broad scale conservation action, with this result varying significantly among taxa (Figure 11.9). They also asked why each of these species required broad scale conservation. This yielded the surprising finding that while only 43% of these 793 species were "area-demanding" and so required corridors for movement, no less than 72% were dependent on broad scale ecological processes acting across the landscape (15% require both). In this light, recent work in South Africa to pioneer techniques for incorporating ecosystem processes into conservation planning is likely to be particularly important (Rouget *et al.* 2003, 2006).

*Current challenges and future directions*

As at the species and site levels, the incorporation of broad scale targets into conservation planning in aquatic systems lags behind the terrestrial environment. Given the regimes of flows and currents inherent in rivers and oceans, the expectation is that broad scale conservation will be even more important in freshwater (Bunn and Arthington 2002) and in the sea (Roberts 1997) than it is on land. Boyd *et al.*'s (2008) results are consistent with this, with 74% of threatened marine tetrapods requiring broad scale conservation, and 38% in freshwater, and only 8% on land (Figure 11.9). This said, some recent work suggests that marine larval dispersal occurs over much narrower scales than previously assumed (Jones *et al.* 1999) and so there is no doubt that site level conservation will remain of great importance in the water as well as on land (Cowen *et al.* 2006).

A second research front for sea/landscape conservation planning concerns dynamic threats. Recent work has demonstrated that changes in the nature and intensity of threats over time have important consequences for the prioritization of conservation actions among sites (Turner and Wilcove 2006). Such dynamism introduces particular complications when considered at the landscape scale, the implications of which are only just beginning to be addressed (Pressey *et al.* 2007). Climate change is one such threat that will very likely require extensive landscape scale response (Hannah *et al.* 2002), and may be even

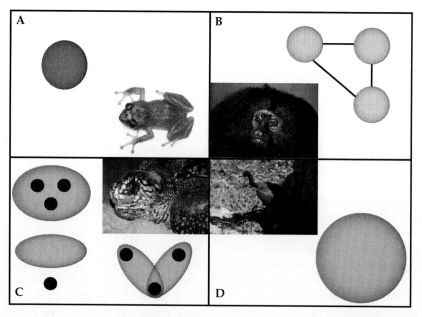

**Figure 11.8** Scale requirements for the conservation of globally threatened species in the short- to medium term (reprinted from Boyd *et al.* 2008). (A, dark green) Species best conserved at a single site (e.g. *Eleutherodactylus corona*); (B, pale green) Species best conserved at a network of sites (e.g. black lion-tamarin *Leontopithecus chrysopygus*); (C, dark blue) Species best conserved at a network of sites complemented by broad-scale conservation action (e.g. leatherback turtle *Dermochelys cariacea*); (D, pale blue) Species best conserved through broad-scale conservation action (e.g. Indian vulture *Gyps indicus*). Photographs by S. B. Hedges (A), R.A. Mittermeier (B), O. Langrand (C), and A. Rahmani (D).

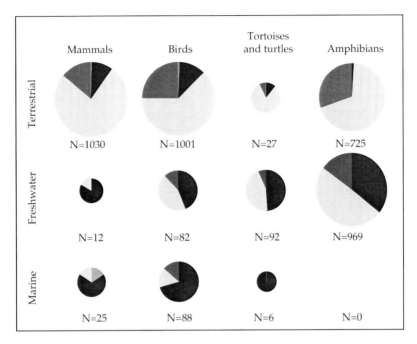

**Figure 11.9** Percentages of globally threatened species requiring different scales of conservation in the short- to medium term (reprinted from Boyd *et al.* 2008). Dark green = species best conserved at a single site; pale green = species best conserved at a network of sites; dark blue = species best conserved at a network of sites complemented by broad-scale conservation action; pale blue = species best conserved through broad-scale conservation action. The totals exclude species insufficiently known to assess the appropriate scale required. Relative size of pies corresponds to the number of species in each taxon/biome combination.

more serious in freshwater (Roessig *et al.* 2004) and the ocean (Xenopoulos *et al.* 2005) (see Chapter 8).

Maybe the largest open research challenge for sea/landscape conservation planning is to move from maintaining current biodiversity towards restoring biodiversity that has already been lost (Hobbs and Norton 1996). Natural processes of succession provide models of how this can proceed most effectively (Dobson *et al.* 1997). However, restoration is much more expensive and much less likely to succeed than is preservation of biodiversity before impacts occur, and so explicit planning towards the specific biodiversity targeted to be restored is essential (Miller and Hobbs 2007). Given these costs and challenges, most efforts to date target very tightly constrained ecosystems that, as restoration proceeds, are then managed at site scales—wetlands are the best example (Zedler 2000). A few ambitious plans for landscape level restoration have already been developed (Stokstad 2008). Moreover, the current explosive growth in markets for carbon as mechanisms for climate change mitigation will likely make the restoration of forest landscapes increasingly viable in the near future (Laurance 2008). Ultimately, planning should move from simple restoration to designing landscapes that allow the sustainability of both biodiversity and human land uses, envisioned as "countryside biogeography" (Daily *et al.* 2001) or "reconciliation ecology" (Rosenzweig 2003).

## 11.3 Coda: the completion of conservation planning

The research frontiers outlined in this chapter are formidable, but conservation planning is nevertheless a discipline with its completion in sight. It is not too far of a stretch to imagine a day where top-down global prioritization and bottom-up conservation planning come together. Such a vision would encompass:

- The completion and continuous updating of IUCN Red List assessments of all vertebrate and plant species, plus selected invertebrate groups.
- Iterative identification of key biodiversity areas, based on these data, representing the full set of sites of global biodiversity conservation significance.
- Measurement and mapping of the continuous global surface of seascape and landscape scale ecological processes necessary to retain these species and sites into the future.
- Continuous measurement and mapping of the threats to these species, sites, and sea/landscapes, and of the costs and benefits of conserving them.
- Free, electronic, continuously updated access to these datasets, and to tools for their interpretation, planning, and prioritization.

A particularly important characteristic of such a vision is its iterative nature. As knowledge of biodiversity increases, threats and costs change, and conservation is implemented successfully (or not) it is crucial that mechanisms exist to capture these changing data, because changes to any one of these parameters will likely impinge on conservation planning across the board.

Under such a vision, it would be possible, at any given point in time, to maximize the overall benefits of a conservation investment at any scale, from *ex situ* management of a particular species, through gap analysis by a national protected areas agency, to investment of globally flexible resources by institutions like the GEF. Given the pace of advance in conservation planning over the last 20 years, it is possible that such a vision is achievable within the coming few decades. Its realization will provide great hope for maintaining as much of the life with which we share our planet as possible.

## Summary

- Conservation planning and prioritization are essential, because both biodiversity and human population (and hence threats to biodiversity and costs and benefits of conservation) are distributed highly unevenly.
- Great attention has been invested into global biodiversity conservation prioritization on land over the last two decades, producing a broad consensus that reactive priority regions are concentrated in the tropical mountains and islands, and proactive priorities in the lowland tropical forests.
- Major remaining research fronts for global biodiversity conservation prioritization include the examination of cross-taxon surrogacy, aquatic priorities, phylogenetic history, evolutionary process, ecosystem services, and costs of conservation.
- Maybe the most important tool for guiding conservation on the ground is the IUCN Red List of Threatened Species, which assesses the extinction risk of 41 415 species against quantitative categories and criteria, and provides data on their distributions, habitats, threats, and conservation responses.
- The predominant threat to biodiversity is the destruction of habitats (Chapter 4), and so the primary conservation response must be to protect these places through safeguarding key biodiversity areas.
- While protecting sites is essential for biodiversity conservation, persistence in the long term also requires the conservation of those landscape and seascape level ecological processes that maintain biodiversity.

## Suggested reading

- Boyd, C., Brooks, T. M., Butchart, S. H. M., *et al.* (2008). Scale and the conservation of threatened species. *Conservation Letters*, **1**, 37–43.
- Brooks, T. M., Mittermeier, R. A., da Fonseca, G. A. B., *et al.* (2006). Global biodiversity conservation priorities. *Science*, **313**, 58–61.
- Eken, G., Bennun, L., Brooks, T. M., *et al.* (2004). Key biodiversity areas as site conservation targets. *BioScience*, **54**, 1110–1118.
- Margules, C. R. and Pressey, R. L. (2000). Systematic conservation planning. *Nature*, **405**, 243–253.
- Rodrigues, A. S. L., Pilgrim, J. D., Lamoreux, J. F., Hoffmann, M., and Brooks, T. M. (2006). The value of the IUCN Red List for conservation. *Trends in Ecology and Evolution*, **21**, 71–76.

## Relevant websites

- BirdLife International Datazone: http://www.birdlife.org/datazone.
- IUCN Red List of Threatened Species: http://www.iucnredlist.org.

- World Database on Protected Areas: http://www.wdpa.org.
- Alliance for Zero Extinction: http://www.zeroextinction.org.

## Acknowledgments

Many thanks to Güven Eken, Murat Ataol, Murat Bozgoğan, Özge Balkiz, Süreyya İsfendiyaroğlu, Dicle Tuba Kiliç, and Yildiray Lise for contributing the box, Dave Knox, Ana Rodrigues, and Will Turner for help, and Navjot Sodhi and Paul Ehrlich for inviting me to write this chapter.

## REFERENCES

Akçakaya, H. R., Ferson, S., Burgman, M. A., Keith, D. A., Mace, G. M., and Todd, C. R. (2000). Making consistent IUCN classifications under uncertainty. *Conservation Biology*, **14**, 1001–1013.

Akçakaya, H. R., Butchart, S. H. M., Mace, G. M., Stuart, S. N., and Hilton-Taylor, C. (2006). Use and misuse of the IUCN Red List Criteria in projecting climate change impacts on biodiversity. *Global Change Biology*, **12**, 2037–2043.

Ando, A., Camm, J., Polasky, S., and Solow, A. (1998). Species distributions, land values, and efficient conservation. *Science*, **279**, 2126–2128.

Baillie, J. E. M., Bennun, L. A., Brooks, T. M. *et al.* (2004). *A global species assessment*. IUCN-The World Conservation Union, Gland, Switzerland.

Baillie, J. E. M., Collen, B., Amin, R., *et al.* (2008). Towards monitoring global biodiversity. *Conservation Letters*, **1**, 18–26.

Balmford, A. and Gaston, K. J. (1999). Why biodiversity surveys are good value. *Nature*, **398**, 204–205.

Balmford, A., Gaston, K. J., Rodrigues, A. S. L., and James, A. (2000). Integrating costs of conservation into international priority setting. *Conservation Biology*, **14**, 597–605.

Balmford, A., Moore, J., Brooks, T., *et al.* (2001). Conservation conflicts across Africa. *Science*, **291**, 2616–2619.

Balmford, A., Gaston, G. J., Blyth, S., James, A., and Kapos, V. (2003). Global variation in conservation costs, conservation benefits, and unmet conservation needs. *Proceedings of the National Academy of Sciences of the United States of America*, **100**, 1046–1050.

Bennun, L., Matiku, P., Mulwa, R., Mwangi, S., and Buckley, P. (2005). Monitoring Important Bird Areas in Africa: towards a sustainable and scaleable system. *Biodiversity and Conservation*, **14**, 2575–2590.

Bennun, L., Bakarr, M., Eken, G., and Fonseca, G. A. B. da (2007). Clarifying the Key Biodiversity Areas approach. *BioScience*, **57**, 645.

Bierregaard, R., Lovejoy, T. E., Gascon, C., and Mesquita, R., eds (2001). Lessons from Amazonia: The Ecology and Conservation of a Fragmented Forest. Yale University Press, Newhaven, CT.

BirdLife International (2004). *State of the world's birds 2004—indicators for our changing world*. BirdLife International, Cambridge, UK.

Bode, M., Wilson, K. A., Brooks, T. M., *et al.* (2008). Cost-effective global conservation spending is robust to taxonomic group. *Proceedings of the National Academy of Sciences of the United States of America*, **105**, 6498–6501.

Boyd, C., Brooks, T. M., Butchart, S. H. M., *et al.* (2008). Scale and the conservation of threatened species. *Conservation Letters*, **1**, 37–43.

Brook, B. W., O'Grady, J. J., Chapman, A. P., Burgman, M. A., Akçakaya, H. R., and Frankham, R. (2000). Predictive accuracy of population viability analysis in conservation biology. *Nature*, **404**, 385–387.

Brook, B. W., Sodhi, N. S., and Ng, P. K. L. (2003). Catastrophic extinctions follow deforestation in Singapore. *Nature*, **424**, 420–423.

Brooke, M. de L., Butchart, S. H. M., Garnett, S. T., Crowley, G. M., Mantilla-Beniers, N. B. and Stattersfield, A. J. (2008). Rates of movement of threatened bird species between IUCN Red List categories and toward extinction. *Conservation Biology*, **22**, 417–427.

Brooks, T. and Kennedy, E. (2004). Biodiversity barometers. *Nature*, **431**, 1046–1047.

Brooks, T. M., Pimm, S. L., and Oyugi, J. O. (1999). Time lag between deforestation and bird extinction in tropical forest fragments. *Conservation Biology*, **13**, 1140–1150.

Brooks, T. M., Mittermeier, R. A., Mittermeier, C. G., *et al.* (2002). Habitat loss and extinction in the hotspots of biodiversity. *Conservation Biology*, **16**, 909–923.

Brooks, T. M., Mittermeier, R. A., da Fonseca, G. A. B., *et al.* (2006). Global biodiversity conservation priorities. *Science*, **313**, 58–61.

Brummitt, N. and Lughadha, E. N. (2003). Biodiversity: where's hot and where's not? *Conservation Biology*, **17**, 1442–1448.

Bruner, A. G., Gullison, R. E., Rice, R. E., and da Fonseca, G.A.B. (2001). Effectiveness of parks in protecting tropical biodiversity. *Science*, **291**, 125–128.

Bruner, A. G., Gullison, R. E., and Balmford, A. (2004). Financial costs and shortfalls of managing and expanding protected-area systems in developing countries. *BioScience*, **54**, 1119–1126.

Bryant, D., Nielsen, D., and Tangley, L. (1997). *Last frontier forests: ecosystems and economies on the edge*. World Resources Institute, Washington, DC.

Bunn, S. and Arthington, A. (2002). Basic principles and ecological consequences of altered flow regimes for aquatic biodiversity. *Environmental Management*, **30**, 492–507.

Butchart, S. H. M., Stattersfield, A. J., Bennun, L. A., *et al.* (2004). Measuring Global Trends in the Status of Biodiversity: Red List Indices for Birds. *PLoS Biology*, **2**(12), e383.

Carwardine, J., Wilson, K. A., Ceballos, G., *et al.* (2008). Cost-effective priorities for global mammal conservation. *Proceedings of the National Academy of Sciences of the United States of America*, **105**, 11446–11450.

Ceballos, G. and Ehrlich, P. R. (2002). Mammal population losses and the extinction crisis. *Science*, **296**, 904–907.

Chape, S., Harrison, J., Spalding, M., and Lysenko, I. (2005). Measuring the extent and effectiveness of protected areas as an indicator for meeting global biodiversity targets. *Philosophical Transactions of the Royal Society of London B*, **360**, 443–455.

Cincotta, R. P., Wisnewski, J., and Engelman, R. (2000). Human population in the biodiversity hotspots. *Nature*, **404**, 990–992.

Collar, N. J. (1993–4). Red Data Books, action plans, and the need for site-specific synthesis. *Species*, **21–22**, 132–133.

Costanza, R., d'Arge, R., Groot, R. d., *et al.* (1997). The value of the world's ecosystem services and natural capital. *Nature*, **387**, 253–260.

Cowen, R. K., Paris, C. B., and Srinivasan, A. (2006). Scaling of connectivity in marine populations. *Science*, **311**, 522–527.

Cowling, R. M. and Heijnis, C. E. (2001). Broad Habitat Units as biodiversity entities for conservation planning in the Cape Floristic Region. *South African Journal of Botany*, **67**, 15–38.

Cowling, R. M., Pressey, R. L., Rouget, M., and Lombard, A. T. (2003). A conservation plan for a global biodiversity hotspot – the Cape Floristic Region, South Africa. *Biological Conservation*, **112**, 191–216.

Crooks, K. R. and Sanjayan, M. (2006). *Connectivity conservation*. Cambridge University Press, Cambridge, UK.

Daily, G. C., Ehrlich, P. R., and Sanchez-Azofeifa, A. (2001). Countryside biogeography: Utilization of human-dominated habitats by the avifauna of southern Costa Rica. *Ecological Applications*, **11**, 1–13.

Darwall, W. R. T. and Vié, J. C. (2005). Identifying important sites for conservation of freshwater biodiversity: extending the species-based approach. *Fisheries Management and Ecology*, **12**, 287–293.

da Fonseca, G. A. B., Balmford, A., Bibby, C., *et al.* (2000). Following Africa's lead in setting priorities. *Nature*, **405**, 393–394.

De Grammont, P. C. and Cuarón, A. D. (2006). An evaluation of threatened species categorization systems used on the American continent. *Conservation Biology*, **20**, 14–27.

Diamond, J. M. (1975). The island dilemma: lessons of modern biogeographic studies for the design of natural reserves. *Biological Conservation*, **7**, 129–146.

Dobson, A. P., Bradshaw, A. D., and Baker, A. J. M. (1997). Hopes for the future: restoration ecology and conservation biology. *Science*, **277**, 515–522.

Edgar, G. J., Langhammer, P. F., Allen, G., *et al.* (2008a). Key Biodiversity Areas as globally significant target sites for the conservation of marine biological diversity. *Aquatic Conservation: Marine and Freshwater Ecosystems*, **18**, 955–968.

Edgar, G. J., Banks, S., Bensted-Smith, R., *et al.* (2008b). Conservation of threatened species in the Galapagos Marine Reserve through identification and protection of marine key biodiversity areas. *Aquatic Conservation: Marine and Freshwater Ecosystems*, **18**, 969–983.

Eken, G., Bennun, L., Brooks, T. M., *et al.* (2004). Key biodiversity areas as site conservation targets. *BioScience*, **54**, 1110–1118.

Erwin, T. L. (1991). An evolutionary basis for conservation strategies. *Science*, **253**, 750–752.

Fitter, R. and Fitter, M. (1987). *The road to extinction: problems of categorizing the status of taxa threatened with extinction*. IUCN, Gland, Switzerland.

Fjeldså, J. and Lovett, J. C. (1997). Geographical patterns of old and young species in African forest biota: the significance of specific montane areas as evolutionary centres. *Biodiversity and Conservation*, **6**, 325–346.

Forest, F., Grenyer, R., Rouget, M., *et al.* (2007). Preserving the evolutionary potential of floras in biodiversity hotspots. *Nature*, **445**, 757–760.

Gärdenfors, U., Hilton-Taylor, C., Mace, G. M., and Rodriguez, J. P. (2001). The application of IUCN Red List criteria at regional levels. *Conservation Biology*, **15**, 1206–1212.

Gaston, K. J. (2000). Global patterns in biodiversity. *Nature*, **405**, 222–227.

Gaston, K. J. and May, R. M. (1992). Taxonomy of taxonomists. *Nature*, **356**, 281–282.

Grenyer, R., Orme, C. D. L., Jackson, S. F., *et al.* (2006). Global distribution and conservation of rare and threatened vertebrates. *Nature*, **444**, 93–96.

Halpern, B. S., Walbridge, S., Selkoe, K. A., *et al.* (2008). A global map of human impact on marine ecosystems. *Science*, **319**, 948–952.

Hannah, L., Rakotosamimanana, B., Ganzhorn, J., *et al.* (1998). Participatory planning, scientific priorities, and landscape conservation in Madagascar. *Environmental Conservation*, **25**, 30–36.

Hannah, L., Midgely, G. F., Lovejoy, T., *et al.* (2002). Conservation of biodiversity in a changing climate. *Conservation Biology*, **16**, 264–268.

Hobbs, R. J. (1992). The role of corridors in conservation—solution or bandwagon. *Trends in Ecology and Evolution*, **7**, 389–392.

Hobbs, R. J. and Norton, D. A. (1996). Towards a conceptual framework for restoration ecology. *Restoration Ecology*, **4**, 93–110.

Hoekstra, J. M., Boucher, T. M., Ricketts, T. H., and Roberts, C. (2005). Confronting a biome crisis: global disparities of habitat loss and protection. *Ecology Letters*, **8**, 23–29.

Hoffmann, M., Brooks, T. M., da Fonseca, G. A. B., *et al.* (2008). The IUCN Red List and conservation planning. *Endangered Species Research*, doi:10.3354/esr99987.

Hunter Jr., M. L. and Hutchinson, A. (1994). The virtues and shortcomings of parochialism: conserving species that are locally rare, but globally common. *Conservation Biology*, **8**, 1163–1165.

Innes, J. L. and Er, K. B. H. (2002). The questionable utility of the frontier forest concept. *BioScience*, **52**, 1095–1109.

IUCN (International Union for the Conservation of Nature) (2001). *IUCN Red List categories and criteria – version 3.1.* IUCN, Gland, Switzerland.

IUCN (International Union for the Conservation of Nature) (2007). *2007 IUCN Red List of threatened species.* IUCN, Gland, Switzerland.

Isaac, N. J. B., Turvey, S. T., Collen, B., Waterman, C., and Baillie, J. E. M. (2007). Mammals on the EDGE: conservation priorities based on threat and phylogeny. *PLoS One*, **2**, e296.

James, A., Gaston, K., and Balmford, A. (1999). Balancing the Earth's accounts. *Nature*, **401**, 323–324.

James, A., Gaston, K., and Balmford, A. (2001). Can we afford to conserve biodiversity? *BioScience*, **51**, 43–52.

Jepson, P. and Whittaker, R. J. (2002). Ecoregions in context: a critique with special reference to Indonesia. *Conservation Biology*, **16**, 42–57.

Jetz, W. and Rahbek, C. (2002). Geographic range size and determinants of avian species richness. *Science*, **297**, 1548–1551.

Jones, G. P., Milicich, M. J., Emslie, M. J., and Lunow, C. (1999). Self-recruitment in a coral reef fish population. *Nature*, **402**, 802–804.

Kaiser, J. (2001). Bold corridor project confronts political reality. *Science*, **293**, 2196–2199.

Kareiva, P. and Marvier, M. (2003). Conserving biodiversity coldspots. *American Scientist*, **91**, 344–351.

Kirkpatrick, J. B. (1983). An iterative method for establishing priorities for the selection of nature reserves – an example from Tasmania. *Biological Conservation*, **25**, 127–134.

Knight, A. T., Smith, R. J., Cowling, R. M., *et al.* (2007). Improving the Key Biodiversity Areas approach for effective conservation planning. *BioScience*, **57**, 256–261.

Knight, A. T., Cowling, R. M., Rouget, M., Balmford, A., Lombard, A. T., and Campbell, B.M. (2008). Knowing but not doing: selecting conservation priority areas and the research–implementation gap. *Conservation Biology*, **22**, 610–617.

Krupnick, G. A. and Kress, W. J. (2003). Hotspots and ecoregions: a test of conservation priorities using taxonomic data. *Biodiversity and Conservation*, **12**, 2237–2253.

Küper, W., Sommer, J. H., Lovett, J. C., *et al.* (2004). Africa's hotspots of biodiversity redefined. *Annals of the Missouri Botanical Garden*, **91**, 525–535.

Lamoreux, J. F., Morrison, J. C., Ricketts, T. H., *et al.* (2006). Global tests of biodiversity concordance and the importance of endemism. *Nature*, **440**, 221–214.

Langhammer, P. F., Bakarr, M. I., Bennun, L. A., *et al.* (2007). *Identification and gap analysis of key biodiversity areas: targets for comprehensive protected area systems.* IUCN World Commission on Protected Areas Best Practice Protected Area Guidelines Series No. 15. IUCN, Gland, Switzerland.

Laurance, W. F. (2008) Can carbon trading save vanishing forests? *BioScience*, **58**, 286–287.

Lens, L., Van Dongen, S., Norris, K., Githiru, M., and Matthysen, E. (2002). Avian persistence in fragmented rainforest. *Science*, **298**, 1236–1238.

Mace, G. M., Balmford, A., Boitani, L., *et al.* (2000). It's time to work together and stop duplicating conservation efforts. *Nature*, **405**, 393.

Margules, C. R. and Pressey, R. L. (2000). Systematic conservation planning. *Nature*, **405**, 243–253.

Meynell, P.-J. (2005). Use of IUCN Red Listing process as a basis for assessing biodiversity threats and impacts in environmental impact assessment. *Impact Assessment and Project Appraisal*, **23**, 65–72.

Miller, J. R. and Hobbs, R. J. (2007). Habitat restoration—do we know what we're doing? *Restoration Ecology*, **15**, 382–390.

Mittermeier, R. A., Robles Gil, P., and Mittermeier, C. G. (1997). *Megadiversity*. Cemex, Mexico.

Mittermeier, R. A., Mittermeier, C. G., Brooks, T. M., *et al.* (2003). Wilderness and biodiversity conservation. *Proceedings of the National Academy of Sciences of the United States of America*, **100**, 10309–10313.

Mittermeier, R. A., Robles Gil, P., Hoffmann, M., *et al.* (2004). *Hotspots: revisited*. Cemex, Mexico.

Myers, N. (1988). Threatened biotas: "hot spots" in tropical forests. *The Environmentalist*, **8**, 187–208.

Myers, N. (2003). Biodiversity hotspots revisited. *BioScience*, **53**, 916–917.

Myers, N., Mittermeier, R. A., Mittermeier, C. G., da Fonseca, G. A. B., and Kent, J. (2000). Biodiversity hotspots for conservation priorities. *Nature*, **403**, 853–858.

Newmark, W. D. (1987). Mammalian extinctions in western North American parks: a land-bridge island perspective. *Nature*, **325**, 430–432.

Olson, D. M. and Dinerstein, E. (1998). The Global 200: a representation approach to conserving the Earth's most biologically valuable ecoregions. *Conservation Biology*, **12**, 502–515.

Olson, D. M., Dinerstein, E., Wikramanayake, E. D., et al. (2001). Terrestrial ecoregions of the world: A new map of life on Earth. *BioScience*, **51**, 933–938.

Orme, C. D. L, Davies, R. G., Burgess, M., et al. (2005). Global hotspots of species richness are not congruent with endemism or threat. *Nature*, **436**, 1016–1019.

Osieck, E. R. and Mörzer Bruyns, M. F. (1981). *Important bird areas in the European community.* International Council for Bird Preservation, Cambridge, UK.

Peres, C. A. (2005). Why we need megareserves in Amazonia. *Conservation Biology*, **19**, 728–733.

Plantlife International (2004). *Identifying and protecting the world's most important plant areas: a guide to implementing target 5 of the global strategy for plant conservation.* Plantlife International, Salisbury, UK.

Polasky, S. (2008). Why conservation planning needs socioeconomic data. *Proceedings of the National Academy of Sciences of the United States of America*, **105**, 6505–6506.

Pressey, R. L. and Taffs, K. H. (2001). Scheduling conservation action in production landscapes: priority areas in western New South Wales defined by irreplaceability and vulnerability to vegetation loss. *Biological Conservation*, **100**, 355–376.

Pressey, R. L. and Tully, S. L. (1994). The cost of ad hoc reservation—a case-study in western New South Wales. *Australian Journal of Ecology*, **19**, 375–384.

Pressey, R. L., Johnson, I. R., and Wilson, P. D. (1994). Shades of irreplaceability—towards a measure of the contribution of sites to a reservation goal. *Biodiversity and Conservation*, **3**, 242–262.

Pressey, R. L., Cabeza, M., Watts, M. E., Cowling, R. M., and Wilson, K. (2007). Conservation planning for a changing world. *Trends in Ecology and Evolution*, **22**, 583–592.

Raimer, F. and Ford, T. (2005). Yellowstone to Yukon (Y2Y)— one of the largest international wildlife corridors. *Gaia-Ecological Perspectives for Science and Society*, **14**, 182–185.

Raxworthy, C. J., Martinez-Meyer, E., Horning, N., et al. (2003). Predicting distributions of known and unknown reptile species in Madagascar. *Nature*, **426**, 837–841.

Reid, W. V. (1998). Biodiversity hotspots. *Trends in Ecology and Evolution*, **13**, 275–280.

Ricketts, T. H., Dinerstein, E., Boucher, T., et al. (2005). Pinpointing and preventing imminent extinctions. *Proceedings of the National Academy of Sciences of the United States of America*, **102**, 18497–18501.

Roberts, C. M. (1997). Connectivity and management of Caribbean coral reefs. *Science*, **278**, 1454–1457.

Roberts, C. M., McClean, C. J., Veron, J. E. N., et al. (2002). Marine biodiversity hotspots and conservation priorities for tropical reefs. *Science*, **295**, 1280–1284.

Robinson, W. D. (1999). Long-term changes in the avifauna of Barro Colorado Island, Panama, a tropical forest isolate. *Conservation Biology*, **13**, 85–97.

Rodrigues, A. S. L. (2007). Effective global conservation strategies. *Nature*, **450**, e19.

Rodrigues, A. S. L. and Brooks, T. M. (2007). Shortcuts for biodiversity conservation planning: the effectiveness of surrogates. *Annual Review of Ecology, Evolution, and Systematics*, **38**, 713–737.

Rodrigues, A. S. L., Andelman, S. J., Bakarr, M. I., et al. (2004a). Effectiveness of the global protected area network in representing species diversity. *Nature*, **428**, 640–643.

Rodrigues, A. S. L., Akçakaya, H. R., Andelman, S. J., et al. (2004b). Global gap analysis – priority regions for expanding the global protected area network. *BioScience*, **54**, 1092–1100.

Rodrigues, A. S. L., Brooks, T. M., and Gaston, K. J. (2005). Integrating phylogenetic diversity in the selection of priority areas for conservation: does it make a difference? In A. Purvis, J. L. Gittleman, and T. M. Brooks, eds *Phylogeny and conservation*, pp. 101–119. Cambridge University Press, Cambridge, UK.

Rodrigues, A. S. L., Pilgrim, J. D., Lamoreux, J. F., Hoffmann, M., and Brooks, T. M. (2006). The value of the IUCN Red List for conservation. *Trends in Ecology and Evolution*, **21**, 71–76.

Rodríguez, J. P., Ashenfelter, G., Rojas-Suárez, F., Fernández, J. J. G., Suárez, L., and Dobson, A. P. (2000). Local data are vital to worldwide conservation. *Nature*, **403**, 241.

Roessig, J. M., Woodley, C. M., Cech, J. J., and Hansen, L. J. (2004). Effects of global climate change on marine and estuarine fishes and fisheries. *Reviews in Fish Biology and Fisheries*, **14**, 251–275.

Rondinini, C., Wilson, K. A., Boitani, L., Grantham, H., and Possingham, H. P. (2006). Tradeoffs of different types of species occurrence data for use in systematic conservation planning. *Ecology Letters*, **9**, 1136–1145.

Rosenzweig, M. L. (2003). *Win-win ecology*. Oxford University Press, Oxford, UK.

Rouget, M., Cowling, R. M., Pressey, R. L., and Richardson, D. M. (2003). Identifying spatial components of ecological and evolutionary processes for regional conservation planning in the Cape Floristic Region, South Africa. *Diversity and Distributions*, **9**, 191–210.

Rouget, M., Cowling, R. M., Lombard, A. T., Knight, A. T., and Kerley, G. I. H. (2006). Designing regional-scale corridors for pattern and process. *Conservation Biology*, **20**, 549–561.

Saccheri, I., Kuussaari, M., Kankare, M., Vikman, P., Fortelius, W., and Hanski, I. (1998). Inbreeding and extinction in a butterfly metapopulation. *Nature*, **392**, 491–494.

Sanderson, E. W., Jaiteh, M., Levy, M. A., Redford, K. H., Wannebo, A. V., and Woolmer, G. (2002a). The human footprint and the last of the wild. *BioScience*, **52**, 891–904.

Sanderson, E. W., Redford, K. H., Vedder, A., Coppolillo, P. B., and Ward, S. E. (2002b). A conceptual model for conservation planning based on landscape species requirements. *Landscape and Urban Planning*, **58**, 41–56.

Saunders, D. A., Hobbs, R. J., and Margules, C. R. (1991). Biological consequences of ecosystem fragmentation: a review. *Conservation Biology*, **5**, 18–32.

Schipper, J., Chanson, J. S., Chiozza, F., *et al.* (2008). The status of the world's land and marine mammals: diversity, threat, and knowledge. *Science*, **322**, 225–230.

Sechrest, W., Brooks, T. M., da Fonseca, G. A. B., *et al.* (2002). Hotspots and the conservation of evolutionary history. *Proceedings of the National Academy of Sciences of the United States of America*, **99**, 2067–2071.

Simberloff, D., Farr, J. A., Cox, J., and Mehlman, D. W. (1992). Movement corridors – conservation bargains or poor investments? *Conservation Biology*, **6**, 493–504.

Smith, T. B., Wayne, R. K., Girman, D. J., and Bruford, M. W. (1997). A role for ecotones in generating rainforest biodiversity. *Science*, **276**, 1855–1857.

Smith, T. B., Kark, S., Schneider, C. J., Wayne, R. K., and Moritz, C. (2001). Biodiversity hotspots and beyond: the need for preserving environmental transitions. *Trends in Ecology and Evolution*, **16**, 431.

Spathelf, M. and Waite, T. A. (2007). Will hotspots conserve extra primate and carnivore evolutionary history? *Diversity and Distributions*, **13**, 746–751.

Stattersfield, A. J., Crosby, M. J., Long, A. J., and Wege, D. C. (1998). *Endemic bird areas of the world: priorities for biodiversity conservation*. BirdLife International, Cambridge, UK.

Stokstad, E. (2008). Big land purchase triggers review of plans to restore Everglades. *Science*, **321**, 22.

Stuart, S. N., Chanson, J. S., Cox, N. A., *et al.* (2004). Status and trends of amphibian declines and extinctions worldwide. *Science*, **306**, 1783–1786.

Terborgh, J., Lopez, L., Nuñez, P. *et al.* (2001). Ecological meltdown in predator-free forest fragments. *Science*, **294**, 1923–1926.

Tewksbury, J. J., Levey, D. J., Haddad, N. M., *et al.* (2002). Corridors affect plants, animals, and their interactions in fragmented landscapes. *Proceedings of the National Academy of Sciences of the United States of America*, **99**, 12923–12926.

Thomas, C. D., Cameron, A., Green, R. E., *et al.* (2004). Extinction risk from climate change. *Nature*, **427**, 145–148.

Turner, M. G. (2005). Landscape ecology: what is the state of the science? *Annual Review of Ecology, Evolution, and Systematics*, **36**, 319–344.

Turner, W. R. and Wilcove, D. S. (2006). Adaptive decision rules for the acquisition of nature reserves. *Conservation Biology*, **20**, 527–537.

Turner, W. R., Brandon, K., Brooks, T. M., Costanza, R., da Fonseca, G. A. B., and Portela, R. (2007). Global conservation of biodiversity and ecosystem services. *BioScience*, **57**, 868–873.

Whittaker, R. J., Araújo, M. B., Jepson, P., Ladle, R. J., Watson, J. E. M., and Willis, K. J. (2005). Conservation biogeography: assessment and prospect. *Diversity and Distributions*, **11**, 3–23.

Wilson, E. O. (1992). *The diversity of life*. Belknap, Cambridge, Massachusetts.

Wilson, K., Pressey, R. L., Newton, A., Burgman, M., Possingham, H., and Weston, C. (2005). Measuring and incorporating vulnerability in conservation planning. *Environmental Management*, **35**, 527–543.

Wilson, K. A., McBride, M. F., Bode, M., and Possingham, H. P. (2006). Prioritizing global conservation efforts. *Nature*, **440**, 337–340.

WWF (World Wildlife Fund) and IUCN (International Union for the Conservation of Nature) (1994–7). *Centres of plant diversity: a guide and strategy for their conservation*. World Wide Fund for Nature, Gland, Switzerland.

Xenopoulos, M. A., Lodge, D. M., Alcamo, J., Marker, M., Schulze, K., and Van Vuuren, D. P. (2005). Scenarios of freshwater fish extinctions from climate change and water withdrawal. *Global Change Biology*, **11**, 571–564.

Zedler, J. B. (2000). Progress in wetland restoration ecology. *Trends in Ecology and Evolution*, **15**, 402–407.

# CHAPTER 12

# Endangered species management: the US experience

David S. Wilcove

To many people around the world, the conservation of endangered species is synonymous with the conservation of biodiversity. Ecologists, of course, understand that biodiversity encompasses far more than endangered species, but it is nonetheless true that endangered species are among the most visible and easily understood symbols of the ongoing loss of biodiversity (see Chapter 10). The protection of such species is a popular and important part of efforts to sustain the earth's natural diversity (see Box 12.1).

The process of conserving endangered species can be divided into three phases: (i) identification—determining which species are in danger of extinction; (ii) protection—determining and implementing the short-term measures necessary to halt a species' slide to extinction; and (iii) recovery—determining and implementing the longer-term measures necessary to rebuild the population of the species to the point at which it is no longer in danger of extinction.

Many countries today have laws or programs designed to protect endangered species, although the efficacy of these efforts varies widely. Most follow the identification/protection/recovery paradigm. One of the oldest and strongest laws is the United States' Endangered Species Act (ESA), which was passed in 1973 and has served as a template for many other nations. In this chapter, I shall focus on the three phases of endangered species management, emphasizing the US experience. My reason for emphasizing the US is not because I believe it has done a better job of protecting its endangered species than other countries. Rather, I am most familiar with conservation efforts in the US. Moreover, because the US has one of the oldest and strongest laws on the books to protect endangered species, it provides a useful case history.

My discussion is admittedly incomplete and, to some extent, idiosyncratic. Endangered species programs, especially those that impose restrictions on human activities, are invariably controversial, and that controversy results in much discussion and debate. The ESA, for example, has been the subject of many books, scientific articles, and popular articles; it has been debated in the halls of Congress and in town halls across the nation; and it has been litigated numerous times in the courts. Complete coverage of all of the issues associated with endangered species in the US or any other large country is simply not possible in a single book chapter. For that reason, I have chosen to review a subset of issues that are likely to be of interest to both scientists and decision-makers in countries with active programs to conserve endangered species.

## 12.1 Identification

### 12.1.1 What to protect

A fundamental question that quickly arises when scientists and decision-makers discuss endangered wildlife is what exactly should be conserved (see Box 12.2). Protection efforts can be directed at species, subspecies, or populations, with important tradeoffs. If, for example, protection is extended to subspecies and populations, the total number of plants and animals that are deemed in need of protection is likely to increase dramatically, resulting in greater

## Box 12.1 Rare and threatened species and conservation planning in Madagascar
### Claire Kremen, Alison Cameron, Tom Allnutt, and Andriamandimbisoa Razafimpahanana

The fundamental challenge of reserve design is how to maximize biodiversity conservation given area constraints, competing land uses and that extinction risk is already high for many species, even without further habitat loss. Madagascar is one of the world's highest priorities for conservation (Brooks et al. 2006) with endemism exceeding 90% for many plant and animal groups (Goodman and Benstead 2005). Recently, the President of Madagascar set the target for habitat protection at 10% of the land surface, representing a tripling of the region to be protected. This provided an unparalleled opportunity to protect Madagascar's biodiversity. To aid the government in site selection, we used a "systematic conservation planning" approach (Margules and Pressey 2000) to identify regions that would protect as many species as possible, especially geographically rare and threatened species, within that 10% target.

We obtained occurrence data for 2315 endemic species of plants, lemurs, frogs, geckos, butterflies and ants (see Box 12.1 Figure 1). We utilized a spatial prioritization decision-support tool (Zonation: Moilanen et al. 2005), and input models of species distributions (for 829 species) and point data for the remaining species (too rare to model, designated RTS for rare target species). The Zonation algorithm preferentially selects the best habitat for geographically rare (range-restricted) species. In addition, by supplying weights based on past habitat loss, we instructed Zonation to favor species that had suffered large range loss within the past 50 years (threatened species). In this manner, our decision support tool picked regions that not only represented all of the species in our analysis, but also identified the habitats most important to geographically rare and/or threatened species.

We ran Zonation in three ways: (i) for each of the six taxonomic groups alone; (ii) for all groups together; and (iii) for all groups together, after first selecting existing protected areas, totaling 6.3% of the country. We then assessed how well the selected regions for each Zonation run protected rare and threatened species by determining what proportions of their habitats (for modeled species) or occurrence points (for RTS species) were included. We also compared Zonation's selections based on all taxa (run ii

**Box 12.1 Figure 1** *Mantella cowanii*, a critically endangered frog of Madagascar. It is one of the species that was used by Kremen et al. (2008) to determine priority sites for protection in Madagascar. Photograph by F. Andreone.

above) against the actual protected areas, from 2.9% area in 2002 to 6.3% area in 2006.

When individual taxonomic groups were utilized to define priority regions (run i), the regions selected by Zonation provided superior protection for members of the taxon itself, but relatively poor protection for species in other groups. It was therefore more efficient to utilize an analysis based on all taxonomic groups together (run ii). Comparing this analysis to the regions that had already been set aside showed that, on an area by area basis, Zonation selected regions that significantly increased the inclusion of habitat for geographically rare and threatened species. In addition, we found that the trajectory for accumulating species and habitat areas from 2002 to 2006 would be insufficient to protect all species within the area target, but that careful selection of the last 3.7% (Run iii) could greatly improve both representation of all species and the selection of habitat for the geographically rare and threatened species (Kremen et al. 2008).

Subsequently, this analysis was used along with other conservation inputs (Key Biodiversity Analyses, Important Bird Areas, and others; see Chapter 11) to justify the final regions for protection totaling 6.4 million hectares (Box 12.1 Figure 2, black zones totaling just over 10%), and served to designate an additional 5.3 million

*continues*

### Box 12.1 (Continued)

hectares as important conservation regions subject to an inter-ministerial decree limiting mining activities. No new mining permits will be issued in the highest priority zones (grey zones), and the remaining areas (light grey zones) will be subject to strict control (e.g. following Environmental Impact Assessment). The rare target species, in particular, were utilized to define these zones, in particular the 505 species currently known from only a single site. Furthermore, as a significant proportion of these priority zones contain existing mining permits (14% of the existing parks and highest priority areas), the Zonation result is an ideal tool for negotiating trade-offs or swaps between mining and protected areas.

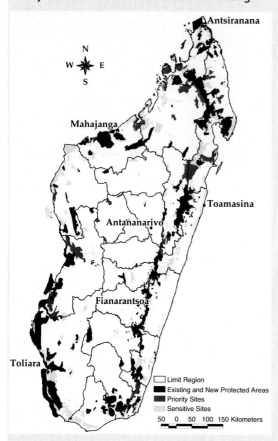

**Box 12.1 Figure 2** This map portrays the Inter-Ministry decree of October 2008 delineating the new and existing protected areas (black), the priority biodiversity areas where no new mining permits may be issued (grey) and the sensitive biodiversity sites (light grey), which will be subject to environmental impact assessment prior to permission of forestry or mining activities. See also Figure 11.6.

### REFERENCES

Brooks, T. M., Mittermeier, R. A., da Fonseca, G. A. B., et al. (2006). Global biodiversity conservation priorities. *Science*, **313**, 58–61.

Goodman, S. and Benstead, J. (2005). Updated estimates of biotic diversity and endemism for Madagascar. *Oryx*, **39**, 73–77.

Kremen, C., Cameron, A., Moilanen, A., et al. (2008). Aligning conservation priorities across taxa in Madagascar with high-resolution planning tools. *Science*, **320**, 222–226.

Margules, C.R. and Pressey, R. L. (2000). Systematic conservation planning. *Nature*, **405**, 243–253.

Moilanen, A., Franco, A. M. A., Early, R. I., et al. (2005). Prioritizing multiple-use landscapes for conservation: methods for large multi-species planning problems. *Proceedings of the Royal Society of London B*, **272**, 1885–1891.

demands for funding and, potentially, more conflicts with landowners, developers, and other resource users. On the other hand, it has been argued that populations should be the fundamental unit of biodiversity protection (see Box 10.1), since it is populations of plants and animals that provide the ecosystem services essential to human welfare (Hughes et al. 1997; Chapter 3).

A second consideration relates to geographic scale. Should the frame of reference for deciding whether or not a species is endangered be the entire world (the species' global status), a particular country (its national status), or a particular

### Box 12.2 Flagship species create Pride
### Peter Vaughan

**Rare:** Rare is a non-governmental organization whose mission is "to conserve imperiled species and ecosystems around the world by inspiring people to care about and protect nature" (see Chapter 15). Rare's *Pride* program utilizes social marketing to educate and motivate people who live in, or adjacent to, areas of high biodiversity to adopt new behaviors that either protect, or are less damaging to, the local environment.

**Social marketing:** Many commercial marketers "brand" their companies and/or products using symbols, such as Pillsbury's "doughboy", or Apple Computer's "bitten apple." Similarly, *Pride* brands its social marketing campaigns using "flagship" species. While concepts such as ecosystem and biodiversity are central to Rare's overall conservation strategies, they are complex and fail to evoke the emotional response that is required to motivate behavior change among most people. The purpose of a flagship is to create a simple, instantly recognizable symbol that evokes a positive emotional response among members of the target audience. As Mckenzie-Mohr (2008) states "All persuasion depends upon capturing attention. Without attention, persuasion is impossible. Communications can be made more effective by ensuring that they are vivid, personal and concrete." A good flagship evokes feelings of trust, affection, and above all for Rare, a sense of *Pride* in the local environment. Pride of place is a powerful emotion that can motivate people to change their behaviors and empower them to take environmental action.

**What makes a good flagship?** Unlike the concepts of "keystone", "indicator", "umbrella", and "endangered" species, which all have ecological or conservation implications, flagship species are chosen for their marketing potential (Walpole and Leader-Williams 2002). The key characteristics of flagship species are (based on Karavanov 2008):

• Be charismatic or appealing to the target audience; no slugs, worms, or mosquitoes!
• Be local or endemic to symbolize the uniqueness of the conservation target area to foster a sense of local pride.
• Be representative of the conservation target area by living in its habitat or ecosystem.
• Have no negative perceptions among local people, such as being a crop pest, being dangerous, or have existing cultural connotations that detract from or compete with the campaign's conservation messages.

**How are flagships chosen?** Flagship species are chosen through a lengthy process that includes input from local stakeholders, interviews with local experts, and results from surveys of the local human population. This process ensures that flagships have the requisite characteristics outlined above.

**How are flagships used?** Flagship species are used in most of the marketing materials produced during a *Pride* campaign, including billboards, posters, puppet shows, songs, videos, etc. such that they become ubiquitous in the community. Although flagship species are non-human, they become symbolic members of the local community, which

*continues*

### Box 12.2 (Continued)

confers on them the credibility they need in order to be perceived as trustworthy sources of information. The flagship species serves as both the "face" of the campaign and as a "spokesperson" for the campaign's messages. This "opinion leadership role" activates the social diffusion networks that exist in all societies by stimulating interpersonal communication among members of the target audience, a key step in the behavior change process (Rogers 1995, Vaughan and Rogers 2000).

**Rare's flagship species:** Among Rare's first 71 campaigns, 59% chose a bird, 16% chose a mammal, and 11% chose a reptile to be their flagships species, but campaigns have also used fish, insects, crustaceans, amphibians, and plants. About half of the chosen species were endemic to the country or region, but only about 8% have been listed as endangered or critically endangered by IUCN. Because flagship species play such a prominent role in *Pride* campaigns, knowledge about them can serve as markers for campaign exposure and impact. For example, during the *Pride* campaign in Laos (Vannalath 2006), awareness among the campaign's target audience of the great hornbill (*Buceros bicornis*; Box 12.2 Figure) increased from 61% to 100%; the percentage of respondents who know that the hornbill is in danger of extinction increased from 22% to 77%; the percentage who knew that hunting or capturing the hornbill is prohibited increased from 31% to 90%; and the percentage that identified "cutting down the forest" as one of the greatest threats to the hornbill increased from 17% to 65%. In addition to increasing knowledge, improving attitudes, and changing personal behavior, *Pride* campaigns have been credited with contributing to the creation of protected areas, enactment of new laws and regulations, and the preservation of endangered species (Jenks *et al.* 2010). Central to all of these efforts has been the use of flagship species.

**Box 12.2 Figure** Pride campaign flagship mascot representing the great hornbill in Laos. Photograph by R. Godfrey.

### REFERENCES

Jenks, B., Vaughan, P. W., and Butler, P. J. (2010). The Evolution of Rare Pride: Using Evaluation to Drive Adaptive Management in a Biodiversity Conservation Organization. *Journal of Evaluation and Program Planning Special Edition on Environmental Education Evaluation*, in press.

Karavanov, A. (2008). *Campaign design - Including work planning and monitoring*. Rare Pride Leadership Development Program, Rare, Arlington, VA.

Mckenzie-Mohr, D. (2008). http://www.graduationpledge.org/Downloads/CBSM.pdf (page 4) accessed December 15, 2008.

Rogers, E. M. (1995). *Diffusion of innovations*, 4th ed. The Free Press, New York, NY.

Vannalath, S. (2006). *Final Report, Rare Pride Campaign in Nam Kading National Protected Area, Lao Peoples Democratic Republic*, Rare Diploma in Conservation Education, University of Kent at Canterbury, United Kingdom. Wildlife Conservation Society and Rare, Arlington, VA.

Vaughan, P. W. and Rogers, E. M. (2000). A staged model of communication effects: Evidence from an entertainment-education radio drama in Tanzania. *The Journal of Health Communication*, **5**, 203–227.

Walpole, M. J. and Leader-Williams, N. (2002). Tourism and flagship species in conservation. *Biodiversity and Conservation*, **11**, 543–547.

state, county, or municipality (its local status)? For example, the northern saw-whet owl (*Aegolius acadicus*) is widely distributed across the northern and western United States and in parts of Mexico. It is not in danger of extinction. But within the US, the State of Maryland considers the northern saw-whet owl to be an endangered species; Maryland is at the southeastern periphery of the owl's range and the bird is quite rare there. Conservationists continue to debate the wisdom of expending scarce resources on the protection of peripheral or isolated populations of otherwise common species. Yet such populations are often a source of pride to the citizens of a given region, and they may contain unique alleles that contribute to the overall genetic diversity of the species.

A third consideration is whether to extend protection to all types of endangered organisms or to limit such efforts to particular groups, such as vertebrates or vascular plants. Proponents of exclusion argue that it is impossible to identify and protect all of the imperiled species in any large area (see below), and that by targeting a few, select groups, it should be possible to protect the habitats of many other species. Although some studies have supported this notion, others have not.

Within the US, the ESA addresses these issues in the following ways: it allows for the protection of species and subspecies of plants and animals (including invertebrate animals). In the case of vertebrates only, it also allows for the protection of distinct population segments. In the early years of the ESA, the US Fish and Wildlife Service, the agency charged with protecting imperiled wildlife, allowed populations to be defined on the basis of political borders. Thus, bald eagles (*Haliaeetus leucocephalus*) in the coterminous 48 states (but not those living in Alaska or Canada) were added to the endangered list when their numbers plummeted due to pesticide poisoning. More recently, the Fish and Wildlife Service has turned away from using political borders to delineate vertebrate populations and has insisted that such populations be discrete ecological entities in order to be eligible for inclusion on the endangered list. An example of the latter would be some of the salmon runs in the Pacific Northwest that have been added to the endangered species list in recent years. To qualify for listing, a given run must show significant genetic, demographic, or behavioral differences from other runs of the same species.

One aspect of the ESA's identification process merits special attention. The law explicitly states that the decision to add a plant or animal to the endangered species list must be based "solely on the basis of the best scientific and commercial data..." (Endangered Species Act, Section 4(b)(1)(A)). In other words, whether or not a species is endangered is treated as a purely scientific question. Political considerations are not allowed to interfere with the identification phase (although in practice they sometimes do, leading to nasty legal battles).

### 12.1.2 Criteria for determining whether a species is endangered

How does one know that a given species is in danger of extinction? Biologists typically look for data that indicate vulnerability: a small population size, a declining population, ongoing losses of habitat (see Chapter 4), etc. In some cases, those data are combined with models to yield short and long-term projections of population viability (see Chapter 16); in other cases, where not enough data exist to construct good models, the determination is based on expert opinion.

Needless to say, different experts weighing different factors are likely to come to different conclusions as to which species are in trouble. Resources may be wasted on plants and animals that are not really endangered, while other, gravely imperiled species go unprotected. The need for a more transparent, standardized way to assess the status of species led the International Union for Conservation of Nature (IUCN) to develop a set of quantitative guidelines in 1994, now known as the Red List categories and criteria. These guidelines enable scientists to assign any plant or animal to one of six categories (Extinct, Extinct in the Wild, Critically Endangered, Endangered, Vulnerable, Near Threatened) based on factors such as range size, amount of occupied habitat, population size, trends in population size, or trends in the amount of habitat (www.iucnredlist.org/static/categories_crtiteria). The original Red List categories and criteria were designed to determine the global status of species, but conservation biologists subsequently have developed guidelines for applying those criteria to individual nations, states, provinces, etc.

The ESA, however, is notably vague in defining what constitutes a species at risk of extinction. It establishes two categories of risk, endangered and threatened, and defines an endangered species as "any species which is in danger of extinction throughout all or a significant portion of its range" and a threatened species as "any species which is likely to become an endangered species within the foreseeable future throughout all or a significant portion of its range" (Endangered Species Act, Sections 3(6) and 3(19)). In practice, most plants and animals have not been added to the US endangered species list until they were close to extinction. A study published in 1993 (Wilcove et al. 1993) showed that the median total population size of a vertebrate at time of listing was 1075 individuals; the median number of surviving populations was two. For invertebrate animals, the median total population size was less than 1000 individuals, and the number of surviving populations was three. In the case of plants, the median total population size was less than 120 individuals, and the number of surviving populations was four. One obvious consequence of waiting until species are so rare before protecting them is that recovery becomes far more difficult, if not impossible, to achieve.

## 12.2 Protection

In order to develop an effective protection plan for endangered species, one needs to know a minimum of two things: (i) What threats do the species in question face?; and (ii) Where do those species occur? Knowledge of the threats will determine protection and recovery efforts, while knowledge of the location and, in particular, the land ownership, will guide the choice of conservation strategy.

### 12.2.1 What are the threats?

Understanding the threats facing endangered species is complicated due to four factors: (i) threats may vary from taxon to taxon; the things that imperil freshwater fish, for example, may not necessarily be the things that imperil terrestrial mammals; (ii) threats may vary geographically, depending on economics, technology, human demography, land-use patterns and social customs in different areas; (iii) threats may change over time, again in response to technological, economic, social, or demographic factors; and (iv) for all but a handful of groups (e.g. birds, mammals, amphibians), scientists simply do not know enough about most species to determine which ones are imperiled and why they are imperiled.

For three groups—birds, mammals, and amphibians—the IUCN has determined the conservation status of virtually all extant and recently extinct species (Baillie et al. 2004). These data provide the best global overview of threats to endangered species (Figure 12.1). With respect to birds and amphibians, habitat destruction is by far the most pervasive threat: over 86% of birds and 88% of amphibians classified by IUCN as globally imperiled are threatened to some degree by habitat destruction. Agriculture and logging are the most widespread forms of habitat destruction (see Chapter 4). Overexploitation for subsistence or commerce contributed to the endangerment of 30% of imperiled birds but only 6% of amphibians (see Chapter 6). Alien species were a factor in the decline of 30% of imperiled birds and 11% of amphibians (see Chapter 7). Pollution affected 12% of imperiled birds and 4% of amphibians (see Box 13.1). Disease, which is often linked to pollution or habitat destruction, was a threat to 5% of birds and 17% of amphibians. Surprisingly, few species were identified as being threatened by human-caused climate change, perhaps because most threats are identified after the fact (see Chapter 8). However, Thomas et al. (2004) modeled the response of localized species of various taxa to climate change and concluded that 15–37% of them could be destined for extinction by 2050, making climate change potentially a grave threat.

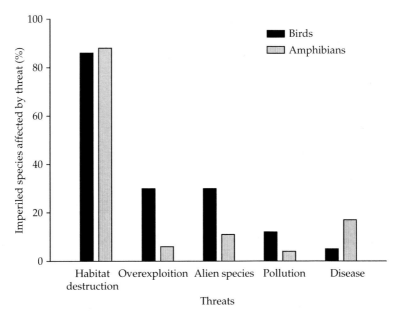

**Figure 12.1** Percentage of the world's imperiled birds (black bars) and amphibians (grey bars) threatened by different factors, based on global analyses performed by the IUCN/World Conservation Union. The total percentages for each group exceed 100% because many species are threatened by more than one factor. Data from Baille et al. (2004).

A comprehensive status assessment of the world's mammals was published in 2008 (Schipper et al. 200; Figure 12.2). Unlike the analyses of birds and amphibians, the mammal assessment did not separate imperiled from non-threatened species in its breakdown of threats. Habitat destruction is the most widespread threat to mammals, affecting 37% of all extant and recently extinct species, followed by overexploitation (17%), invasive species (6%), pollution (4%), and diseases (2%). (The lower percentages compared to birds and amphibians reflect the fact that the mammal assessment covered both imperiled and non-threatened species). Accidental mortality, usually associated with bycatch in fisheries, affects 5% of the world's mammals; in the special case of marine mammals, it affects a staggering 83% of species (see Schipper et al. 2008).

These global analyses of threats mask some important regional differences that could influence conservation decisions. For example, in the US, the most pervasive threat to vertebrates is habitat destruction, affecting over 92% of imperiled mammals, birds, reptiles, amphibians, and fish. This was followed by alien species (affecting 47% of imperiled vertebrates), pollution (46%), overexploitation (27%), and disease (11%) (Wilcove et al. 1998). In contrast, the most pervasive threat to imperiled vertebrates in China is overexploitation, affecting 78% of species, followed by habitat destruction (70%), pollution (20%), alien species (3%), and disease (<1%) (Li and Wilcove 2005; Figure 12.3).

Ecologists have long recognized that island ecosystems are more vulnerable to alien species than most continental ecosystems. In the Hawaiian archipelago, for example, 98% of imperiled birds and 99% of imperiled plants are threatened at least in part by alien species (Figure 12.4 and Plate 14). Comparable percentages for imperiled birds and plants in the continental US are 48% and 30%, respectively (Wilcove et al. 1998).

## 12.2.2 Where do endangered species live?

There is now a burgeoning literature that aspires to identify key sites for endangered species, typically by developing sophisticated algorithms that optimize the number of rare species protected per acre or per dollar (see Dobson et al. 2007; see

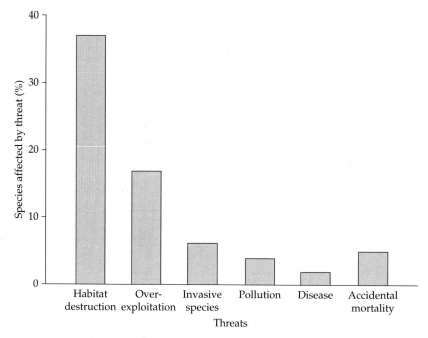

**Figure 12.2** Percentage of the world's mammals threatened by different factors, based on a global analysis by Schipper *et al.* (2008). Note that this analysis covered both threatened and unthreatened species; as such, the data include threats to species that are not yet at risk of extinction, unlike Figure 12.1.

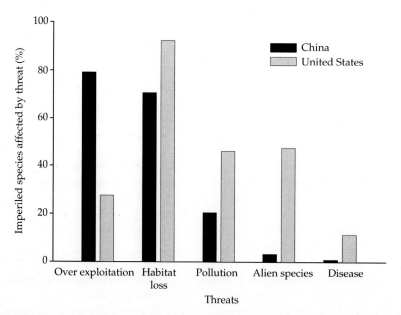

**Figure 12.3** Percentage of imperiled vertebrates in China and the USA threatened by different factors. Reprinted from Li and Wilcove (2005) © American Institute of Biological Sciences.

**Figure 12.4** Akohekohe (*Palmeria dolei*), an endangered Hawaiian honeycreeper. Like many Hawaiian honeycreepers, it is endangered by a combination of habitat destruction and diseases transmitted by introduced mosquitoes. Photograph by Jaan Lepson.

Chapter 11). In this section I shall focus on the simpler issue of land ownership: does the species in question occur on publicly owned (federal or state government) land or private land? In the US, at least, land ownership patterns are a prime consideration in devising effective protection and recovery strategies, given that approximately 60% of land in the US is privately owned.

In the most authoritative assessment of land ownership and endangered species in the US, Groves *et al.* (2000) estimate that private lands harbor populations of more than half of the nation's imperiled species; if one focuses exclusively on those imperiled species that have made it onto the official federal list, that value rises to two-thirds. Approximately one-quarter of all documented populations of federally protected endangered species occur on privately owned land. This figure almost certainly underestimates the degree to which private lands are important to endangered species because many landowners are reluctant to allow biologists to come onto their property to look for rare plants and animals.

### 12.2.3 Protection under the ESA

An effective law or program for endangered species must, at a minimum, be capable of protecting essential habitat, halting overexploitation, and slowing the spread of harmful alien species. In the US at least, it must also extend to both public and private lands.

In the US, once a species has been added to the official list of threatened and endangered species (making it a "listed species"), it is protected to varying degrees on both publicly-owned and privately-owned lands. Federal agencies, for example, are prohibited from engaging in, authorizing, or funding any activities that may jeopardize the survival and recovery of a listed species, including activities that damage or destroy important habitats. Depending on circumstances, such activities can range from timber cutting in the national forests to the construction of federally-funded dams or the allocation of funds for the construction of interstate highways. Federal agencies are required to consult with the US Fish and Wildlife Service, the agency charged with administering the ESA, prior to undertaking any activities that may harm listed species. This consultation requirement minimizes the risk that these other agencies will ignore the needs of imperiled species in the course of their day-to-day operations. Typically, the US Fish and Wildlife Service will work with other government agencies to modify projects so they no longer pose a threat to listed species or, if such modifications are impractical, to develop a mitigation plan that compensates for any harm to a listed species.

Private citizens are prohibited from harming listed animals. This includes direct harm, such as shooting or trapping, as well as indirect harm, such as habitat destruction. Listed plants, on the other hand, are not afforded protection on private lands unless the activity in question (e.g. filling a wetland) requires a federal permit for some other reason. This distinction between animals and plants dates back to English common law and does not have any ecological basis.

The decision to extend the ESA's reach to the activities of private citizens was revolutionary at the time, and it has been the source of considerable controversy ever since. When the ESA originally was passed in 1973, the prohibition on harming a listed species was absolute. But this rigid requirement had an unfortunate consequence: Landowners refused to discuss their endangered-species issues with the US Fish and Wildlife Service because they knew the agency could only say "no," and the US Fish and Wildlife

Service turned a blind eye to the activities of private landowners because it feared a political backlash if it slavishly enforced the law. Thus, paradoxically, the law was too strong to protect endangered species effectively. In 1982, the US Congress modified the ESA so that private landowners could obtain permits from the US Fish and Wildlife Service to engage in activities harmful to listed species provided the landowners developed a plan to minimize and mitigate the impacts of those activities, "to the maximum extent practicable." This change to the law, while controversial, probably averted a much greater weakening of the ESA down the road.

For both federal agencies and private citizens there is also an exemption process that permits important activities to go forward notwithstanding their impact on endangered species. It is reserved for cases where the project in question cannot be modified or mitigated so as to avoid jeopardizing the survival and recovery of a listed species. Because the exemption process is complicated, time-consuming, and politically charged, it has been very rarely used. Instead, the vast majority of conflicts are resolved through consultations with the US Fish and Wildlife Service and modifications to the proposed projects.

Finally, it should be noted that while the ESA can prevent a landowner from undertaking activities that are harmful to a listed species (e.g. habitat destruction), it is doubtful that it can compel an individual to take affirmative steps to improve the well-being of a listed species, for example by removing an invasive plant that is choking out the habitat of an endangered bird. This is an important limitation of laws, such as the ESA, that focus on prohibiting harmful activities; they may not be effective at dealing with more passive threats, such as invasive species or diseases. I return to this issue in my discussion of recovery programs (see below).

## 12.3 Recovery

### 12.3.1 Recovery planning

Recovery aims to secure the long-term future of the species, to rebuild its populations, restore its habitat, or reduce the threats such that it no longer is in danger of extinction and no longer requires extraordinary conservation measures. That process demands a careful balancing of science, economics, and sociology (see Chapter 14). For example, scientific tools like population viability analysis can be used to figure out how many populations must be protected, how large those populations should be, and how they must be distributed across the landscape in order to sustain the species in question (Chapter 16). Restoration ecology can be used to determine how to rehabilitate degraded habitats so as to increase the numbers and distributions of endangered species (see Chapter 13). But securing the cooperation of landowners in the targeted areas or obtaining the necessary funding to implement the restoration plan requires careful consideration of economics, politics, and social customs. All these steps need to be integrated in order to recover an endangered species.

In the US, the ESA requires that recovery plans be developed for all listed species. Those plans should, in theory, spell out the steps necessary to ensure that a given species is no longer in danger of extinction as well as provide a budget for achieving that goal. One might assume that recovery plans play a pivotal role in endangered species management in the US but, in fact, they rarely do. Part of the problem is that the plans are not legally binding documents. Moreover, according to several studies (Clark *et al.* 2002; Hoekstra *et al.* 2002), the plans often fail to make good use of available biological data for the purposes of developing quantitative recovery goals and outlining recovery actions. In addition, many plans lack adequate information on the threats facing endangered species or fail to link recovery actions to specific threats. And still others fail to set out a scientifically sound monitoring protocol for detecting changes in the status of species or assessing the impacts of recovery actions. In short, the recovery planning process has failed to deliver the sort of guidance needed to move species back from the brink of extinction.

### 12.3.2 The management challenge

In theory, the goal of endangered species management is to undertake a series of steps that eliminate the threats to the species in question and result in healthy populations that no longer require special protection or attention. And yet these sorts of success stories—sometimes termed "walk-away-species" because conservationists are able to walk away from them—will be few and far between. Instead, most endangered species are likely to require intensive management and protection for the indefinite future. The reasons are three-fold.

First, the leading cause of species endangerment worldwide is habitat loss (Chapter 4). If, as a result of this problem, species are reduced to living in small, fragmented patches of habitat, they are likely to remain at high risk of extinction until such time as more suitable habitat is created via ecological restoration. In places where human demands for land are great (e.g. southern California), there may be no practical way for conservation organizations or government agencies to acquire land for restoration. Moreover, even if the land is available, it can take decades, even centuries, to restore certain types of ecosystems, such as old-growth forests—if those ecosystems can be restored to anything resembling their pre-industrial state [see Hobbs and Harris (2001) for a discussion of key conceptual issues in ecological restoration; also Box 5.3].

Second, many species live in ecosystems that are maintained by natural disturbances such as fires and floods. Examples of such ecosystems include longleaf pine forests in the Southeastern United States and riparian forests in the Southwestern United States. As people dam rivers, clear native vegetation to build homes and farms, and settle those ecosystems, they disrupt or eliminate the natural disturbances. The result is a growing roster of endangered species for which overt habitat destruction is compounded by the elimination of the natural disturbances that were essential to maintaining the habitat. Given that people are unlikely to allow wildfires or floods to reappear in places where these forces have been "tamed," the only way to ensure the survival of disturbance-dependent species is to mimic the disturbances by using techniques such as prescribed fire, controlled releases of water from dams, or direct manipulation of the vegetation. In short, a growing number of species will not survive without constant human intervention.

Third, more and more species are becoming endangered by the spread of alien, invasive species. In most cases, scientists have no way to eliminate or permanently control the invasive species. Indeed, most attempts at biological control, such as introducing a predator or pathogen of the harmful alien, prove unsuccessful or, worse yet, end up harming other native species (Simberloff and Stiling 1996). Consequently, the usual recourse is to control invasive species by pulling them up, poisoning them, hunting them, or trapping them. Since these activities must be repeated whenever the population of the alien species rebounds, there is little prospect of declaring victory and "walking away."

Wilcove and Chen (1998) estimate that 60% of the species protected or proposed for protection under the ESA are threatened to some degree by alien species or fire suppression. For virtually all of these species, ongoing management of their habitats will be necessary to ensure their long-term survival. Wilcove and Chen (1998) further note that the longer the necessary management is delayed, the greater the risk of extinction of rare species and the greater the cost when the necessary management is finally performed. For example, *Tamarix*, an invasive woody plant, dominates riparian areas in the Southwestern US unless it is controlled via herbicides and cutting. In places where *Tamarix* has been allowed to grow for many years, the cost of removal can be as high as US$675 per acre in the first year, dropping below US$10 per acre in the second year. Subsequent maintenance requires an expenditure of under US$10 per acre every two to three years. We can think of endangered species management as having two cost components: an accrued debt reflecting a deferred maintenance problem that arises from inadequate management efforts in the past and an annual payment reflecting the necessary upkeep of properly managed habitats.

Scott *et al.* (2005) recommend that recovery be viewed as a continuum of states. At one extreme are the species that can survive in the wild with essentially no active management once key threats have been eliminated or enough habitat has been protected. At the other extreme are species that can persist in the wild, but only if people actively manage their habitats or control their competitors, predators, etc. A simple recovered/ not recovered dichotomy, as exists under the ESA, does not reflect the complexity of contemporary conservation.

## 12.4 Incentives and disincentives

Policy tools to conserve endangered species can be divided into two categories: incentives and disincentives. An example of an incentive would be a cash payment to a landowner for maintaining the habitat of an endangered species. An example of a disincentive would be a fine or jail sentence for harming an endangered species; this latter approach is the one taken by the ESA.

Conservationists have long debated the merits of the two approaches. Theoretically, with unlimited financial resources, it should be possible to protect and restore endangered species without incurring much opposition. Landowners or resource users who stand to lose money or opportunities due to restrictions on development could be "bought off" at whatever price they demand. It's an appealing scenario but also a deeply unrealistic one. Conservation programs are chronically under-funded. Moreover, at least in the US, some of the regions of the country with the highest concentrations of imperiled species are also regions with some of the highest real estate prices (e.g. San Francisco Bay region; Ando et al. 1998), a congruence that would quickly break the budget of any incentives program. Fines and jail sentences are thus used to deter developers from destroying the habitat of endangered golden-cheeked warblers (*Dendroica chrysoparia*) in the US or poachers from killing black rhinoceroses (*Diceros bicornis*) in many African countries. These types of laws, however, are effective only if they are enforced, i.e., if violators feel there is a non-trivial chance they will be caught and punished.

Unfortunately, penalties sometime force people to engage in activities that are counterproductive for conservation. Consider the case of the red-cockaded woodpecker (*Picoides borealis*). This woodpecker is restricted to mature, open pine forests in the southeastern US. A combination of residential development and short-rotation forestry resulted in the elimination of most of the old-growth pine forests in the Southeast and led the US Fish and Wildlife Service to place the woodpecker on the endangered list in 1970. This action ultimately resulted in protection of much of the woodpecker's remaining habitat. However, reports began to trickle in of landowners cutting down stands of young pine trees because they were afraid that red-cockaded woodpeckers would colonize their property if the trees got much older. The landowners knew that once the woodpeckers arrived, their ability to cut down the trees at a later date could be severely restricted; they reasoned that cutting the trees now would ensure the woodpeckers never arrived. Similar fears prevented some landowners from participating in recovery efforts for red-cockaded woodpeckers and other endangered species. Why go out of one's way to restore habitat for endangered species if doing so could result in restrictions on the use of one's property?

To remedy this situation, the federal government implemented a program known as "safe harbor" in 1995. Under this program, the government assures landowners who engage in voluntary activities that benefit endangered species that they will not incur additional regulatory restrictions as a result of their good deeds. In other words, a landowner who restores a part of her property to benefit an endangered species—and agrees to maintain the restored habitat for a certain period of time—will be given permission to undo those improvements (i.e. develop the property) at a later date, notwithstanding the fact that endangered species may now reside there. The reasoning is that without such assurances, the landowner would never engage in the beneficial action in the first place. In some cases, government agencies or private conservation organizations have provided financial assistance to landowners to cover some or all of the costs of habitat restoration. To date, landowners have enrolled over 1.5 million hectares in the safe harbor program (www.edf.org), benefiting a wide variety of endangered species, from Houston toads (*Bufo houstonensis*) to northern aplomado falcons (*Falco femoralis septentrionalis*) to Utah prairie dogs (*Cynomys parvidens*).

Fee-hunting is another interesting and controversial incentives program that has been used in parts of Africa to raise revenues and build local support for wildlife conservation. A limited number of licenses to hunt game animals are sold, with a portion of the revenues being returned to the local communities on whose land the hunting occurs. The goal of such programs is to give these communities an economic incentive to conserve wildlife, including animals such as lions (*Panthera leo*) and African elephants (*Loxodonta africana*) that can be harmful to crops or dangerous to people (Corn and Fletcher 1997).

Both disincentives and incentives play important roles in endangered species conservation. Disincentives are most useful in the protection phase as a means to discourage killing of endangered species or further destruction of their habitats. Incentives are useful in the recovery phase as a means to encourage landowners to restore habitats.

## 12.5 Limitations of endangered species programs

Many conservation biologists believe that a focus on endangered species is misplaced. They argue that the sheer number of species at risk makes a species-by-species approach impractical or even futile. Thus, conservation efforts would be more efficient and successful if they were focused at the level of whole ecosystems and landscapes, rather than individual species.

The US experience highlights the extreme difficulty of identifying and protecting even a fraction of a country's imperiled species, even when that country is wealthy. To date, only about 15% of the known species in the US have been studied in sufficient detail to determine their conservation status (i.e. which species are in danger of extinction). Embedded in this figure is a tremendous variance between groups, reflecting a predictable bias in favor of vertebrates. Thus, the status of almost 100% of the mammals, birds, reptiles, amphibians, and freshwater fishes is known; in contrast, fewer than 4% of invertebrate species have been assessed (Wilcove and Master 2005).

Among the species that have been assessed by experts, over 4800 are considered possibly extinct, critically imperiled, or imperiled; a strong case can be made that all of them merit federal protection under the ESA. Yet as of November 2008, less than a third of these species had been added to the federal endangered species list. Adding a species to the federal list is a time-consuming and often controversial process. Moreover, the US Fish and Wildlife Service is chronically under-funded and under-staffed. One can only imagine how much more difficult the situation must be in most of the developing countries in the tropics, where the total number of species at risk is far greater, yet resources for conservation are far fewer. Hence, it does seem reasonable to conclude that a species-by-species approach to conservation inevitably will leave many imperiled plants and animals unprotected and vulnerable to further losses.

Nonetheless, it would be dangerous to assume that endangered species conservation is a poor use of conservation resources. First, efforts to protect particular endangered species, especially those with large territories or home ranges (e.g. northern spotted owl, *Strix occidentalis caurina*), often result in de facto protection for other endangered species that share the same ecosystem. By choosing the right species to focus on, conservationists can improve the efficiency of their efforts. Second, many conservationists would argue that an essential goal of ecosystem or landscape conservation should be to protect all of the constituent species within that system, including the endangered ones. Moreover, certain ecosystems, such as the Florida scrub or Hawaiian rainforests, have such high concentrations of endangered species that there is little practical difference between conservation programs aimed at endangered species and those aimed at the ecosystem as a whole. Finally, and perhaps most importantly, endangered species have always enjoyed tremendous support from the public. Species such as the whooping crane (*Grus americana*), giant panda (*Ailuropoda melanoleuca*), golden lion tamarin (*Leontopithacus rosalia*), and black rhinoceros have inspired millions of people around the world to care about biodiversity. While it may be impossible to identify and protect each and every species that humanity has brought to the brink of extinction, there will always be many that we care deeply about and cannot afford to lose.

## Summary

- Endangered species conservation has three phases: identification, protection, and recovery.
- Protection can be directed toward species, subspecies, or populations. There are important economic and ecological trade-offs associated with protecting subspecies and populations.
- Consistent, quantitative criteria for determining the status of species have been developed by IUCN.
- Protection of endangered species requires accurate knowledge of the threats to those species, the location of existing populations, and land ownership patterns.
- Recovery of many endangered species will require continual, active management of the habitat or continual efforts to control populations of alien species.
- Incentives may be needed to entice people to participate in recovery programs.

## Suggested reading

Goble, D. D., Scott, J. M., and Davis, F. W. (2006). *The endangered species act at thirty*. Volume 1. Island Press, Washington, DC.
Wilcove, D. S. and Chen, L. Y. (1998). Management costs for endangered species. *Conservation Biology*, **12**, 1405–1407.
Wilcove, D. S., Rothstein, D., Dubow, J. A., *et al.* (1998). Quantifying threats to imperiled species in the United States. *BioScience*, **48**, 607–615.

## Relevant websites

- US Fish and Wildlife Service, Endangered Species Program: http://www.fws.gov/endangered/.
- A list of endangered species: http://www.iucnredlist.org/.

## REFERENCES

Ando, A., Camm, J., Polasky, S., and Solow, A. (1998). Species distributions, land values, and efficient conservation. *Science*, **279**, 2126–2128.
Baille, J. E. M., Hilton-Taylor, C., and Stuart, S. N., eds (2004). IUCN Red List of threatened species: a global species assessment. IUCN, Gland, Switzerland.
Clark, J. A., Hoekstra, J. M., Boersma, P. D., and Kareiva, P. (2002). Improving U.S. Endangered Species Act recovery plans: key findings and recommendations of the SCB recovery plan project. *Conservation Biology*, **16**, 1510–1519.
Corn, M. L. and Fletcher, S. R. (1997). African elephant issues: CITES and CAMPFIRE. Congressional Research Service Report 97–752 ENR. Available online at http://digital.library.unt.edu/govdocs/crs/permalink/meta-crs-388:1 (accessed 9 November 2008).
Dobson, A., Turner, W. R., and Wilcove, D. S. (2007). Conservation biology: unsolved problems and their policy implications. In R. May and A. McLean, eds. *Theoretical ecology: principles and applications*, pp. 172–189. Third edition. Oxford University Press, Oxford, UK.
Groves, C.R., Kutner, L. S., Stoms, D. M., *et al.* (2000). Owning up to our responsibilities. In B.A. Stein, L.S. Kutner, and J.S. Adams, eds *Precious heritage: the status of biodiversity in the United States*, pp. 275–300. Oxford University Press, Oxford, UK.
Hobbs, R. J. and Harris, J. A. (2001) Restoration ecology: repairing the earth's ecosystems in the new millennium. *Restoration Ecology*, **9**, 239–246.
Hoekstra, J. M., Clark, J. A., Fagan, W. F., and Boersma, P. D. (2002). A comprehensive review of Endangered Species Act recovery plans. *Ecological Applications*, **12**, 630–640.
Hughes, J. B., Daily, G. C., and Ehrlich, P. R. (1997). Population diversity: its extent and extinction. *Science*, **278**, 689–692.
Li, Y. and Wilcove, D. S. (2005). Threats to vertebrate species in China and the United States. *BioScience*, **55**, 147–153.
Schipper, J., Chanson, J. S., Chiozza, F. *et al.* (2008). The status of the world's land and marine mammals: diversity, threat, and knowledge. *Science*, **322**, 225–230.
Scott, J. M., Goble, D. D., Wiens, J. A., *et al.* (2005). Recovery of imperiled species under the Endangered Species Act: the need for a new approach. *Frontiers in Ecology and the Environment*, **3**, 383–389.
Simberloff, D. and Stiling, P. (1996). Risks of species introduced for biological control. *Biological Conservation*, **78**, 185–192.

Thomas, C.D., Cameron, A., Green, R. E., *et al.* (2004). Extinction risk from climate change. *Nature*, **427**, 145–148.

Wilcove, D. S. and Chen, L. Y. (1998). Management costs for endangered species. *Conservation Biology*, **12**, 1405–1407.

Wilcove, D. S. and Master, L. L. (2005). How many endangered species are there in the United States? *Frontiers in Ecology and the Environment*, **3**, 414–420.

Wilcove, D. S., McMillan, M., and Winston, K. C. (1993). What exactly is an endangered species? An analysis of the U.S. endangered species list: 1985–1991. *Conservation Biology*, **7**, 87–93.

Wilcove, D. S., Rothstein, D., Dubow, J. A., *et al.* (1998). Quantifying threats to imperiled species in the United States. *BioScience*, **48**, 607–615.

# CHAPTER 13

# Conservation in human-modified landscapes

Lian Pin Koh and Toby A. Gardner

In the previous two chapters, we learn about the importance and difficulties of prioritizing areas for conservation (Chapter 11), and the management of endangered species in these habitats (Chapter 12). In this chapter, we discuss the challenges of conserving biodiversity in degraded and modified landscapes with a focus on the tropical terrestrial biome, which is undergoing rapid deforestation and habitat degradation (Chapter 4) and contains an untold diversity of rare and endemic species that are in urgent need of conservation attention. We first highlight the extent to which human activities have modified natural ecosystems, and how these changes are fundamental in defining ongoing conservation efforts around the world. We then outline opportunities for conserving biodiversity within the dominant types of human land-use, including logged forests, agroforestry systems, monoculture plantations, agricultural lands, urban areas, and regenerating land. We also highlight the highly dynamic nature of modified landscapes and the need to recognize important human development benefits that can be derived from conservation action in these areas.

## 13.1 A history of human modification and the concept of "wild nature"

Efforts to improve human welfare have led to landscapes and ecosystems worldwide being domesticated to enhance food supplies and reduce exposure to natural dangers (Kareiva *et al.* 2007). As a consequence there are few places left on earth that have escaped some form of obvious human impact (see Chapter 4) that can have negative effects on biodiversity. This is especially so because human beings have released toxic synthetic organic chemicals, many of which are endocrine disrupters (Box 13.1), that are now distributed from pole to pole.

Although few data are available on changes to the extent and condition of many habitats, regions and ecosystems, what we do know is that, with few exceptions, changes that are currently underway are negative, anthropogenic in origin, ominously large and often accelerating (Balmford and Bond 2005). For example, the conversion of forests to agricultural land continues at a rate of approximately 13 million hectares per year, and the last global assessment classified a full two-thirds of the world's forests as having been modified by human impacts (FAO 2006).

Some ecologists have gone so far as to consider that the traditional concept of an intact ecosystem is obsolete, and instead propose a classification system based on global patterns of human interaction with ecosystems, demonstrating that much of the world currently exists in the form of different "anthropogenic biomes" (Figure 13.1 and Plate 15; Ellis and Ramankutty 2008). For many types of ecosystems, large areas of intact vegetation simply no longer exist, as is the case of the Atlantic forest hotspot of Brazil which has been reduced, except for a few conservation units, to a fragmented network of very small remnants (< 100 ha), mainly composed of secondary forest, and immersed in agricultural or urban matrices (Ribeiro *et al.* 2009).

Even when we turn to areas that at first appear to be undisturbed by human impact, the boundaries between "pristine" and "degraded" can

## Box 13.1 Endocrine disruption and biological diversity
### J. P. Myers

Since the beginning of the Industrial Revolution, over 80 000 new chemicals have entered commerce and hence the biosphere. These are compounds for which no organism has any evolutionary history and hence no opportunity to evolve over generations any metabolic protections against potential harm.

Depending upon how they are used and upon their chemical characteristics, they have dispersed widely, many globally. For example, whales feeding hundreds of feet beneath the surface of the mid-Atlantic accumulate brominated flame retardants from their prey. Bark of mature trees from virtually any forest in the world contains pesticides and industrial pollutants, even though they may be thousands of miles from the source. Penguins in the Antarctic store persistent organic pollutants that have been carried to the Antarctic by atmospheric transport and stored for decades in glacial snow but that are now being liberated by global warming. Seemingly pristine cloud forest in Costa Rica is more contaminated by the pesticides used on lowland banana plantations than forest adjacent to the bananas, because the pesticides volatilize in the lowland but are carried downwind and upward into the mountains, where they condense because of lower temperatures.

Decades of toxicological research focused on the effects of high exposures, which unquestionably can be serious, indeed directly lethal. Over the past 20 years, however, research has emerged revealing that this approach to toxicology was blind to serious effects that stem from the ability of some contaminants to interfere with hormones, altering gene expression, even at extremely low doses. These effects, deemed 'endocrine disruption' have forced toxicologists to rethink how they assess risk and have raised a wide array of questions about how contaminants may be affecting the biosphere in unexpected ways, since hormones regulate a wide array of biological functions in both plants and animals. Moreover, the signaling systems used by the endocrine system are highly conserved evolutionarily, operating in essentially the same ways in fish and mammals despite 300 million years of evolutionary separation. Hence the sudden and unprecedented arrival of hundreds, if not thousands, of chemicals capable of disrupting hormone action and novel to body chemistry is a source of concern.

Three key discoveries lie at the center of this revolution in toxicology. First, hormones – and contaminants that behave like hormones – can cause completely different effects at different levels of exposure. This is because the suite of genes up- or down-regulated by a hormone can vary dramatically as the concentration of the hormone varies. And at high levels, the hormone (or a hormone-like contaminant) can be overtly toxic, shutting down gene expression altogether. Hence all of the tests that toxicologists have run that assume high dose testing will catch low dose effects are invalid. Compounds judged to be safe based on data from high dose testing may not be. Some, widely used in commerce, clearly are not.

Second, changes in gene expression as an organism is developing—in the womb, as an egg, as a larvae or a tadpole, etc—can have lifelong consequences, affecting virtually every system of the body, including altering fertility, immune system function, neurological competency (and thus behavior), etc. Frogs in suburban Florida are less likely to be feminized than frogs in agricultural Florida, where endocrine-disrupting agricultural chemicals are used. Frogs exposed as tadpoles to a mixture of pesticides die from bacterial meningitis when adult, from a common bacteria easily resisted by control animals.

Third, individuals vary significantly in their capacity to metabolize these compounds and resist their effects. Specific variants of genes are more, or less, effective at safely metabolizing a contaminant and rendering it harmless. In people, for example, there is at least a 40-fold difference in capacity to metabolize organophosphate pesticides.

This is the stuff of Darwin...heritable differences among individuals that alter reproductive success...but it is happening to people and biodiversity at a pace that may be unprecedented in the history of most, if not all species. Hundreds, if not more, of compounds capable of altering gene expression at low levels

*continues*

### Box 13.1 (Continued)

of exposure have been introduced into the biosphere in fewer than 200 years. They alter fertility, cognition, immune and cardiovascular function, and more. The inescapable prediction, clearly speculative but highly plausible, is that this past 200 years has been a period of remarkable, if not unprecedented speed in the molecular evolution of life on earth.

Documented effects extend to interactions among species as well. For example, several environmental estrogens decrease the efficacy of communication between *Rhizobium* bacteria and their leguminaceous hosts, reducing nitrogen fixation. One widely used herbicide, atrazine, both increases the likelihood that ponds will contain large numbers of trematode parasites, which cause limb deformities in frogs, it also undermines the frog's immune defenses against trematode infections.

These emerging discoveries have come as surprises to traditional toxicology, because they raise questions about many chemicals in common use that based on traditional approaches had been deemed safe. For conservation biologists, they offer competing hypotheses to test against other interpretations. For example, is the disappearance of the golden toad (*Bufo periglenes*) from Costa Rica a result of global warming? Or have the pesticides now known to be present in significant concentrations in Costa Rican cloud forests undermined their viability? What is the role of contaminant-reduced immune system function in fungal-caused deaths in frogs, clearly an important factor in amphibian extinctions? Is the chytrid fungus new? Or are frogs less able to withstand infestation? Was the lake trout extinction in the Great Lakes the result of lampreys and over-fishing, or because dioxin sediment loads became so heavy that 100% of fry died? Have impairments by endocrine disrupters in the ability of young salmon to switch their osmoregulation from fresh water to salt water when they reach the ocean in their first downstream migration contributed to salmon population declines along the Pacific coast? Are declines in Chesapeake Bay oysters and crabs a result of invertebrate vulnerability to endocrine-disrupting contaminants? Is the relationship between coral and their symbiotic algae disrupted by contamination? Does this contribute to coral bleaching?

In the most elegant experimental field test to date of population-level effects of endocrine disruptors, Kidd *et al.* (2007) contaminated a lake in western Ontario with an active ingredient of birth control pills (17alpha-ethynylestradiol), maintaining the contaminant's concentration at 5–6 parts per trillion for two years. This concentration is just above levels typically found in sewage effluent and also in surface waters. The treatment led initially to delayed sexual development of fathead minnows in the lake. By the second year they observed that some males had eggs in their testes (ova-testis). And by the end of the seventh year, long after the treatments were halted, very few individuals were left. The population had crashed. There are many reports of ova-testis in fresh water fish populations from around the world.

How large a role endocrine disruption plays in biodiversity declines isn't yet clear, because few conservation biologists have included these mechanisms in the suite of hypotheses their studies are designed to test. The solutions to biodiversity declines caused by endocrine disruption will contrast sharply with those from more conventional forces. No harvest zones and artificial reefs, for example, will prove futile if shellfish declines are caused by chemical contamination. Hence in the search for tools to maintain biodiversity, it is imperative that conservation biologists' science widens to incorporate these effects.

### Relevant website

- Synopses of new studies on endocrine disruption: http://tinyurl.com/a6puq7.

### REFERENCE AND SUGGESTED READING

Colborn, T., Dumanoski, D., and Myers, J. P. (1996). *Our stolen future*. Dutton, New York, NY.

Cook, P. M., Robbins, J. A., Endicott, D. D., *et al.* (2003). Effects of aryl hydrocarbon receptor-mediated early life stage toxicity on Lake Trout populations in Lake Ontario during the 20th century. *Environmental Science and Technology*, **37**, 3864–3877.

Dally, G. L., Lei, Y. D., Teixeira, C., *et al.* (2007). Accumulation of current-use pesticides in Neotropical montane

*continues*

> **Box 13.1 (Continued)**
>
> forests. *Environmental Science and Technology*, **41**, 1118–1123.
>
> Gore, A. C. (2007). Introduction to endocrine-disrupting chemicals. In A.C. Gore, ed. *Endocrine-disrupting chemicals: From basic research to clinical practice*, pp. 3–8. Humana Press, New Jersey.
>
> Kidd, K. A., Blanchfield, P. J., Mills, K. H., *et al.* (2007). Collapse of a fish population after exposure to a synthetic estrogen. *Proceedings of the National Academy of Science of the United States of America*, **104**, 8897–8901.
>
> Welshons, W. V., Nagel, S. C., and vom Saal, F. S. (2006). Large effects from small exposures: III. Endocrine mechanisms mediating effects of bisphenol A at levels of human exposure. *Endocrinology*, **147**, S56–S69.

quickly become blurred on closer inspection. Archaeological and paleoecological studies over the last two decades suggest that many contemporary pristine habitats have in fact undergone some form of human disturbance in the past (Figure 13.2 and Plate 16; Willis *et al.* 2005; Willis and Birks 2006; see Chapter 14).

For example, the Upper Xingu region of Brazil comprises one of the largest contiguous tracts of tropical rainforest in the Amazon today. Emerging archaeological evidence suggests that parts of this region had been densely populated with pre-European human settlements (circa ∼1250 to ∼1600 A.D.), and that extensive forests underwent large-scale transformation to agricultural areas and urbanized centres (Heckenberger *et al.* 2003; Willis *et al.* 2004). Much of the lowland rainforests of the Congo basin had similarly experienced extensive human habitation, forest clearance, and agricultural activities between ∼3000 and ∼1600 years ago, as evidenced by extensive finds of stone tools, oil palm nuts, charcoal horizons (subsoil layers of charcoal), banana phytoliths (silica bodies found in plants preserved in sediments), and pottery fragments (Mbida *et al.* 2000; White 2001). Many further examples of extensive pre-European disturbance have been found in areas that conservationists today frequently describe as "pristine" or "intact", including Southeast Asia, Papua New Guinea and Central America (Willis *et al.* 2004).

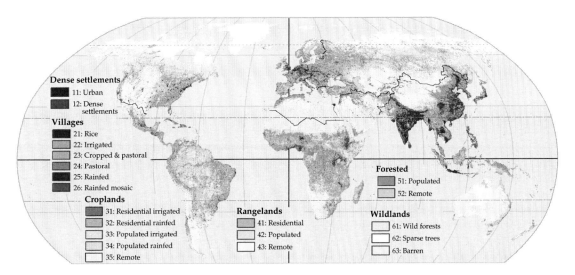

**Figure 13.1** Anthropogenic biomes. Global land-cover analysis reveals that that less than a quarter of the Earth's ice-free land can still be considered as wild. Biomes displayed on the map are organized into groups and are ranked according to human population density. Reprinted from Ellis and Ramankutty (2008).

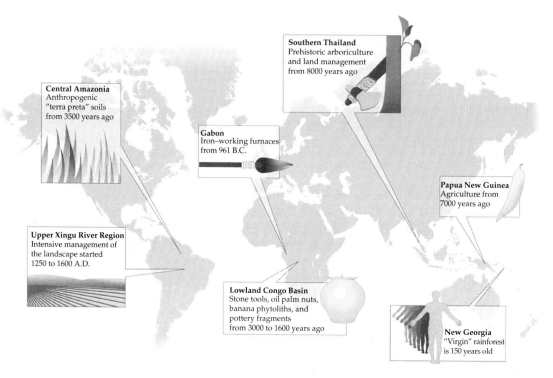

**Figure 13.2** Evidence of human modification of "pristine" tropical rainforest. Archaeological and paleoecological studies suggest that rainforests in the Amazon basin, the Congo basin, and Southeast Asia have regenerated from disturbance by prehistoric human settlements. Reprinted from Willis *et al.* (2004) with permission from AAAS (American Association for the Advancement of Science).

In most of these cases, forest regeneration followed the abandonment of human settlements and agricultural activities resulting in the old-growth stands that are regarded as pristine today.

## 13.2 Conservation in a human-modified world

How does all this evidence of historical and ongoing human modification of the natural world relate to efforts to conserve biological diversity today? There are at least two very profound implications.

First, the sheer extent to which we have dominated the biosphere (terrestrial, freshwater, and marine) (Ehrlich and Ehrlich 2008) means that we have no choice but to integrate conservation efforts with other human activities. It is broadly accepted that strictly protected areas provide a necessary yet grossly inadequate component of a broader strategy to safeguard the future of the world's biota. Gap analyses show that approximately one quarter of the world's threatened species live outside protected areas (Rodrigues *et al.* 2004; Chapter 11), and that most of the world's terrestrial ecoregions fall significantly short of the 10% protection target proposed by the IUCN (Figure 13.3 and Plate 17; Schmitt *et al.* 2009). Even where they exist, the integrity of protected areas is often threatened by encroachment and illegal extraction in areas that are undergoing widespread deforestation (Pedlowski *et al.* 2005), and management of neighboring areas is vital to ensuring their long-term viability (Wittemyer *et al.* 2008; Sodhi *et al.* 2008).

Second, evidence of historical recovery in areas that once hosted high levels of human activity illustrates that while long-time scales are often involved, the biotic impacts of many types of disturbance might not be completely irreversible.

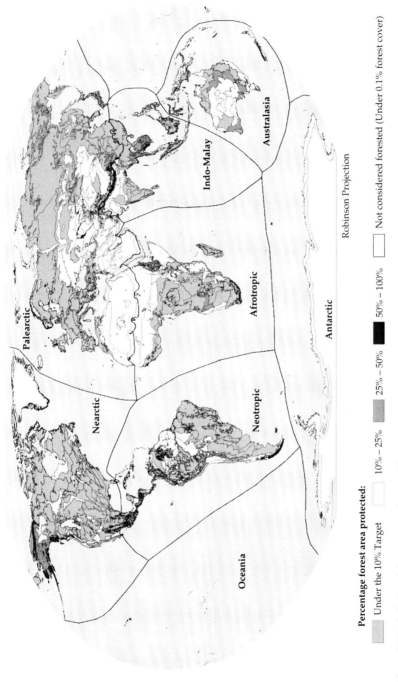

**Figure 13.3** Distribution of the percentage of protected forest area within WWF ecoregions. The highest levels of protection can be seen in parts of Australia, the Amazon, Southeast Asia, and Alaska. Notable areas of low protection include the Congo Basin in Central Africa and Northern Boreal forests. Black lines indicate biogeographic realms. White areas indicate no forest cover. Reprinted from Schmitt et al. (2009).

It is clear therefore, that partially modified landscapes are an important and valuable asset for biodiversity conservation, and should not be overlooked by biologists and conservationists, and abandoned to yet further levels of intensification.

Against this backdrop of necessity and hope, it is self-evident that the future of much of the world's biodiversity depends on the effective management of human-modified systems (Daily 2001; Lindenmayer and Franklin 2002; Bawa *et al.* 2004). To face up to this challenge conservation biology needs to adopt a research perspective that incorporates human activities as integral components of ecosystems, and place a strong emphasis on understanding the coupled social-ecological dynamics of modified lands (Palmer *et al.* 2004; Sayer and Maginnis 2005).

Ultimately conservation biologists need to improve their understanding of how different types of human land-use may confer different benefits for conservation. To what extent can modified land-uses support viable populations of native species, and help ensure the long-term viability of isolated remnants of undisturbed vegetation? Understanding which native species can maintain viable populations in modified landscapes, and under what management regimes, is one of the greatest challenges currently facing conservation biologists (Fischer and Lindenmayer 2007; Sekercioglu *et al.*, 2007; Sodhi 2008; Chazdon *et al.*2009a). While it is generally accepted that the conversion of primary habitat for intensive agriculture inevitably leads to dramatic losses in biodiversity (Donald 2004; Sodhi *et al.* 2009), more information is certainly needed. Conservation biologists are particularly uncertain of the extent to which more structurally and floristically complex land-uses such as secondary and agroforests can conserve native biotas (e.g. Dunn 2004; Gardner *et al.* 2007), although mixed agricultural landscapes can be more hospitable to forest birds than once suspected (Daily *et al.* 2001; Ranganathan *et al.* 2008). In the rest of this chapter we briefly outline the biodiversity prospects that exist within different land-use systems, focusing in particular on forested landscapes in the tropics.

## 13.3 Selectively logged forests

As of 2005, approximately one third of the world's forests—a total of 1.3 billion hectares—were designated primarily for timber production (FAO 2006). In 2006, member nations of the International Tropical Timber Organization (ITTO) exported over 13 million cubic meters of tropical non-coniferous logs worth US$2.1 billion, making a substantial contribution to the economies of these nations (ITTO 2007). Logging activity on this massive scale has resulted in huge areas of forest being degraded following the selective removal of high-value trees, and the collateral damage associated with tree felling and extraction. Asner *et al.* (2005) estimated that in the Brazilian Amazon between 1999 and 2002 the area of rainforest annually degraded by logging is approximately the same as that which is clear-felled for agriculture (between 12 and 19 million hecatres).

Although all logging activity has a negative impact on the structure and composition of the forest, the severity of this impact depends on the logging intensity, including the number of trees removed per ha, length of the rotation time, and site management practices. The density of felled trees varies among regions and management regimes from as few as one tree every several has (e.g. mahogany, *Swietenia macrophylla* in South America) to more than 15/ha in lowland dipterocarp forests of Southeast Asia (Fimbel *et al.* 2001). In the last few decades Reduced Impact Logging (RIL) techniques have been developed that involve careful planning and controlled harvesting (e.g. preliminary inventories, road planning, directional felling) to greatly minimize deleterious impacts (Fimbel *et al.* 2001; Putz *et al.* 2008).

Differences in how forests are managed determine the extent to which logging negatively affects wildlife, with impacts felt through changes to the structure and composition of the forest environment, including alterations in tree size structure, a shift towards early successional vegetation, changes in composition of fruiting trees, fragmentation of the canopy, soil compaction, and alteration of aquatic environments. In general, broad patterns of wildlife response can be explained by differences in the intensity of logging activity as well as the amount of recovery time elapsed before a study was conducted (Putz *et al.* 2001).

While there is no available evidence of any species having been driven extinct by selective logging there are abundant data showing marked population declines and local extinctions in a wide range of species groups (Fimbel *et al.* 2001; Meijaard and Sheil 2008). Arboreal vertebrates appear to be particularly badly affected through the loss of nesting and food resources. Both Thiollay (1995) and Sekercioglu (2002) reported losses of approximately 30% of forest dependent birds from logged areas in Sumatra and Uganda, respectively. Felton *et al.* (2003) reported depleted numbers of adult orangutans (*Pongo borneo*) in selectively logged peat forest in Kalimantan, Borneo, compared to neighboring intact sites. Bats also appear to be especially sensitive to even low levels of logging as changes in canopy cover and understory foliage density have knock-on effects on foraging and echolocation strategies (e.g. Peters *et al.* 2006).

Nevertheless, for many taxa the impacts of selective logging are far less severe, even under conventional management regimes. For example, Lewis (2001) found that logging at a density of six stems per hectare had little effect on the diversity and structure of butterfly assemblages in Belize, while Meijaard and Sheil (2008) concluded that only a few terrestrial mammal species have shown marked population declines following logging in Borneo. These studies suggest that different species groups exhibit significantly different responses to logging impacts depending on their life-history strategies and resource requirements. Within any one group it is invariably the forest dependent and specialist species that decline, while generalist and omnivorous species are unaffected or even increase in abundance and diversity.

For most of the world we lack detailed information on the extent to which specific management practices can enhance levels of biodiversity in managed natural forests. Nevertheless, many best practice general guidelines do exist, which, if implemented more broadly, could greatly improve the value of logged forests for wildlife (Fimbel *et al.* 2001; Lindenmayer *et al.* 2006; Meijaard and Sheil 2008). These guidelines include stand-level practices such as the retention of structural complexity (including dead wood), long-rotation times, maintenance of canopy cover, and fire control and timber removal techniques. In addition many landscape scale measures can greatly improve the value of logged forests for conservation, including the designation of no-take areas, careful road design and maintenance of landscape connectivity with intact corridors and riparian buffers (Gillies and St Clair 2008).

More work is urgently needed to prescribe strategies for effective biodiversity conservation in managed forests. Despite receiving criticism from conservation biologists on the adequacy of criteria to support conservation, timber certification authorities such as the Forest Stewardship Council (www.fsc.org) offer a promising approach to improving the responsibility of forest management standards.

## 13.4 Agroforestry systems

Agroforestry is a summary term for practices that involve the integration of trees and other woody perennials into crop farming systems through the conservation of existing trees, their active planting and tending, or the tolerance of natural regeneration in fallow areas (Schroth *et al.* 2004). Its main purpose is to diversify production for increased social, economic and environmental benefits, and has attracted increasing attention from scientists working at the interface between integrated natural resource management and biodiversity conservation, especially in tropical countries (Schroth *et al.* 2004; Scherr and McNeely 2007). Farmers in many traditional agricultural systems have maintained or actively included trees as parts of the landscape for thousands of years to provide benefits such as shade, shelter, animal and human food (McNeely 2004).

Although many different definitions exist to define different agroforestry systems, here we highlight two broad categories; complex agroforestry and home-gardens (Scales and Marsden

2008). Complex agroforestry is an extension of the swidden agriculture system where tree seedlings are co-planted with annual crops and left in fallow (e.g. rattan), or maintained in an annual-perennial association (e.g. damar-coffee). After 25–50 years the trees are felled and the cycle is repeated. Home-gardens are small areas of agricultural land located near to houses that are cultivated with a mixture of annuals and perennials, including trees and shrubs. They are semi-permanent and typically more intensively managed than complex agroforests. Because of their high levels of floristic diversity and complex vegetation, agroforests represent a mid-point in forest structural integrity between monoculture plantations and primary forest (Figure 13.4; Schroth and Harvey 2007).

Agroforestry can benefit biodiversity conservation in three ways; the provision of suitable habitat for forest species in areas that have suffered significant historical deforestation, the provision of a landscape matrix that permits the movement of species among forest remnants, and the provision of livelihoods for local people which may in turn relieve pressure on remaining areas of primary forest (see also Chapter 14). In areas of the tropics that have lost the majority of old-growth forest the dominant near-forest vegetation is frequently comprised of some form of agro-forestry, highlighting the importance of these systems for conservation in some regions, including shade-coffee in Central America, shade-cacao in the Atlantic Forest of Brazil, jungle rubber in the Sumatran lowlands, and home-gardens in countries across the world.

The majority of studies that have examined the biodiversity value of agroforestry systems have found that although some species are invariably lost following conversion of native habitat, a large proportion of the original fauna and flora is maintained when compared to more intensified agricultural land-uses (Ranganathan *et al.* 2008). In reviewing the results of 36 studies Bhagwat *et al.* (2008) found that agroforestry systems consistently hosted more than two-thirds of the species found in reserves, while patterns of similarity in species composition between agroforest plots and areas of native forest ranged from 25% (herbaceous plants) to 65% (mammals). Although existing studies have not revealed any clear pattern regarding which groups of species are unlikely to be conserved within agroforestry systems, it appears that rare and range-restricted species are often those that suffer the greatest declines following forest conversion, while those that increase in abundance are often open-habitat and generalist taxa (Scales and Marsden 2008). However, even species that are usually only found in areas of native vegetation may use agroforests to move between forest remnants, as is the case for two species of sloth in Costa Rica that frequently use shade-cacao plantations as a source of food and resting sites (Vaughan *et al.* 2007).

Differences in the amount of biodiversity that is retained in different agroforestry systems can often be explained by differences in the intensity of past and present management regimes (Bhagwat *et al.* 2008). For example, the effect of

**Figure 13.4** Shade-coffee plantation in the Western Ghats, India. Photograph by M. O. Anand.

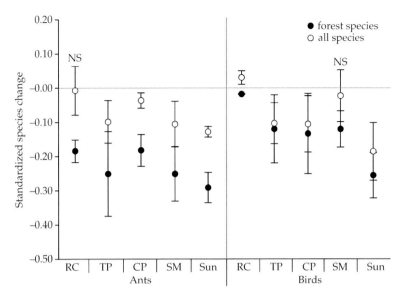

**Figure 13.5** Standardized change in species richness for ants and birds in coffee sites compared with nearby forests from 18 datasets in the Neotropics. Error bars are bootstrapped 95% CIs (Confidence Intervals). Points below zero show species loss relative to forests, and points above zero show significant increases in species richness compared with forests. Error bars that do not overlap zero show significantly higher or lower richness in coffee habitats compared with forests (NS, points not significantly different from zero). Habitat abbreviations: RC, rustic coffee; TP, traditional polyculture coffee; CP, commercial polyculture coffee; SM, shade monoculture coffee; Sun, sun coffee. Reprinted from Philpott et al. (2008).

management intensification on biodiversity is clearly demonstrated by the marked loss of forest species following the simplification of shade-coffee plantations and a decrease in the density and diversity of shade trees (Figure 13.5; reviewed by Philpott et al. 2008).

Despite the potential value of agroforestry systems for biodiversity, it is important to recognize key limitations in their contribution towards long-term conservation strategies. First, the ability of agroforestry systems to maintain a significant proportion of the regional biota depends on the maintenance of sufficient areas of natural habitat, both to support highly sensitive species (Schroth and Harvey 2007) and to provide source populations (Anand et al. 2008). By encompassing sufficient areas of native forest within an agroforestry landscape it is possible to ensure the persistence of a large number of species for very long time periods, as recently demonstrated by Ranganathan et al. (2008) who reported the presence of more than 90% of the regional forest avifauna in arecanut (*Areca catechu*) production systems that have been cultivated for more than 2000 years in the Western Ghats, India. Second, appropriate regulations on hunting and resource extraction are vital to ensure that keystone vertebrate and plant species are not depleted from otherwise diverse systems. Finally, and most importantly, agroforestry systems can only survive with the support of market incentives and favorable land-use policies that maintain viable livelihoods of local people, and prevent conversion to more intensified land-uses (Steffan-Dewenter et al. 2007).

## 13.5 Tree plantations

As for agroforestry systems, tree plantations have the potential to make an important contribution to biodiversity conservation for two key reasons: (i) they may more closely reflect the structural complexity of native forest than many more intensive production land-uses; and (ii) they occupy a large area of once-forested land in many parts of the world. The total area of the plantation forest estate in 2005 was about 109 million hectares, and is continuing to increase by approximately 2.5 million hectares per year (FAO 2006). In the tropics alone, the total coverage of plantation forestry increased from approximately 17 million hectares in 1980 to 70 million hectares in 2000 (FAO 2006). As demands for timber and wood fiber continue to increase around the world, it is highly likely that these upward trends will persist or even accelerate.

Many tree plantations have been traditionally labeled as "green deserts", and are presumed or found to be hostile to native species and largely devoid of wildlife (Kanowski et al. 2005; Sodhi et al. 2009). However, closer inspection of available data indicates that while it is certainly true that some intensively managed plantation monocultures offer very little value to biodiversity (e.g. oil palm in Southeast Asia; Koh and Wilcove 2007, 2008, 2009; Koh 2008a, b), other plantation systems may provide valuable species habitat, even for some threatened and endangered taxa (Hartley 2002; Carnus et al. 2006). This apparent contradiction is explained in part by marked differences in the levels of biodiversity that can be supported by different types of plantation. For example there is a stark contrast in the conservation value of industrial monocultures of exotic species that often have little or no intrinsic value for native forest species, compared with complex multi-species plantations that encompass remnants of native vegetation and are managed as a mosaic of differently aged stands (Hartley 2002; Lindenmayer and Hobbs 2004; Kanowski et al. 2005). However, a second reason why many plantations are incorrectly presumed to be biological deserts is that human perceptions of habitat quality are often distinct from how native species themselves perceive the landscape (Lindenmayer et al. 2003). Although few comprehensive and robust field studies have been conducted to examine the conservation value of plantations, those that exist suggest that under certain conditions the numbers of species inhabiting these areas may be greater than expected. For example, a very thorough study in north-east Brazilian Amazonia found that *Eucalyptus* plantations contained nearly half of the regional forest fauna, although it is very unlikely that all of these taxa could maintain viable populations in the absence of large areas of neighboring primary forest (Barlow et al. 2007; see Box 13.2).

The value of a given plantation forest for conservation is partly determined by how it is managed. For example, at the stand level, many studies have found that faunal diversity in tree plantations is strongly influenced by the maintenance of structural attributes such as snags and dead wood, and the tolerance of succession by native plant species in the understory (Hartley 2002). More floristically and structurally complex plantations provide more resources for many forest species (e.g. fruit feeding butterflies; Barlow et al. 2008). At the landscape scale, spatial heterogeneity in stand management and age has been shown to be a key factor in determining the overall level of diversity within a given plantation forest (Lindenmayer and Hobbs 2004; Lindenmayer et al. 2006).

However, the true conservation value of a plantation depends upon the comparison with alternative land-uses that may otherwise exist in its place (Kanowski et al. 2005; Brockerhoff et al. 2008). Clearly there is a net loss of biodiversity if plantations replace native forest. There is also a net loss of regional biodiversity if plantations are grown on areas of natural grassland, as seen in many areas of southern Africa. However, if plantations represent the "lesser evil" and prevent land from being converted to croplands or pasture, or have been grown on areas of degraded land, then their importance for biodiversity may be significant. In areas where very little native vegetation remains plantation forests may provide the last refuge for endemic species, such as the case of the critically endangered ground beetle (*Holcaspis brevicula*) in New Zealand which is only known from *Pinus* plantations (Brockerhoff et al. 2005).

Ultimately, the extent to which plantations can be managed to enhance biodiversity depends upon the level of economic cost incurred by responsible management strategies, and the availability of market incentives to offset such costs. Some minor improvements in management technique may generate some conservation benefits with little loss in productivity (Hartley 2002) but our knowledge of the economic-conservation trade-offs implicit in major changes to stand and

**Box 13.2 Quantifying the biodiversity value of tropical secondary forests and exotic tree plantations**
**Jos Barlow**

Ecologists and conservation scientists have found it difficult to make an accurate assessment of the conservation value of secondary and plantation forests in the tropics. Many studies have been conducted in small forest blocks, and may be influenced by the presence of transient species moving between patches of adjacent old-growth forest. Furthermore, studies are often beset by a variety of methodological shortcomings. As a result, there is little consistency in their results, and studies may systematically overestimate the conservation value of non-primary forests (Gardner et al. 2007).

Many of these potential methodological shortcomings were addressed by a recent comprehensive study that utilized a quasi-experimental landscape mosaic that resulted from a large-scale attempt to implement fast-growing tree monocultures in the Brazilian Amazon in the 1970s. In 2004, a large international team of researchers attempted to quantify the biodiversity that persists in primary forests, 4–5 year old Eucalyptus plantations and 14–19 year old native second-growth (Barlow et al. 2007). They sampled 15 different groups of biodiversity, including most of the terrestrial vertebrates, a wide range of invertebrates, and

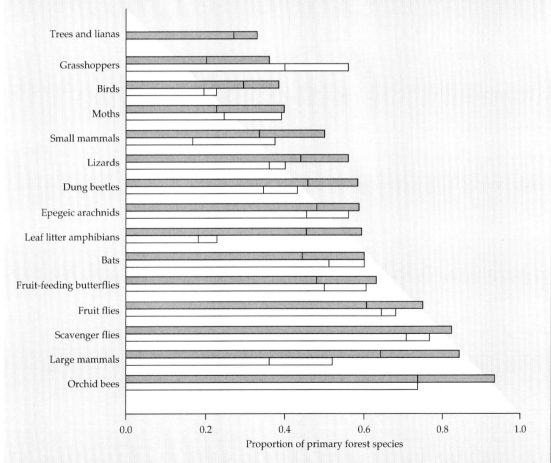

**Box 13.2 Figure** The proportion of primary forest species that were also recorded in 14-19 year old native second growth (grey bars) and 4–6 year old *Eucalyptus* plantations (white bars) in the Brazilian Amazon. The bars are split by a line that indicates the decrease in the proportion of primary forest species when occasional species (those that were recorded only once in each of the non-primary forests) are removed from the comparison.

*continues*

> **Box 13.2 (Continued)**
>
> the trees and lianas (see Box 13.2 Figure). The researchers spent >18 200 person hours collecting specimens in the field and identifying them in the laboratory, and recorded 61 325 individuals and identified 1442 species.
>
> Their results provide a clear message regarding the unique value of primary or old-growth forests. Averaging across all taxa, secondary forests and Eucalyptus plantations held only 59% and 47% of the species that were recorded in the old-growth forests, respectively. These results should be interpreted as a best case scenario, as the wider landscape was dominated by old-growth forests, maximizing recolonization opportunities for primary forest species. Furthermore, many primary forest species were recorded just once within the non-primary habitats, and the presence of single individuals is unlikely to represent a species ability to persist in these regenerating forests. Removing these occasional species from the results reduces the estimated value of non-primary habitats for most taxa (Box 13.2 Figure) to an average of 46% of species for second-growth and 39% for plantations.
>
> This research was unique as it allows us to make a robust comparison between the responses of different taxa across the same land-use gradient. This shows that the estimated value of non-primary forests is much higher for highly mobile taxa such as orchid bees, large mammals, and bats (see Box 13.2 Figure), which include many mobile species that fly tens of kilometers each day, and perceive landscape and habitat quality at a very large-scale. There was also a marked difference among taxa in the kinds of species that come to dominate these non-primary forests. For example, more than 60% of the species of birds, grasshoppers and moths that were recorded in secondary forests were never recorded in old-growth forests. These taxa contrast with the orchid bees, fruit flies and large mammals, for which most of the species recorded in secondary forests (more than 75%) were also recorded in primary forests. These data illustrate an important point about the consequences of land-use change; the species persisting in anthropogenic habitats can be either composed of a subset of the species found in primary forests, or like birds, they may be wide-ranging generalists that have invaded from open habitats, riparian vegetation, and even urban areas.
>
> **REFERENCES AND SUGGESTED READING**
>
> Barlow, J., Gardner, T. A., Araujo, I. S., et al. (2007). Quantifying the biodiversity value of tropical primary, secondary and plantation forests. *Proceedings of the National Academy of Science of the United States of America*, **104**, 18555–18560.
>
> Gardner, T. A., Barlow, J., Parry, L. T. W., and Peres, C. A. (2007). Predicting the Uncertain Future of Tropical Forest Species in a Data Vacuum. *Biotropica*, **39**, 25–30.

landscape management regimes is poor. An alternative to more ecologically sensitive management within individual plantations which also deserves further research attention is to adopt a land sparing approach, where intensified silviculture in one area generates sufficient revenue to "spare" other lands for conservation (e.g. Cyranoski 2007; see next section).

## 13.6 Agricultural land

The human population is expected to increase from 6 billion today to 8–10 billion by 2050 (Cohen 2003). Global demand for agricultural products is predicted to grow even faster due to rising demand for food and higher quality food (e.g. meat), as well as for bioenergy crops used in biofuel production (UN 2005; Scherr and McNeely 2008). It has been estimated that feeding a population of 9 billion people would require the conversion of another billion hectares of natural habitats to croplands (Tilman et al. 2001), which will almost certainly increase the risks of extinction already faced by numerous species worldwide (see Boxes 13.3 and Introduction Box 1).

## Box 13.3 Conservation in the face of oil palm expansion
### Matthew Struebig, Ben Phalan, and Emily Fitzherbert

The African oil palm (*Elaeis guineensis*) is one of the world's most rapidly expanding crops, and has the highest yields and largest market share of all oil crops. While cultivation has historically focused in Malaysia and Indonesia, oil palm is increasingly grown across the lowlands of other countries in Southeast Asia, Latin America and Central Africa. Expansion is driven by large companies and smallholders responding to global demand for vegetable oil (mainly from Indonesia, India and China), and the growing biofuel markets of the European Union. With high demand, and strong overlap between areas suitable for oil palm and those of endemic-rich tropical forests, expansion poses an increasing threat to biodiversity.

The few studies available show that oil palm is a poor substitute habitat for the majority of tropical forest species, particularly those of conservation concern. On average only 15% of species recorded in primary forest are found in oil palm plantations (Box 13.3 Figure), even fewer than in most other tree crops. Plantation assemblages are typically dominated by a few abundant generalists (e.g. macaques), alien invasives (e.g. crazy ants), pests (e.g. rats), and their predators (e.g. pythons). Oil palm is a major contributor to deforestation in a few countries, although its role is sometimes obscured by ambiguous land-tenure laws and its links with other enterprises (e.g. timber profits are used to offset plantation establishment costs in Indonesia).

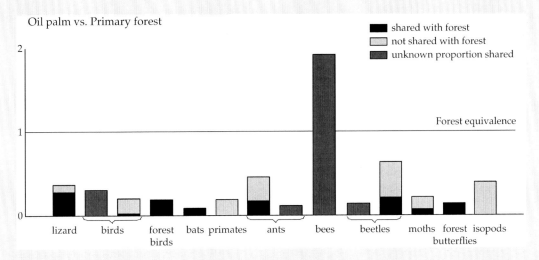

**Box 13.3 Figure** The biodiversity impact of converting forests to plantations is shown by comparing species richness in oil palm relative to primary forest. The species richness of oil palm is presented as a proportion of forest richness such that equal species richness is 1. Each column contains a study of one taxon and shows the proportion of oil palm species shared and those not shared with forest. One study of bees found fewer species in forests than oil palm, but might have underestimated forest species richness because the canopy was not sampled.

In response to consumer concerns about deforestation, the Roundtable for Sustainable Palm Oil (RSPO) was formed from industry-NGO (non-governmental organization) collaboration in 2003. Under this scheme members commit to environmental and social standards for responsible palm oil production, including an assurance that no forests of High

*continues*

> **Box 13.3 (Continued)**
>
> Conservation Value (http://www.hcvnetwork.org/) will be cleared for plantations. However, certification is not yet a panacea. Unless land planning is expanded to regional assessments, biodiversity losses outside of RSPO-member plantations will continue; certification risks remaining a niche market, with mainly older plantations exporting to responsible buyers, while demand from others is filled by newer plantations pushing into forests.
>
> Conservation science is needed to inform oil palm policies, but it is not enough to understand only the biodiversity impacts of plantations. The real challenge is for conservation scientists to translate their findings into better land planning and forest protection strategies, whilst accounting for social, economic and political realities.
>
> **SUGGESTED READING**
>
> Fitzherbert, E. B., Struebig, M. J., Morel, A., et al. (2008). How will oil palm expansion affect biodiversity? *Trends in Ecology and Evolution*, **23**, 538–545.
>
> Koh, L. P. and Wilcove, D. S. (2008). Is oil palm agriculture really destroying tropical biodiversity? *Conservation Letters*, **1**, 60–64.

What can conservation biologists do to mitigate the threat from agricultural expansion? This problem has traditionally been framed as a zero-sum game—agricultural production will take away land that would otherwise be used for biodiversity conservation, and vice versa. More recently however, researchers have suggested that "countryside biogeography" (also known as "win-win ecology" or "reconciliation ecology") should be a key consideration in practical conservation (Dale *et al.* 2000; Daily *et al.* 2001; Miller and Hobbs 2002; Daily 2003; Rosenzweig 2003).

Proponents of countryside biogeography argue that because a large proportion of the planet is already dominated by humans and what little remains of pristine habitats will not be sufficient for the long-term survival of many species, conservation planning should include mitigation measures that enable human activities to proceed with minimum displacement of native species (Rosenzweig 2003, see Box 13.4).

In the context of agricultural expansion, it is often the case that after natural habitats have been converted, what remains is an agricultural mosaic—forest fragments in a matrix of production systems (Vandermeer and Perfecto 2007). Both theoretical and empirical ecological research over the past decade has shown that species survival in such fragmented landscapes depends on the size and isolation of fragments, as well as the permeability of the intervening matrix to the movement of organisms (Hanski 1999; Stratford and Stouffer 1999; Vandermeer and Carvajal 2001; Perfecto and Vandermeer 2002; Chapter 5).

To enhance the survivability of native species in an agricultural mosaic, two approaches may be pursued. The first approach is to intensify agricultural production to increase overall yield while avoiding further cropland expansion and deforestation (Balmford *et al.* 2005; Green *et al.* 2005). This "land sparing" approach, though conceptually straightforward, remains controversial among the conservation community. Critics have argued that the ecological impacts of intensive farming often extend over a wider area than the land so farmed (Matson and Vitousek 2006). Intensive farming would require more irrigation, and fertilizer and pesticide inputs, which would divert water away from downstream ecosystems and species, and result in greater pollution. Furthermore, intensifying agricultural production could lead to extensive land use by displacing people to other forested areas or by providing the economic incentives for migration into the area (Matson and Vitousek 2006).

A second approach is to focus on improving the quality of the matrix to make it more hospitable for habitat generalist species that are able to utilize it, and be less of a barrier to the migration

## Box 13.4 Countryside biogeography: harmonizing biodiversity and agriculture
### Jai Ranganathan and Gretchen C. Daily

With human impacts expected to intensify rapidly (e.g. Tilman et al. 2001), the future of biodiversity cannot be separated from the future of people. Although protected areas are central to conservation strategy, they alone are unlikely to ensure survival of more than a tiny fraction of Earth's biodiversity (e.g. Rosenweig 2003). Here we discuss the scope for expanding conservation strategy to include the countryside: active and fallow agricultural plots, gardens and pasture, plantation or managed forest, and remnants of native vegetation in landscapes otherwise devoted primarily to human activities (Daily et al. 2001). Little is known about the capacity of the countryside to support native species, particularly in the tropics, where the majority of the Earth's species are found (Wilson and Peter 1988).

We summarize information on the best-studied groups—birds, mammals, and insects—in well-studied systems in Mesoamerica. On the question of what fraction of native species can survive in countryside, the answer appears to be about 50% or more, though abundance of many species is low (Estrada et al. 1997; Daily et al. 2001; Daily et al. 2003; Horner-Devine et al. 2003). Three landscape characteristics stand out as important in conferring a survival advantage to native species in the countryside. First, species richness is considerably higher in the vicinity of large remnants of relatively intact forest, suggesting that many species that occur in the countryside can persist only in the nearby presence of that native forest (Estrada et al. 1997; Rickets et al. 2001; Perfecto and Vandermeer 2002; Sekercioglu et al. 2007). Second, the presence of native vegetation in human-dominated habitat (in the form of living fences, windbreaks, and remnant trees) facilitates persistence (Estrada et al. 1994; Estrada et al. 2000; Hughes et al. 2002; Harvey et al. 2004). Third, the intensity of agriculture in a landscape is negatively correlated with that landscape's conservation potential (Bignal and McCraken 1996; Green et al. 2005).

The question of which attributes of native species confer an advantage in the countryside has perhaps been best studied in birds, where a high population growth rate and the ability to disperse through open habitat greatly increases the chance of occurrence in the countryside (Sekercioglu et al. 2002; Pereira et al. 2004). Additionally, the conversion of forest to agriculture severely impacts forest-interior bird species; in one case the cause seemed to be a decrease in available nesting habitat (Lindell and Smith 2003).

It is uncertain if high levels of native diversity can be maintained over the long term (centuries to millennia), as almost all of the countryside under study has been under cultivation for less than a century (at least in recent centuries). A possible indication of the long-term prospects can be found within the Western Ghats mountain range, India, where high levels of bird diversity have been maintained in a low-intensity agricultural landscape, despite >2000 years of continuous agricultural use (Ranganathan et al. 2008, see Box 13.4 Figures 1 and 2). Though tentative, these results show that conservation investments in countryside may pay off for biodiversity in the long term.

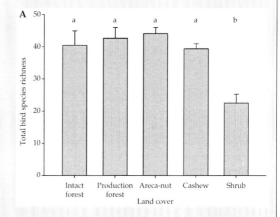

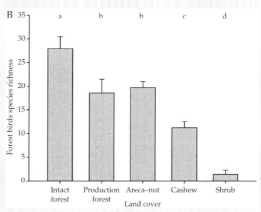

**Box 13.4 Figure 1** Patterns of bird species richness within an agricultural landscape on the fringes of the Western Ghats, India, where land use patterns help to maintain avian diversity (reprinted from Ranganathan et al. 2008). There are five major land covers in the landscape: forest (itself divided into relatively pristine "intact

*continues*

## Box 13.4 (Continued)

forest" and "production forest", within which the extraction of non-timber forest products is permitted), arecanut plantations, cashew plantations, shrubland, and rice/peanut farms. The last land cover was omitted from analysis due to the fact that they are seasonally devoid of vegetation and, thus, wildlife. With the exception of the depauperate shrublands, the land covers contained a similar richness of birds (A). However, when just birds associated with forest habitat ("forest species") are examined, much larger differences can be seen, with production forest and arecanut plantations second only to intact forest in richness (B). Thus, it can be seen that arecanut plantations are important for maintaining forest species across the landscape. Their importance is all the greater because the production forests serve primarily as a source of agricultural inputs for the arecanut plantations, thereby providing a powerful economic incentive to maintain those areas as forest. © National Academy of Sciences, USA.

**Box 13.4 Figure 2** Biodiversity of birds, and likely other taxa, is especially rich in the low-intensity agricultural landscapes on the fringes of the Western Ghats, India. Photograph by J. Ranganathan.

The time is ripe for developing and promoting best management practices for farmers—and, similarly, best conservation practices for conservation organizations—that integrate biodiversity and human well-being in meaningful, effective ways globally.

## REFERENCES

Bignal, E. M., and McCracken, D. I. (1996). Low-intensity farming systems in the conservation of the countryside. *Journal of Applied Ecology*, **33**, 413–424.

Daily, G. C., Ehrlich, P. R., and Sanchez-Azofeifa, G. A. (2001). Countryside biogeography: Use of human-dominated habitats by the avifauna of southern Costa Rica. *Ecological Applications*, **11**, 1–13.

Daily, G. C., Ceballos, G., Pacheco, J. Suzan, G., and Sanchez-Azofeifa, A. (2003). Countryside biogeography of neotropical mammals: Conservation opportunities in agricultural landscapes of Costa Rica. *Conservation Biology*, **17**, 1814–1826.

Estrada, A., Coates-Estrada, R., and Meritt, D. (1994). Non-flying Mammals and Landscape Changes in the Tropical Rain-Forest Region of Los-Tuxtlas, Mexico. *Ecography*, **17**, 229–241.

Estrada, A., Coates-Estrada, R., and Meritt, D. A. (1997). Anthropogenic landscape changes and avian diversity at Los Tuxtlas, Mexico. *Biodiversity and Conservation*, **6**, 19–43.

Estrada, A., Cammarano, P., and Coates-Estrada, R. (2000). Bird species richness in vegetation fences and in strips of residual rain forest vegetation at Los Tuxtlas, Mexico. *Biodiversity and Conservation*, **9**, 1399–1416.

Green, R. E., Cornell, S. J., Scharlemann, J. P. W., and Balmford, A. (2005). Farming and the fate of wild nature. *Science*, **307**, 550–555.

Harvey, C. A., Tucker, N. I. J., and Estrada, A. (2004). Live fences, isolated trees, and windbreaks: tools for conserving biodiversity in fragmented tropical landscapes. In G. Schroth, G. A. B. da Fonseca, C. A. Harvey, C. Gascon, H. L. Vasconcelos, and A.-M. N. Izac, eds *Agroforestry and biodiversity conservation in tropical landscapes*, pp. 261–289. Island Press, Washington, DC.

Horner-Devine, M. C., Daily, G. C., Ehrlich, P. R., and Boggs, C. L. (2003). Countryside biogeography of tropical butterflies. *Conservation Biology*, **17**, 168–177.

Hughes, J. B., Daily, G. C., and Ehrlich, P. R. (2002). Conservation of tropical forest birds in countryside habitats. *Ecology Letters*, **5**, 121–129.

Lindell, C. and Smith, M. (2003). Nesting bird species in sun coffee, pasture, and understory forest in southern Costa Rica. *Biodiversity and Conservation*, **12**, 423–440.

Pereira, H. M., Daily, G. C., and Roughgarden, J. (2004). A framework for assessing the relative vulnerability of species to land-use change. *Ecological Applications*, **14**, 730–742.

Perfecto, I. and Vandermeer, J. (2002). Quality of agro-ecological matrix in a tropical montane landscape: Ants in coffee plantations in southern Mexico. *Conservation Biology*, **16**, 174–182.

Ranganathan, J., Daniels, R. J. R., Chandran, M. D. S., Ehrlich, P. R., and Daily, G. C. (2008). How biodiversity can live with agriculture. *Proceedings of the National*

*continues*

> **Box 13.4 (Continued)**
>
> *Academy of Sciences of the United States of America*, **105**, 17852–17854.
>
> Ricketts, T. H., Daily, G. C., Ehrlich, P. R., and Fay, J. P. (2001). Countryside biogeography of moths in a fragmented landscape: Biodiversity in native and agricultural habitats. *Conservation Biology*, **15**, 378–388.
>
> Rosenzweig, M. L. (2003). *Win-win ecology: how the earth's species can survive in the midst of human enterprise*. Oxford University Press, Oxford, UK.
>
> Sekercioglu, C. H., Ehrlich, P. R., Daily, G. C., Aygen, D., Goehring, D., and Sandi, R. F. (2002). Disap-
>
> pearance of insectivorous birds from tropical forest fragments. *Proceedings of the National Academy of Sciences of the United States of America*, **99**, 263–267.
>
> Sekercioglu, C. H., Loarie, S. R., Brenes, F. O., Ehrlich, P. R., and Daily, G. C. (2007). Persistence of forest birds in the Costa Rican agricultural countryside. *Conservation Biology*, **21**, 482–494.
>
> Tilman, D., Fargione, J., Wolff, B., *et al.* (2001). Forecasting agriculturally driven global environmental change. *Science*, **292**, 281–284.
>
> Wilson, E. O. and Peter, F. M., eds (1988). *Biodiversity*. National Academy Press, Washington, DC.

of forest specialist species between forest fragments (Vandermeer and Perfecto 2007). The goal is to increase the permeability of the matrix, which is critical for the long-term persistence of metapopulations and metacommunities (Hanski 1999; Stratford and Stouffer 1999; Vandermeer and Carvajal 2001; Chapter 5).

## 13.7 Urban areas

Urban areas represent an extreme case in the spectrum of human-modified land uses. Unlike the other forms of habitat modification discussed above, urbanization often irreversibly replaces natural habitats with persistent artificial ones, resulting in long-term impacts on many native species (Stein *et al.* 2000). Despite the rapid rate at which urban sprawl is occurring worldwide, urban ecology has received relatively little attention from conservation biologists (Miller and Hobbs 2002). This can be attributed to the traditional focus of conservation research on "natural" ecosystems such as old-growth forests (Fazey *et al.* 2005).

As the trend of rapid economic growth continues in the tropics, urban areas will likely be increasingly ubiquitous in the tropics. An obvious research agenda, therefore, is to understand the response of tropical species to urbanization and to develop effective measures for their conservation. We ideally would want to be able to excise a tropical country, allow it to fulfill its economic potential and experience the associated landscape changes within a greatly accelerated time frame, and use this natural laboratory to study what species survive, where they persist and how they are able to do so. The island nation of Singapore in tropical Southeast Asia represents just such an ecological worst case scenario (Sodhi *et al.* 2004).

Koh and Sodhi (2004) studied butterfly diversity in Singapore, and found that forest reserves had higher species richness than secondary forest fragments and urban manmade parks (Figure 13.6). They attributed this to the larger areas of forest reserves and greater floristic complexity (compared to the other habitats they studied), which can sustain larger populations of species with lower risks of extinction, and contain greater diversities of microhabitats with myriad ecological niches that can support more species (MacArthur and Wilson 1963, 1967; Simberloff 1974; Laurance *et al.* 2002). Koh and Sodhi further explained that the last remaining tracts of old-growth vegetation in forest reserves can provide the unique microclimatic conditions such as a closed canopy, and specific larval host plants vital to the persistence of specialist butterfly species.

**Figure 13.6** Urban manmade park in Singapore. Photograph by Lian Pin Koh.

A second important finding of Koh and Sodhi's study was that urban parks adjoining forests were more diverse than secondary forest fragments. This was likely due to the prevalence of numerous ornamental flowering plants cultivated in these urban parks, which can support resident butterfly species adapted to an open canopy, as well as species from adjacent forests that forage in these parks. Indeed, the authors reported that both the number of potential larval host-plant species and the amount of surrounding forest cover were statistically significant predictors of butterfly species richness in urban parks.

Koh and Sodhi's study has two key conservation lessons: first, in highly urbanized tropical landscapes the least human-disturbed land uses are likely also most valuable for preserving the native biodiversity, and should therefore be given the highest conservation priority; second, with a good understanding of the biology of organisms, it is possible to enhance the conservation value of manmade habitats within human-modified landscapes. Although urban landscapes represent the worst case scenario in ecosystem management we are increasingly faced with the task of conserving species in such "unnatural" environments. Therefore, it is crucial that more research be focused on developing viable strategies for the effective conservation of biodiversity in urban landscapes.

## 13.8 Regenerating forests on degraded land

In most areas of the world, secondary forests regenerate naturally on abandoned agricultural land if human disturbance declines. Following centuries of human disturbance, the total area of regenerating forest is now enormous (millions of hectares). Indeed, for parts of the world that have suffered widespread historical deforestation secondary forests comprise the majority of remaining forest area (e.g. east coast of the USA, much of Western Europe, and areas of high human population density like Singapore). In the tropics secondary regrowth together with degraded old-growth forests (e.g. through logging, fire, fragmentation) comprise roughly half of the world's remaining tropical forest area (ITTO 2002).

Understanding the potential importance of these large areas of secondary forest for conservation has attracted much research attention from ecologists and conservation biologists, as well as considerable controversy. For example, Wright and Muller-Landau (2006) recently proposed that the regeneration of secondary forests in degraded tropical landscapes is likely to avert the widely anticipated mass extinction of native forest species. However, other researchers have highlighted serious inadequacies in the quantity and quality of species data that underpin this claim, casting doubt on the potential for secondary forest to serve as a "safety net" for tropical biodiversity (Brook *et al.* 2006; Gardner *et al.* 2007).

In perhaps the only quantitative summary of biodiversity responses to forest regeneration to date, Dunn (2004) analyzed data from 39 tropical data sets and concluded that species richness of some faunal assemblages can recover to levels similar to mature forest within 20–40 years, but that recovery of species composition is likely to take substantially longer. The recovery of biodiversity in secondary forests varies strongly between different species groups depending on their life histories with species responses generally falling into three categories (Bowen *et al.* 2007): (i) species that decline in abundance or are absent from regrowth due to specialist habitat requirements; (ii) old-growth forest species that benefit from altered conditions in regenerating forest and increase in abundance or distribution; and (iii) open-area species that invade regenerating areas to exploit newly available resources. These conclusions are mirrored by the results the comprehensive Jari study in north-east Brazil that found that 41% of old-growth vertebrate and invertebrate species were lacking from secondary forests of 12–18 years of age, and that species responses varied strongly among and within taxonomic groups (see Box 13.2).

The general lack of data and the context dependent nature of existing studies on biodiversity recovery in secondary forests severely limit our ability to make general predictions about the potential for species conservation in tropical secondary forests (Chazdon *et al.* 2009b). However, we can conclude that secondary forests are likely to be more diverse the more closely they reflect the structural, functional, and compositional properties of mature forest and are set within a favorable landscape context (Chazdon 2003; Bowen *et al.* 2007). In particular, the conservation of old-growth species in secondary forests will be maximized in areas where extensive tracts of old-growth forest remain within the wider region, older secondary forests have persisted, post-conversion land-use was of limited duration and low intensity, post-abandonment anthropogenic disturbance is relatively low, seed dispersing fauna are protected, and old-growth forests are close to abandoned sites (Chazdon *et al.* 2009b).

The conservation value of a secondary forest should increase over time as old-growth species accumulate during forest recovery, but older secondary forests are poorly studied and long-term datasets are lacking. Existing chronosequence studies of regenerating forests demonstrate that biotic recovery occurs over considerably longer time scales than structural recovery, and that re-establishment of certain species and functional group composition can take centuries or millennia (de Walt *et al.* 2003; Liebsch *et al.* 2008). However, for much of the world, secondary forests exist in highly dynamic landscape mosaics and are invariably clear-felled within one or two decades, thereby greatly limiting the opportunity for these forests to develop into older successional stands that are of higher value for conservation (Chazdon *et al.* 2009b).

Despite this uncertainty, regeneration represents the only remaining conservation option for many regions of the world that have suffered severe historical deforestation. An estimated 350 million hectares of the tropics are classified as degraded due to poor management (Maginnis and Jackson 2005). While the natural recovery of this land is not inevitable there is encouraging evidence that judicious approaches to reforestation can greatly facilitate the regeneration process and enhance the prospects of biodiversity in modified landscapes (Chazdon 2008).

## 13.9 Conservation and human livelihoods in modified landscapes

Modified and degraded landscapes around the world are not only of vital importance for biodiversity conservation, but are also home to millions of the world's poorest people. This is especially true in tropical countries where areas of high species richness and endemism frequently overlap with centers of human population density (e.g. sub-Saharan Africa; Balmford *et al.*

2001). It is estimated that the livelihoods of at least 300 million rural poor in tropical countries depend upon degraded or secondary forests (ITTO 2002). For impoverished communities biodiversity is about the basic human needs of eating, staying healthy, and finding shelter (see Chapter 14). Furthermore, it is local people that ultimately decide the fate of their local environments, even if the decisions they make fall within a wider political, social and economic context (Sodhi *et al.* 2006, 2008; Ghazoul 2007; Chapter 14).

These facts make it clear that human livelihoods and poverty concerns need to receive high priority in the conservation agenda if we are to develop management strategies for agricultural and modified landscapes that are not only viable into the long term, but are also socially just (Perfecto and Vandermeer 2008; Chapter 14). Recognition of this broader challenge has led to calls for a "pro-poor" approach to conservation (Kaimowitz and Sheil 2007). However, developing such an approach and successfully reconciling the interdependent objectives of poverty alleviation and biodiversity conservation is far from trivial. Opportunities for much-sought after "win-win" solutions (Rosenzweig 2003) are often hard if not impossible to achieve when faced by real-world trade-offs between economic and conservation goals, especially in the short-term. However, with careful planning and good science there is significant potential for synergies in achieving development and biodiversity benefits in the management of modified landscapes (Figure 13.7; Lamb *et al.* 2005).

The greatest difficulty in developing a pro-poor approach to biodiversity conservation lies in the fact that the structure and dynamics of human communities, and their interactions with the local environment, varies significantly across different parts of the world. There are no silver bullet, "off the shelf" solutions that can be successfully applied to any situation. Instead individual management strategies for individual landscapes need to be developed with explicit recognition of the socioeconomic, political, and ecological context within which they are embedded (Ostrom 2007). Furthermore, it is not enough to accommodate development considerations that do no more than secure livelihood levels at subsistence levels. Local guardians of modified landscapes have the right to develop management strategies that generate higher economic returns that can raise them out of poverty (Ghazoul 2007).

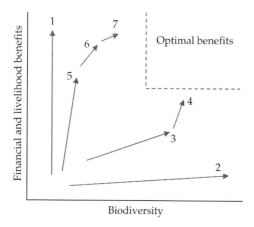

**Figure 13.7** Balancing trade-offs between human livelihoods and biodiversity conservation in reforestation projects. Arrows represent alterative reforestation methods. Traditional monoculture plantations of exotic species (arrow 1) mostly generate just financial benefits, whereas restoration using methods that maximize diversity and enhance biodiversity (arrow 2) yields few direct financial benefits to landowners, at least in the short term. Protecting forest regrowth (arrow 3) generates improvements in both biodiversity and livelihoods, although the magnitude of the benefits depends on the population density of commercially or socially important species; these can be increased by enrichment of secondary forest with commercially attractive species (arrow 4). Restoration in landscapes where poverty is common necessitates attempting both objectives simultaneously. But, in many situations, it may be necessary to give initial priority to forms of reforestation that improve financial benefits, such as woodlots and agroforestry systems (arrow 5). In subsequent rotations, this balance might change over time (moving to arrow 6 and later to arrow 7 by using a greater variety of species). There may be greater scope for achieving multiple objectives by using several of these options at different locations across the landscape. Reproduced with permission from Lamb *et al.* (2005).

## 13.10 Conclusion

The challenge of safeguarding the future of tropical forest species is daunting. Spatial and temporal patterns of biodiversity in modified landscapes are the product of interacting human and ecological processes that vary strongly between different land-use systems and among regions, and have effects that may take years to become fully manifest (Gardner *et al.* 2009). Conservation biologists have little option but to tackle this challenge head-on as very few, if any tropical forest species exist in isolation from human interference. Perhaps the most important conclusion to emerge from biodiversity research in modified landscapes is that different human land-uses can have enormously different implications for conservation. In this chapter we have shown that a broad gradient of structural complexity and species diversity exists from lightly logged production forests at one end to intensive arable and pastoral systems and semi-urban landscapes at the other. We have also highlighted how responsible management strategies at local and landscape scales can greatly enhance opportunities for biodiversity conservation in these systems. Throughout we have drawn attention to some of the real world economic and social considerations that will determine the success of any attempts to implement improved conservation strategies in the real world.

To truly understand the prospects for conservation in modified landscapes, we need to increase our emphasis on the study of biodiversity in managed land-use systems (Chazdon *et al.* 2009a). Key knowledge gaps remain in our understanding of the long-term viability of native species in different land-uses (Sodhi 2008), and how patterns of species persistence are influenced by differences in the composition and configuration of entire landscapes. Increasingly severe levels of environmental degradation in modified landscapes across the world means that the costs and benefits of ecological restoration are deserving of particular research attention. There is also an urgent need for an improved understanding of the interaction between people and their local environment in human-modified systems, including the importance of ecosystem services (see Chapter 3) and opportunities for generating livelihood benefits from conservation activities.

If it is to be successful, the conservation research agenda in modified landscapes needs to be effective at incorporating new tools and approaches, both conceptual and analytical, that have the potential to bridge the divide between theory and practice and translate policies into effective field implementation (Chazdon *et al.* 2009a; Gardner *et al.* 2009). Key to achieving success and developing sustainable management strategies is the ability to build participatory and multidisciplinary approaches to research and management that involve not only conservation biologists, but also agroecologists, agronomists, farmers, indigenous peoples, rural social movements, foresters, social scientists, and land managers (see Chapter 14).

## Summary

- Given that approximately one quarter of the world's threatened species live outside protected areas, and that the integrity of protected areas where they exist is often threatened, we need to integrate conservation efforts with other human activities.
- Recent studies demonstrate there are important opportunities for conserving biodiversity within the dominant types of human land-use, including logged forests, agroforestry systems, monoculture plantations, agricultural lands, urban areas, and regenerating land.
- It is the local people that ultimately decide the fate of their local environments, even if the decisions they make fall within a wider political, social, and economic context.
- Key to achieving success and developing sustainable management strategies is the ability to build participatory and multidisciplinary approaches to research and management that involve not only conservation biologists, but also agroecologists, agronomists, farmers, indigenous peoples, rural social movements, foresters, social scientists, and land managers.

## Suggested reading

Barlow, J., Gardner, T. A., Araujo, I. S., *et al.* (2007). Quantifying the biodiversity value of tropical primary, secondary and plantation forests. *Proceedings of the National Academy of Sciences of the United States of America*, **104**, 18555–18560.

Dunn, R. R. (2004). Recovery of faunal communities during tropical forest regeneration. *Conservation Biology*, **18**, 302–309.

Gardner, T., Barlow, J., Chazdon, R., *et al.* (2009). Prospects for tropical forest biodiversity in a human-modified world. *Ecology Letters*, **12**, 561–582.

Koh, L. P. (2008). Can oil palm plantations be made more hospitable for forest butterflies and birds? *Journal of Applied Ecology* **45**, 1002–1009.

## Relevant website

- Mongabay: http://www.mongabay.com

## REFERENCES

Anand, M., Krishnaswamy, J., and Das, A. (2008). Proximity to forests drives bird conservation value of coffee plantations: implications for certification. *Ecological Applications*, **18**, 1754–1763.

Asner, G. P., Knapp, D. E., Broadbent, E. N., Oliveira, P. J. C., Keller, M., and Silva, J. N. (2005). Selective logging in the Brazilian Amazon. *Science*, **310**, 480–482.

Balmford, A. and Bond, W. (2005). Trends in the state of nature and their implications for human well-being. *Ecology Letters*, **8**, 1218–1234.

Balmford, A., Moore, J. L., Brooks, T. *et al.* (2001). Conservation conflicts across Africa. *Science*, **291**, 2616–2619.

Balmford, A., Green, R. E., and Scharlemann, J. P. W. (2005). Sparing land for nature: exploring the potential impact of changes in agricultural yield on the area needed for crop production. *Global Change Biology*, **11**, 1594–1605.

Barlow, J., Gardner, T. A., Araujo, I. S., *et al.* (2007). Quantifying the biodiversity value of tropical primary, secondary and plantation forests. *Proceedings of the National Academy of Sciences of the United States of America*, **104**, 18555–18560.

Bawa, K. S., Kress, W. J., Nadkarni, N. M., *et al.* (2004). Tropical ecosystems into the 21st century. *Science*, **306**, 227–228.

Bhagwat, S. A., Willis, K. J., Birks, H. J. B., and Whittaker, R. J. (2008). Agroforestry: a refuge for tropical biodiversity? *Trends in Ecology and Evolution*, **23**, 261–267.

Bowen, M. E., McAlpine, C. A., House, A. P. N., and Smith, G. C. (2007). Regrowth forests on abandoned agricultural land: A review of their habitat values for recovering forest fauna. *Biological Conservation*, **140**, 273–296.

Brockerhoff, E. G., Berndt, L. A., and Jactel, H. (2005). Role of exotic pine forests in the conservation of the critically endangered New Zealand ground beetle *Holcaspis brevicula* (Coleoptera: Carabidae). *New Zealand Journal of Ecology*, **29**, 37–43.

Brockerhoff, E. G., Jactel, H., Parrotta, J. A., Quine, C. P., and Sayer, J. (2008). Plantation forests and biodiversity: oxymoron or opportunity? *Biodiversity and Conservation*, **17**, 925–951.

Brook, B. W., Bradshaw, C. J. A., Koh, L. P., and Sodhi, N. S. (2006). Momentum drives the crash: mass extinction in the tropics. *Biotropica*, **38**, 302–305.

Carnus, J. M., Parrotta, J., Brockerhoff, E., *et al.* (2006). Planted forests and biodiversity. *Journal of Forestry*, **104**, 65–77.

Chazdon, R. L. (2003). Tropical forest recovery: legacies of human impact and natural disturbances. *Perspectives in Plant Ecology Evolution and Systematics*, **6**, 51–71.

Chazdon, R. L. (2008). Beyond Deforestation: Restoring Forests and Ecosystem Services on Degraded Lands. *Science*, **320**, 1458–1460.

Chazdon, R. L., Harvey, C. A., Oliver, K., *et al.* (2009a). Beyond Reserves: A Research Agenda for Conserving Biodiversity in Human-modified Tropical Landscapes. *Biotropica*, **41**, 142–153.

Chazdon, R. L., Peres, C. A., Dent, A. (2009b). Where are the wild things? Assessing the potential for species conservation in tropical secondary forests *Conservation Biology* (in Press).

Cohen, J. E. (2003). Human population: the next half century. *Science*, **302**, 1172–1175.

Cyranoski, D. (2007). Logging: the new conservation. *Nature*, **446**, 608–610.

Daily, G. C. (2001). Ecological forecasts. *Nature*, **411**, 245–245.

Daily, G. C. (2003). Time to rethink conservation strategy. *Science*, **300**, 1508–1509.

Daily, G. C., Ehrlich, P. R., and Sanchez-Azofeifa, A. (2001). Countryside biogeography: Utilization of human-dominated habitats by the avifauna of southern Costa Rica. *Ecological Applications*, **11**, 1–13.

Dale, V. H., Brown, S., Haeuber, R. A., *et al.* (2000). Ecological principles and guidelines for managing the use of land. *Ecological Applications*, **10**, 639–670.

DeWalt, S. J., Maliakal, S. K., and Denslow, J. S. (2003). Changes in vegetation structure and composition along a tropical forest chronosequence: implications for wildlife. *Forest Ecology and Management*, **182**, 139–151.

Donald, P. F. (2004). Biodiversity impacts of some agricultural commodity production systems. *Conservation Biology*, **18**, 17–37.

Dunn, R. R. (2004). Recovery of faunal communities during tropical forest regeneration. *Conservation Biology*, 18, 302–309.

Ehrlich, P. R. and Ehrlich, A. H. (2008). *The Dominant Animal: Human Evolution and the Environment*. Island Press, Washington, DC.

Ellis, E. C. and Ramankutty, N. (2008). Putting people in the map: anthropogenic biomes of the world. *Frontiers in Ecology and the Environment*, 6, 439–447.

FAO (Food and Agriculture Organization of the United Nations) (2006). *Global forest resources assessment 2005: progress towards sustainable forest management*. FAO, Rome, Italy.

Fazey, I., Fischer, J., and Lindenmayer, D. B., (2005). What do conservation biologists publish? *Biological Conservation*, 124, 63–73.

Felton, A. M., Engstrom, L. M., Felton, A., and Knott, C. D. (2003). Orangutan population density, forest structure and fruit availability in hand-logged and unlogged peat swamp forests in West Kalimantan, Indonesia. *Biological Conservation*, 114, 91–101.

Fimbel, R. A., Grajal, A., and Robinson, J. G., eds (2001). *The cutting edge. Conserving wildlife in logged tropical forests*. Colombia University, New York, NY.

Fischer, J. and Lindenmayer, B. D. 2007. Landscape modification and habitat fragmentation: a synthesis. *Global Ecology and Biogeography*, 16, 265–280.

Gardner, T. A., Barlow, J., Parry, L. T. W., and Peres, C. A. (2007). Predicting the uncertain future of tropical forest species in a data vacuum. *Biotropica*, 39, 25–30.

Gardner, T. A., Barlow, J., Chazdon, R. L., *et al.* (2009). Prospects for tropical forest biodiversity in a human-modified world. *Ecology Letters*, 12, 561–582.

Ghazoul, J. (2007). Placing humans at the heart of conservation. *Biotropica*, 39, 565–566.

Gillies, C. S. and St Clair, C. C. (2008). Riparian corridors enhance movement of a forest specialist bird in fragmented tropical forest. *Proceedings of the National Academy of Sciences of the United States of America*, 105, 19774–19779.

Green, R. E., Cornell, S. J., Scharlemann, J. P. W., and Balmford, A. (2005). Farming and the fate of wild nature. *Science*, 307, 550–555.

Hanski, I. (1999). *Metapopulation ecology*. Oxford University Press, Oxford, UK.

Hartley, M. J. (2002). Rationale and methods for conserving biodiversity in plantation forests. *Forest Ecology and Management*, 155, 81–95.

Heckenberger, M. J., Kuikuro, A., Kuikuro, U. T., *et al.* (2003). Amazonia 1492: Pristine forest or cultural parkland? *Science*, 301, 1710–1714.

ITTO (International Tropical Timber Association) (2002). *ITTO guidelines for the restoration, management and rehabilitation of degraded and secondary tropical forests*. ITTO Policy Development Series No. 13, International Tropical Timber Organization.

ITTO (International Tropical Timber Association) (2007). *Annual Review and Assessment of the World Timber Situation 2007*. ITTO, Yokohama, Japan.

Kaimowitz, D. and Sheil, D. (2007). Conserving what and for whom? Why conservation should help meet basic human needs in the tropics. *Biotropica*, 39, 567–574.

Kanowski, J., Catterall, C. P., and Wardell-Johnson, G. W. (2005). Consequences of broadscale timber plantations for biodiversity in cleared rainforest landscapes of tropical and subtropical Australia. *Forest Ecology and Management*, 208, 359–372.

Kareiva, P., Watts, S., McDonald, R., and Boucher, T. (2007). Domesticated nature: shaping landscapes and ecosystems for human welfare. *Science*, 316, 1866–1869.

Koh, L. P. (2008a). Birds defend oil palms from herbivorous insects. *Ecological Applications* 18, 821–825.

Koh, L. P. (2008b). Can oil palm plantations be made more hospitable for forest butterflies and birds? *Journal of Applied Ecology* 45, 1002–1009.

Koh, L. P., and Sodhi, N. S. (2004). Importance of reserves, fragments and parks for butterfly conservation in a tropical urban landscape. *Ecological Applications*, 14, 1695–1708.

Koh, L. P. and Wilcove, D. S. (2007). Cashing in palm oil for conservation. *Nature* 448, 993–994.

Koh, L. P. and Wilcove, D. S. (2008). Is oil palm agriculture really destroying tropical biodiversity? *Conservation Letters* 1, 60–64.

Koh, L. P. and Wilcove, D. S. (2009). Oil palm: disinformation enables deforestation. *Trends in Ecology and Evolution* 24, 67–68.

Lamb, D., Erskine, P. D., and Parrotta, J. A. (2005). Restoration of degraded tropical forest landscapes. *Science*, 310, 1628–1632.

Laurance, W. F., Lovejoy, T. E., Vasconcelos, H. L., *et al.* (2002). Ecosystem decay of Amazonian forest fragments: a 22-year investigation. *Conservation Biology*, 16, 605–618.

Lewis, O. T. (2001). Effect of experimental selective logging on tropical butterflies. *Conservation Biology*, 15, 389–400.

Liebsch, D., Marques, M. C. M., and Goldenberg, R. (2008). How long does the Atlantic Rain Forest take to recover after a disturbance? Changes in species composition and ecological features during secondary succession. *Biological Conservation*, 141, 1717–1725.

Lindenmayer, D. B. and Franklin, J. F. (2002). *Conserving biodiversity: a comprehensive multiscaled approach*. Island Press, Washington, DC.

Lindenmayer, D. B. and Hobbs, R. J. (2004). Fauna conservation in Australian plantation forests – a review. *Biological Conservation*, 119, 151–168.

Lindenmayer, D. B., McIntyre, S., and Fischer, J. (2003). Birds in eucalypt and pine forests: landscape alteration and its implications for research models of faunal habitat use. *Biological Conservation*, **110**, 45–53.

Lindenmayer, D. B., Franklin, J. F., and Fischer, J. (2006). General management principles and a checklist of strategies to guide forest biodiversity conservation. *Biological Conservation*, **131**, 433–445.

MacArthur, R. H. and Wilson, E. O. (1963). An equilibrium theory of insular zoogeography. *Evolution*, **17**, 373–387.

MacArthur, R. H. and Wilson, E. O. (1967). *The theory of island biogeography*. Princeton University Press, Princeton, New Jersey.

Maginnis, S. and Jackson, W. (2005). *Balancing restoration and development*. ITTO Tropical Forest Update, International Tropical Timber Organization. 15/2, 4–6.

Matson, P. A. and Vitousek, P. M. (2006). Agricultural intensification: will land spared from farming be land spared for nature? *Conservation Biology*, **20**, 709–710.

Mbida, C. M., Van Neer, W., Doutrelepont, H., and Vrydaghs, L. (2000). Evidence for banana cultivation and animal husbandry during the First Millennium BC in the forest of Southern Cameroon. *Journal of Archaeological Science*, **27**, 151–162.

McNeely, J. A. (2004). Nature vs. nurture: managing relationships between forests, agroforestry and wild biodiversity. *Agroforestry Systems*, **61**, 155–165.

Meijaard, E. and Sheil, D. (2008). The persistence and conservation of Borneo's mammals in lowland rain forests managed for timber: observations, overviews and opportunities. *Ecological Restoration*, **23**, 21–34.

Miller, J. R. and Hobbs, R. J. (2002). Conservation where people live and work. *Conservation Biology*, **16**, 330–337.

Ostrom, E. (2007). A diagnostic approach for going beyond panaceas. *Proceedings of the National Academy of Sciences of the United States of America*, **104**, 15181–15187.

Palmer, M., Bernhardt, E., Chornesky, E., et al. (2004). Ecology for a crowded planet. *Science*, **304**, 1251–1252.

Pedlowski, M. A., Matricardi, E. A. T., Skole, D., et al. (2005). Conservation units: a new deforestation frontier in the Amazonian state of Rondonia, Brazil. *Environmental Conservation*, **32**, 149–155.

Perfecto, I. and Vandermeer, J. (2002). The quality of agroecological matrix in a tropical montane landscape: ants in coffee plantations in southern Mexico. *Conservation Biology*, **16**, 174–182.

Perfecto, I. and Vandermeer, J. (2008). Biodiversity conservation in tropical agroecosystems. *Annals of the New York Academy of Science*, **1134**, 173–200.

Peters, S. L., Malcolm, J. R., and Zimmerman, B. L. (2006). Effects of selective logging on bat communities in the southeastern Amazon. *Conservation Biology*, **20**, 1410–1421.

Philpott, S. M., Arendt, W. J., Armbrecht, I., et al. I (2008). Biodiversity loss in Latin American coffee landscapes: review of the evidence on ants, birds, and trees. *Conservation Biology*, **22**, 1093–1105.

Putz, F. E., Blate, G. M., Redford, K. H. Fimbel, R., and Robinson, J. (2001). Tropical forest management and conservation of biodiversity: an overview. *Conservation Biology*, **15**, 7–20.

Putz, F. E., Sisk, P., Fredericksen, T., and Dykstra, D. (2008). Reduced-impact logging: Challenges and opportunities. *Forest Ecology and Management*, **256**, 1427–1433.

Ranganathan, J., Daniels, R. J. R., Chandran, M. D. S., Ehrlich, P., and Daily. G. C. (2008). Sustaining biodiversity in ancient tropical countryside. *Proceedings of the National Academy of Sciences of the United States of America*, **105**, 17852–17854.

Ribeiro, M. C., Metzger, J. P., Martensen, A. C., Ponzoni, F., and Hirota, M. M. (2009). Brazilian Atlantic forest: how much is left and how is the remaining forest distributed? Implications for conservation. *Biological Conservation*, **142**, 1141–1153.

Rodrigues, A. S. L., Andelman, S. J., Bakarr, M. I., et al. (2004). Effectiveness of the global protected area network in representing species diversity. *Nature*, **428**, 640–643.

Rosenzweig, M. L. (2003). *Win-win ecology: how the earth's species can survive in the midst of human enterprise*. Oxford University Press, New York.

Sayer, J. and Maginnis, S., eds (2005). *Forests in landscapes: ecosystem approaches to sustainability*. Earthscan, London.

Scales, B. R. and Marsden, S. J. (2008). Biodiversity in small-scale tropical agroforests: a review of species richness and abundance shifts and the factors influencing them. *Environmental Conservation*, **35**, 160–172.

Scherr, S. J. and McNeely, J. A., eds (2007). *Farming with nature: the science and practice of ecoagriculture*. Island Press, Washington, DC.

Scherr, S. J. and McNeely, J. A. (2008). Biodiversity conservation and agricultural sustainability: towards a new paradigm of 'ecoagriculture' landscapes. *Philosophical Transactions of the Royal Society of London B*, **363**, 477–494.

Schmitt, C. B., Belokurov, A., Besançon C., et al. (2009). *Global Ecological Forest Classification and Forest Protected Area Gap Analysis. Analyses and recommendations in view of the 10% target for forest protection under the Convention on Biological Diversity (CBD)*. 2nd revised edn Freiburg University Press, Freiburg, Germany.

Schroth, G. and Harvey, C. A. (2007). Biodiversity conservation in cocoa production landscapes: an overview. *Biodiversity and Conservation*, **16**, 2237–2244.

Schroth, G., da Fonseca, G. A. B., Harvey, C. A., Gascon, C., Vasconcelos, and Izac, A.-M. N., eds (2004).

*Agroforestry and biodiversity conservation in tropical landscapes*. Island Press, Washington, DC.

Sekercioglu, C. H. (2002). Effects of forestry practices on vegetation structure and bird community of Kibale National Park, Uganda. *Biological Conservation*, **107**, 229–240.

Sekercioglu, C. H., Loarie, S. R., Ruiz-Gutierrez, V., Oviedo Brenes, F., Daily, G. C., and Ehrlich, P. R. (2007). Persistence of forest birds in tropical countryside *Conservation Biology* **21**, 482–494.

Simberloff, D. S. (1974). Equilibrium theory of island biogeography and ecology. *Annual Review of Ecology and Systematics*, **5**, 161–182.

Sodhi, N. S. (2008). Tropical biodiversity loss and people – a brief review. *Basic and Applied Ecology*, **9**, 93–99.

Sodhi, N. S., Koh, L. P., Brook. B. W., and Ng, P. K. L. (2004). Southeast Asian biodiversity: an impending disaster. *Trends in Ecology and Evolution*, **19**, 654–660.

Sodhi, N. S., Brooks, T. M., Koh, L. P. Koh *et al.* (2006). Biodiversity and human livelihood crises in the Malay Archipelago. *Conservation Biology*, **20**, 1811–1813.

Sodhi, N. S., Acciaioli, G., Erb, M., and Tan, A. K.-J., eds (2008). *Biodiversity and human livelihoods in protected areas: case studies from the Malay Archipelago*. Cambridge University Press, Cambridge, UK.

Sodhi, N. S., Lee, T. M., Koh, L. P., and & Brook, B. W. (2009). A meta-analysis of the impact of anthropogenic forest disturbance on Southeast Asia's biotas. *Biotropica*, **41**, 103–109.

Steffan-Dewenter, I., Kessler, M., Barkmann, J., *et al.* (2007). Tradeoffs between income, biodiversity, and ecosystem functioning during tropical rainforest conversion and agroforestry intensification. *Proceedings of the National Academy of Sciences of the United States of America*, **104**, 4973–4978.

Stein, B.A., Kutner, L. S., and Adams, J. S. (2000). *Precious heritage*. Oxford University Press, Oxford, UK.

Stratford, J. A. and Stouffer, P. C. (1999). Local extinctions of terrestrial insectivorous birds in a fragmented landscape near Manaus, Brazil. *Conservation Biology*, **13**, 1416–1423.

Thiollay, J. M. (1995). The role of traditional agroforests in the conservation of rain-forest bird diversity in Sumatra. *Conservation Biology*, **9**, 335–353.

Tilman, D., Fargione, J., and Wolff, B., *et al.* (2001). Forecasting agriculturally driven global environmental change. *Science*, **292**, 281–284.

UN (United Nations) (2005). *Halving hunger: it can be done*. Earthscan, London, UK.

Vandermeer, J. and Carvajal, R. (2001). Metapopulation dynamics and the quality of the matrix. *The American Naturalist*, **158**, 211–220.

Vandermeer, J. and Perfecto, I. (2007). The agricultural matrix and a future paradigm for conservation. *Conservation Biology*, **21**, 274–277.

Vaughan, C., Ramirez, O., Herrera, G., and Guries, R. (2007). Spatial ecology and conservation of two sloth species in a cacao landscape in limon, Costa Rica. *Biodiversity and Conservation*, **16**, 2293–2310.

White, L. J. T. (2001). The African rain forest: climate and vegetation. In W. Weber, L. J. T. White, A. Vedder, and L. Naughton-Treves, eds *African rain forest ecology and conservation: an interdisciplinary perspective*, pp. 3–29. Yale University Press, New Haven, CT.

Willis, K. J. and Birks, H. J. B. (2006). What is natural? The need for a long-term perspective in biodiversity conservation. *Science*, **314**, 1261–1265.

Willis, K. J., Gillson, L., and Brncic, T. M. (2004). How "virgin" is virgin rainforest? *Science*, **304**, 402–403.

Willis, K. J., Gillson, L., Brncic, T. M., and Figueroa-Rangel, B. L. (2005). Providing baselines for biodiversity measurement. *Trends in Ecology and Evolution*, **20**, 107–108.

Wittemyer, G., Elsen, P, Bean, W. T., Burton, C. O., and Brashares, J. S. (2008). Accelerated human population growth at protected area edges. *Science*, **321**, 123–126.

Wright, S. J. and Muller-Landau, H. C. (2006). The future of tropical forest species. *Biotropica*, **38**, 207–301.

CHAPTER 14

# The roles of people in conservation

C. Anne Claus, Kai M. A. Chan, and Terre Satterfield

The study of human beings in conservation is often eclipsed by the study of threatened species and their environments. This is surprising given that conservation activities are human activities, and that the very need for conservation arises out of human actions. In this chapter, we begin with the premise that understanding human activities and human roles in conservation is fundamental to effective conservation. Specifically, we address the following:

- Conservation history: how has conservation changed since its inception?
- Common conservation perceptions: how do conservationists characterize the relationship between human beings and the environment, and how have these perceptions influenced the trajectory of conservation?
- Organizational institutions: What factors mediate the relationship between human beings and their environments? What implications do these have for conservation?
- Biodiversity conservation and local resource use: in what ways do we conserve our environments?
- Equity, rights, and resources: how do we understand conservation-induced change?
- Social research in conservation: how do social science and humanities studies inform conservation practice?

## 14.1 A brief history of humanity's influence on ecosystems

Human beings have influenced Earth's ecosystems for many millennia (see Chapter 13). Since *Homo sapiens* migrated out of East Africa in the late Pleistocene, we have subsequently fanned out to inhabit virtually every terrestrial environment on this planet. From high altitudes to high latitudes, people have adapted culturally, technologically, and biologically to diverse landscapes. Just as coevolution and coadaptation occur among plants and animals in ecosystems, so too do they occur between humans and other components of ecosystems around the world. We are crucial elements of ecosystems, and for better or worse, we help shape the environment of which we are a part.

## 14.2 A brief history of conservation

Indigenous and local people have practiced conservation possibly for hundreds of thousands of years (see Box 1.1). The Western conservation movement, however, has arisen over the past 150 years. We briefly address the history of the modern conservation movement here (see also Chapter 1). In its earliest period, a concern for biodiversity was not a dominant motivating factor of this movement. Rather most historians link modern conservation to writings of romantic and transcendentalist philosophers, and to the often violent colonizing of indigenous peoples in the Americas (White and Findley 1999), Africa (Neumann 2004), and worldwide (Grove 1996).

Environmental historians in the United States locate the origins of conservation with the writings of early ecologists and the advocacy of key thinkers in the latter half of the nineteenth century. As early as 1864, George Perkins Marsh published a remarkable book, *Man and Nature,* based in part on his observations of the depletion of the woods near his American home. Criticizing the cultivated gardens idealized by the Jeffersonian tradition, and deeming them an agent of destruction, he outlined the impact of logging on watersheds, water supply, salmon runs, and flooding (Robbins 1985). At the same time, John Muir, Ralph Waldo Emerson, and Henry David Thoreau came to be known for their highly influential transcendentalist philosophy, which contemplated nature's capacity for spiritual healing. This philosophy in particular is closely associated with early efforts at wilderness preservation. Vast tracks of land were integral to this view of nature, and this idea sparked the establishment of preservationist nature parks worldwide.

Several key policy initiatives ensured both a large legacy of public lands and a national park system in the United States and elsewhere. That so much "public" land was available for national parks was the product of two often ignored facts in the history of conservation. The first was the reservation system or the forced removal of aboriginal populations onto vastly reduced and parceled "reservation" lands, and the second was the rise in sedentary settlements. Much of this forced removal from what would become public and park land was made possible by the epidemics of disease amongst aboriginal populations that followed contact with Europeans (Stevens 1997). In the Americas, virtually all groups succumbed to successive waves of disease outbreaks, especially measles and small pox, introduced by "discoverers" (possibly as early as the Vikings and certainly by Western European explorers of Christopher Columbus' time) and by early settlers. Where disease did not decimate populations, people were forcibly removed from conservation areas in the Americas, Australia, Africa, and Asia. The conservation movement became more complex in the early 1900s with the advocacy of forester Gifford Pinchot, who insisted that conservation shift from primarily preservationist to that of resource management, or "sustained yield".

In 1960, the first of a set of legislative acts meant to represent both conservation and industrial interests was introduced. Under pressure from environmentalists and recreationists, the US federal government came out with a new mission statement: The Multiple Use Sustained Yield Act, 1960. Multiple uses incorporated outdoor recreation, range, timber, watershed, wildlife, and fishing interests. This early inscription of multiple uses for multiple people followed two events singularly important to modern environmentalism: Aldo Leopold's promotion of his "land ethic", which emphasized the biota's role in ethics (e.g. "A thing is right when it tends to preserve the integrity, stability and beauty of the biotic community. It is wrong when it tends otherwise"; from the *Sand County Almanac*, 1949); and the work of Rachel Carson (1962). A marine biologist, Carson published what has been called the basic book of North America's environmental revolution—*Silent Spring*. Its stirring argument exposed the actual and potential consequences of using the insecticide DDT (dichloro-diphenyl-trichloroethane), although even with DDT the social, environmental, and medical landscape is a complicated one. Regardless, Carson's work continues to be highly relevant to our understanding of biological processes, is cited in the inspirational biographies of environmentalists, and has spurred dozens of environmental groups into action. In retrospect, it is evident that what are today called "environmentalist ideas" coalesced around this time. While we have used the US as an example, it is the case that environmentalist ideas appeared independently but nonetheless concurrently in many parts of the world (see Box 1.2). This has led in part to the establishment of international conservation organizations, some of which originated in the developed world but all of which act in conjunction with partners worldwide. Examples of the larger and more well known such organizations include the World Wide Fund for Nature (WWF), The Nature Conservancy, and the International Union for Conservation of Nature (IUCN).

International conservation organizations have thus become particularly active in the advocacy for and the establishment and management of conservation areas worldwide. All of these philosophies of conservation are now evident in the multitude of conservation interventions across the planet. Box 14.1 illustrates how customary management and Western conservation are integrated to achieve conservation goals in the Pacific.

## Box 14.1 Customary management and marine conservation
### C. Anne Claus, Kai M. A. Chan, and Terre Satterfield

Can traditional management strategies and marine conservation be integrated? Cinner and Aswani (2007) set out to uncover the commensurability of these divergent resource management strategies in the Pacific. Customary management in marine systems refers to a generational, culturally embedded, dynamic system for regulating natural resource use. Cinner and Aswani first review studies on the ecological impacts of customary management. While more research is clearly needed on this topic, the available literature points to species-specific benefits, often on a small spatial scale. Viewing the smaller scale of customary practices in light of current social and economic threats, Cinner and Aswani suggest that customary management must be paired with marine conservation in order to produce ecological successes.

There are similarities in customary and marine conservation traditions. Cinner and Aswani define six types of restrictions present in both systems:

- Spatial (such as temporary ritualistic reef closures, or marine protected areas).
- Temporal (fishing bans on the Sabbath, or closed seasons).
- Gear (bans on harvesting technologies, or gear prohibitions).
- Effort (gender restriction on access to specific areas, or licensing).
- Species (class restrictions on particular species consumption, or species-specific bans).
- Catch (avoidance of waste, or quotas).

In spite of these resemblances, there are differences in the scale, concept, and intent of these two types of marine resource management. For example, in customary management, fishing bans may regularly be lifted to provide food for feasts. Therefore fish may be conserved but they are also harvested at regular intervals, pointing to a difference in concept between the two systems (in marine conservation fishing bans are generally considered permanent). Additionally, customary management in the Pacific is often embedded in ceremonies and traditions.

"Although resources may be consciously improved by these practices, conservation in the Western sense may be simply a by-product of other economic, spiritual, or social needs" (Ruttan in Cinner and Aswani).

Cinner and Aswani point out how hybrid systems have been socially successful in Vanuatu and the Western Solomon Islands. They summarize some principles for hybrid customary and marine conservation management systems:

- Approaches should echo local socioeconomic and cultural conditions.
- Planning and implementation should integrate both scientific and local knowledge systems.
- Strategies should be appropriate for varying social and ecological processes
- Management should provide flexible legal capacity.
- Planners and implementers should recognize that hybrid systems may not always be appropriate.
- Hybrid management should embrace the utilitarian nature of customary systems as well as its ecosystem benefits.

Finally, Cinner and Aswani caution that socioeconomic transformations such as population increase, technological change, urbanization, and the adoption of new legal systems can drastically and rapidly change customary management systems. How these systems are impacted by such changes varies depending on the heterogeneity of customary institutions and the scale of socioeconomic change. Cinner and Aswani conclude by endorsing hybrid management systems for their potential to encourage compliance amongst communities involved in creating them.

### REFERENCE

Cinner, J. E. and Aswani, S. (2007). Integrating customary management into marine conservation. *Biological Conservation*, 140, 201–216.

## 14.3 Common conservation perceptions

Observe Figure 14.1. You may see either a young woman or an old woman (hint: the chin of the young woman is the nose of the old woman). The lines remain the same, but their meaning changes based on your perception, the process by which you translate information into organized understanding. People construct meaning based on perceptions arising through their experiences. Since human beings have a broad range of experience, perception is also highly variable, and it is based on these perceptions that people act. In conservation, people comport themselves in accordance with observations they make about the state of a given ecosystem (for example, by hunting species they perceive as abundant and avoiding species that seem scarce). Similarly, conservationists base resource management strategies on their perceptions of local resource use. This whirlwind of perceptions can often lead to misperceptions. These misperceptions can also be enhanced by unequal relations of power within and between international organizations and local people. This can result, for instance, in such things as unnecessary burdens placed on local peoples. Box 14.2 contains a case study showing how such burdens can be placed on livelihoods as a result of both misperceptions and inequities.

**Figure 14.1** An optical illusion illustrates how the human brain perceives objects differently.

That is, the necessity for conservation often arises out of a misperception about the abundance of resources, which leads to excessive extraction. For conservationists who seek to alter

---

**Box 14.2 Historical ecology and conservation effectiveness in West Africa**
**C. Anne Claus, Kai M. A. Chan, and Terre Satterfield**

How does faulty perception lead to misguided conservation policies? Fairhead and Leach (1996) explore this question in the forest-savanna transition zone of Guinea. This landscape is unique because amidst the open woodland savanna exist patches of dense semi-deciduous rain forest. Conservationists and policy makers viewed these forest patches as either relics of a more extensive original forest or as a relatively stable pattern of vegetation. Regardless of the viewpoint taken on the forest patches, policy makers agreed that local people were contributing to their destruction. This supposed deforestation encouraged strict fire restrictions. At one point the punishment for setting a fire was the death penalty!

By using new historical data sets combined with oral histories of vegetation use, remote sensing data, archival research, and ethnographic fieldwork, Fairhead and Leach demonstrate that local human activities actually encouraged the formation of forest patches. Originally created around villages to provide fire and wind protection, the forest patches also provided resources for consumption and use. People enriched the

*continues*

> **Box 14.2 (Continued)**
>
> forest patches by managing soil fertility and fire. While the focus of local interest in the forest patches has changed since the nineteenth century from village defense to coffee production to timber for logging, they have been cultivated consciously and unconsciously by local resource users. And, contrary to the commonly accepted perception, remote sensing and photo analysis demonstrate that forest cover actually increased during the past century.
>
> So what were the consequences of the misguided forest policy? Because they assumed that locals were deforesting the landscape, policy makers excluded local resource users from resource management. Policies curbed early season grass burning, creating the potential for destructive natural fires in the dry season. The perception that locals were to blame for deforestation ultimately impacted their livelihoods and created an acrimonious conservation climate.
>
> Fairhead and Leach point out that the policy makers, due to their initial assumptions about the role of local resource users in deforestation, did not question the accuracy of historical vegetation records. The authors therefore advocate mixed historical and satellite data collection methods for reconstructing historically accurate pictures of vegetation patterns on which to base conservation policy. Their study illustrates how perceptions can negatively impact society and the environment.
>
> **REFERENCE**
>
> Fairhead, J. and Leach, M. (1996). Enriching the landscape: social history and the management of transition ecology in the forest-savanna mosaic of the Republic of Guinea. *Africa*, **66**, 14–36.

human action, understanding that perceptions and power differ is critical.

Here we discuss conservation in the sense of conservation biology, the science of understanding Earth's biological diversity for the sake of its protection. We refer to conservationists as people who identify themselves as practitioners or advocates of wild living resource conservation. Local resource users are people who live in close proximity to, and derive their livelihoods from, natural resources. Local resource users may be indigenous people, long-standing immigrant communities, or new residents. Like conservationists, they may represent homogeneous communities or encompass diverse ethnic groups. It is of course possible that conservationists may also be local resource users and vice-versa. As conservationists interact with local resource users around the world, they make considered judgments, as well as erroneous assumptions, about the relationship that human beings have with their environment (e.g. Sundberg 1998). In the past, conservationists have broadly characterized local resource users alternately as both enemies of and saviors of the environment – and the complexity of those ethical relationships are explored in Box 14.3.

Fundamental to these binary depictions are ideas of nature as a pristine wilderness. Images of this sort helped spur the modern conservation movement, and are still pervasive in conservation marketing. Wilderness is imagined "as a remnant of the world as it was before man appeared, as it was when water was fit to drink and air was fit to breathe" (Caufield 1990). These ideas rest on a perceived separation between humans and nature, a sentiment that appeals to many North Americans (Cronon 1995). Some conservationists assume that in order to conserve a system it should be restored to this idealized human-free state. Anthropologists, archaeologists, and historical ecologists have increasingly found that even landscapes that were once considered pristine have had considerable human influence (see Chapter 13). North America at the time of European contact, for example, has been depicted in literature and films as a vast wilderness. In reality, archaeological evidence and historical

## Box 14.3 Elephants, animal rights, and *Campfire*
### Paul R. Ehrlich

A conservation success story is that elephant populations have recently rebounded over much of Africa. That has fueled a heated debate over whether or not it is ethical to cull the herds (http://news.bbc.co.uk/2/hi/africa/7262951.stm). On one hand the giant beasts can be serious agricultural pests; on the other, animal rights activists and many other nature lovers are offended by the killing of these charismatic and intelligent animals. Like many of today's ecoethical dilemmas, this one is not easy to resolve (I do not wish to get into the animal rights debate here. For intelligent discussion of these issues, see Singer 1975; Midgley 1983; Jamieson 1999. Although I sometimes disagree with Peter Singer's conclusions on a variety of issues, sometimes emotionally rather than intellectually, I always find him a clear thinker). There are ways to attain needed population reductions other than culling, including relocation and contraception. But suitable areas into which to introduce elephants are growing scarce, and using contraceptives is difficult except in small parks and is more complicated and expensive than shooting. Animal rights groups are properly (in my view) concerned about cruelty to elephants, and the plight of young elephants orphaned when their mothers are killed is especially heart-rending. Furthermore immature elephants who have witnessed culling seem to suffer from something resembling Post-Traumatic Stress Disorder that frequently causes them to become very violent. But overpopulation of elephants can lead both to problems of sustainability for them and to collisions with another overpopulated species that has the capability of destroying them.

A similar elephant controversy took place in the 1990s – one demonstrating the extreme complexity of the ecoethical issues in conservation – centered around the Zimbabwean *Campfire* (Communal Areas Management Program for Indigenous Resources) program, partially funded by USAID (United States Agency for International Development) (Smith and Duffy 2003). The *Campfire* program was designed to build the capacity of local populations to manage natural resources, including game for hides, meat, sport hunting, and photographic tourism. The situation can be briefly summarized. Elephant herds outside of parks and reserves were capable of decimating a family's livelihood in an hour by destroying its garden plot. That led to defensive killing of marauding animals by local people. Rogue elephants were also responsible for hundreds of human deaths each year (http://findarticles.com/p/articles/mi_m1594/is_n4_v9/ai_20942049/pg_1). Defensive killing was accelerating a decline already under way in elephant herds because of poaching.

*Campfire* supported the return of elephant herds to the control of local communities and the issuing of some 100–150 elephant hunting licenses per year for community lands. The licenses to shoot an elephant were sold to sport hunters for US$12 000–15 000 each. Rural District Councils determined how the funds were spent. Herds grew dramatically in the hunting areas because poaching was suppressed by the elephants' new "owners," local people got more money and suffered less damage because marauders were targeted, and it seemed to many that it was a win-win situation. But the Humane Society of the U.S. (HSUS) objected, saying that the intelligent and charismatic elephants should never be killed by hunters, and animal rights groups lobbied to get funding stopped (http://digital.library.unt.edu/govdocs/crs/permalink/meta-crs-388:1).

The issue was further clouded by arguments over how much of *Campfire*'s motivation was centered on reopening the ivory trade (partly sanctioned by CITES) and its impact on elephants outside of Zimbabwe, and on whether a switch to entirely photographic safaris (a trend then well under way) would not be equally effective in protecting herds.

More recently, despite the shocks of a cessation of international funding and the deterioration of the political situation in Zimbabwe, the conservation benefits of

*continues*

> **Box 14.3 (Continued)**
>
> *Campfire* remained remarkably robust (Balint and Mashinya 2008) – although their present status is in doubt. The situation emphasizes the need to keep the ethics of the "big picture" always in mind, and to pay attention to factors such as "political endemism" – organisms found in only a single nation which, if poor, may not be able to adequately protect them (Ceballos and Ehrlich 2002).
>
> The *Campfire* controversy highlights the ethical conflict between those who believe the key conservation issue is maintaining healthy wildlife populations and those concerned primarily about the rights of individual animals and who decry the "utilization" or "commodification" of nature – "wise use" or "multiple use" as discussed in the text. Much as I personally hate to see elephants hunted, in this case I tend to come down on the side of the *Campfire* program. It seems more ethical to give local people a beneficial stake in maintaining the herds instead of permitting their extermination than it does to avoid the "unethical" killing of individuals by rich hunting enthusiasts. I also think it is more ethical to consider the non-charismatic animals and plants that, as I have seen in the field, can be laid waste by elephant overpopulation, even while some organisms can be dependent on normal elephant activities (e.g. Pringle 2008). Others may, of course, have a different ethical compass.
>
> **REFERENCES**
>
> Balint, P. and Mashinya, J. (2008). Campfire during Zimbabwe's national crisis: local impacts and broader implications for community-based wildlife management. *Society and Natural Resources*, 21, 783–796.
>
> Ceballos, G. and Ehrlich, P. R. (2002). Mammal population losses and the extinction crisis. *Science*, 296, 904–907.
>
> Jamieson, D., ed. (1999). *Singer and his critics*. Blackwell, Oxford, UK.
>
> Midgley, M. (1983). *Animals and why they matter*. University of Georgia Press, Athens, GA.
>
> Pringle, R. M. (2008). Elephants as agents of habitat creation for small vertebrates at the patch scale. *Ecology*, 89, 26–33.
>
> Singer, P. (1975). *Animal liberation: a new ethics for our treatment of animals*. New York Review Books, New York, NY.
>
> Smith, M. and Duffy, R. (2003). *The ethics of tourism development*. Routledge, London, UK.

accounts reveal that the Americas were extensively populated by millions of indigenous peoples who extensively altered their surroundings (Denevan 1993; Ruddiman 2005). In fact, "scientific findings indicate that virtually every part of the globe, from the boreal forests to the humid tropics, has been inhabited, modified, or managed throughout our human past... Although they may appear untouched, many of the last refuges of wilderness our society wishes to protect are inhabited and have been so for millennia" (Gomez-Pompa and Kaus 1992). And historical ecologists have demonstrated how these changes have had profound and lasting effects on populations and ecosystems, which should influence our current conservation strategies (e.g. Janzen and Martin 1982; Jackson *et al.* 2001). In short, there is no such thing as wilderness.

Did the activities of indigenous people threaten the environment? Conservationists' perception of people has long been that they are largely threats to biodiversity. Mitigating those threats is viewed as important to maintaining and recovering biodiversity. Often conservation organizations systematically identify threats long before their social causes are identified. Many social scientists see environmentally destructive behavior as symptomatic of broader societal issues, which can be obscured by the hasty labeling of local resource users as threats to biodiversity. While human activities can indeed threaten biodiversity, an exaggerated emphasis on curbing behaviors that are harmful can stand in the way of promoting those that are beneficial to conservation. Ultimately, local resource users are also conservation agents.

An opposing view suggests that local or indigenous people live in harmony with nature. As Redford (1991) points out, some researchers and conservationists have idealized the relationship indigenous people have with their environments. They have subscribed to the myth of the "ecologically noble savage", which asserts that indigenous people naturally live in harmony with the environment and have developed superior systems of resource management (or "traditional ecological knowledge") that should be adopted by conservationists.

> "Indians walked softly and hurt the landscape hardly more than the birds and squirrels, and their brush and bark huts last hardly longer than those of wood rats, while their enduring monuments, excepting those wrought on the forests by fires they made to improve their hunting grounds, vanish in a few centuries".

This quote by John Muir, an American naturalist, exemplifies this attitude. In reality, the relationship of local or indigenous people with their environments is variable. The Miwok whom Muir refers to above burned, pruned, and selectively harvested their lands to create the highly managed Yosemite landscapes that Muir saw (Anderson in Nabhan 1998:160). Another example comes from a closed tropical forest zone in South Asia. For centuries, the practice of swidden cultivation (alternately known as shifting, or slash and burn) brought about ideal habitat conditions for herbivores that do not typically inhabit this forest zone. Deer, elephants, and rhinos were drawn to grasslands and edge habitats created by swidden fallows. More generally, by altering terrestrial vegetation, human activity has changed soil structure, water availability, wildlife, and possibly the global climate system for hundreds of millennia (Westbroek *et al.* 1993; see Chapter 8). Critiques of the ecologically noble savage myth point out that some indigenous cultures have reverent environmental behaviors, and others have eroded their resource base. Such "good user/bad user" judgment is often counter-productive, especially as standards are more a product of popular imaginations than they are true to the human and ecological histories involved (cf. Fairhead and Leach 1996). But this is not to say that there are not practices we might learn from as well as those that have turned out to be destructive. Swidden agriculture, for example, *can* be environmentally destructive if practiced partially (Conklin 1975). It would also be a mistake to assume that all indigenous people are naturally stewards of their environments, any more than are any peoples. Primarily, then, it is important to remember that the ecologically-noble savage myth is, more often than not, reductionist and potentially misleads conservation activities (Buege 1996).

The final important point is to recognize and understand practices on the ground, in their historical context. To critique the ecologically noble savage myth is not to say that long-term indigenous or local residents do not develop an extensive body of knowledge related to species and ecosystem relationships. They certainly do. Knowledge borne of sustained practice and trial and error is often instructive to conservation. Not all indigenous or local people have developed or retained these bodies of knowledge, but where this knowledge does exist it can be critical to, and in effect be, the conservation effort most needed.

## 14.4 Factors mediating human-environment relations

Perceptions also arise from, and concurrently shape, our worldviews. Often, institutions direct or mediate those worldviews. Cultural, political, and economic institutions are powerful social forces that dynamically impact the environment, as coevolution of social institutions and ecological systems occurs in interesting and often unpredictable ways. For example, agriculture in North America traditionally involved cultivation of many crops. The advent of mechanical agriculture made monoculture agriculture more efficient. This *social change* led to increased use of pesticides, since fewer natural predators visit single variety crop fields. Monoculture fields produced less fertile soil and increased soil erosion. These *ecosystem changes* required specialized upkeep and changes in the organization of human labor on large, consolidated farms. This ecosystem change led to further social changes in economic institutions. Subsistence-based agriculture increasingly gave way to a surplus market-based system. These social and ecological changes also impacted worldviews as agriculturalists believed that technological advances would continue to increase crop yields and make food production more efficient.

Social institutions operate at multiple scales. They are dynamic and mutually reinforcing, as the discussion of one agricultural ecosystem illustrates. A general understanding of the following three institutions provides a framework for understanding human behavior.

### 14.4.1 Cultural

Culture is a dynamic system of collectively shared symbols, meanings, and norms – the nongenetic information possessed by a society. People are born into cultural settings, which help shape their perceptions of the world around them. For example, societies that believe guardian spirits reside in forests will often take measures to protect those forests; likewise societies that believe that ecosystems are naturally held in balance might do little to actively conserve their resources. These belief systems are often called "worldviews". Societies share some worldviews, or systems through which they interpret information and then consequently act. Within societies many interpretations of reality coexist, depending on an individual's gender, age, occupation, or education level. Understanding these perceptions helps to explain why individuals act in particular ways in their environments (Marten 2001).

Many traditional societies and human ecologists share a common perception of nature, that everything on earth is connected (for example, the Nuu Chah Nulth of Vancouver Island say "Hishuk ish tswalk," which means, "Everything is one and all is interconnected"). This worldview asserts that the actions humans take have consequences in nature. Human ecologists are inclined to focus on the details of these consequences. Another common worldview sees nature as benign, and presumes that as long as people do not alter the environment too much, she will not harm them (Marten 2001). It is not uncommon to hold contradictory worldviews at the same time. For example, one's religious beliefs may encourage human domination over nature, while one's academic field may view humans as just one part of a broader environment (Box 14.4 further explicates the role of religion in conservation biology). Together with values and norms, worldviews

---

**Box 14.4 Conservation, biology, and religion**
**Kyle S. Van Houtan**

Conservation is said to be a worthy cause for a variety of reasons. The great wilderness evangelist John Muir advocated nature preservation by describing its majesty. Forests were "sparkling and shimmering, covering hills and swamps, rocky headlands and domes, ever bravely aspiring and seeking the sky" (Muir and Cronon 1997). When she warned of the threats of pesticide pollution, Rachel Carson invoked peaceableness. Her landmark *Silent Spring* (1962) opens with: "There was once a town… where all life seemed to live in harmony with its surroundings." And in describing the perils of human overpopulation Paul Ehrlich pleads for justice. His *The Population Bomb* (1968) asserts that enjoying nature and breathing clean air are "inalienable rights." Often, however, such arguments forget their deep roots in religious traditions. For centuries, religious practices carried the torch of virtue and moral guidance. So it seems appropriate that science and religion might partner in the work of conservation (see Clements *et al.* 2009). Yet today both religion and science face a number of complaints from conservation.

Some Christians, for example, rationalize environmental destruction based on their interpretation of human dominion. Their view

*continues*

## Box 14.4 (Continued)

holds humans to be superior above all the Earth's creatures and therefore baptizes industrial development and economic growth. Any environmental regulation that limits human enterprise then becomes the enemy of divine order. Other religious beliefs maintain that the path of history culminates in apocalyptic fury. Such sects do not regard ecological preservation because they believe the planet is destined for destruction.

But science is not safe from ecological criticism either. The scientific revolution, some argue, institutionalized ecological destruction by linking experimentation, knowledge, and political power. Scientists then claimed their craft to be the new means to master human limits. It was the great empiricist Francis Bacon, after all, who dreamed a society where nature's secrets were tortured from her. Critics contend that modern science has inherited an insatiable curiosity and lacks the capacity to restrain itself, working alongside government agencies and economic corporations in a united program to exploit the biosphere. Mountains become "natural resources," ancient forests are seen as "agriculture," rivers of fish are "stocks," and human communities become the "labor force."

The question then should not be how religion and conservation biology can combine forces (Box 14.4 Figure). This might forget the ecological complaints against science and religion, which are very real and must be taken seriously. A different approach would be to cultivate the virtues conservation requires. The wisdom to know the virtues from the counterfeits that have been passed down to us requires the practical intelligence and witness of those who practice them. It is the scientists who know science best. And it is from within religious traditions where religions are most faithfully judged. Knowledge in both traditions is social and must be vigorously encouraged. People who have a foot in both a scientific and a religious tradition might be especially important here. They may see more clearly the transgressions that produce the ecological crisis. They may know more than most the virtues that conservation requires.

**Box 14.4 Figure** Nature is the context for virtue in many religious traditions. Saint Jerome, a father of the early Christian Church, is commonly depicted as a desert ascetic, pulling a thorn from a lion's paw. The upper left image in the painting suggests Jerome is drawn to the wilderness for healing and renewal, the same reason the lion is drawn to him. (Saint Jerome and the Lion. Roger van der Weyden. Reprinted with permission from The Detroit Institute of Arts.)

## REFERENCES AND SUGGESTED READING

Berry, W. (1977). *The unsettling of America: culture and agriculture*. Sierra Club Books. San Francisco, CA.

Carson, R. (1962). *Silent spring*. Houghton Mifflin, Boston, MA.

Clements, R., Foo, R., Othman, S., et al. (2009). Islam, turtle conservation and coastal communites. *Conservation Biology*, **23**, 516–517.

Ehrlich, P. R. (1968). *The population bomb*. A Sierra Club-Ballantine Book, New York, NY.

Lodge, D. M. and Hamlin C. (2006). *Religion and the new ecology*. University of Notre Dame Press, Notre Dame, IN.

Muir, J. and Cronon, W. (1997). *Nature writings*. Library of America, New York, NY.

Northcott, M. S. (1996). *The environment and Christian ethics*. Cambridge University Press, Cambridge UK.

*continues*

> **Box 14.4 (Continued)**
>
> Tucker, M. E. and Grim J., eds (1997–2003). *Religions of the world and ecology* (Vols. I–IX). Harvard University Press, Cambridge, MA.
>
> Van Houtan, K. S. (2006). Conservation as virtue: a scientific and social process for conservation ethics. *Conservation Biology*, **20**, 1367–1372.

underlie resource management systems, and form the basis for decision-making and action.

An example from Papua New Guinea illustrates how culture impacts conservation interventions (West 2006). Noticing the decline in birds of paradise, an international conservation organization set out to save these species from extinction. Since conservationists saw the importance of these birds for the ecosystem and local community, they believed the local resource users would readily comply with their project. Unbeknownst to the conservationists, when they asked the local villagers to engage in conservation actions, they were entering into a complex exchange relationship. Villagers expected not only medicine, technology, and tourism development, but an ongoing reciprocal relationship by which the villagers would continue to protect species in exchange for ongoing assistance in any number of areas that are usually the purview of government. Fundamental cultural misunderstandings such as this undermine conservation interventions, leading to disappointment and project disintegration.

### 14.4.2 Political

Political systems are a set of institutions that govern a particular territory or population. These systems are not to be confused with politics, or the maneuvering for power (though politics heavily influence whether conservation initiatives will be carried out). Political institutions can be distinguished by degree of power concentration, level of formality, global to local scale, and normative characteristics. Conservation interventions often require the reinforcement of policy by multiple political systems at different scales. A small-scale conservation intervention, for example, may draw on traditional authority, the national environmental ministry, an international NGO (non-governmental organization), and global trade policies to achieve its goals. While there are a range of political systems that impact conservation efforts, we focus on political processes, or governance here. Governance refers to a set of regulatory processes and mechanisms through which the state, communities, businesses and NGOs act (Lemos and Agrawal 2006).

In addition to compensation and clarification on land tenure and access (see Equity, Resource Rights, and Conservation section below), participation in governance has been critical to establishing good relationships between conservationists and local resource users (Zerner 2003). This sharing of resource management, sometimes referred to as co-management, more equitably distributes authority between local people, stakeholders, state-level political systems, and conservation organizations involved (Brechin *et al.* 2002). There is considerable controversy over when, where, and to what extent co-management should be endorsed. Some worry that co-management and consideration of local concerns are dangerous, over-riding the maintenance of biodiversity, whereas others call for increased equity for indigenous and local communities. These debates should be contemplated in light of the fact that political and economic institutions are the products of a contestation for power between various sectors of a population, and that historically this struggle has resulted in entrenchment of institutions that favor the powerful. That is, conservation must often arise through institutions that are themselves considerably inequitable.

Further, the degree to which co-management is in fact shared management, with equitable distributions of positions of power and decision making across the governing body (e.g. a park's management board) is hotly debated. Generally, however, it is assumed that genuine or bona fide systems of co-management include multiple opportunities for participation; equitable control over decisions and outcomes; the existence of governing bodies that are truly representative of those with recognized rights and/or local populations more broadly; and capacity building for the realization of tasks and responsibilities (McKean 2000).

Where co-management is not desirable or feasible, political scientists and other social researchers have uncovered a number of principles for designing effective conservation governance (Ostrom 2008). These vary greatly depending on local circumstances. Governance processes that have been effective in a range of case studies include participatory decision-making, the presence of enforcement and conflict resolution mechanisms, and flexible management. Cultural variations in implementing these design principles lead to innumerable governance arrangements.

In the Comoros Islands, devolution of management rights to the local resource users increased participation in marine protected area (MPA) decision-making. Ultimately, the resource users decided to limit outside use of the MPA by restricting certain types of fishing gear (Granek and Brown 2005). Local resource users set limits on resource extraction through the regulatory process of governance.

### 14.4.3 Economic

In every society goods and services have values. These goods are distributed in networks that range from household to international scales. It would be difficult to find a society today that is not affected by the worldwide economy, however. The structure of the global economy impacts resource use in surprising ways. We illustrate this here via the "hamburger connection".

The 1980s brought worldwide consciousness of Amazonian deforestation. Clearing land for cattle was the main driver of local deforestation, yet cattle ranches in this area were less lucrative, and less destructive, than cash cropping. So why did people engage in this labor-intensive activity? Researchers traced the political economy of Amazonian deforestation and found that international consumers were implicated in this destruction (Leduc 1985). The Brazilian government and the World Bank blamed deforestation on local swidden cultivators, yet Leduc illustrates how Brazilian and international development policy was actually at fault. These institutions encouraged the conversion of rainforest to pasture land by providing tax cuts and perverse incentives for cattle ranchers. Leduc finds that it was the *wealth* of European consumers and Brazilian cattle ranchers, not the *poverty* of local resource users, which drove Amazonian deforestation.

The global economy supplied goods to feed Europe's desire for hamburger, impacting environmental degradation thousands of miles away.

Increasing population, decreasing household size, and especially the associated rise in consumption, negatively impact resource use (see Introduction Box 1). Yet some researchers have highlighted how poorly executed global distribution networks are equally implicated in resource destruction. Awareness of these large-scale economic forces is important for unraveling local resource use patterns. Household economic characteristics are also significant in designing conservation interventions. Figure 14.2 relates household economic factors to conservation strategies.

Given the diversity of cultural, political, and economic institutions and their variable local manifestations, there can be no worldwide conservation program. However, there are similar conservation strategies that have been implemented around the world. We turn now to the role that local resource users play in these common conservation strategies.

## 14.5 Biodiversity conservation and local resource use

Conservation interventions that focus more exclusively on biodiversity are now pervasive, and most focus on ameliorating threats to species and

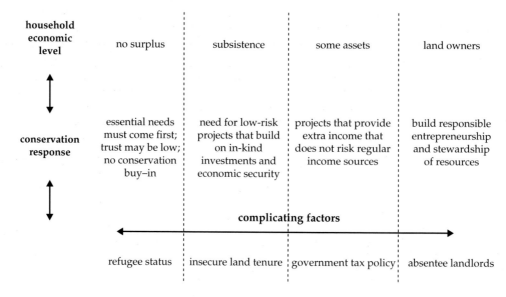

**Figure 14.2** Economic site characteristics and conservation initiatives: points to take into consideration. Adapted from Russell and Harshbarger (2003). Teasing out relevant factors that determine one conservation intervention over another is challenging. This figure presents one element considered in conservation planning: income. Where income is a defining factor in conservation planning, the conservation response may derive from the household economic level. We also present a few factors that complicate this simplistic view of an appropriate conservation response.

systems (see Chapter 15). According to analyses based on identified threats for species listed as threatened or endangered, the five greatest threats to imperiled terrestrial species are habitat loss (Chapter 4), overexploitation (Chapter 6), pollution, non-native species (Chapter 7), and disease, with the order of importance varying greatly between USA and China (Wilcove et al. 1998; Li and Wilcove 2005). Marine, estuarine, and coastal species in the USA are threatened by many of the same and some additional threats, such as water diversion, increased human presence, vessel interaction, and climate change (Kappel 2005). Although climate change was missing from the threats to terrestrial species in the studies above, it is increasingly recognized as a rising direct and indirect threat in all systems (Chapter 8).

Local resource users may be positively and/or negatively impacted by conservation. They may see, for instance, an increase in income due to ecotourism, or a decrease in fish catch concurrent with the opening of a marine protected area. Their health may suffer with the decline of resource access, or the gazetting of a protected area may privilege their resource rights over those of regional loggers. In short, there is no simple answer to the question of how local resource users are integrated into and impacted by conservation.

Conservation interventions, too, can be undertaken by a variety of actors from local to global scales. Some actors include governments and government agencies, NGOs, corporations, communities, and individual stakeholders. These actors address conservation from a diversity of angles, such as law, policy, management of wildlife and ecosystems, or individual actions [for example, through direct payments for conservation-friendly behavior (Ferarro and Kiss 2002)]. Table 14.1 outlines some conservation interventions that involve local resource users, and the potential positive and negative impacts of those interventions. Anticipating potential impacts of conservation interventions may also make them more robust and, ultimately, more sustainable (Chan et al. 2007).

Conservation that involves local resource users does not adopt a similar form worldwide. Table 14.2 compares protected area policies from Nepal, Brazil, and Australia to illustrate the

Table 14.1 This table presents the orthodox view of how local resource users impact their environments, and how they in turn are impacted by conservation interventions. A deeper analysis might ask questions such as; who is responsible for the degradation, overexploitation, etc. in column three? Does the conservation strategy disproportionately affect local resource users? Are the strategies in column one responsive to the threats presented in column three?

| Strategy | Impact on local resource users | Response to... |
|---|---|---|
| Protected area | Limits entry, extraction, and use of designated area | Habitat loss and degradation, overexploitation |
| Zoning | Designates areas where local resources may be extracted | Habitat loss and degradation, overexploitation |
| Purchase of water/land rights | Transfers ownership or use rights to/from locals | Water diversions, overexploitation |
| Ecotourism | Brings outside investment to local businesses, employment in service industry | Multiple threats |
| Community-based natural resource management | Formal encoding of local monitoring and managing resource extraction | Multiple threats |
| Direct payments for ecosystem services | Payment received for successfully maintaining local resources | Overexploitation, habitat loss and degradation, pollution |
| Integrated conservation and development projects | Development of small-scale economic initiatives that incorporate sustainable resource use | Multiple threats |

diversity of local involvement in one particular conservation strategy, protected areas. Local resource users may not want to be involved in all aspects of conservation management (as was the case in Kakadu National Park, Australia). Ethical imperatives direct the extent and manner of consulting local resource users. We turn to these now.

## 14.6 Equity, resource rights, and conservation

Questions of rights and equity have recently emerged as paramount to the practice of conservation and are in part the by-product of several years of debate between social scientists and

Table 14.2 Three examples of variations in local involvement in protected areas (from Claus, unpublished data). The three national parks presented here involve local resource users in conservation policy and implementation. As park policies are implemented in myriad ways, this empirical comparison accentuates the differing degree and nature of local conservation involvement.

| Do policies allow... | Sagarmatha National Park, Nepal | Rio Ouro Preto Extractive Reserve, Brazil | Kakadu National Park, Australia |
|---|---|---|---|
| Continued subsistence use | ✓ | ✓ | ✓ |
| Formalized policy consultation | | | ✓ |
| Sharing of entrance fees | ✓ | | ✓ |
| Co-management | ✓ | | ✓ |
| Integration of local resource management regimes | ✓ | ✓ | ✓ |
| Local land ownership | ✓ via enclaves | | ✓ |
| Training/integration into management structure | ✓ Limited | ✓ in local organizations only | ✓ |
| Establishment of organizations with substantial local representation | ✓ | ✓ | ✓ |
| Power to determine land use | | | |
| Technical assistance regarding resource management | ✓ | ✓ | ✓ |

conservation biologists (Chapin 2004). Critical to these debates is both a recognition that rights heretofore ignored need be recognized, addressed, and integrated into conservation (including those pertaining to access, use rights, and compensation in the event of their loss), and that justice and equity more broadly need to become a cornerstone principle in the advancement of conservation and the maintenance of biodiversity.

Rights refer to a bundle of entitlements or permissions assigned to or affiliated with a group or population. These rights may be individually or collectively held, and they include the right to tenure and/or ownership, the maintenance of livelihood security and resource access (such as the ability, for instance, to reduce the impact of damage-causing animals on the periphery of protected areas), and the right to be involved in the governance of both these rights and the lands or waters with which they are associated. Box 14.5 provides an example of how asserting rights can lead to the protection of forests.

It is often difficult for conservation officials to understand the histories of peoples that may have preceded a park or protected area. It is nonetheless important to recognize that many rights-based systems are formalized in treaties; in titles

---

### Box 14.5 Empowering women: the Chipko movement in India
### Priya Davidar

Rural societies depend on natural resources for fuel, fodder, food, medicine, and construction materials. Women play an important role in the collection, use and sale of forest products such as fuel-wood and fodder to meet household requirements, thereby enhancing the economic security of their households (Wickramasinghe 2005). Fuel-wood is an important source of domestic energy for rural households in tropical countries and women are often involved in the harvesting and sale of fuel-wood collected from the forest (Gera 2002). Commercialization of forestry operations and deforestation therefore adversely affect the livelihoods of poorer forest dependent households. The Chipko movement is one case in point, where rural women and children fought back against timber operations. Chipko means "to embrace" or "to hug" and the concept of saving trees from felling by embracing them is old in Indian culture. The first recorded instance of such action was in 1604 among the Bishnois community in Rajastan when two Bishnoi women, Karma and Gora, sacrificed their lives in an effort to prevent the felling of Khejri (*Prosopis cineraria*) trees.

The Chipko agitation began in 1971 in the Uttaranchal in the Himalayas, as a grass roots Gandhian movement to assert the rights of the local communities over forest produce (Berreman 1985; Joshi 1982). During British rule, the Himalayan forests were heavily exploited for timber, particularly during the two world wars. Commercial exploitation continued with India's independence. Deforestation led to soil erosion and landslides, destroying crops and houses. Women had to walk longer distances to collect fuel-wood and fodder. Women's participation in the movement can be traced to a remote hill town called Reni where a contractor in 1973 was given a permit to fell about 3000 trees for a sporting goods store. When the woodcutters appeared, the men had been called away for other tasks. The alarm was sounded and a widow in her 50s, Gaura Devi, collected twenty-seven women and children and rushed into the forest to protect the trees. After threats and altercations from the woodcutters, the women would not back off, and then embraced the trees, as a consequence of which the woodcutters backed off (see Box 14.5 Figure). This movement spread to many areas in this region, and village women saved around 100 000 trees from being cut. The movement was characterized by de-centralized and locally autonomous activism by local communities, led mostly by village women. Following this, the government was forced to abolish the private contract system of felling and in 1975 the Uttar Pradesh Forest Corporation was set up to perform this function.

*continues*

> **Box 14.5 (Continued)**
>
>
>
> Box 14.5 Figure  A demonstration of the Chipko movement. Photograph courtesy of The Right Livelihood Award.
>
> The Chipko Movement had two elements: one section concentrated on protecting existing forests from being logged and the other focused on promoting reforestation and developing sustainable village production systems based on forests and agroforestry. The latter were led by Shri Chand Prasad Bhatt, one of the original organizers of the Chipko movement who provided a unified vision and leadership to the movement. Bhatt worked closely with the village women and encouraged them to assert their environmental rights. In 1987 Chipko was chosen for a "Right Livelihood Award," known as the "alternate Nobel" prize honor. The honor was rightly deserved for this small movement dominated by women that became a call to save forests.
>
> **REFERENCES**
>
> Berreman, G. (1985). Chipko: Nonviolent direct action to save the Himalayas. *South Asian Bulletin*, 5, 8–13.
> Gera, P. (2002). *Women's role and contribution to forest based livelihoods*. HRDC (Human Rights and Documentation Center), UNESCO, New Delhi, India.
> Joshi, G. (1982). *The Chipko movement and women*. PUCL bulletin, September.
> Wickramasinghe, A. (2005). Gender relations beyond farm fences: reframing the spatial context of local forest livelihoods. In L. Nelson and J. Seager, eds *A female companion to feminist geography*, pp. 434–445. Blackwell Publishing, Oxford, UK.

bought and exchanged in markets; or informal to the extent that they are based in traditional or indigenous systems wherein some flexibility of rules is to be expected. Treaties, however formal, may be recognized and intrinsic to rights or they may be largely ignored. Great variability of treaty or claim-based rights exists across nation states, and most of those with significant implications for protected areas follow recent court decisions in their "home" nation states such as those in Australia (the Mabo decision) and Canada (Delgamuukw; Haida-Taku; Tsilhqot'in). Conversely, informal systems of rights tend to be oral (though not exclusively so) and tied to local systems of kinship, governance and decision-making. For instance, a group may distribute rights to traditional territories or specific fishing grounds to a lineage (rights passed through the matriline, patriline, or both), but that same system may allow for considerable room to negotiate temporary rights. Overall, it is crucial to understand formal and informal as well as long-standing and ephemeral systems of rights in any community with which conservation organizations engage.

Given the above-mentioned history of colonialisms closely linked to the establishment of many protected areas, in many cases peoples' rights have been lost or dispossessed. Recognition, restitution, equitable compensation, and the settling of land claims noted above, is necessary for the (re-) establishment of good relations between parks and people (Colchester 2004).

The principle of equity (fair distribution of benefits) permeates many of the above considerations regarding eviction or restrictions on resource use. Debates about compensation for the dispossession of lands or co-management suggest that a more equitable world is necessary for effective conservation. Further, the struggle to sustain biodiversity may involve justice (the principle or ideal of right action) or equity as assigned to extant peoples, future generations, and nonhuman organisms. In order to consider justice, one must also ask: justice for whom? Utilitarian approaches consider justice done when the greatest good is done, in the aggregate across a number of stakeholders, and many of these stakeholders might be very distant from the protected areas in question. Whereas much of the above discussion suggests that it is proximate peoples, with standing or legitimate claims to rights, whom are the appropriate focus for conservation. This is particularly so as those who enjoy the benefits of conservation may not be those who suffer its costs. Considering justice then requires us to ask the identities and condition of peoples most affected or impacted by a decision (Rawls 1999). For the purposes of conservation, rights, *and* justice, the point here is not to defend a particular approach, but to assert that justice should involve a fair distribution of rights, responsibilities, costs, and benefits. "And when actions have impacts with unfair distribution, justice requires appropriate restitution" (Chan and Satterfield 2007). At the heart of much of the controversy about justice and conservation is the incongruence amongst the intended stakeholders (those who are affected by a decision in a morally relevant way) (Chan *et al.* 2007). Whereas conservationists consider non-human organisms—and sometimes also species, ecosystems, and inanimate components of the environment—to have rights or to be deserving of moral consideration, critics of conservation frequently focus first and foremost or exclusively on human beings. If we are to make headway in controversial settings for conservation, we will require an ethical framework that allows us to consider our obligations to the non-human world alongside those to fellow human beings (Sodhi *et al.* 2008). Such an applied "global" environmental ethic is elusive, however, and so *in-situ* ones can and should suffice.

Conservation often also requires changing local behavior, resource access or livelihoods. This social engineering serves biodiversity, yet biodiversity itself is not a value-neutral concept. The biodiversity concept is rooted in what a particular group of people view as "ideal nature", and places value on what cultural practices are good or bad. Where biodiversity conservation clashes with local environmental values, it is necessary to consider the implicit prescriptions of ideal conservation-oriented behavior that underlie the distribution of benefits from conservation projects. An extreme example is the complete exclusion of local people from protected areas, but subtler measures such as the acceptance of hunting with spears instead of guns displays underlying assumptions about how people should interact with their environments. Acting justly requires recognizing one's assumptions and the behavior judgments that arise from them. Only then is it possible to prescribe appropriate conservation behavior (Chan and Satterfield 2007). Biodiversity conservation depends on solutions that are socially just, and attentive to rights where they exist formally and informally. As a society, we should strive for sustainable economic development and socially as well as ecologically sustainable conservation—for its global and future benefits and for its own sake—in harmony with the cultural, social, and economic well-being of local peoples.

## 14.7 Social research and conservation

One of the anomalies of modern ecology is that it is the creation of two groups, each of which seems barely aware of the existence of the other. The one studies the human community almost as if it were a separate entity, and calls its findings sociology, economics and history. The other studies the plant and animal community and comfortably relegates the hodge-podge of politics to the liberal arts. The inevitable fusion of the two lines of thought will, perhaps, constitute the outstanding advance of the present century

—Aldo Leopold (1935 in Meine and Knight 1999)

As Aldo Leopold suggests, integrating understanding of human and other natural systems is crucial for conservation success. Social disciplines like history, ethics, policy and business studies, and the social sciences provide insights into conservation implementation, from formulating plans to enacting them on the ground (see Figure 14.3 for a conceptual diagram of how these disciplines interact). Examples of social research in conservation include clarifying resource use patterns, mapping socio-political territories, and uncovering regional resource tenure institutions.

Yet, social research in conservation is undervalued. There are a number of reasons conservation fails to appreciate social research. Firstly, social research takes time, and conservation moves at a rapid pace. Secondly, when funds are limited, as they nearly always are, biological research takes precedence. Thirdly, conservation organizations are most often staffed by natural scientists, who may feel that simply because they are human they understand human behavior. As Russell and Harshbarger (2003) point out, anyone with field experience can collect social data. However, it takes considerable training and practice to collect good social data and to interpret those data in meaningful ways. Finally, disciplinary tensions between natural and social scientists complicate cross-disciplinary work. They frequently differ in worldviews, with natural scientists more likely to see (other) people as threats to biodiversity and social researchers more likely to see (local) people as autonomous agents worthy of respect and sovereignty.

Given the serious consequences of failed conservation projects—for the environment as well as for future conservation initiatives—conservation organizations are increasingly turning to social researchers to answer social problems. Social researchers are well placed to answer questions surrounding lack of buy-in to conservation initiatives, why people engage in particular environmental behaviors, or what

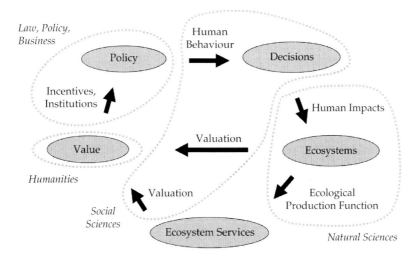

**Figure 14.3** The study of conservation of biodiversity in functioning ecosystems has been largely pursued separately in diverse academic disciplines. Effective conservation requires an integrated understanding of how people's decisions influence ecosystems, how ecosystems produce services for people, how those services are valued by people, how those values translate into policies, and how those policies result in human actions. This requires an integration of diverse fields, natural and social.

**Table 14.3** Social research and conservation. This table draws on resources available on the Society for Conservation Biology's Social Science Working Group webpage (www.conbio.org/sswg).

| Discipline | Definition | Prevalent conservation-relevant methods | Traditional unit of analysis | Sample conservation contribution and case study |
|---|---|---|---|---|
| Anthropology | The scientific and humanistic study of the human species: present and past biological, linguistic, and cultural variations. | Ethnography, discourse analysis, participant observation, excavation/paleological coring | Community | Analyzing cultural context of conservation intervention determines differences between local and outsider perceptions of conservation projects, and associated biodiversity implications (West 2006). |
| Business, Management | The study of corporate action and the effective operation of organizations. | Simulation and analytical modeling | Individual, Firm, NGO | Understanding how corporations and other organizations respond to circumstances fosters more effective conservation interventions; effective management approaches improve operation of conservation organizations (Stoneham et al. 2003). |
| Economics | The study of the allocation of resources under scarcity: how we behave when using resources (e.g. time, money) with insufficient quantity to satisfy all users. | Econometrics, simulation and analytical modeling | Individual, Firm & State | Incorporating conservation costs into strategy planning phase results in larger biological gains than when costs are ignored (Naidoo et al. 2006). |
| Ethics | The study of right and wrong actions based on normative premises and logical argument. | Inductive/deductive reasoning | Individual - Earth | Exploring the competing values underlying people's actions and potential policies can foster mutual understanding among stakeholders and agreement over appropriate decisions (Chan and Satterfield 2007). |
| Geography | The study of human activity, culture, politics and economics within its spatial and environmental context. | GIS, remote sensing, spatial analysis, geostatistics | Community - Earth | Using participatory GIS integrates diverse stakeholder knowledge to clarify spatial aspects of landscape level environmental change (Balram et al. 2004). |
| History | The reconstruction and analysis of past events of importance to the human race. | Text and media analysis | Individual - Earth | Understanding sequences of past events and their possible causes (both events and social contexts) suggests how present-day conservation actions may unfold and be received (Turner 2006). |
| Law | The study of laws and policies, their origins, implementation, judicial interpretation, and enforcement. | Policy and law analysis | Region, State, Nation, and between these | Understanding legal and judicial processes improves efforts to implement new legislation and regulations, and to use existing ones in court processes (e.g. lawsuits) (Thompson 2001). |

| | | | | |
|---|---|---|---|---|
| Political Science | The study of governments, public policies, and political processes, systems, and behavior. | Text analysis, scenario modeling, comparative statistics | State | Presenting framework for understanding and designing compensation schemes for resource rights acquisition/loss due to protected area establishment (Mascia and Claus 2009). |
| Psychology | The study of human thought, feeling and behavior in order to understand behavior and promote human welfare. | Controlled experimentation, psychoanalysis, brain scanning, computational modeling | Individual - Small group | Researching the relationship between values and environmental behavior to inform environmental message framing in the USA (Schultz and Zelezny 2003). |
| Sociology | The study of societies, particularly social relations, stratification, and interaction. | Social network analysis, content analysis, longitudinal studies | Community - Nation | Researching a collaborative watershed planning effort that, through creating social capital, led to cooperative conservation amongst participants (Salamon et al. 1998). |

social pressures encourage environmental degradation. As social research disciplines vary in methodology, scale, and scope of data collection, they have different contributions to the field of conservation. Table 14.3 details these disciplines and research they undertake that informs conservation action.

## Summary

- Conservation is inherently a social process operating in a social context. As such, conservationists will benefit from a nuanced understanding of people's perceptions and behaviors as individuals and in organizations and institutions.
- While there is no easy recipe for how local resource users should participate in modern conservation initiatives, attentiveness to resource rights and equity are critical in every conservation project.
- A successful conservation movement will effectively integrate the natural sciences and diverse fields of social research.

## Relevant websites

- Society for Conservation Biology's Social Science Working Group: http://www.conbio.org/workinggroups/sswg/.
- Advancing Conservation in a Social Context: http://www.tradeoffs.org/static/index.php.
- Conservation and Society Interdisciplinary Journal: http://www.conservationandsociety.org/.

## Suggested reading

Chan, K. M. A. and Satterfield, T. (2007). Justice, equity, and biodiversity. In S. Levin, G. C., Daily, and R. K. Colwell, eds *The Encyclopedia of Biodiversity Online Update 1.* Elsevier, Oxford, UK.

Colchester, M. (2004). Conservation policy and indigenous peoples. *Environmental Science and Policy,* 7, 145–153.

## REFERENCES

Balram, S., Suzana, D. E., and Dragicevic, S. (2004). A collaborative GIS method for integrating local and technical knowledge in establishing biodiversity conservation priorities. *Biodiversity and Conservation*, **13**, 1195–1208.

Brechin, S., Wilshusen, P., Fortwangler, C., and West, P. C. (2002). Beyond the square wheel: Toward a more comprehensive understanding of biodiversity conservation as social and political process. *Society and Natural Resources*, **15**, 41–64.

Buege, D. J. (1996). The ecologically noble savage. *Environmental Ethics*, **18**, 71–88.

Carson, R. (1962). *Silent Spring*. Houghton Mifflin, Boston, MA.

Caufield, C. (1990). The ancient forest. *The New Yorker*, 14 May 1990, 46–84.

Chan, K. M. A. and Satterfield, T. (2007). Justice, equity, and biodiversity. In S. Levin, G. C., Daily, and R. K. Colwell, eds *The Encyclopedia of Biodiversity Online Update 1*. Elsevier Ltd, Oxford.

Chan, K. M. A., Pringle, R. M., Janganathan, J., et al. (2007). When agendas collide: Human welfare and biological conservation. *Conservation Biology*, **21**, 59–68.

Chapin, M. (2004). A Challenge to Conservationists. *World Watch Magazine*, November/December 2004, 17–31.

Colchester, M. (2004). Conservation policy and indigenous peoples. *Environmental Science and Policy*, **7**, 145–153.

Conklin, H. (1975). *Hanunoo agriculture*. Forestry Development Paper. Food and Agriculture Organization of the United Nations, Rome, Italy.

Cronon, W. (1995). The trouble with wilderness; or getting back to the wrong nature. In W. Cronon, ed. *Uncommon ground: Rethinking the human place in nature*. W. W. Norton and Company, New York, NY.

Denevan, W. (1992). The pristine myth—the landscape of America in 1492. *Annals of the Association of American Geographers*, **82**, 369–385.

Fairhead, J. and Leach, M. (1996). Enriching the landscape: social history and the management of transition ecology in the forest-savanna mosaic of the Republic of Guinea. *Africa*, **66**, 14–36.

Ferraro, P. J., and Kiss, A. (2002). Direct payments to conserve biodiversity. *Science*, **298**, 1718–1719.

Gomez-pompa, A. and Kaus, A. (1992). Taming the wilderness myth. *BioScience*, **42**, 271–279.

Granek, E. F. and Brown, M. A. (2005). Co-management approach to marine conservation in Moheli, Comoros Islands. *Conservation biology*, **19**, 1724–1732.

Grove, R. H. (1996). *Green imperialism, colonial expansion, tropical island edens and the origins of environmentalism 1600–1860*. Cambridge University Press, Cambridge, UK.

Jackson, J. B. C., Kirby, M. S., and Berger, W. H. (2001). Historical overfishing and the recent collapse of coastal ecosystems. *Science*, **293**, 629–638.

Janzen, D. H. and Martin, D. H. (1982). Neotropical anachronisms: the fruits the gomphotheres ate. *Science*, **215**, 19–27.

Kappel, C. V. (2005). Losing pieces of the puzzle: threats to marine, estuarine, and diadromous species. *Frontiers in Ecology and the Environment*, **3**, 275–282.

Leduc, G. (1985). The political economy of tropical deforestation. In H. J. Leonard, ed. *The political economy of environmental abuse in the third world*. Holmes and Meier, New York, NY.

Lemos, M. C., and Agrawal, A. (2006). Environmental governance. *Annual Review of Environment and Resources*, **31**, 297–325.

Leopold, A. (1949). *A sand county almanac and sketches here and there*. Oxford University Press, New York, NY.

Li, Y. M. and Wilcove, D. S. (2005). Threats to vertebrate species in China and the United States. *Bioscience*, **55**, 147–153.

Mascia, M. and Claus, C. A. (2009). A property rights approach to understanding human displacement from protected areas: the case of marine protected areas. *Conservation Biology*, **23**, 16–23.

Marsh, G. P. (1864). *Man and nature*. Harvard University Press, Cambridge, MA (1965; annotated reprint of the 1864 original).

Marten, G. (2001). *Human ecology*, Earthscan Publications Ltd., UK.

McKean, M. A. (2000). Common property: what is it, what is it good for, and what makes it work. In C. C. Gibson, M. A. McKean and E. Ostrom, eds *People and forests: communities, institutions, and governance*. MIT Press, Cambridge, MA.

Meine, C. and Knight, R. L., eds (1999). *The essential Aldo Leopold: quotations and commentaries*. University of Wisconsin Press, Madison, WI.

Nabhan G. (1998). *Cultures of habitat*, Counterpoint, Washington, DC.

Naidoo, R., Balmford, A., Ferraro, P. J., Polasky, S., Ricketts, T. H., and Rouget, M. (2006). Integrating economic costs into conservation planning. *Trends in Ecology and Evolution*, **21**, 681–687.

Neumann, R. (2004). Moral and discursive geographies in the war for biodiversity in Africa. *Political Geography*, **23**, 813–837.

Ostrom, E. (2008). The challenge of common-pool resources. *Environment*, **50**, 8–20.

Rawls, J. (1999). *A theory of justice*. Harvard University Press, Cambridge, MA.

Redford, K. (1991). The ecologically noble savage. *Cultural Survival Quarterly*, **15**, 46–48.

Robbins, W. (1985). The social context of forestry: the Pacific Northwest in the Twentieth Century. *Western Historical Quarterly,* **16**, 413–427.

Ruddiman W. F. (2005). *Plows, plagues, and petroleum: how humans took control of climate.* Princeton University Press, Princeton, NJ.

Russell, D. and Harshbarger, C. (2003). *Groundwork for community-based conservation,* Altamira Press, Lanham, ML.

Salamon, S., Farnsworth, R. L., and Rendziak, J. A. (1998). Is locally led conservation planning working? A farm town case study. *Rural Sociology,* **63**, 214–234.

Schultz, P. W. and Zelezny, L. (2003). Reframing environmental messages to be congruent with American values. *Human Ecology Review,* **10**, 126–136.

Stevens, S. (1997). *Conservation through cultural survival: Indigenous peoples and protected areas.* Island Press, Washington, DC.

Stoneham, G., Chaudhri, V., Ha, A., and Strappazzon, L. (2003). Auctions for conservation contracts: an empirical examination of Victoria's BushTender trial. *Australian Journal of Agricultural and Resource Economics,* **47**, 477–500.

Sodhi, N. S., Acciaioli, G., Erb, M. and Tan, A. K.-J., eds (2008). *Biodiversity and human livelihoods in protected areas: case studies from the Malay Archipelago.* Cambridge University Press, Cambridge, UK.

Sundberg, J. (1998). NGO landscapes in the Maya Biosphere Reserve, Guatemala. *Geographical Review,* **88**, 388–412.

Thompson, B. H. J. (2001). Providing biodiversiy through policy diversity. *Idaho Law Review,* **38**, 355–384.

Turner, J. M. (2006). Conservation science and forest service policy for roadless areas. *Conservation Biology,* **20**, 713–722.

West, P. (2006). *Conservation is our government now: the politics of ecology in Papua New Guinea,* Duke University Press, Durham, NC.

Westbroek, P., Collins, M. J., Jansen, J. H. F., and Talbot, L. M. (1993). World archaeology and global change: did our ancestors ignite the Ice Age? *World Archaeology,* **25**, 122–133.

White, R. and Findley, J., eds (1999). *Power and place in the North American West.* University of Washington Press, Seattle, WA.

Wilcove, D. S., Rothstein, D., Dubow, J., Phillips, A., and Losos, E. (1998). Quantifying threats to imperiled species in the United States. *BioScience,* **48**, 607–615.

Zerner, C, ed. (2003). *Culture and the question of rights: forests, coasts, and seas in Southeast Asia.* Duke University Press, Durham, NC.

# CHAPTER 15

# From conservation theory to practice: crossing the divide

Madhu Rao and Joshua Ginsberg

Conservation biology is continually developing new tools and concepts that contribute to our understanding of populations, species and ecosystems (Chapter 1). The science underpinning the field has undoubtedly made rapid strides generating more effective methods to document biodiversity, monitor species and habitats. Scientists have developed comprehensive priority setting exercises to help determine where and what to conserve in on-going attempts to identify which factors would best serve as the basis for triage for species and ecosystems (Wilson et al. 2007; Chapter 11). They are well positioned to track the loss of species and ecosystems in broad patterns even if precise details are not always available (Chapter 10). However applying the science effectively requires the efforts of conservation biologists combined with a diversity of other actors, most of whom are non-biologists and include local and indigenous communities, civil servants at all levels of government, environmental consultants, park managers, environmental lobbyists, private industry, and even the military (Box 15.1; Chapter 14). This amorphous group of practitioners will pursue a diverse set of activities which include putting up or taking down fences (literal and metaphorical), lobbying politicians, buying land, negotiating with members of local and indigenous communities, tackling invasive species problems, guarding against poachers and managing off-take of plants and animals.

There are many pressing challenges facing practical conservation. Forces affecting biodiversity in different ecosystems have altered over the past two decades. For instance, the nature of tropical forest destruction has changed from being dominated by rural farmers to currently being driven substantially by major industries and economic globalization, with timber operations, oil and gas development, large-scale farming and exotic-tree plantations being the most frequent causes of forest loss (see Chapter 4). A direct result of these changes is the need for engaging not just conservation minded individuals and organizations, but those in the largest, and most influential, of the world's corporations and multilateral institutions (Box 15.2). In addition, the changes in those factors driving loss– and in the scale of loss – requires that we diversify our approaches, and focus not just on biodiversity, but on the whole issue of those goods and services that natural systems provide for us (Daily 1997, Woodwell 2002; Box 15.3). Global threats, and opportunities, such as climate change (Chapter 8), are forcing conservation practitioners to work at a variety of scales to better integrate these challenges (Bonan 2008). Conservation science must meet the continually changing nature of threats to biodiversity (Butler and Laurance 2008); conservation biologists and practitioners need to design and leverage solutions in response to these global changes in threat.

Not only is the practice of conservation getting more complicated, but it has a stronger global presence, and increasingly large expenditures (Cobb et al. 2007). As a result, implementing agencies and specifically conservation organizations are being held to a higher standard in monitoring and evaluating their conservation success, and failure (Wells et al. 1999; Ferraro and Pattanayak 2006; see also Box 15.4). Another issue that

## Box 15.1 Swords into ploughshares: reducing military demand for wildlife products
### Lisa Hickey, Heidi Kretser, Elizabeth Bennett, and McKenzie Johnson

Illegal trade in wild animals and plants is one of the greatest threats to populations of many species. The impacts are diverse, and the direct impact in reducing wildlife populations is well studied, and often noted (Robinson and Bennett 2000; Bennett 2005). Indirect effects – including the global movement of emerging infectious diseases (Karesh et al. 2005) pose a different, but equally compelling case for better management of such trade. The economic imperatives are great as well, with current estimates of the value of illicit trade (estimated at US$6 billion; Warchol 2004) second only to narcotics and arms trafficking. Legal trade is clearly occurring at a much higher level (on the order of US$150 billion per year) if trade in commodities such as timber and ocean fish are included in these studies (Warchol 2004), but this also produces a significant threat since legal trade in many species is unsustainable.

US military personnel have a long-term presence abroad, including in countries of great biodiversity importance. These personnel and affiliates have significant buying power that influences local markets, including the ability to drive the demand for wildlife products. The Afghanistan Biodiversity Project funded by USAID (United States Agency for International Development) and implemented by the Wildlife Conservation Society (WCS) has found that US soldiers serving in Afghanistan are primary buyers of illegal wildlife products there, including big cat skins and other types of trophies. WCS has initiated a program focused on education and awareness-raising to reduce purchasing of wildlife products by the US military, and protect American soldiers from serious penalties related to the import of illegal wildlife. WCS, in conjunction with the Department of State, traveled to Bagram Air Base, the largest military base in Afghanistan, to educate soldiers on issues related to illegal trade in wild species. Military Police (MPs) received instruction on issues of biodiversity in Afghanistan and how to identify threatened and endangered Afghan species. The partnership between Bagram customs officials and WCS aims to reduce illegal buying of wildlife products by soldiers, and MPs have already shown an adept ability to identify and seize prohibited wild species before they leave the base, as well as enthusiasm to collaborate on the program.

To further address the demand for wildlife products by US military personnel, WCS is complementing its work in Afghanistan by working with the military in the US. As part of this effort WCS ran a booth at Safety Day, in Fort Drum, to raise awareness about illegal wildlife trade for both pre and post-deployment troops. A survey conducted at Fort Drum as part of this effort indicated that fewer than 12% of soldiers ($n = 371$) had heard of CITES, yet more than 40% had either purchased wildlife products while overseas or seen other members of the military purchase these items (Kretser, unpublished data).

To increase its effectiveness in working on wildlife trade issues with the military, WCS is planning to develop a template approach to begin addressing wildlife trade within all branches of the military. Activities include the development of pocket cards and playing cards for soldiers as well as handouts and power point slides for incorporation into military-run environmental training including officer training, pre-departure briefings, and in-theater briefings. The playing cards will communicate information about wildlife, wildlife products, and legal concerns pertaining to wildlife of Iraq and Afghanistan.

### REFERENCES

Bennett, E. L. (2005). Consuming wildlife in the tropics. In S. Guynup, ed. *State of the Wild 2006: a global portrait of wildlife, wildlands, and oceans*, pp. 106–113. Island Press, Washington, DC.

Karesh, W. B., Cook, R. A., Bennett, E. L., and Newcomb, J. (2005). Wildlife trade and global disease emergence. *Emerging Infectious Diseases*, 11, 1000–1002.

Robinson, J. G. and Bennett, E. L., eds (2000). *Hunting for sustainability in tropical forests*. Columbia University Press, New York, NY.

Warchol, G. L. (2004). The Transnational Illegal Wildlife Trade. *Criminal Justice Studies*, 17, 57–73.

## Box 15.2 The World Bank and biodiversity conservation
### Tony Whitten

The World Bank is well known as a development agency providing both concessionary credits and commercial-rate loans to governments to reduce poverty, but is less well known as a leader in biodiversity conservation. In fact, the biodiversity portfolio has grown steadily, especially since 1992 when funding from the Global Environment Facility (GEF) became available. In the last ten years the World Bank approved 598 projects that fully or partially supported biodiversity conservation and sustainable use (see Box 15.2 Figure). These are being executed in 122 countries and through 52 multi-country efforts and include activities in almost all terrestrial and coastal habitats, although more than half of all projects are directed towards the conservation of different types of forests. Many of these habitats provide critical ecosystem services and can be an important buffer to climate change, providing low-cost options for adaptation and mitigation actions. During the last 20 years, the World Bank has committed almost US$3.5 billion in loans and GEF resources, and leveraged US$2.7 billion in co-financing, resulting in a total investment portfolio for biodiversity exceeding US$6 billion. Protected-area projects account for more than half of the investments, but the Bank is increasingly seeking to mainstream biodiversity in production landscapes, especially where GEF-funded activities can be integrated within Bank lending.

Partner governments have borrowed just over 31% of the US$6 billion, whereas grants comprise 25%, mostly facilitated through Bank-executed GEF projects, as well as through trust funds, and carbon financing. The remaining 44% represents co-financing and parallel financing, and global initiatives, such as the IFC Small and Medium Enterprise Fund, the Critical Ecosystems Partnership Fund, Coral Reef Targeted Research, and projects funded under the World Bank-Netherlands Partnership Program's Forests and Biodiversity windows.

**Box 15.2 Figure** These villagers on Buton, Sulawesi (Indonesia), are members of a cooperative within a village which has developed a conservation agreement vowing not to encroach into the natural forest and not to hunt wildlife, with sanctions for members who go against the agreement. In return they get access to high prices for their cashews (*Anacardium* sp.) which became the world's first Fairtrade cashews. This World Bank project is executed by Operation Wallacea - see www.opwall.com and www.lambusango.com.

The scale and variety of Bank financing mechanisms provide many opportunities to integrate biodiversity concerns into development assistance, to address the root causes of biodiversity loss, and to develop local capacity and interest. The Bank's leadership and coordinating role within the donor community can help to promote biodiversity conservation within national sustainable development agendas. As well as being a major funding source for biodiversity projects in developing countries, the Bank is also a source of technical knowledge and expertise, and has the convening power to facilitate participatory dialogue between governments and other relevant stakeholders.

In addition to the biodiversity projects themselves, each and every World Bank project is subjected to a 'safeguard review' to ensure that they meet the requirements of the various policies it has on, for example, environmental assessment, resettlement, indigenous peoples, international waterways, physical cultural

*continues*

> **Box 15.2** (Continued)
>
> property – and natural habitats (World Bank 1998). The last of these is an important tool by which biodiversity concerns are integrated into improved project design because the policy forbids the Bank supporting projects involving the significant conversion of natural habitats unless there are no feasible alternatives for the project and its sites, and unless comprehensive analysis demonstrates that overall benefits from the project outweigh the environmental costs. Likewise the Bank will not approve a project that would involve the significant conversion or degradation of a gazetted or approved protected area. Mitigation for anticipated project impacts on biodiversity might include conservation offsets or additional species protection.
>
> For further information and details of projects, see Mackinnon *et al.* (2008) and www.worldbank.org/biodiversity.
>
> **REFERENCES**
>
> Mackinnon, K., Sobrevila, C. and Hickey, V. (2008). *Biodiversity, climate change, and adaptation: nature-based solutions from the World Bank portfolio*. The World Bank, Washington, DC.
>
> World Bank (1998). *Operational Policy 4.04: natural habitats*. The World Bank, Washington, DC.

is increasingly taking the forefront in the application of conservation science to conservation practice is the often real, and sometimes, perceived, conflicting mandates of biodiversity conservation and poverty alleviation. While there are, clearly, situations in which development can facilitate conservation efforts, it cannot be assumed that economic development will automatically lead to conservation benefits (Redford and Sanderson 2003). Furthermore, we cannot impose the world's development needs on the relatively small (approximately 10%) part of the land surface that constitutes protected areas and doing so poses significant, and perhaps insurmountable, challenges to the effective management of these areas to achieve global biodiversity goals. The value of protected areas – and their costs to local and indigenous people – has often been framed as one of opposition – with protected areas seen by some as depriving local and indigenous peoples of resources, by others as potentially beneficial (Sodhi *et al.* 2008). As one would expect, the reality is that such relationships are complex, and often locally specific (Upton *et al.* 2008) and the problem is more subtle (see for instance West and Brockington (2006) for a more detailed discussion of some of the effects). That parks may actually benefit the rural poor and serve as an attractant with human growth at their boundaries is both an argument for such areas, and flags a concern for their future conservation. The much contested relationship between parks and people will continue to stimulate both better analysis of the reality of such conflict, and provoke the design of innovative approaches for reconciliation between human needs and biodiversity conservation (Sodhi *et al.* 2006).

The technical and financial capacity for biodiversity conservation is significantly limited in developing economies harboring high levels of biodiversity (for example, most tropical countries). Such human resource deficits have been at the root of the changes in the way that conservation NGOs (Non-governmental Organizations), local governments, and international donors have implemented conservation projects over the last four decades (Cobb *et al.* 2007). Effectively tackling this issue – and empowering both local and national governments and institutions – will require visionary and far-sighted approaches that are able to justify investment of scarce resources to long-term capacity building objectives in the face of immediate conservation problems.

The gap between conservation science and its application has been long acknowledged (Balmford *et al.* 1998) and there are numerous efforts directed at bridging it (Sutherland *et al.* 2004).

## Box 15.3 The Natural Capital Project
**Heather Tallis, Joshua H. Goldstein, and Gretechen C. Daily**

The vision of the Millennium Ecosystem Assessment is a world in which people and institutions appreciate natural systems and the biodiversity that constitutes their principal working parts as vital assets, recognize the central roles these assets play in supporting human well-being, and routinely incorporate their material and intangible values into decision-making. This vision has now caught fire, fueled by innovations worldwide – from pioneering local leaders to the belly of government bureaucracy, and from traditional cultures to a new experimental wing of Goldman Sachs – a giant investment banking firm (Daily and Ellison 2002; Bhagwat and Rutte 2006; Kareiva and Marvier 2007; Ostrom 2007; Goldman et al. 2008). China, for instance, is investing over 700 billion yuan in ecosystem service payments over 1998–2010 (in early 2009, US$ 1.0 = 6.85 yuan) (Liu et al. 2008).

The aim of the Natural Capital Project is to act on this vision and mainstream ecosystem services into everyday decisions around the world. Launched in October 2006, the Project is a unique partnership among Stanford University, The Nature Conservancy, and World Wildlife Fund, working together with many other institutions (www.naturalcapitalproject.org). Its core mission is to align economic forces with conservation by: (i) developing tools that make incorporating natural capital into decisions easy; (ii) demonstrating the power of these tools in important, contrasting places; and (iii) engaging leaders globally.

Making conservation mainstream requires turning the valuation of ecosystem services into effective policy and finance mechanisms – a problem no one has solved on a large scale. A key challenge remains that, relative to other forms of capital, assets embodied in ecosystems are often poorly understood, scarcely monitored, typically undervalued, and undergoing rapid degradation (Daily et al. 2000; Heal 2000; Balmford et al. 2002; MEA 2003; NRC 2005; Mäler et al. 2008). Often the importance of ecosystem services is recognized only upon their loss, such as in the wake of Hurricane Katrina (Chambers et al. 2007).

To help address this challenge, we have developed a software system for integrated valuation of ecosystem services and tradeoffs (InVEST; Nelson et al. 2009). This tool informs managers and policy makers about the impacts of alternative resource management choices on the economy, human well-being and the environment, in an integrated way.

Examples of urgent questions that InVEST can help answer include:

- Which parts of a landscape provide the greatest carbon sequestration, biodiversity, and tourism values?
- Where would reforestation achieve the greatest downstream water quality benefits?
- How would agricultural expansion, climate change and population growth affect a downstream city's drinking water supply or flood risk?

InVEST is designed for use as part of an active decision-making process. The first phase of the approach involves working with decision makers and other stakeholders to identify critical management decisions and to develop scenarios to project how the provision of services might change in response to those decisions as well as to changing climate or population. Based on these scenarios, a modular set of models quantifies and maps ecosystem services in a flexible way. The outputs of these models provide decision makers with maps and other information about costs, benefits, tradeoffs, and synergies of alternative investments in ecosystem service provision.

InVEST is now being used in major resource decisions in Bolivia, Brazil, China, Colombia, Ecuador, Mexico, Peru, Tanzania, and the United States (California, Hawaii, Oregon, and Washington; see Box 15.3 Figure). The tool has proven useful with stakeholders as diverse as national governments, private landowners and corporations, and increasing demand for the tool indicates that the time is ripe for ecosystem service thinking to change the face of management across sectors and around the globe.

*continues*

### Box 15.3 (Continued)

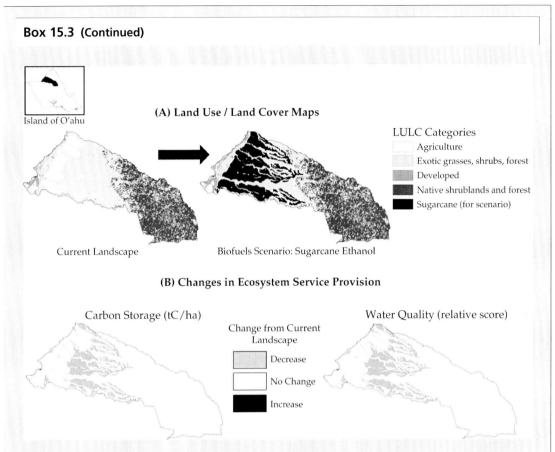

**Box 15.3 Figure** Application of InVEST to a planning region on the Island of O'ahu, Hawaii. The parcel covers approximately 10 500 ha from mountaintop to the sea, including 800 ha of developed rural community lands along the coast, 3600 ha of agricultural lands further inland, and 6100 ha of rugged forested lands in the upper part of the watershed. While many of the agricultural fields have been fallow for over a decade, stakeholders are exploring using the fields to grow sugarcane for ethanol biofuel (among other options). InVEST was used to assess how this land-use change scenario would affect the ecosystem services of water quality and carbon storage. Part (A) shows land use/land cover (LULC) maps for the current landscape and the sugarcane ethanol scenario. Part (B) shows the projected changes for water quality and carbon stock. The dominant effect is a decrease in service provision relative to the current landscape. Water quality decreases by 44.2%, driven by increased fertilizer application on the fallow fields returned to crop production. Taking advantage of next-generation sugarcane breeds, however, could greatly reduce these impacts. Carbon stock decreases by 12.6%, which is due to clearing of woody exotic species that grew while the fields were not in production. This "carbon debt" (Fargione *et al.* 2008) could be repaid through time by using sugarcane ethanol to offset more carbon intensive fuel sources. The information generated from this InVEST analysis elucidates ecosystem service tradeoffs apparent in undertaking biofuel production, which can inform land use decisions alongside economic and other benefits not shown here. Furthermore, the analysis helps land managers identify where to focus efforts, spatially for each ecosystem service, to improve management practices. See also Figure 3.1.

*continues*

## Box 15.3 (Continued)

### REFERENCES

Balmford, A., Bruner, A., Cooper, P., et al. (2002). Economic reasons for conserving wild nature. *Science*, **297**, 950–953.

Bhagwat, S. A. and Rutte, C. (2006). Sacred groves: potential for biodiversity management *Frontiers in Ecology and the Environment*, **4**, 519–524.

Chambers, J. Q., Fisher, J. I., Zeng, H., et al. (2007). Hurricane Katrina's carbon footprint on U.S. Gulf Coast forests. Science, **318**, 1107.

Daily, G. C. and Ellison, K. (2002). *The new economy of nature: the quest to make conservation profitable*. Island Press, Washington, DC.

Daily, G. C., Söderqvist, T., Aniyar, S., et al. (2000). The value of nature and the nature of value. *Science*, **289**, 395–396.

Fargione, J., Hill, J., Tilman, D., Polasky, S., and Hawthorne, P. (2008). Land clearing and the biofuel carbon debt. *Science*, **319**, 1235–1238.

Goldman, R. L., Tallis, H., Kareiva, P., and Daily, G. C. (2008). Field evidence that ecosystem service projects support biodiversity and diversify options. *Proceedings of the National Academy of Sciences of the United States of America*, **105**, 9445–9448.

Heal, G. (2000). *Nature and the marketplace: capturing the value of ecosystem services*. Island Press, Washington, DC.

Kareiva, P. and Marvier, M. (2007). Conservation for the people. *Scientific American*, **297**, 50–57.

Liu, J., Li, S., Ouyang, Z., et al. (2008). Ecological and socioeconomic effects of China's policies for ecosystem services. *Proceedings of the National Academy of Sciences of the United States of America*, **105**, 9489–9494.

MEA (millennium Ecosystem Assessment). (2003). *Ecosystems and human well-being: a framework for assessment*. Island Press, Washington, DC.

Mäler, K-G., Aniyar, S., and Jansson, Å. (2008). Accounting for ecosystem services as a way to understand the requirements for sustainable development. *Proceedings of the National Academy of Sciences of the United States of America*, **105**, 9501–9506.

Nelson, E., Mendoza, G., Regetz, J., et al. (2009). Modeling multiple ecosystem services, biodiversity conservation, commodity production, and tradeoffs at landscape scales. *Frontiers in Ecology and the Environment*, **7**, 4–11.

NRC (National Research Council). (2005). *Valuing ecosystem services: toward better environmental decision-making*. National Academies Press, Washington, DC.

Ostrom, E. (2007). A diagnostic approach for going beyond panceas. *Proceedings of the National Academy of Sciences of the United States of America*, **104**, 15181–15187.

These efforts are based on the assumption that effective conservation is dependent not only on science catching up with the dynamic aspects of a changing world (Chapter 13) but also on conservation practice catching up with science (Pressey et al. 2007). There is a recognized need to integrate the activities of conservation biologists (and other conservation minded scientists) with those of practitioners, with conservation biologists interacting more frequently with practitioners and the latter better documenting their actions (Sutherland et al. 2004). This chapter provides a glimpse into the realm of practical conservation with examples and case studies to illustrate some of the diverse approaches that are being implemented to conserve biodiversity and how these approaches benefit from, and offer opportunities to, the science that underlies them.

## 15.1 Integration of Science and Conservation Implementation

A good example of integrating conservation science with implementation is a project that is being undertaken in South Africa (Balmford 2003). Richard Cowling and his colleagues have successfully attempted to build the input of decision-makers and local people into scientifically rigorous conservation planning for the Cape Floristic Region in South Africa (Cowling and Pressey 2003; Cowling et al. 2003; Pressey et al.

## Box 15.4 Measuring the effectiveness of conservation spending
### Matthew Linkie and Robert J. Smith

Conservationists can only develop cost-effective strategies by evaluating the success of their past efforts. However, few programs measure project performance adequately: most carry out no assessment at all or rely on descriptive analyses that cannot distinguish between the confounding effects of different covariates. In response, Ferraro and Pattanayak (2006) have presented a counterfactual design for determining conservation success. This involves comparing similar sampling units, e.g. villages, people or forest patches, which receive conservation intervention (the treatment group) with those that do not (the control group). Here, we describe two studies that have used this approach to evaluate conservation effectiveness.

### Case study 1

Linkie et al. (2008) studied a US$19 million project that ran from 1997–2002 in and around Kerinci Seblat National Park, Sumatra. Part of this project involved spending US$1.5 million on development schemes within 65 villages (the treatment) that border the park, in return for the villagers signing agreements to stop the illegal clearance of their forest (see Box 15.4 Figure). Thus, determining the success of this strategy involved measuring subsequent village deforestation rates. However, deforestation patterns are often explained by covariates relating to accessibility, such as proximity to roads, and some project villages were chosen for logistical or political reasons. Linkie et al. accounted for the influence of these different factors by using a propensity score matching technique. This approach used data on ten socio-economic and biophysical covariates from a village profile dataset to identify the factors that best predicted forest loss, and to identify the 65 non-project villages (the control) that most closely matched the project villages in terms of these factors. Deforestation rates between these two groups were then compared and no difference was found, showing that project participation had no effect. In contrast, a questionnaire survey conducted by the project found stronger conservation support in project villages than non-project villages, and on this basis alone the project might have been considered a success.

**Box 15.4 Figure** Small scale logging in Sumatra (Indonesia). Photograph by Jeremy Holden.

### Case study 2

Andam et al. (2008) evaluated the effectiveness of protected areas (PAs) in avoiding deforestation in Costa Rica. They also used a propensity score matching technique to identify similar unprotected areas (the control) that most closely matched the PAs (the treatment), based on similarities of accessibility and land use opportunities. From 1960–1997, the PAs were found to avoid about 10% of the deforestation that was predicted to have occurred if they had not been present. In addition, Andam et al. tested a commonly used method for evaluating PA effectiveness, which compares deforestation in PAs against that in adjacent unprotected areas. Such comparisons can be problematic because PAs tend to be located on land that is less

*continues*

> **Box 15.4 (Continued)**
>
> accessible and less suitable for agriculture and therefore has a lower risk of clearance. This was illustrated by their results, which showed that not controlling for these confounding effects led to a threefold over-estimation of deforestation reduction within the PAs.
>
> These two case studies illustrate the importance of using statistically robust approaches for measuring conservation success. Such an approach should be widely adopted, as it provides vital information for donors, policy developers and managers. However, this will depend in part on developing a conservation culture that discusses and learns from failure, instead of hiding it from scrutiny (Knight 2006).
>
> **REFERENCES**
>
> Andam, K. S., Ferraro, P. J., Pfaff, A., Sanchez-Azofeifa, G. A., and Robalino, J. A. (2008). Measuring the effectiveness of protected area networks in reducing deforestation. *Proceedings of the National Academy of Sciences of the United States of America*, **105**, 16089–10694.
>
> Ferraro, P. J. and Pattanayak, S. K. (2006). Money for nothing? A call for empirical evaluation of biodiversity conservation investments. *PLoS Biology*, **4**, 482–488.
>
> Knight, A. T. (2006). Failing but learning: writing the wrongs after Redford and Taber. *Conservation Biology*, **20**, 1312–1314.
>
> Linkie, M., Smith, R. J., Zhu, Y., *et al.* (2008). Evaluating biodiversity conservation around a large Sumatran protected area. *Conservation Biology*, **22**, 683–690.

2007). The Cape Floristic Region, covering 90 000 km$^2$ of the south-west tip of Africa, contains over 9000 species of plants and is globally recognized for its biological significance (Davis *et al.* 1994; Olson and Dinerstein 1998; Stattersfield *et al.* 1998; Myers *et al.* 2000). Over 1400 of the plant species found here are Red Data Book listed and nearly 70% are endemic to the region. Conversion to intensive agriculture, forestry, urbanization, infestation with alien plants and widespread grazing are key threats in the region with 22% of all land protected in conservation areas (only half in statutory reserves) and 75% in private ownership (Balmford 2003).

Against this backdrop of escalating threats, declining institutional capacity, and a biologically unrepresentative reserve system, a project known as the Cape Action Plan for the Environment (CAPE) was launched (Cowling and Pressey 2003). The project has since expanded into a 20-year implementation program addressing three broad themes: (i) the protection of biodiversity in priority areas; (ii) the promotion of its sustainable use; and (iii) the strengthening of local institutions and capacity. From its inception, the project engaged not only the statutory agencies that would ultimately be responsible for implementation, but also land-owners, local communities and the non-governmental sector. Building these partnerships early on enabled a diversity of local actors and external practitioners to work with planners in developing broad project goals and strategies. The approach of integrating the involvement of stakeholders and practitioners with scientifically rigorous planning not only earned the project credibility with external donors but the resulting wide ownership of the conservation plan has been crucial to its ongoing implementation (Balmford 2003).

## 15.2 Looking beyond protected areas

During the past century, the standard practice for safeguarding biodiversity (Chapter 2) and reducing the rate of biodiversity loss has been the establishment of protected areas (Lovejoy 2006). The steady and significant increase in the area protected and number of protected areas created over the past three to four decades has been accompanied by an evolution of protected areas from being small refuges for particular species to the protection of entire ecosystems. But even large protected areas can be inadequate to ensure the persistence of some wildlife populations, particularly large carnivores (Woodroffe and Ginsberg 1998). Furthermore, biodiversity conservation, or the preservation of ecological integrity, are only two reasons for establishing and maintaining protected areas. Other goals may relate to sustainable development, poverty alleviation, peace and social equity. The disparate and often conflicting global mandates for protected areas pose the greatest challenge for the design and implementation of effective conservation strategies. The need for reconciliation of conflicting mandates will drive the design and implementation of innovative approaches to management, governance, financing and monitoring of protected areas, all of which will directly and indirectly impact their effectiveness in conserving biodiversity.

One such approach involves the design of strategies aimed at managing protected areas as components of a larger landscape. Given that wildlife, ecological processes and human activities often spill across the boundaries of protected areas, conservation that is focused solely within the limits of protected areas is often faced with difficult challenges. The management of protected areas therefore cannot occur in isolation from the surrounding human-dominated landscapes. Box 15.5 provides a description of a landscape approach to conservation where protected areas are managed as one component of a larger conservation landscape that is traversed by land uses where biodiversity conservation is not the primary objective (see Box 5.3). The entire field of countryside biogeography, of course, focuses on this key issue (see Box 13.4).

## 15.3 Biodiversity and human poverty

There is a considerable degree of spatial overlap of poverty, inequality and biodiversity with high levels of biodiversity occurring in some of the world's poorest countries (McNeely and Scherr 2001). The creation of protected areas in order to restrict the use of biodiversity in such countries therefore has impacts on communities and other user groups who benefit economically from directly utilizing biodiversity or converting the land to a more profitable form of use such as oil palm plantations. Protected areas established to conserve biodiversity in regions of high poverty are under tremendous pressure to serve the dual purpose of economic development and biodiversity conservation. Consequently, there is much contention surrounding the relationship between protected areas, people and economic

---

**Box 15.5 From managing protected areas to conserving landscapes**
**Karl Didier**

The Ewaso Ecosystem is a vast (40 000 km$^2$) and diverse savanna region in central Kenya. It is relatively intact, with most of its biodiversity and all of its megafauna still present, including elephants (*Loxodonta africana*), lions (*Panthera leo*), giraffe (*Giraffa camelopardalis reticulate*), the endangered African wild dog (*Lycaon pictus*), and the last populations of the critically endangered Grevy's zebra (*Equus grevyi*). The relative intactness of the Ewaso is owed, in large part, to a large network of protected areas covering 6000 km$^2$ (~15% of the region), including national parks and reserves, and provincial forest reserves (see Box 15.5 Figure and Plate 18). However, even with so much of the land in protected areas, conservation goals have yet to be met: populations of some species remain dangerously low (e.g., <300 wild dogs), many other biological species and communities are threatened with imminent decline due to increasing habitat fragmentation (Chapter 5) and conflict beyond the boundaries of the protected area network (e.g., elephants; see Box 14.3), and basic ecosystem services (Chapter 3), such as production of clean water, are threatened by land development (e.g., logging and agriculture) (Chapter 4) and climate change (Chapter 8).

*continues*

## Box 15.5 (Continued)

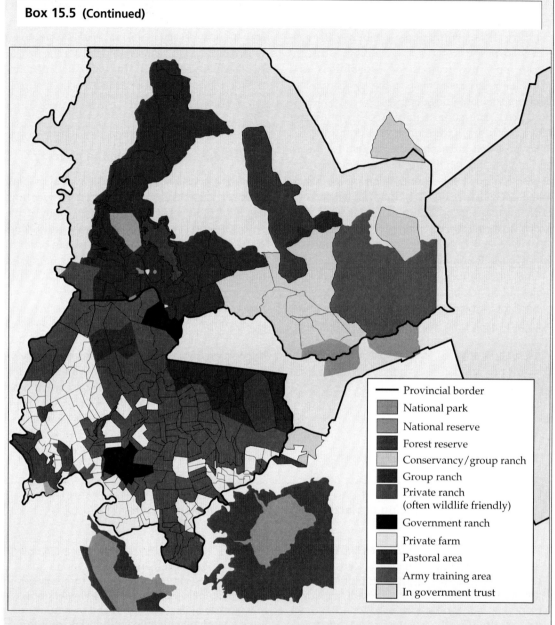

**Box 15.5 Figure** The biodiversity of the Ewaso ecosystem in central Kenya is relatively intact due in large part to a strong set of protected areas. However, even these are not sufficient to preserve the patterns and processes of biodiversity and to reach conservation objectives. To do so, conservationists are working in the complex matrix of land uses beyond the protected areas, with a vast array of stakeholders, and using actions that benefit both people and biodiversity.

### Why protected areas are not enough?

In the Ewaso and in most areas around the globe, there are two reasons why protected area creation is an incomplete strategy to meet the conservation objectives. First, protected areas, whether they cover 5 or 50% of a region, simply cannot represent the enormous array of biodiversity out there. Existing protected area networks tend to be biased toward representing a small subset of species,

*continues*

**Box 15.5 (Continued)**

such as large mammals, and fail to represent other taxa well, such as plants. This is especially true in western Africa (see Gardner et al. 2007).Second, even for the elements of biodiversity that are represented (i.e. occur at least once) in protected areas, their long-term persistence is rarely ensured by management of the protected area alone. The problem is that both biodiversity and the threats to biodiversity move freely across protected area boundaries. For example, elephants and wild dogs in the Ewaso rely on habitats and corridors well beyond protected areas, bringing them into conflict with humans. Also, although many threats have their source outside of protected areas, like pollution added to the Ewaso River by flower farmers or wandering livestock, they manage to directly impact biodiversity inside parks. Mitigation of such threats cannot be achieved by park management alone, and expansion of protected areas is untenable. To ensure that ecosystem services are maintained and that viable and functional populations (i.e., at appropriate densities) of species persist, conservation practitioners need to work beyond park boundaries, into the surrounding human-dominated matrix.

### Defining a "landscape" for conservation practitioners

The term "landscape" has been defined as "a heterogeneous land area composed of a cluster of interacting ecosystems that is repeated in similar form throughout" (Forman and Godron 1986) or "an area that is spatially heterogeneous in at least one factor of interest" (Turner et al. 2001). These are interesting from a theoretical perspective, but are not very useful for a park manager or conservation practitioner. An alternative definition of a "landscape" for conservation practitioners could be 'an area sufficient in size, composition, and configuration of land elements (e.g., habitats, management types) to support the long-term persistence and functioning of all conservation features of interest, including ecological communities and processes, ecosystem services, and functional populations of species'.

Most frequently, this kind of landscape will be heterogeneous in many aspects, including human land uses, ecosystems or ecological communities, political units, and management units. In the Ewaso, the "landscape" includes protected areas, private lands, villages, community-owned lands, untenured lands, parts of at least 10 districts, and a diversity of habitats that include rivers, montane forests, acacia savanna and moorlands (see Box 15.5 Figure). A typical landscape will also include a diversity of stakeholders. In the Ewaso, this includes local ranch owners and farmers, non-governmental development organizations [e.g. CARE (Cooperative for Assistance and Relief Everywhere)], powerful "county councils" who control large community-owned areas, industrial-scale flower farmers often from Europe, and poor, nomadic pastoralists who graze their livestock on tracts of government-owned land. While defining the boundaries and users of a landscape are difficult tasks, implementation of conservation activities at the landscape scale presents an enormous challenge.

### Implementing landscape conservation

Conservation at landscape scales requires, first and foremost, that practitioners engage communities and landowners and implement activities that meet their needs while improving the situation for biodiversity. In the Ewaso, several organizations such as the Laikipia Wildlife Forum (LWF) and the Northern Rangelands Trust (NRT) spend much of their resources working outside the boundaries of protected areas, with community-owned ranches and conservancies. For example, NRT helps communities obtain formal land ownership from the Kenyan government. Once this occurs, they implement a suite of activities to help communities generate sustainable income and improve conditions for biodiversity. For example,

*continues*

> **Box 15.5 (Continued)**
>
> NRT helps pastoralists on community-run ranches improve their access to livestock markets. Improved market access means that owners receive a higher price per head, can reduce total number of livestock on their lands, and, therefore, improve rangeland quality for wildlife. NRT and LWF also help local communities develop ecotourism enterprises, which supplement local incomes, make owners less susceptible to the vagaries of livestock management, and gives them incentive to conserve biodiversity. A further example of conservation action outside protected area boundaries is the work being done by organizations under the banner of the Laikipia Elephant Project (see also Boxes 5.3 and 13.4). This project aims to decrease incidents of crop raiding by elephants in several ways, including providing farmers with "early warning systems", training them how to plant and sell chili peppers (a crop that elephants hate and which is valuable on international markets), or even training people to make paper out of elephant dung. As conservation in the Ewaso demonstrates, to implement landscape-scale conservation practitioners need an expanded set of tools and skills. Just to name a few, they need skills in the ecological and social sciences, law, business and finance, facilitation and negotiation, conservation planning, zoning, geographic information systems, remote sensing, and fund raising. While the creation and management of protected areas will remain a cornerstone strategy for biodiversity conservation, there is an increasing need for traditional strategies to be augmented with new tools and approaches to implement landscape scale conservation.
>
> **REFERENCES AND SUGGESTED READING**
>
> Forman, R. T. T. and Gordon, M. (1986). *Landscape ecology*. John Wiley, New York, NY.
>
> Gardner, T. A., Caro, T., Fitzherbert, E.B., Banda, T., and Lalbhai, P. (2007). Conservation value of multiple use areas in East Africa. *Conservation Biology*, **21**, 1516–1525.
>
> Gaston, K. J., Pressey, R. L., and Margules, C. M. (2002). Persistence and vulnerability: retaining biodiversity in the landscape and in protected areas. *Journal of Biosciences*, **27**, 361–384.
>
> Poiani, K. A., Richter, B. D., Anderson, M. G., and Richter, H. E. (2000). Biodiversity conservation at multiple scales: Functional sites, landscapes, and networks. *BioScience*, **50**, 133–146.
>
> Turner, M. G., Gardner, R. H., and O'Neill, R. V. (2001). *Landscape ecology in theory and practice*. Springer-Verlag, New York, NY.

development, with conservationists and those concerned with human welfare locked in debate (West and Brockington 2006; Vermeulen and Sheil 2007; Robinson 2007). Conservationists argue that environmental regulations are essential to ensure the sustainability of the planet's biological systems and the health and welfare of people, especially local people, and that protected areas are an indispensable tool in that regulatory toolbox (Peres 1995; Kramer et al. 1997; Brandon et al. 1998; Terborgh 1999). Some social advocates, on the other end of the spectrum, contest the establishment and management of protected areas, and support the beliefs that: (i) only initiatives related to poverty alleviation will lead to successful biodiversity conservation since poverty is the root cause of environmental destruction (Duraiappah 1998; Ravnborg 2003); and (ii) Protected areas have been frequently established at the expense of local communities (in and around protected areas) through displacement and dispossession, and are responsible for perpetuating poverty by the continued denial of access to land and other resources (Ghimire and Pimbert 1997; Colchester 2004). In addition, others contend that even if parks do generate economic value, the distribution of these benefits is so skewed against poor rural people that the role of parks in local development is negligible, and they neither justly compensate for lost property and rights nor contribute to poverty alleviation (Brockington 2003; McShane 2003).

In an analysis of programmatic interventions aimed at achieving both biodiversity conservation and poverty alleviation, Agrawal and Redford (2006) indicate that there is basic lack of evidence on the extent to which the two goals can be simultaneously achieved. While the role of poverty in destroying biodiversity in poor countries is indisputable, one should never lose sight of the overwhelming role that the rich, through their overconsumption, play in extinguishing life forms all over the Earth (Ehrlich and Ehrlich 2005).

Identifying win-win strategies that simultaneously benefit biodiversity and people continues to dominate the agenda of researchers and practitioners alike and the integration of poverty alleviation and biodiversity conservation goals has been approached in various ways. Biodiversity use may not be able to alleviate poverty, but may have an important role in sustaining the livelihoods of the poor, and preventing further impoverishment (Angelsen and Wunder 2003). Furthermore, while the vast majority of the world's poor live in semi urban areas, significant progress in poverty alleviation will not be affected by most conservation activities (Redford *et al.* 2008). Biodiversity-rich tropical forests subject to high deforestation rates nonetheless harbor some of the poorest, most remote and politically disenfranchised forest dwellers offering distinct opportunities for joint conservation and development initiatives, and have drawn advocates for new approaches to "pro-poor conservation" (Kaimowitz and Sheil 2007).

### Box 15.6 Bird nest protection in the Northern Plains of Cambodia
**Tom Clements**

Cambodia is identified by many global assessments as a conservation priority: for example it lies within the Indo-Burma hotspot (Myers *et al.* 2000) and contains four of the Global 200 Ecoregions (Olson and Dinerstein 1998). Although it does not support high species diversity, Cambodia is of particular importance for conservation because it contains some of the largest remaining examples of habitats that previously spread across much of Indochina and Thailand, which still support almost intact species assemblages. Many of these species are listed as Globally Threatened by IUCN due to significant declines elsewhere in their range. Following the restoration of peace in Cambodia in 1993, conservation strategies have primarily focused on the establishment of Protected Areas (PAs). These PAs generally have a small number of poorly paid staff with limited capacity or infrastructure, i.e. they are 'paper parks' (Wilkie *et al.* 2001). Moreover, PAs usually contain existing human settlements, in some cases with >10 000 people, whose rights are respected under law but with varying degrees of implementation. Such a situation is not uncommon: 70% of a non-random sample of global PAs contained people, and 54% had residents who contested the ownership of some percentage of the PA area (Bruner *et al.* 2001). Since limited site information was available when PAs were declared many areas of importance for biodiversity conservation lie outside the system, emphasizing the importance of adopting a landscape approach. This requires tools to engage local communities in conservation (see Chapter 14).

In the 1980s and 90s Integrated Conservation and Development Projects (ICDPs) were a popular methodology for combining the needs of local communities with conservation, both inside and outside of PAs. However, there is very little evidence of conservation success (Wells *et al.* 1999; Chape 2001; Ferraro and Kiss 2002; Linkie *et al.* 2008). One of the principle reasons suggested for this failure is that the linkages between project activities (benefits) and biodiversity conservation were weak, i.e. benefits were not contingent on conservation outcomes. Ferraro and Kiss (2002) have therefore proposed that community conservation interventions would be more effective if they concentrated on initiatives where these linkages are much stronger. 'Direct

*continues*

**Box 15.6 (Continued)**

payments' and 'conservation easements' are actually much more accepted in the USA and Europe and have been recently established in other countries such as Costa Rica (Zbinden and Lee 2004). This section describes a direct payment scheme established by the Wildlife Conservation Society (WCS), an international non-governmental organization, in Cambodia. The scheme is evaluated against some of the original claims made by Ferraro and Kiss (2002), specifically that direct payments schemes would be simpler to implement and therefore have: (i) efficient institutional arrangements; (ii) be cost-effective; and (iii) deliver substantial development benefits, in addition to the conservation benefits.

### Methods

The Northern Plains support probably the largest breeding global population of giant ibis (scientific names in tables) (Critically Endangered), a species known from only a handful of records in the 1900s until it was rediscovered in 2000 by WCS in the area. Some of the only known nesting sites in mainland Asia of another Critically Endangered species – white-shouldered ibis – are also located in the Northern Plains. These two ibises are amongst the most endangered bird species in the world. In addition, the Northern Plains supports breeding populations of three Critically Endangered vulture species – white-rumped, slender-billed, and red-headed vultures – and eight species of large waterbirds: greater adjutant, lesser adjutant, white-winged duck, sarus crane, Oriental darter, black-necked stork, and woolly-necked stork. This unique assemblage of nine globally threatened large bird species means that the Northern Plains is of exceptional importance for conservation. The primary immediate threat to all these birds is collection of nest contents by local people, often for sale to middlemen who trade with Thai and Lao border markets. This is especially true for both adjutant species and the sarus cranes – the latter is known to fetch a high market price (>US$100 in Thailand). The collection is mostly done by people from local communities, who then re-sell the eggs and chicks on to middlemen. The Bird nest Protection Program was launched in 2002 by WCS in order to locate, monitor and protect the nesting sites. Initially the research, protection and monitoring was undertaken by WCS staff and rangers. However increasingly it has been discovered that a much greater number of nests can be found and successfully protected by working in cooperation with the local communities, who were originally the principal threat. Under the program, local people are offered a reward of up to US$5 for reporting nests, and are then employed to monitor and protect the birds until the chicks successfully fledge. The protection teams are regularly visited every one to two weeks by community rangers employed by WCS and WCS monitoring staff to check on the status of the nests and for the purposes of research and data collection. The program operates year-round, as some species nest in the dry season and others during the wet season. It started in four pilot villages in 2002 at one site and was extended to a second site in 2004. By 2007 it was operating in >15 villages. In 2003 and 2004 nest protectors were paid US$2/day at the end of the month, assuming that the nest went undisturbed during that period. In 2005 the payment system was changed following community consultations to US$1/day for protecting the nest with a bonus $1/day provided if the chick(s) successfully fledged. The payment values were based on an acceptable daily wage, rather than compensating for the opportunity cost of not collecting, which would be much greater. Local people were concerned about natural predation, and it was decided that payments would still be made in these cases.

### Results and Conclusions

The scheme has been extremely successful (see Box 15.6 Table 1), protecting over 1200 nests of globally threatened or near-threatened species since 2002, including 416 nests in 2007–8. The numbers of nests monitored and protected have increased by an average of 36% each year since 2004. Most of this increase is due to greater numbers of sarus crane, vultures (three

*continues*

## Box 15.6 (Continued)

**Box 15.6 Table 1** Bird Nest Protection Program: Nests Protected, 2002-2008. In some cases nests were protected but there is no data available. '-' indicates species that were probably present, but were not protected in that year. Initially the program started at one sites and operated in two sites from 2004. Numbers found have grown by 36% per year since 2004.

| Species | Global Status | 2002-3 (1 site only) Nests (Colonies) | 2003-4 (1 site only) Nests (Colonies) | 2004-5 (2 sites) Nests (Colonies) | 2005-6 (2 sites) Nests (Colonies) | 2006-7 (2 sites) Nests (Colonies) | 2007-8 (2 sites) Nests (Colonies) |
|---|---|---|---|---|---|---|---|
| White shouldered Ibis *Pseudibis davisonii* | Critical | 1 | 1 | 2 | 3 | 4 | 6 |
| Giant Ibis *Pseudibis gigantea* | Critical | - | 5 | 27 | 28 | 28 | 29 |
| Sarus Crane *Grus antigone* | Vulnerable | - | 6 | 19 | 29 | 37 | 54 |
| Vulture spp. (*Sarcogyps calvus* & *Gyps spp.*) | Critical | - | - | 1 | 4 | 5 | 5 |
| Black-necked Stork *Ephippiorhynchus asiaticus* | Near-threatened | - | - | - | 2 | 3 | 2 |
| Oriental Darter *Anhinga melanogaster* | Near-threatened | 13 | - | - | - | 26(1) | 33(1) |
| Greater Adjutant *Leptoptilus dubius* | Endangered | - | (present, no data) | 21(2) | 17(2) | 18(2) | 10(2) |
| Lesser Adjutant *Leptoptilus javanicus* | Vulnerable | - | 34(5) | 97(16) | 134(15) | 221(22) | 277(27) |
| Totals, both sites | | 14 | 46+ | 166 | 219 | 342 | 416 |

species), white-shouldered ibis, Oriental darter, and lesser adjutant being found, suggesting that persecution and nest collection were the main factors limiting populations of these species. Local awareness regarding the importance of bird conservation has substantially improved, with an almost complete cessation of collection activity at one site, and significant reductions at the other. The direct payments scheme has therefore been very effective at delivering conservation results.

Reviewing the first of the claims of Ferraro and Kiss (2002), the scheme involves a very simple institutional arrangement: with contracts made directly between WCS and the protectors without involving any other institution. Under Cambodian Law collection of bird nests contents is actually strictly illegal, but Government authorities are not directly involved in the scheme, although they do participate in regular reviews of results. The scheme therefore reinforces national law by providing an incentive to villagers not to collect bird nests, but not fully compensating for the opportunity cost.

A detailed breakdown of the payments made in the 2005–6, 2006–7 and 2007–8 seasons is given in Box 15.6 Table 2. The total cost to WCS of the program is around US$25 000 per year, with an average cost of $60–$120 per nest protected. The average cost has declined as the number of nests has increased, partly because monitoring costs can be shared between adjacent sites and also due to the

*continues*

## Box 15.6 (Continued)

**Box 15.6 Table 2** Bird Nest Protection Program: Costs, 2005-2008. WCS, Wildlife Conservation Society. Currency in US dollars.

|  | 2005-6 | 2006-7 | 2007-8 |
|---|---|---|---|
| Local Payments (%) | $ 19850 (78%) | $ 19119 (74%) | $ 17434 (69%) |
| Nest Protection Payments | $ 12597 | $ 11248 | $ 9786 |
| Community Rangers | $ 7253 | $ 7871 | $ 7648 |
| WCS Monitoring (%) | $ 5603 (22%) | $ 6800 (26%) | $ 7747 (31%) |
| Expenses | $ 2506 | $ 3640 | $ 4192 |
| Salaries | $ 3098 | $ 3160 | $ 3555 |
| Total | $ 25453 | $ 25918 | $ 25180 |
| Nests Protected | 219 | 342 | 416 |
| Average Cost/Nest | $ 116.22 | $ 75.78 | $ 60.53 |

greater number of nests at colonies. Of the cost of the program, 69–78% of payments went directly to local people, with the remaining expenditure being monitoring costs incurred by WCS. The program is therefore very cost-effective, with an overhead of only 22–31%, substantially less than other conservation approaches (Ferraro and Kiss 2002). Average payments per family are around US$120/year, with considerable variation depending upon how long people were employed. Some individuals are specialist protectors, switching species depending on the season and receiving continual employment for several months. The amounts paid, sometimes >US$400/individual, are substantial in villages where annual cash incomes are $200–$350/year. Evaluations have shown that this money is used to pay for clothes, schooling, housing improvements and to enhance food security. The scheme therefore does provide substantial development benefits, although these are not a primary objective of the program. It is also very popular with villagers because they are able to decide for themselves how to spend the money (i.e. benefits are not in-kind).

The initial scheme was based upon 'payments for work' (i.e. US$2/day) rather than 'payments for success'. This led to perverse situations where WCS was perceived as an employer with responsibility for protectors' well-being, whilst the protectors shared little of the risk and were not responsible for the final outcome. In 2005 the payment system was changed to increase the risk shared by the protectors. That is, they are paid $1/day for their work and $1/day for results upon successful fledging. This revised payment system delegates decision-making to local people, who are probably more familiar with the situation and more aware of threats.

Payments are also entirely dependent on money raised annually by WCS, although the scheme is relatively inexpensive in comparison with the substantial conservation benefits. However, given the extreme level of threat to many of these species, with average population sizes <20 pairs per site when the scheme was initiated, these were judged acceptable risks. In the longer-term financing could become more sustainable through direct sponsorship, for example through websites or exhibits in zoos. One risk is that collection would resume if the payment scheme was stopped.

The bird nests protection scheme is linked to a community-based ecotourism program. Under this, communities receive rights to locally manage ecotourism enterprises in exchange for active protection of the biodiversity that tourists come to see. The ecotourism enterprises employ additional groups within the communities, including more marginal groups such as women and poorer households, reinforcing the value of the birds. In addition, as the community enterprises become more empowered they have begun to take over local payments for bird nest protection, funded from tourism receipts. This provides a long-term sustainable financing mechanism for the initiative.

## REFERENCES

Bruner, A. G., Gullison, R. E., and Rice, R. E. (2001). Effectiveness of parks in protecting tropical biodiversity. *Science*, **291**, 125–128.

Chape, S. (2001). An overview of integrated approaches to conservation and community development in the Lao People's Democratic Republic. *Parks*, **11**, 24–32.

*continues*

### Box 15.6 (Continued)

Ferraro, P. J., and Kiss, A. (2002). Direct payments to conserve biodiversity. *Science*, **298**, 1718–1719.

Linkie, M., Smith, R. J., Zhu, Y., *et al.* (2008). Evaluating biodiversity conservation around a larger Sumatran protected area. *Conservation Biology*, **22**, 683–690.

Myers, N., Mittermeier, R. A., Mittermeier, C. G., *et al.* (2000). Biodiversity hotspots for conservation priorities. *Nature*, **403**, 853–858.

Olson, D. M. and Dinerstein, E. (1998). The Global 200: a representation approach to conserving the Earth's most biologically valuable ecoregions. *Conservation Biology*, **12**, 502–515.

Wells, M., Guggenheim, S., Khan, A., Wardojo, W., and Jepson, P. (1999). *Investing in biodiversity: a review of Indonesia's integrated conservation and development projects*. Directions in development series. World Bank, Indonesia and Pacific Islands Country Department, Washington, DC.

Wilkie, D. S., Carpenter, J. F., and Zhang, Q. (2001). The under-financing of protected areas in the Congo Basin: so many parks and so little willingness-to-pay. *Biodiversity and Conservation*, **10**, 691–709.

Zbinden, S., and Lee, D. R. (2004). Paying for Environmental Services: an analysis of participation in Costa Rica's PSA program. *Word Development*, **33**, 255–272.

---

Increasingly, conservation practitioners try to provide incentives to individuals and user groups to prevent the degradation of biodiversity. These incentives lie on a spectrum from indirect to direct with respect to their link with conservation objectives (Ferraro and Kiss 2002). The least direct approaches include support for the use and marketing of extracted biological products (e.g. logging, non-timber forest product extraction, hunting) and subsidies for reduced impact land and resource use (e.g. sustainable agriculture). Performance based payments for biodiversity conservation represents one of the most direct approaches of providing incentives. Box 15.6 outlines an example of this approach that has been implemented to conserve endangered bird species in Cambodia.

The evolving relationship between parks and people will continue to dominate international and national dialogues on biodiversity conservation and stimulate the evolution of innovative

### Box 15.7 International activities of the Missouri Botanical Garden
Peter H. Raven

The Missouri Botanical Garden (MBG) is the oldest botanical garden in the United States, established in 1859. Modern botanical gardens were first developed in Europe in the early 1500s as adjuncts to schools of medicine, since the physicians the medical schools trained had to be able to recognize those kinds of plants that would be effective in treating their patients. Consequently, botanical gardens are often associated with universities: they have carried out research on plants over the years, as they still do at the present time. During the era of colonization, the colonial powers often established botanical gardens as places where they could grow and investigate what crops of economic value might be useful in that particular area. The botanical gardens in Sydney, Singapore and Bogor are examples of institutions of this kind that have survived from the nineteenth century. Botanical gardens came from very different beginnings from zoos, which started as carnivals and displays, became permanent facilities under first royal and later municipal or state patronage, and are not historically connected with universities. In the modern era, both botanical gardens and zoos have recognized their common interest in conservation, since the organisms in their care often are becoming increasingly rare in nature. The kinds of research collections, herbaria, libraries, and associated databases that are associated with comprehensive botanical gardens are not mirrored in the holdings of zoos. Such research collections of both plants and animals are found as part of the holdings of natural history museums, including those in universities.

*continues*

### Box 15.7 (Continued)

The research program of the Missouri Botanical Garden, which initially was centered on the central United States and eventually spread to the Pacific Coast and into Mexico, has since the first part of the twentieth century been largely devoted to the tropics. A comprehensive account of the plants of Panamá begun in 1927 was completed in 1981. From this base, the research program of the garden spread north to southern Mexico and south throughout South America, to Africa, especially Madagascar; to China, Vietnam, Lao, and Cambodia, and to New Caledonia. Our style has often resulted in the preparation of comprehensive databases, and we are pushing increasingly towards a state in which all of the information about plants would be on the web and available for use or revision directly. Over a third of the plants of the world, more than 100 000 species, are being treated through one or more of the projects of the Missouri Botanical Garden.

Since the 1970s, the Garden's program has been organized around the activities of botanists resident in individual countries whose plants we are studying. We decided early on that it would not be possible to investigate the plants of any area thoroughly enough by means of intermittent expeditions and that we would be far more able to help in building institutions and training people if we lived on the ground with them. Thus our work in Nicaragua was based on Doug Stevens' residence of 11 years in the country, starting in the 1970s, that lead to the formation of substantial library and herbarium resources, and has, then and subsequently, resulted in the training of dozens of Nicaraguan botanists and conservationists. Through our continuing interactions with the government and many visits since, we have been able to do a great deal not only in technical botany but more importantly in building institutions through collaboration and by keeping in touch with individuals in our fields of study. Conservation and sustainability have become landmarks of our long-term intentions. In Peru, for example, empowering the Yanesha, indigenous people who want to use their resources sustainably, has been a major effort that continues to the present. Similar efforts are underway in Ecuador and Bolivia, and of course they are complementary in building knowledge of the plants of a particular region. In Costa Rica, resident MBG botanist Barry Hammel collaborates with the National Institute of Biodiversity (INBio) and the Museo Nacional in the production of a Manual Flora of the Plants of Costa Rica, one of the countries in the Neotropics where the most varied and comprehensive biological research is being conducted – we are sure that our manual will fill a gap by providing complete and up-to-date information on all kinds of plants found in the country.

In Madagascar, where MBG has been active for more than 30 years, we maintain a staff of more than 50 people, all but one Malagasy, and many trained in our joint Masters' degree programs with the University of Antananarivo. We are preparing a comprehensive, highly revised database on all the plants of the country, and finding about a third more kinds than had been recorded earlier, so that this island, which is about 50% larger than California, may be home to more than 13 000 species of plants. More than 90% of these are found nowhere else, and more than 80% of the natural vegetation in Madagascar has been destroyed, so that our team is literally engaged in a race against time, finding the places where plants grow and determining which are most critical for conservation. By Presidential Decree, the amount of preserved land in the country is being greatly increased at present, and it is of key importance to make the best choices concerning what should be set aside. The sustainability of certain communities, such as Mahabo, is being enhanced through collaboration with the Scandellaris Center of the Business School at Washington University in St. Louis, so that poor people may have alternatives to simply taking products unsustainably from an ever-diminishing forest – the key to biological conservation on a large scale.

The world will achieve sustainability only if efforts of this kind are repeated everywhere and the local efforts are united as a basis for common action. Along with sister institutions such as the Royal Botanic Gardens, Kew, and The New York Botanical Garden, we are contributing what we can toward the solution of our common challenge.

approaches for reconciliation (Koh and Wilcove 2007). Solutions for capturing opportunities that simultaneously protect biodiversity and reduce poverty, often boil down to improving institutions and governance, but there are no easy generalizations (Chomitz 2007).

## 15.4 Capacity needs for practical conservation in developing countries

In many developing economies with rich tropical biodiversity, government agencies responsible for the management of protected areas lack the necessary technical capacity to stem biodiversity loss effectively. These gaps in capacity occur at all levels, from the need for direct management of natural resources, to the compliance requirements of multilateral agreements (Steiner *et al.* 2003). At the ground level, managers of natural resources including biodiversity within protected areas often have limited access to the vast and dynamic body of knowledge and tools in conservation science. There is an urgent and critical need to transfer the advances in conservation science to individuals and institutions in biodiversity-rich countries. Building the capacity needed to implement conservation strategies and apply conservation principles represents one of the greatest challenges facing the field of conservation biology (Rodriguez *et al.* 2006).

Increasing capacity in applied conservation is complex: it involves not only the training of in-service conservation professionals but also the enhancement of university graduate and undergraduate programs that will generate a cadre of future conservation professionals. In order to be effective in the field of conservation, graduates of such training programs need relevant multidisciplinary knowledge and practical skills such as problem-solving and conflict resolution to tackle the complexities of biological and societal issues that characterize applied conservation (Noss 1997).

The urgency of the biodiversity crisis coupled with the general scarcity of funds and short project timelines make on-the-job training of individuals the most common approach to tackle the lack of capacity. NGOs for instance, work with individuals on specific projects and attempt to build capacity that is often quite specialized. However, a longer-term approach to building capacity would necessarily involve targeting relevant programs at universities and professional training institutions. Lack of financial resources and educational infrastructure are key limitations facing universities with regard to training for conservation. Addressing these issues will require concerted investment in financial and human capacity, but important initiatives are underway to begin this process.

Strong linkages between international NGOs and academic/professional institutions in countries such as Lao PDR are often key to provide field training opportunities in applied conservation research and management. Organizations such as zoological societies, natural history museums, and botanical gardens (see Box 15.7) are increasingly engaged in long-term conservation and capacity building efforts. In certain situations, such linkages may be the only means for students as well as staff of natural resource management agencies to gain valuable field experience in project design and management to complement theoretical knowledge and skills they may have acquired in the classroom.

For instance, the Network of Conservation Educators and Practitioners (NCEP, http://ncep.amnh.org), a project led by the Center for Biodiversity and Conservation of the American Museum of Natural History, aims to improve training in conservation biology through innovative educational materials and methods that directly target teachers of conservation biology. NCEP is a global initiative, currently active in Bolivia, Lao PDR, Madagascar, Mexico, Myanmar, Peru, Rwanda, the United States and Vietnam. The project seeks to create and make widely available a variety of resources to teach biodiversity conservation, and develop networks and resource centers to increase mentoring and training opportunities in biodiversity conservation worldwide. A central goal of the project is to increase teachers' and trainers' access to high quality and free of cost teaching materials. To meet this goal, NCEP develops collaborations with partner institutions and individuals including conservation practitioners to develop a series of multi-component teaching resources called modules adapted for local use. For example, in the Lao People's Democratic Republic (Lao PDR), a densely forested, land-locked country with high levels of biodiversity in Southeast Asia, NCEP established a partnership with The Wildlife Conservation Society and the National University of Laos (NUoL) to help develop the capacity of trainers in the science and forestry faculties to teach topics in conservation biology to undergraduates. Most young professionals employed in natural resource research and management agencies in the country today have graduated from the science or forestry faculties at the NUoL. These faculties have a critical need for up-to-date relevant materials in the Lao language for teaching biodiversity conservation principles.

Local adaptation is an important feature of the NCEP project, empowering in-country partners and making the materials immediately useful for faculty, students, and professionals who are already working in or associated with the field of biodiversity conservation. The project also found it useful to couple module training with applied research for students and faculty at field sites. The applied research served to reinforce learning and comprehension of new biodiversity conservation topics and terms in addition to providing critical exposure to real-world conservation.

A second phase of the NCEP project in Lao PDR involves building the capacity of university trainers to teach relevant aspects of applied conservation to protected area managers from seven National Protected Areas across the country. During this process, conservation science principles and case studies of applied conservation approaches will be adapted to make them more accessible to instructors to use as training materials for protected area managers who could apply those principles to achieve conservation results on the ground.

Capacity building activities can consume vast resources, potentially diverting already limited conservation funds away from other, more immediate conservation problems that involve direct actions at the site-level to reduce threats (for example, monitoring and enforcement). Moreover, justifying investment in capacity building activities is sometimes challenged by the difficulties involved in measuring success in the short-term. Yet, building capacity is vital to a longer-term vision of enabling responsible stewardship of biodiversity.

## 15.5 Beyond the science: reaching out for conservation

Globally, a key challenge to achieving conservation goals is the need to capture the interest of local people in a manner that stimulates cooperation and positive conservation actions (Brewer 2002). This need, sometimes defined as a form of social marketing, is a compelling reason for conservation biologists to work more closely with local communities to mobilize support for conservation through better informed and carefully designed outreach (Johns 2003). The process of involving local communities living adjacent to threatened species and their habitats helps build a constituency that is more aware of its role either as part of the problem or sometimes, as part of the solution, in a protected area (Steinmetz et al. 2006). This awareness is crucial to the effective implementation of conservation strategies. Field-based research outreach and partnership programs facilitate a two-way dialogue: local participants learn firsthand what scientists do, how they do it, and why they do it and by working with local communities, scientists can learn how local residents relate to the threatened species and habitats they study.

In the Thung Yai Naresuan Wildlife Sanctuary ($3622$ km$^2$) in western Thailand, commercial hunting contributed heavily to extensive population declines for most species and subsistence hunting was locally significant for some carnivores, leaf monkeys (*Presbytis* sp.), and deer. Workshops with local communities clarified which species were at highest risk of local extinction, where the most threatened populations were, and the causes of these patterns. Scientists, protected area managers and local people worked together to assess wildlife declines and jointly define and understand the scale of the problem during workshops. As a result, local people and sanctuary managers increased communication, initiated joint monitoring and patrolling, and established wildlife recovery zones.

While conflict between local people and the park authorities has not completely disappeared, there is interest to work together on wildlife issues (Steinmetz et al. 2006).

## 15.6 People making a difference: A Rare approach

Recognizing the important role that communities can play in conservation, a US based conservation organization known as Rare has adopted a mission to "conserve imperiled species and ecosystems around the world by inspiring people to care about and protect nature" (www.rareconservation.org; see Box 12.2). Rare fulfills this mission by addressing some of the most pressing needs of the global conservation movement. Rare trains and mentors local conservation leaders in the use of proven outreach tools, builds partnerships to leverage their investments and evaluate lessons learned to continuously improve the practice of conservation.

Rare's flagship program for constituency building is known as the Rare Pride campaign (Box 12.2). A hybrid of traditional education and private sector marketing strategies, Rare Pride campaigns inspire people who live in the world's most biodiverse places to take pride in their natural heritage and embrace conservation. Pride campaign managers are local conservationists who make two year commitments to inspire environmental protection at every level in their communities. Campaign managers are trained by the organization during a university-based program in social marketing culminating in a Master's degree in Communications for Conservation from the University of Texas (El Paso).

Pride campaigns utilize a charismatic flagship species, like the Saint Lucia parrot or the Philippine cockatoo, which becomes a symbol of local pride and acts as a messenger to build support for needed behavior changes for habitat and wildlife protection. Marketing tools such as billboards, posters, songs, music videos, sermons, comic books, and puppet shows make conservation messages positive, compelling, relevant, and fun for the community. Campaigns aim to generate an increased sense of pride and public stewardship that goes beyond mere awareness-raising. Pride campaigns involve and engage several segments of the community: teachers, business and religious leaders, elected officials, and the average citizen. Rare Pride is currently being employed on a global scale, and has been successfully replicated by partner organizations in over 40 countries.

## 15.7 Pride in the La Amistad Biosphere Reserve, Panama

The farming town of Cerro Punta, with a population of 7000, lies at the gateway to a forest corridor between Barú Volcano National Park in Panama and La Amistad Biosphere Reserve shared with Costa Rica that encompasses one of the largest tracts of undisturbed rainforest in one of the most biologically diverse regions in the world. The corridor between the two parks is important for the movement of globally significant species including ocelot (*Leopardus pardalis*), puma (*Puma concolor*), Baird's tapir (*Tapirus bairdii*), white-faced capuchin monkey (*Cebus capucinus*), and the Resplendent Quetzal (*Pharomachrus mocinno*). The land is under threat. The mild climate and rich volcanic soil creates fertile conditions that include four growing seasons a year for agricultural crops. Consequently, Cerro Punta produces 80% of all the vegetables grown in Panama (population 3.2 million). Crops are cultivated on the steep mountainsides without any terraces causing heavy erosion during the rainy season. Given the farmers' heavy reliance on synthetic chemical pesticides and fertilizers, erosion and run-off from the cultivated slopes leads to downstream water pollution with deleterious health impacts for residents. Furthermore, the erosion slowly forces farmers to clear more land for new fields, closer and closer to the two parks and the corridor between them. In addition to the threat of agricultural expansion, there is persistent pressure to build roads or highways through the La Amistad Biosphere reserve as exploitation for coal and minerals increases. Deforestation, cattle ranching, hunting, and commercial extraction are also serious threats to the Park's rich flora and fauna.

La Amistad needed a strong constituency lobbying for conservation, as well as significant change in community farming methods. Luis Olmedo Sanchez Samudio, a Sunday school teacher from a farming family in Cerro Punta, knew that creating real change in his community would require a dramatically different approach. Sanchez Samudio completed Rare's program at the University of Guadalajara in Mexico to learn how to implement a full scale Pride social marketing campaign in La Amistad. The Fundacion para el Desarrollo Integral del Corregimiento de Cerro Punta (FUNDICCEP), with Sanchez Samudio on their staff, allied with Rare and one of the biggest International NGOs, The Nature Conservancy, in this effort. Sanchez undertook the formidable task of reaching out to radio stations, schools, fairs, and the farmers themselves in a relentless effort to change decades-old customs and attitudes. Panama's Resplendent Quetzal was chosen to serve as the campaign's flagship species and used to talk about a range of conservation issues. Named "Quelly", an image of the Resplendent Quetzal appears on all campaign materials, reminding people of the importance of habitat protection. After several months of formative research, including surveys and focus groups with local farmers, Sanchez Samudio launched his campaign with over 30 outreach vehicles including posters, advertisements, bumper stickers, radio shows, mascots, classroom visits, sermons, workshops, festivals, and much more. Sanchez encouraged farmers to adopt sustainable agricultural practices while garnering support from clergymen, legislators and other relevant sources. Post campaign survey data to measure effectiveness showed that 52% of the respondents were aware of the benefits of living near a protected area, up from just 15% at the beginning of the campaign; 85% said they were ready to petition the government for better controls of agricultural chemicals, up from 61% at the beginning. Other indicators, such as whether respondents knew of alternatives to agricultural chemicals, remained flat at around 30%. Promoting alternatives became the central focus of Sanchez' follow-up efforts to conserve the La Amistad Biosphere Reserve—his local pride. To learn more about Rare's social marketing methodologies for conservation, visit www.rareconservation.org.

## 15.8 Outreach for policy

While local communities and protected area officials are important targets for outreach activities, an equally challenging need is for scientists and practitioners to engage in outreach that influences policy goals (Noss 2007). However there is acknowledged lack of clarity regarding advocacy in conservation biology which influences the ability of conservation biologists to effectively direct their expertise to policy decisions (Chan *et al.* 2005, 2008). At the core of this debate is the degree to which conservation biologists honor their commitment to the inherent value of biodiversity.

Given that scientists are still trained almost entirely in research methods, not public communication or policy intervention (Lovejoy 1989), there is some fear that engagement in public education and policy intervention can reduce credibility (Blockstein 2002). One thread of this debate is based on the need to relinquish commitment to the inherent value of biodiversity while another thread suggests that conservation biologists should explicitly advocate for values (e.g. biodiversity) and are obligated to step well beyond research to recommend solutions to policy goals (Chan 2008).

## 15.9 Monitoring of Biodiversity at Local and Global Scales

Monitoring is critically essential to determine the extent to which protected areas are effective in conserving biodiversity or achieving other management objectives. Monitoring that provides assessment of threats in a manner that allows managers to respond effectively, is central to good conservation management (see Chapter 16). Danielsen *et al.* (2000) define 'monitoring' as data sampling which is: (i) repeated at certain intervals of time for management purpose; (ii) replicable over an extended time frame; and (iii) focuses on rates and magnitude of change. Monitoring helps identify priority areas for research and conservation, and to quantify the response of plant and animal populations to disturbance and management interventions. Countries contracting to the Convention on Biological Diversity are obliged to monitor

## Box 15.8 Hunter self-monitoring by the Isoseño-Guaraní in the Bolivian Chaco
### Andrew Noss

The 34 400 km² Kaa-Iya del Gran Chaco National Park (KINP) in Bolivia was created in 1995 to protect the Gran Chaco's natural resources and the traditional use areas of the indigenous residents surrounding it, the Isoseño-Guaraní, Chiquitano and Ayoreo (and a group of non-contacted Ayoreo living within it). It is the largest dry forest protected area in the world, and contains high levels of biological diversity, particularly mammals, with at least 10 endemic mammal taxa, most notably the Chacoan guanaco (*Lama guanicoe voglii*) and the Chacoan peccary (*Catagonus wagneri*) (Ibisch and Me´rida 2003). KINP is the first protected area in South America co-managed by an indigenous organization, the Capitanı´a del Alto y Bajo Isoso (CABI) which is the political authority representing the 10 000 Isoseño-Guaraní inhabitants of the Isoso. Isoseño livelihoods are based on agriculture, livestock, hunting, fishing and permanent and seasonal wage labor. Prior to the creation of the KINP, most of the 23 Isoso communities had legal titles of their lands as community lands covering an area of 650 km², encompassing settlements, farming, and livestock lands. In 1997, based on their historical occupation of the area over the past 300 years, CABI formally demanded 19 000 km² as a 'Tierra Comunitaria de Orı´gen' or TCO adjacent to, but not overlapping, the KINP. Principal threats to both the TCO and KINP include illegal settlements and inappropriate management of land and natural resources with the conversion of Chaco forests to soybean farms and extensive cattle ranches (overstocking, no management of forage, minimal veterinary care), sport hunting by city-based hunters, and large-scale regional infrastructure programs that include international gas pipelines and highways.

Like other indigenous groups, many traditional beliefs and local practices among the Isoseño influence their hunting behaviors to favor wildlife conservation. A hunter must follow certain rules in order to retain the favor of the spirits that guard wildlife. For example, hunters should not hunt young animals, hunt excessively or beyond family needs, or mistreat animals by wounding them and allowing them to escape. Additional local practices that favor wildlife conservation include seasonal rotation of hunting areas that respond to seasonal movements of animals according to availability of food, as well as the accessibility of different areas, no hunting of certain vulnerable species (primates, guanacos) and the substitution of other activities (such as fishing and farming) to hunting in particular seasons. Seeking to integrate these traditional beliefs and local knowledge of wildlife with political/administrative requirements and scientific management, in 1996 a joint team of an international NGO, the Wildlife Conservation Society, and CABI personnel initiated a wildlife and hunting monitoring program in the 23 Isoseño communities. The principal objectives were to: (i) determine whether subsistence (armadillos, peccaries, brocket deer, tapir) and commercial (parrots, tegu lizards) hunting by Isoseño communities was sustainable; (ii) generate management recommendations to ensure that hunting would be sustainable in the indigenous territory, thereby reducing potential pressure on the KINP; and (iii) consolidate the concepts and practices of wildlife management together with hunters and communities (Painter and Noss 2000). The principal method to estimate hunting offtakes was a hunter self-monitoring program with voluntary participation: hunters carried data sheets with them on hunting excursions to record information on the hunt and on any captured animals, and they collected specimens (skulls/jawbones, stomach contents, fetuses) of hunted animals. Community hunting monitors assisted the hunters to record, collect and analyze the data for the entire community on a monthly basis (Noss *et al.* 2003, 2004).

The communities selected Isoseño parabiologists and hunting monitors, the majority with an elementary and some with high school education. Following an initial six month volunteer period, those who expressed the most interest and initiative were hired by the program. Monitors (seven to ten individuals each living and working in their home community) were hunters hired part-time to support the recording of hunting data in communities (by encouraging hunters to participate in the self-monitoring program, and

*continues*

### Box 15.8 (Continued)

by periodically collecting information from hunters in their community). Parabiologists (six to eight individuals working in their home community or other research sites in the Isoso) were hired full-time to support wildlife research according to their individual specialization. Through field courses and practical experience, these Isoseño technicians began to assume greater responsibility for designing and implementing research programs with hunters.

Hunter self-monitoring (100–150 hunters per month) combined with monthly activity records for potential hunters (7637 observed hunter-months) permitted estimations of total offtakes of subsistence game species for 1996–2003, as well as catch-per-unit-effort over the same time period. These data showed considerable fluctuations from year to year and no declining trends that would suggest over-hunting.

Experience from the monitoring project suggested that even simple approaches such as hunter self-monitoring or line transect surveys required considerable effort by both project staff and volunteers in order to provide sufficient information for management interventions. Thus, precise and detailed population density estimates are difficult to obtain in situations with a large number of species and/or large study areas such as the Isoso with only basic tools and non-professional personnel.

Ideally, adaptive management would include continuous population monitoring over long time periods using selected indicator species assemblages, detailed studies of ecological principles and processes, and studies of population trends in sink and source areas (Kremen et al. 1994; Hill et al. 2003). Such detailed monitoring is prohibitively complex and expensive not only for territories under the jurisdiction of indigenous peoples but also for most protected areas in general.

Instead, it may be more useful to consider adaptive management in a broader context focusing on fundamental requirements for informed decision-making. Assuming that communal decision-making is the key, detailed scientific information and sophisticated analyses may not be as important as ensuring that: (i) information familiar to resource managers is used; and (ii) participatory methods provide the inputs and framework for discussion (see also Danielsen et al. 2005). Hunter self-monitoring provides a means to engage large numbers of community members in data collection. By generating the data themselves, people become conscious of underlying problems, for example perceived or actual over-hunting of a certain species, and can thus think about solutions to address the problems. In turn, reflection processes may lead to preliminary management action that can be consolidated in an adaptive management process. Approaches that integrate traditional customs and knowledge with scientific methods, bringing together community members with specialists can have positive outcomes for conservation (Becker et al. 2005; Townsend et al. 2005). In the Isoso case, this integration took place at several levels. At a first level, community members indicated through discussions the most important game species and described hunting practices and traditions regarding wildlife management. In turn, through hunter self-monitoring and observation of hunting activities, hunters themselves and trained community members (parabiologists and monitors) confirmed and quantified what hunters did in practice. Strong traditional authority structure and community organization, a favorable legal/institutional framework, the ability of government authorities to appropriately implement their responsibilities, and financial and technical support from private partners to the process were all important determinants of effective engagement of communities in this wildlife monitoring program.

This box is adapted from Noss et al. (2005).

### REFERENCES

Becker, C. D., Agreda, A., Astudillo, E., Constantino, M., and Torres P. (2005). Community-based surveys of fog capture and biodiversity monitoring at Loma Alta, Ecuador enhance social capital and

*continues*

> **Box 15.8 (Continued)**
>
> institutional cooperation. *Biodiversity Conservation*, **14**, 2695–2707.
>
> Danielsen, F., Burgess, N., and Balmford, A. (2005). Monitoring matters: examining the potential of locally-based approaches. *Biodiversity Conservation*, **14**, 2507–2542.
>
> Hill, K., McMillan, G., and Farin, A. R. (2003). Hunting-related changes in game encounter rates from 1994 to 2001 in the Mbaracayú Reserve, Paraguay. *Conservation Biology*, **17**, 1312–1323.
>
> Ibisch, P. and Mérida, G. (2003). *Biodiversidad: la riqueza de Bolivia. Estado de conocimien to y conservacion*. Fundación Amigos de la Naturaleza, Santa Cruz, CA.
>
> Kremen, C., Merenlender, A. M., and Murphy, D. D. (1994). Ecological monitoring: a vital need for integrated conservation and development programs in the tropics. *Conservation Biology*, **8**, 388–397.
>
> Noss, A. J., Cuéllar, E., and Cuéllar, R. L. (2003). Hunter self-monitoring as a basis for biological research: data from the Bolivian Chaco. *Mastozoología Neotropical*, **10**, 49–67.
>
> Noss, A. J., Cuéllar, E., and Cuéllar, R.l. (2004). An evaluation of hunter self-monitoring in the Bolivian Chaco. *Human Ecology*, **32**, 685–702.
>
> Noss, A. J., Oetting, I., and Cuéllar, R. L. (2005). Hunter self-monitoring by the Isoseño-Guaraní in the Bolivian Chaco. *Biodiversity Conservation*, **14**, 2679–2693.
>
> Painter, M. and Noss, A. (2000). La conservación de fauna con organizaciones comunales: experiencia con el pueblo Izoceño en Bolivia. In E. Cabrera, C. Mercolli and R. Resquin, eds *Manejo de fauna silvestre en Amazonía y Latinoamérica*. pp. 167–180. CITES Paraguay, Fundación Moises Bertoni, University of Florida, Asunción, Paraguay.
>
> Townsend, W. R., Borman, A. R., Yiyoguaje, E., and Mendua, L. (2005). Cofán Indians' monitoring of freshwater turtles in Zábalo, Ecuador. *Biodiversity and Conservation*, **14**, 2743–2755.

biodiversity (Article 7.b), and donors increasingly demand accountability and quantifiable achievements in return for their assistance. Given that biodiversity conservation is one of the key objectives of protected areas, the development of biodiversity monitoring systems for protected areas now attracts a significant proportion of the international funding for biodiversity conservation.

However, conflicts between the scientific ideals and practical realities of monitoring influence the implementation and effectiveness of monitoring systems. For instance, most practitioners agree that in an ideal world, monitoring programs would always be spatially and temporally comprehensive, rigorous in their treatment of sampling error, and sustainable over the time scales necessary to examine population and community level processes (Yoccoz *et al.* 2001). Nevertheless, monitoring of biodiversity and resource use in the real world is often costly and hard to sustain, especially in developing countries, where financial resources are limited. Moreover, such monitoring can be logistically and technically difficult, and is often perceived to be irrelevant by resource managers and local stakeholders. Many suggest the need to identify some middle ground between the need for scientific rigor and goals for program sustainability. Practitioners disagree about whether such a balance exists, and the issue has become a source of debate. At the centre of this debate is the fact that where suggestions or examples of 'appropriate' monitoring in developing countries exist, they generally are unproven in their ability to detect 'true' trends. On the one hand, poor statistical power and bias may turn overly simplistic monitoring schemes into wastes of time and precious resources – yet equally wasteful are programs so intensive they cannot be sustained long enough to address questions fundamental to effective management (Yoccoz *et al.* 2001, 2003; Danielsen *et al.* 2003; Chapter 16). Box 15.8 examines the issue of biological monitoring within the context of a community wildlife management program in the Kaa Iya Del Gran Chaco in Bolivia.

The technical and statistical problems of monitoring at a local level are relatively benign when compared to the problems of tackling monitoring at a global scale. Under the terms of an agreement signed at the Johannesburg Summit on Sustainable Development in 2002, 190 countries committed to "a significant reduction" in the current rate of loss of biodiversity. But the challenges of estimating global rates of loss are enormous (summarized by Balmford et al. 2005), and as the target of 2010 is approached, most indicators developed inevitably involve the use of indirect or surrogate data on habitat loss, protected area overlays with known patterns of biodiversity or with targeted studies of well known vertebrate taxa.

In this brief review we have touched on a number of the challenges in translating conservation science into practical, field based conservation actions. Conservation action lags behind conservation science for a number of reasons. Inevitably, there will be time lags in the dissemination and application of new ideas to real world situations, and the way in which theory informs practice will not always be clear at the outset. But there will also be gaps between the interests and needs of conservation practitioners, and the issues and areas of intellectual pursuit that are valued by academic departments, and institutional science donors.

## Summary

- Integrating the inputs of decision-makers and local people into scientifically rigorous conservation planning is a critically important aspect of effective conservation implementation.
- Protected areas represent an essential component of approaches designed to conserve biodiversity. However, given that wildlife, ecological processes and human activities often spill across the boundaries of protected areas, designing strategies aimed at managing protected areas as components of larger human-dominated landscapes will be necessary for their successful conservation.
- Identifying strategies that simultaneously benefit biodiversity conservation and economic development is a challenge that remains at the forefront of applied conservation. Biodiversity use may not be able to alleviate poverty, but may have an important role in sustaining the livelihoods of the poor, and preventing further impoverishment. Strong institutions and good governance are prerequisites for successful conservation interventions.
- Capacity needs for practical conservation in developing countries occur at many levels from skills needed for management of natural resources to the compliance requirements of multilateral agreements. Filling gaps in capacity involves a diversity of approaches from on-the-job training of individuals to restructuring academic and professional training programs. Prioritizing capacity needs is vital to a longer-term vision of enabling responsible stewardship of biodiversity.
- The engagement of local communities in planning and implementation is critical for effective conservation. Carefully designed social marketing approaches have proved to be successful in capturing the interest of local people while achieving conservation goals.
- Monitoring is a central tenet of good conservation management. Conflicts between the scientific ideals and practical realities of monitoring influence the implementation and effectiveness of monitoring systems.
- Many of the key issues and barriers to effective conservation that face conservation biologists are inherently political and social, not scientific. Thus efforts to close the gap between conservation biologists and conservation practitioners who take action on the ground will require unprecedented collaboration between ecologists, economists, statisticians, businesses, land managers and policy-makers.

## Suggested reading

Ferraro, P. J. and Pattanayak, S. K. (2006). Money for nothing? A call for empirical evaluation of biodiversity conservation investments. *PLoS Biology*, **4**, 482–488.
Pressey, R. L., Cabeza, M., Watts, M. E., et al. (2007). Conservation planning in a changing world. *Trends in Ecology and Evolution*, **22**, 583–592.
Terborgh, J. (1999). *Requiem for nature*. Island Press, Washington, DC.

## Relevant websites

- Cape Action Plan for the Environment (CAPE): http://www.capeaction.org.za/
- Centre for Evidence-based conservation: http://www.cebc.bangor.ac.uk/ and http://www.conservationevidence.com/
- Network of Conservation Educators and Practitioners: http://ncep.amnh.org

- RARE: www.rareconservation.org
- Living Landscapes: http://www.wcslivinglandscapes.com/
- Natural Capital Project: www.naturalcapitalproject.org

## REFERENCES

Agrawal, A. and Redford, K. H. (2006). Poverty, development, and biodiversity conservation: Shooting in the dark? *Wildlife Conservation Society Working Paper*, **26**, 1-48.

Angelsen, A. and Wunder, S. (2003). *Exploring the forest-poverty link: key concepts, issues, and research implications.* CIFOR (Center for International Forestry Research) Occasional Paper No. 40. Bogor, Indonesia.

Balmford, A., Mace, G. M., and Ginsberg, J. R. (1998). The challenges to conservation in a changing world: putting processes on the map. In G. M. Mace, A. Balmford, and J. R. Ginsberg, eds *Conservation in a Changing World.* pp. 1–28. Cambridge University Press, Cambridge, UK.

Balmford, A. (2003). Conservation planning in the real world: South Africa shows the way. *Trends in Ecology and Evolution*, **18**, 435–438.

Balmford, A., Crane, P., Dobson, A. P., Green, R. E., and Mace, G.M. (2005). The 2010 challenge: data availability, information needs, and extraterrestrial insights. *Philosophical Transactions of the Royal Society B*, **360**, 221–228.

Blockstein, D. E. (2002). How to lose your political virginity while keeping your scientific credibility. *BioScience*, **52**, 91–96.

Bonan, G. B. (2008). Forests and climate change: forcings, feedbacks, and the climate benefits of forests. *Science*, **320**, 1444–1449.

Brandon, K., Redford, K. and Sanderson, S., eds (1998). *Parks in peril: people, politics, and protected areas.* Island Press, Washington, DC.

Brewer, C. (2002). Outreach and partnership programs for conservation education where endangered species conservation and research occur. *Conservation Biology* **16**, 5–6.

Brockington, D. (2003). Injustice and conservation: is local support necessary for sustainable protected areas? *Policy Matters*, **12**, 22–30.

Butler, R. F. and Laurance, W. F. (2008). New strategies for conserving tropical forests. *Trends in Ecology and Evolution*, **23**, 469–472.

Chan, K. M. A. (2008). Value and advocacy in conservation biology: crisis discipline or discipline in crisis? *Conservation Biology*, **22**, 1–3.

Chan, K. M. A., Higgins, P. A. T., and Porder, S. (2005). Protecting science from abuse requires a broader form of outreach. *PLoS Biology*, **3**, 1177–1178.

Chomitz, K. (2007). *At loggerheads? Agricultural expansion, poverty reduction, and environment in the tropical forests.* The World Bank, Washington, DC.

Cobb, S., Ginsberg, J. R., Thomsen, J. (2007). Conservation in the tropics: evolving roles for governments, international donors and non-government organizations. In D. Macdonald and K. Service, eds *Key topics in conservation biology*, pp. 145–155. Blackwell Scientific, Oxford, UK.

Colchester, M. (2004). Conservation policy and indigenous peoples. *Cultural Survival Quarterly*, **28**, 17–22.

Cowling, R. M. and Pressey, R. L. (2003). Introduction to systematic conservation planning in the Cape Floristic Region. *Biological Conservation*, **112**, 1–13.

Cowling, R. M., Pressey, R. L., Rouget, M., and Lombard, A. T. (2003). A conservation plan for a global biodiversity hotspot - the Cape Floristic Region, South Africa. *Biological Conservation*, **112**, 191–216.

Daily, G. C., ed (1997). *Nature's services: societal dependence on natural ecosystems.* Island Press, Washington, DC.

Danielsen F., Balete, D. S., Poulsen, M. K., et al. (2000). A simple system for monitoring biodiversity in protected areas of a developing country. *Biodiversity and Conservation*, **9**, 1671–1705.

Danielsen F., Mendoza, M. M., Alviola, P., et al. (2003). Biodiversity monitoring in developing countries: what are we trying to achieve? *Oryx*, **37**, 407–409.

Davis, S. D., Heywood, V. H., and Hamilton, A. C., eds (1994). *Centres of Plant Diversity. A Guide and Strategy for their Conservation. Vol. 1. Europe, Africa, South West Asia and The Middle East.* WWF and IUCN, IUCN Publications Unit, Cambridge, UK.

Duraiappah, A. K. (1998). Poverty and environmental degradation: a review and analysis of the nexus. *World Development*, **26**, 2169–2179.

Ehrlich, P. R. and Ehrlich, A. H. (2005). *One with Nineveh: politics, consumption, and the human future, (with new afterword).* Island Press, Washington, DC.

Ferraro, P. J. and Kiss, A. (2002). Direct Payments to Conserve Biodiversity. *Science*, **298**, 1718–1719.

Ferraro, P. J. and Pattanayak, S. K. (2006). Money for nothing? A call for empirical evaluation of biodiversity conservation investments. *PLoS Biology*, **4**, 482–488.

Ghimire, K. B. and Pimbert, M. P. (1997). *Social change and conservation: environmental politics and impacts of national parks and protected areas.* Earthscan, London, UK.

ICEM (International Centre for Environmental Management). (2003). Lessons learned in Cambodia, Lao PDR, Thailand and Vietnam. Review of Protected Areas and Development in the Lower Mekong River Region, Indooroopilly, Queensland, Australia.

Johns, D. M. (2003). Growth, Conservation, and the Necessity of New Alliances. *Conservation Biology*, **17**, 1229–1237.

Kaimowitz, D. and D. Sheil. (2007). Conserving what and for whom? Why conservation should help meet basic human needs in the tropics. *Biotropica*, **39**, 567–574.

Koh, L. P., and Wilcove, D. S. (2007). Cashing in palm oil for conservation. *Nature*, **448**, 993–994.

Kramer, R., C. van Schaik, and J. Johnson, eds (1997). *Last stand: protected areas and the defense of tropical biodiversity*. Oxford University Press, Oxford, UK.

Lovejoy, T. (1989). The obligations of a biologist. *Conservation Biology*, **3**, 329–330.

Lovejoy, T. (2006). Protected areas: a prism for a changing world. *Trends in Ecology and Evolution*, **21**, 329–333.

McNeely, J. and Scherr, S. J. (2001). *Common ground common future: how ecoagriculture can help feed the world and save wild biodiversity*. IUCN Washington, DC.

McShane, T. O. (2003). Protected areas and poverty. *Policy Matters*, **12**, 52–53.

Myers, N., Mittermeier, R. A., Mittermeir, C. G., et al. (2000). Biodiversity hotspots for conservation. *Nature*, **403**, 853–858.

Noss, R. F. (1997). The failure of universities to produce conservation biologists. *Conservation Biology*, **11**, 1267–1269.

Noss, R. F. (2007). Values are a good thing in conservation biology. *Conservation Biology*, **21**, 18–20.

Olson, D. M. and Dinerstein, E. (1998). The Global 200: a representation approach to conserving the earth's most biologically valuable ecoregions. *Conservation Biology*, **12**, 502–515.

Peres, C. A. (1995). Indigenous reserves and nature conservation in Amazonian forests. In D. Ehrenfield, ed. *Readings from conservation biology: the social dimension*, pp. 25–57. Blackwell Science, Cambridge, MA.

Pressey, R. L., Cabeza, M., Watts, M. E., et al. (2007). Conservation planning in a changing world. *Trends in Ecology and Evolution*, **22**, 583–592.

Ravnborg, H. M. (2003). Poverty and environmental degradation in the Nicaraguan hillsides. *World Development* **31**, 1933–1946.

Redford, K. H and Sanderson S. E. (2003). Contested relationships between biodiversity conservation and poverty alleviation. *Oryx*, **37**, 389–390.

Redford, K. H., Levy, M. A., Sanderson, E. W., and de Sherbinin, A. (2008). What is the role for conservation organizations in poverty alleviation in the world's wild places? *Oryx*, **42**, 516–528.

Robinson, J. (2007). Recognizing differences and establishing clear-eyed partnerships: a response to Vermeulen and Sheil. *Oryx*, **41**, 443–444.

Rodriguez, J. P., Rodriguez-Clark, K. M., Oliveira-Miranda, M. A., et al. (2006). Professional capacity building: the missing agenda in conservation priority setting. *Conservation Biology*, **20**, 1341.

Sodhi, N. S., Brooks, T. M., Koh, L. P., et al. (2006). Biodiversity and human livelihood crises in the Malay Archipelago. *Conservation Biology*, **20**, 1811–1813.

Sodhi, N. S., Acciaioli, G., Erb, M., and Tan, A. K.-J., eds (2008). *Biodiversity and human livelihoods in protected areas: case studies from the Malay Archipelago*. Cambridge University Press, Cambridge, UK.

Stattersfield, A. J., Crosby, M. J., Long, A. J., and Wedge, D. C. eds (1998). *Endemic Bird Areas of the World. BirdLife Conservation Series 7*. BirdLife International, Cambridge, UK.

Steiner, A., Kimball, L. A., and Scanlon, J. (2003). Global governance for the environment and the role of Multilateral Environmental Agreements in conservation. *Oryx*, **37**, 227–237.

Steinmetz, R., Chutipong, W., and Seuaturien, N. (2006). Collaborating to conserve large mammals in Southeast Asia. *Conservation Biology*, **20**, 1391–1401.

Sutherland, W. J., Pullin, A. S., Dolman, P. M., and Knight, T. M. (2004). The need for evidence-based conservation. *Trends in Ecology and Evolution*, **19**, 305–308.

Terborgh, J. (1999). *Requiem for nature*. Island Press, Washington, DC.

Upton, C., Ladle, R., Hulme, D., Jiang, T., Brockington, D., and Adams, W. M. (2008). Are poverty and protected area establishment linked at a national scale? *Oryx*, **42**, 19–25.

Vermeulen, S. and Sheil, D. (2007). Partnerships for tropical conservation. *Oryx*, **41**, 434–440.

Wells, M., Guggenheim, S., Khan, A., Wardojo, W., and Jepson. P. (1999). *Investing in biodiversity. a review of Indonesia's Integrated Conservation and Development Projects*. World Bank, Washington, DC.

West, P. and Brockington, D. (2006). An anthropological perspective on some unexpected consequences of protected areas. *Conservation Biology*, **20**, 609–616.

Wilson K.A., Underwood E. C., Morrison S. A., et al. (2007). Conserving biodiversity efficiently: What to do, where and when. *PLoS Biology*, **5**, 1851–1861.

Woodroffe, R. and Ginsberg, J. R. (1998). Edge effects and the extinction of populations inside protected areas. *Science*, **280**, 2126–2128.

Woodwell, G. M. (2002). On purpose in science, conservation and government. *Ambio*, **31**, 432–436.

Yoccoz, N. G., Nichols, J. D., and Boulinier, T. (2001). Monitoring of biological diversity in pace and time. *Trends in Ecology and Evolution*, **16**, 446–453.

Yoccoz, N.G., Nichols, J. D., and Boulinier, T. (2003). Monitoring of biological diversity – a response to Danielsen et al. *Oryx*, **37**, 410.

# CHAPTER 16

# The conservation biologist's toolbox – principles for the design and analysis of conservation studies

## Corey J. A. Bradshaw and Barry W. Brook

"Conservation biology" is an integrative branch of biological science in its own right (Chapter 1); yet, it borrows from most disciplines in ecology and Earth systems science; it also embraces genetics, dabbles in physiology and links to veterinary science and human medicine. It is also a mathematical science because nearly all measures are quantified and must be analyzed mathematically to tease out pattern from chaos; probability theory is one of the dominant mathematical disciplines conservation biologists regularly use. As rapid human-induced global climate change (Chapter 8) becomes one of the principal concerns for all biologists charged with securing and restoring biodiversity, climatology is now playing a greater role. Conservation biology is also a social science, touching on everything from anthropology, psychology, sociology, environmental policy, geography, political science, and resource management (Chapter 14). Because conservation biology deals primarily with conserving life in the face of anthropogenically induced changes to the biosphere, it also contains an element of economic decision making (Chapter 14). This is a big toolbox indeed, so we cannot possibly present all aspects here. We therefore focus primarily in this chapter on the ecological components of conservation biology (i.e. we concentrate on the biology *per se*).

Conservation biology, and the natural sciences in particular, require simplified abstractions, or models, of the real world to make inferences regarding the implications of environmental change. This is because ecosystems are inherently complex networks of species interactions, physical constraints and random variation due to stochastic (random) environmental processes. The conservation biologist's analytical toolbox therefore comprises methods that mainly serve to simplify the complexity of the real world such that it is understandable and (partially) predictable. The quantification of these relationships – from the effects of habitat loss on biodiversity (Chapter 4) to the implications of small population size for extinction risk (Chapter 10) – is the backbone of analytical conservation biology and evidence-based decision making. Without quantified relationships and robust measures of associated uncertainty, recommendations to improve biodiversity's plight via management intervention or policy change are doomed to fail.

Even though we have chosen to focus on the techniques dealing with the biological data in the conservation realm, we can by no means be comprehensive; there are simply too many ideas, metrics, tests, paradigms, philosophies and nuances to present within a single chapter of this book. However, we have striven to compile a compendium of the major approaches employed along with a list of the best textbook guides and peer-reviewed scientific papers providing the detail necessary for their implementation. We first present measures of biodiversity patterns followed by a general discussion of experimental design and associated statistical paradigms. We then introduce the analysis of abundance time series followed by assessments of species' fate risks. The final section is a brief introduction to genetic tools used to assess a species' conservation status. Although issues of reserve design and their associated algorithms are an essential part of the conservation biologist's toolbox, they have been discussed in detail elsewhere in this book (Chapter 11) and so do not feature in this chapter.

## 16.1 Measuring and comparing 'biodiversity'

Chapter 2 provides an excellent overview of the somewhat nebulous concept of 'biodiversity' and a brief mention of how it can be measured, and Chapter 11 introduces the concept of 'surrogacy' (simplified measures of biodiversity patterns) in conservation planning. Here we develop these concepts further with particular emphasis on practical ways to obtain comparable and meaningful metrics over space and time. It should be noted that regardless of the logistic constraints, biological consideration and statistical minutiae driving the choice of a particular set of metrics for biodiversity, one must not forget to consider the cost-benefit ratio of any selected method (Box 16.1) or the difficulties and challenges of working across cultures (Box 16.2).

### 16.1.1 Biodiversity indices

It is simply impossible to measure every form of life (Chapter 2), regardless of the chosen metric or focal taxon, due to the sheer number of species and the difficulty of sampling many of the Earth's habitats (e.g. ocean depths and tropical forest canopies). We are therefore required to simplify our measurements into tractable, quantifiable units that can be compared across time and space. The simplest and perhaps easiest way to do this has traditionally been to use organism-based metrics that count, in one way or another, the number of 'distinct' species in a defined area. Species richness is therefore the base currency used for most biodiversity assessments, but it can be complicated by adjusting for relative abundance, uniqueness, representativeness, spatial scale or evolutionary history.

As mentioned above, a direct count of the number of species within a defined area is known as *species richness* (S). Species richness can be corrected for total abundance (number of individuals) to produce the *diversity* index better known as *Simpson's Diversity Index* $(1 - \hat{D})$ (Simpson 1949):

---

**Box 16.1 Cost effectiveness of biodiversity monitoring**
**Toby A. Gardner**

There is a shortage of biological data with which to meet some of the primary challenges facing conservation, including the design of effective protected area systems and the development of responsible approaches to managing agricultural and forestry landscapes. This data shortage is caused by chronic under-funding of conservation science, especially in the species-rich tropics (Balmford and Whitten 2003), and the high financial cost and logistical difficulties of multi-taxa field studies. We must therefore be judicious in identifying the most appropriate species groups for addressing a particular objective. Such focal groups are varyingly termed 'surrogates' or 'indicators'. However, indicators are often chosen subjectively on the basis of anecdotal evidence, 'expert' opinion, and ease of sampling. This common approach has resulted in finite resources being wasted on the collection of superficial (including the 'record everything' mantra) and unrepresentative biodiversity data that may be of only limited value. This failing threatens to erode the credibility of conservation science to funding bodies and policy makers.

To maximize the utility of biodiversity monitoring, it should adhere to the concepts of *return on investment*, and *value for money*. In essence this means that field-workers need to plan around two main criteria in selecting which species to sample: (i) what types of data are needed to tackle the objective in hand; and (ii) feasibility of sampling different candidate species groups. Practical considerations should include the financial cost of surveying, but also the time and expertise needed to conduct a satisfactory job. Species groups that satisfy

*continues*

## Box 16.1 (Continued)

both demands can be thought of as having a 'high performance'.

Using a large database from work in the Brazilian Amazon, Gardner et al. (2008) recently presented a framework and analytical approach for selecting such high performance indicator taxa. The objective of that study was to provide representative and reliable information on the ecological consequences of converting tropical rainforest to *Eucalyptus* plantations or fallow secondary regeneration. An audit was conducted of the cost (in money and time) of sampling 14 groups of animals (vertebrates and invertebrates) across a large, managed, lowland forest landscape. Notably, survey costs varied by three orders of magnitude and comparing standardised costs with the indicator value of each taxonomic group clearly demonstrated that birds and dung beetles (Coleoptera: Scarabaeinae) are high-performance groups – they provide the most amount of valuable information for the least cost. By contrast, other groups like small mammals and large moths required a large investment for little return (see Box 16.1 Figure). The fact that both birds and dung beetles are well-studied and perform important ecological functions gives further support to their value for biodiversity monitoring and evaluation. This important finding will help conservation biologists in prioritising the study of the effects of deforestation on land-use change in the Amazon, allowing them to design cost-effective field expeditions that will deliver the most useful information for the money available.

Finally when planning biodiversity surveys it is also important to consider how the data may be used to address ancillary objectives that may ensure an even greater return on investment.

One example is the opportunity to synthesise information from many small-scale monitoring programs to provide robust nation-wide assessments of the status of biodiversity without needing to implement independent studies. A better understanding of the distribution of species in threatened ecosystems will improve our ability to safeguard the future of biodiversity. We cannot afford to waste the limited resources we have available to achieve this fundamental task.

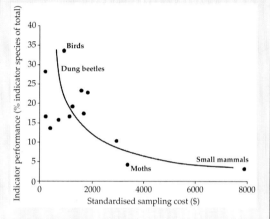

**Box 16.1 Figure** Cost effectiveness of different species groups for indicating habitat change in a multi-purpose forest landscape in Brazilian Amazonia.

## REFERENCES

Balmford, A. and Whitten, T. (2003). Who should pay for tropical conservation, and how could the costs be met? *Oryx*, **37**, 238–250.

Gardner, T. A., Barlow, J., Araujo, I. S., et al. (2008). The cost-effectiveness of biodiversity surveys in tropical forests. *Ecology Letters*, **11**, 139–150.

$$1 - \hat{D} = \frac{\sum_{i=1}^{S} n_i(n_i - 1)}{N(N - 1)}$$

where $S$ = the number of species, $N$ = the total number of individual organisms, and $n_i$ = the number of individuals of species $i$. The unique-ness of species in a sample can be incorporated by using indices of *evenness* (also known as *equitability*), of which *Shannon's Index* ($H$; also known mistakenly as the *Shannon-Weaver* index, or correctly as the *Shannon-Weiner* index) is the most common:

## Box 16.2 Working across cultures
### David Bickford

Establishing conservation projects in countries with cultures and languages that are different from your own can be both daunting and challenging. Without proper thoughtfulness, openness, flexibility, and (most importantly) humour, these projects fail for reasons that are often difficult to distil. All conservation projects involve a mix of stakeholders (local people, scientists, conservation practitioners, governmental and public administrators, educators, community leaders, etc.) that may have widely different expectations and responsibilities for the project. Having worked on both successful and failed projects with a diversity of people in nine countries and six languages, much of what I have learned can be summed up in two simple yet powerful ideas for all stakeholders: *clear communication* and *equity*. The two are intricately linked.

Clear communication is an ideal often sought after, yet rarely achieved. No matter the socio-cultural context, a common denominator of transparency is necessary for a successful conservation project. Having stakeholders explicitly state their intentions, desires and goals is a good start. It also helps elicit traditional or anecdotal knowledge that can be useful in formal analysis (e.g. as Bayesian priors, see Box 16.4). Methods, benefits, and responsibilities should be outlined and agreed upon, as well as limits of what objective(s) each stakeholder perceives as 'bare minimum'. A common pitfall is an inability for leaders to communicate effectively (for many and sundry reasons), re-enforcing top-down stereotypes. Lateral communication (peer-to-peer) can be more effective and avoids many constraints imposed by translating among different languages or cultures, effectively levelling the playing field and enabling everyone to participate (at least for heuristic purposes). Activities that enhance transparent communication include small group discussions, workshops, regular and frequent meetings, project site visits and even informal gatherings such as shared meals or recreational activities.

Almost all social hierarchies involve some component of conflict based around inequity. People want to balance their personal costs and benefits relative to others'. Conservation projects should, wherever possible, bridge gaps and narrow divides by developing equitably among stakeholders. By alleviating large disparities in cost:benefit ratios, responsibilities, and expectations between different stakeholders, the project will become more efficient because there will be less conflict based on inequity. Equity will evolve and change, with stakeholders adapting to behave fairly in a transparent system. In general, teams will reward members who treat others unselfishly and promote the overall goals of the group.

To achieve such a framework of open communication and equity, impartial leadership and long periods of interpersonal relationship building are often required. As hackneyed as they seem, capacity-building exercises, when done correctly, are excellent mechanisms of sharing information and building the competency to use it. Engaging and training local or regional counterparts is an outstanding method for ensuring clearer communication and promoting fairness, instead of forcing information from the top-down and expecting results to emerge from the bottom-up. Further links between transparency and equity can be realised through 'hands-on' applications instead of just talking about concepts. Leaders should participate at all levels, learning the most menial tasks associated with the project (e.g. an administrator should go and catch frogs for a monitoring project).

In the broadest terms, working across cultures is a high risk-high reward system. Although there are complex obstacles, the ultimate litmus for biodiversity conservation might be our ability to learn and work together across cultures to preserve nature.

### SUGGESTED READING

Reed, M.S. (2008). Stakeholder participation for environmental management: a literature review. *Biological Conservation*, 141, 2417–2431.

$$H' = \sum_{i=1}^{s} \frac{n_i}{N} \log_e \left(\frac{n_i}{N}\right)$$

The index provides a measure of the amount of disorder in a system, such that communities with more unique species have higher $H$ (a system with $S = 1$, by this definition, is perfectly ordered but has no diversity). Most of these measures assume a random sampling of species within a community, but this assumption is often violated (Pielou 1966). When sampling is done without replacement, then indices such as Brillouin's $H$ are recommended:

$$H = \frac{1}{N} \log \left(\frac{N!}{n_1! n_2! n_3! \ldots}\right)$$

However, where representativeness is unknown, then *rarefaction* or *resampling* can be used to standardize samples from different areas or periods to a comparable metric (Krebs 1999). This includes inferring the total diversity of a community by using a statistical model to predict unobserved data (unsampled species). Of course, the measures presented here are the basic foundations of species diversity indices, but there are myriad variants thereof, many assumptions that can be tested and adjusted for, and different distributions that may be more or less important under particular circumstances. For an excellent overview of these issues, we recommend the reader refers to Krebs (1999).

## 16.1.2 Scale

Interpretation of the indices and their variants described above depend on the scale of measurement. Whittaker (1972) introduced the concepts of *alpha* ($\alpha$), *beta* ($\beta$), and *gamma* ($\gamma$) *diversity* to measure and compare biodiversity patterns over various spatial scales. $\alpha$ (local) diversity refers to the quantification of species richness, etc. within a particular area or ecosystem, whereas $\beta$ diversity (differentiation) is the difference in the metric between ecosystems. In other words, $\beta$ diversity is a measure of species uniqueness between areas, so as $\beta$ diversity increases, locations differ more from one another and sample a smaller proportion of the total species richness occurring in the wider region (Koleff *et al*. 2003).

Whittaker (1972) sensibly recommended that $\beta$ diversity (Whittaker's $\beta_w$) should be measured as the proportion by which the species richness of a region exceeds the average richness of a single locality within that region:

$$\beta_w = \frac{S}{\bar{\alpha}} = \frac{(a+b+c)}{\frac{(2a+b+c)}{2}}$$

where $S$ = the total number of species recorded for all sites (regional richness) and the average number of species found within sites (local richness), $a$ = the number of species in common in both sites (e.g. for a simple two-site comparison), $b$ = the number of species in site 1, and $c$ = the number of species in site 2. Since then, however, many other variants of the metric have been proposed. These include comparisons along spatial or environmental gradients, between patches of similar habitats, and the degree of similarity between sites (see references in Koleff *et al*. 2003). Indeed, Koleff *et al*. (2003) reviewed 24 different measures of $\beta$ diversity and categorized them into four main groups: measures of (i) continuity (similarity in species composition among sites) and loss (fewer species relative to focal sites); (ii) species richness gradients; (iii) continuity only; and (iv) gain and loss. Not only is there lack of agreement on the most appropriate measure to use, there is also variation in the pattern of scaling applied. As such, Koleff *et al*. (2003) suggested that one should use measures that exhibit the homogeneity property (i.e. the measure is independent of the total number of species as long as the proportions comprising the different components are constant) and that when measures reveal different patterns of variation when based on absolute and proportional species numbers, both types should be examined.

$\gamma$ diversity is otherwise known as "geographic-scale species diversity" (Hunter 2002), which means it is used as a measure of overall diversity for the different constituent ecosystems of a region. This metric becomes particularly valuable to explain broad-scale (regional or continental) patterns of species relative to local (site-specific) indices. Indeed, there are two theoretical types of relationships hypothesized for local versus regional species richness (Figure 16.1). Most datasets support the existence of a proportional relationship between local and regional richness (Type I), albeit local richness always tends to be less than regional (Gaston 2000). It appears that Type II relationships (local richness reaching an asymptote) are rare because local assemblages do not seem to become saturated as one might expect from ecological mechanisms such as density dependence, parasitism and predation (Gaston 2000).

### 16.1.3 Surrogacy

An important goal of conservation biology, which deals with a world of limited resources and options, is to protect areas that have relatively higher biodiversity than surrounding areas. Prioritizing areas for conservation, however, does not always require a complete description of a site's biodiversity, but merely relative measures of differences among them (Margules *et al.* 2002) described using a representative taxonomic subset. The quest for a simple estimator, a *surrogate* (i.e. the number, distribution or pattern of species in a particular taxon in a particular area thought to indicate a much wider array of taxa) that is sufficiently related to the biodiversity parameter of interest is an essential tool in conservation planning (see Chapter 11).

Unfortunately, there is no consensus regarding which surrogates are best for what purposes among ecosystems – many problems with current surrogate approaches remain. For instance, focusing only on a set of species-rich sites may select only a single habitat type with similar species in all areas, thus many rare species may be excluded from protection (Margules and Pressey 2000). Many methods to overcome these problems have been developed based on multivariate measures of biodiversity (e.g. multi-taxa incidence matrices) or reserve-selection algorithms (e.g. Sarkar and Margules. 2002). Advances have been made with recent work (Mellin *et al.* In review) examining surrogate effectiveness in the marine realm. It was shown that higher-taxa surrogates (taxonomic levels such as order, family or genus acting as a surrogate for some lower taxonomic level such as species) outperform cross-taxa (one taxon is used as a surrogate for another at the same taxonomic resolution) and subset-taxa (diversity in one taxonomic group is taken as representative of the entire community) surrogates. Likewise, surrogacy was least effective at broad (> 100 km) spatial scales.

### 16.1.4 Similarity, dissimilarity, and clustering

Although indices of biodiversity take on different aspects of species richness, abundance, evenness and scale, there are many relatively simple techniques available for comparing samples of species and individuals among sites. Most indices of similarity (> 25 types exist – Krebs 1999) are simple descriptors that do not lend themselves easily to measures of uncertainty (e.g. confidence intervals; although resampling methods can provide an index of parameter uncertainty), so their application is generally exploratory. There are two broad classes of similarity: (i) binary; and (ii) quantitative. Binary measures are applied to presence-absence data (i.e. does a species exist in a defined area?) and can be compared among sites using contingency tables using metrics such as Jaccard's similarity, Sorren's similarity, simple matching, or Baroni-Urbani and Buser

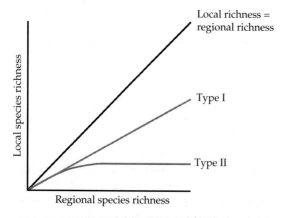

**Figure 16.1** Hypothesized relationship between local and regional species richness (number of species). Type I occurs where local richness is proportional to, but less than, regional richness; Type II demonstrates situations where local richness asymptotes regardless of how much regional richness increases. Reprinted from Gaston (2000).

coefficients (see Krebs 1999). Of course, some method to assess the probability of missing species in presence-absence surveys should also be applied to account for insufficient sampling effort (e.g. MacKenzie et al. 2002).

Quantitative indices require some aspect of individual abundance to be assessed such as the number of individuals, biomass, cover or productivity. Distance dissimilarity indices using abundance data instead of species richness can be applied to the same binary indices listed above. Alternatively Euclidean, Manhattan, Canberra or Bray-Curtis distances between samples can be calculated using relative abundance measures between sites (see Krebs 1999). Simple correlation coefficients such as Pearson product-moment, Spearman's rank and Kendall's $\tau$ can also be used in certain situations to compare sites, but these tend to be insensitive to additive or proportional differences between community samples (Romesburg 1984) and they depend strongly on sample size (generally, $n > 30$ is sufficient for a reliable characterization of the relationship).

When many focal communities are sampled, some form of cluster analysis may be warranted. Cluster analysis refers to any technique that builds classifications, but there is no preferred method given that the choice depends on the type of data being compared. Some considerations for choice include whether the data are: (i) hierarchical (e.g. taxonomic classifications) or reticulate (overlapping classifications); (ii) divisive (sample divided into classes) or agglomerative (fine to coarse resolution); (iii) monothetic (groups distinguished by a single attribute) or polythetic (many attribute-based); or (iv) qualitative (binary) or quantitative (distance measures) (see Krebs 1999 for an overview).

### 16.1.5 Multivariate approaches

When the principal aim of a conservation study is to quantify the relationships between a large number of measurements, whether they be of species, individuals or abiotic predictors of ecological patterns, some form of multivariate analysis is usually required. Over thirty different multivariate techniques have been designed for various applications (Pérez et al. 2008), each with their own particular strengths and weaknesses. Ordination describes those methods that summarize multivariate information in a low-dimensional scatter diagram where points represent samples and distances among them are proportional to their similarity measured, for example, by Euclidean distance, Bray-Curtis or other indices. Common techniques include eigen-based principal components analysis (PCA) or correspondence analysis (CA) and distance-based multidimensional scaling (MDS), cluster analysis or polar ordination that provide coefficients quantifying the relative contribution of component variables to the reduced-dimension principal axes.

Such multivariate approaches are useful for visualizing patterns that would otherwise be difficult or impossible to discern in multidimensional space, such as ecologically related species assemblages or trophic guilds. They can also summarize the principal gradients of variation within and among communities and condense abiotic and other potential explanatory variables (e.g. climate, soil conditions, vegetation structure, chemistry, etc.) into simple gradients themselves that may be used as correlates to explain variation in species or community patterns. Their disadvantage is that they cannot be used to test the relative likelihood of alternative hypotheses, may not appropriately reflect statistical power and effect size, and if applied incautiously, can be misused to mine data for phantom 'patterns' that on closer examination turn out to be random noise or system-specific peculiarities.

## 16.2 Mensurative and manipulative experimental design

Conservation biology typically deals with assessments of previous environmental degradation and the quantification of its effects on biodiversity patterns. Another major aim is to design ways of preserving existing, relatively intact communities through management intervention (e.g. reserve design, control of harvest). Conservation biologists also devote a large proportion of their efforts to quantifying the most efficient and effective methods for restoring degraded habitats to some semblance of previous ecological function. These three principal aims, and the logistical constraints on large-scale system manipulations, generally preclude the use of strict experimental design and control – there are simply too many extenuating variables modifying species patterns to control, and the systems of interest are generally too expensive to apply meaningful manipulations such as those which typify medical experimentation.

There are some notable exceptions to this rule, such as replicated microcosm experiments examining the processes of extinction in rapidly reproducing invertebrate populations. For example, the frequency of extinction times under conditions of low and high environmental variability (Drake 2006), the persistence probability of populations exposed to various spatial configurations of refugia and intensities of harvest (Fryxell et al. 2006) and the implications for extinction risk of chaotic and oscillatory behavior in populations (Belovsky et al. 1999; Hilker and Westerhoff 2007), have all been successfully examined in controlled laboratory settings. Other well-known manipulations at broader spatial scales (albeit with far less experimental control) include examining the effects of forest fragmentation on species diversity (Laurance et al. 2002), controlling the size and configuration of agricultural plots to test bee pollination success (Brosi et al. 2008), examining the effects of landscape composition on the initial dispersal success of juvenile amphibians (Rothermel and Semlitsch 2002), determining the effects of inbreeding depression on individual survival (Jimenez et al. 1994), measuring arthropod responses in tropical savannas exposed to repeated catchment-scale prescribed burning (Andersen and Müller 2000) and the many applications of Before-After-Control-Impact (BACI) experimental designs to detect point-source changes to systems (Underwood 1994).

The above notwithstanding, most conservation studies rely mainly on quantifying existing patterns (observational studies) or take advantage of existing gradients or measurable differences in habitat quality or type to infer mechanisms. This latter category is sometimes referred to as *mensurative* experimentation because it does not explicitly control for confounding variables (Hurlbert 1984). There has been plenty of discussion on this topic over the past twenty or so years (Hurlbert 1984; Krebs 1991; Hargrove and Pickering 1992; Oksanen 2001; Hurlbert 2004; Oksanen 2004), but it is now accepted among most conservation biologists that to make strong inferences on biological patterns and mechanisms, multiple lines of evidence, from observational, mensurative and manipulative experiments, are all required at various spatial and temporal scales (Brook et al. 2008).

### 16.2.1 Hypothesis testing

The classic scientific approach adopts the concept of *falsifiability* (Popper 1959) – that is, demonstrating that a mechanism or phenomenon is not true (null hypothesis) by controlling all other plausible determinants except the one of interest and replicating the experiment sufficiently to avoid spurious patterns that may arise simply by chance (see section below). This is still a core aspect of science because it reduces the chance of making subjective interpretations of the data collected. This is the philosophical basis for the majority of the statistical techniques used by natural scientists; we attempt to discern pattern from the 'noise' in natural systems using theory to estimate the probability that our observations could have been derived merely by chance.

Neyman-Pearson null hypothesis testing (NHT) begins with the assertion that no differences exist between experimental units (null hypothesis), with the implicit view that if the null is unsupported by the data, then one or more 'alternative' hypotheses must therefore be plausible (although these are not explicitly evaluated). Classic statistical theory that has been developed around the NHT approach provides methods to estimate the chance of making an error when rejecting the null hypothesis (*Type I or* $\alpha$ *error*); in other words, this is the probability of concluding that there is a difference (or effect) when in fact, there is none. The flip side to this is that classic NHT tests do not provide an estimate of the probability of making an error when failing to reject the null hypothesis (known as *Type II* or *β* error) – this is essentially the chance one concludes there is no difference (or effect) when in fact, there is. Various *a priori* and *a posteriori* methods exist to estimate Type II errors (more precisely, the *power* of a statistical test taken as 1 – Type II error), with the latter depending on three principal elements: sample size (see below), magnitude of the difference one is attempting to detect (effect size) and the total variance associated with the measure used (see Gerrodette 1987; Osenberg *et al.* 1994; Steidl *et al.* 1997; Thomas 1997; Thomas & Krebs 1997; Thompson *et al.* 2000 for more detail on power analyses).

The disconnect between these two estimates of hypothesis-conclusion error, the implicit conflation of effect size and sample size, as well as the ambiguity related to just *how much chance of making an error is acceptable* (i.e. the moribund and bankrupt concept of statistical 'significance' beyond some arbitrary threshold), have formed for decades some of the main arguments against using NHT (reviewed in Elliott and Brook 2007, see also Burnham and Anderson 2002; Lukacs *et al.* 2007). This is especially true in the ecological and psychological sciences, which are typically restricted to observational studies and subject to extensive variability. The alternative approaches can be classed into the general category of *multiple working hypotheses* (MWH), including best-model selection and multimodal inference (Box 16.3). MWH approaches are now becoming recognized as providing the most logical and objective approaches to assess conservation issues because they explicitly consider uncertainty in the underlying models used to abstract the real world, rather than relying on simple and arbitrarily assessed 'yes-or-no' conclusions typical of the NHT paradigm.

### 16.2.2 Sample size

Regardless of the statistical paradigm invoked or analysis method applied, perhaps the least controversial requirement of good scientific inference

---

**Box 16.3 Multiple working hypotheses**
**Corey J. A. Bradshaw and Barry W. Brook**

Science is, at its core, all about evaluating the support for different ideas – working hypotheses – about how the world works. Because they never reflect the totality of real-world effects, any such hypothesis can be considered a model. But how to decide what ideas have support and which ones should be discarded?

A traditional approach has been to set up some null model (which states that there is no change or measureable effect in a variable of interest), and then proceed to evaluate whether the data conform to this model. This usually involves the arbitrary selection of a threshold probability of making Type I errors (i.e. failing to reject a null hypothesis when it is true) to conclude so-called 'significance' of effect. This line of reasoning still pervades most probabilitistic sciences today. Yet many have called for the abandonment of such subjective statistical practices (Burnham and Anderson 2004; Lukacs *et al.* 2007) in favour of a concept originally forwarded in 1890 by Thomas C. Chamberlin known as *multiple working hypotheses* (Elliott and Brook 2007). The idea is relatively simple – instead of considering a single (null) hypothesis and testing whether the data can falsify it in favour of some alternative (which is not directly tested), the use of multiple working hypotheses does not restrict the number of models considered to abstract the system under investigation. In fact, the approach can specifically accommodate the simultaneous comparison of hypotheses in systems where it is common to find multiple factors influencing the observations made (such as complex ecological systems). This is also particularly applicable to conservation biology because experimental manipulation is often technically difficult or ethically unreasonable.

The basic approach is to construct models (abstractions of complex systems) that

*continues*

### Box 16.3 (Continued)

represent combinations of hypotheses constructed to explain variation in the metric of interest. Models (plausible hypotheses) then can be ranked or compared on the basis of relative evidential support, using methods that tend to reinforce the principle of parsimony (the simplest combination of factors providing the strongest explanatory power) via their bias correction terms. Model comparison based on information theory (usually assessed using Aikaike's information criterion – AIC – when conforming to maximum likelihood approaches – Box 16.4) immediately supposes that all models are false because they represent incomplete approximations of the truth (Elliott and Brook 2007). Weighting AICs then can be used as a means to assess the relative distance to 'truth' by approximating Kullback-Leibler information loss (i.e. measuring the relative distance between conceptual reality and the abstraction under consideration). The Bayesian information criterion (BIC) is a dimension-consistent form of model comparison that provides a measure of the weight of evidence relative to other models (the Bayes factor – see Box 16.4), assuming uninformative prior information. As sample sizes increase, BIC approaches the estimation of the dimension of a 'true' model (not necessarily embedded in the model set) with a probability = 1 (Burnham and Anderson 2004). Here the true model is one which captures main effects but ignores minor (tapering) influences.

It is generally accepted that AIC performs well when sample sizes are small (and AIC itself can be corrected to account for small samples), but it is *a priori* weighted to favour more complex models when tapering effects (biologically important signals that characterise full truth but defy reductionism) are present (Link and Barker 2006). When the aim is to determine the most important variables explaining variation in some measured 'response', BIC is recommended, especially when sample sizes are large (Link and Barker 2006). When prediction is the goal, AIC-based rankings are preferred.

Multimodel inference is gaining increasing popularity in conservation biology because it embraces the concept of multiple working hypotheses to describe complex systems. Rather than choose a single 'best' model (or not even test alternative models, as per null hypothesis testing), multimodel inference is made on the basis of all models in the *a priori* candidate set; here, each model's prediction is weighted by its relative support from the data (e.g. AIC weights or Bayesian posterior probabilities – see Box 16.4) (Burnham and Anderson 2002; Burnham and Anderson 2004; Elliott and Brook 2007). Thus, multimodel inference is advantageous because it accounts for uncertainty in the underlying choice of models used to describe the system of interest, it permits inference from different models simultaneously, and it allows for unconditional ranking of the relative contribution of variables tested (Elliott and Brook 2007). Of course, no inference is made on models/variables not included in the *a priori* model set.

The cases where null hypothesis testing can be justified (see Johnson and Omland 2004; Stephens *et al.* 2005; Stephens *et al.* 2007) are rare in conservation biology for the reasons described above (system complexity, lack of experimentation potential). It is our opinion that the multiple working hypotheses approach, even for relatively simple assessments of effect, should embrace the philosophy of estimating the *strength of evidence* and avoid the pitfalls associated with arbitrary Type I error probability thresholds. This can be usefully done even for a comparison of a null model to a single alternative, using evidence factors (the ratio of AIC or BIC weights of the two models – a concept akin to Bayesian odds ratios) and is preferable to a classic null hypothesis test because the likelihood of the alternative model is explicitly evaluated.

The basic formulae for the most common model-ranking criteria (AIC, $AIC_c$, QAIC and BIC) are provided below:

*continues*

**Box 16.3 (Continued)**

$$\text{AIC} = -2L + 2k$$

where AIC = Akaike's information criterion, $k$ = number of model parameters and $L$ = the maximised log-likelihood function for the estimated model (MLE). Note that the variance term of a statistical model, when estimated (e.g. in a Gaussian model), is a parameter.

$$AIC_c = AIC + \frac{2k(k+1)}{n-k-1}$$

where $AIC_c$ = AIC corrected for small sample size and $n$ = sample size.

$$QAIC = \frac{1}{\hat{c}}2L + 2k$$

where QAIC = quasi-AIC and $\hat{c}$ = the variance inflation factor (when data are over-dispersed). This is commonly used in capture-mark-recapture model assessments (see White and Burnham 1999). The small-sample version of QAIC ($QAIC_c$) is calculated the same way as $AIC_c$. The Bayesian information criterion (BIC) is calculated as:

$$-2\log_e p(xk) \approx \text{BIC} = -2L + k\log_e n$$

where $x$ = observed data and $P(x|k)$ = the likelihood of $x$ given $k$ which is the same as the MLE used in AIC.

**REFERENCES**

Burnham, K. P. and D. R. Anderson. (2002). *Model selection and multimodel inference: a practical information-theoretical approach.* 2nd edn. Springer-Verlag, New York, NY.

Burnham, K. P. and Anderson, D. R. (2004). Understanding AIC and BIC in model selection. *Sociological Methods and Research*, 33, 261–304.

Elliott, L. P. and Brook, B. W. (2007). Revisiting Chamberlain: multiple working hypotheses for the 21st Century. *Bioscience*, 57, 608–614.

Johnson, J. and Omland, K. (2004). Model selection in ecology and evolution. *Trends in Ecology and Evolution*, 19, 101–108.

Link, W. A. and Barker, R. J. (2006). Model weights and the foundations of multimodel inference. *Ecology*, 87, 2626–2635.

Lukacs, P. M., Thompson, W. L., Kendall, W. L., Gould, W. R., Doherty, P. F., Burnham, K. P., and Anderson, D. R. (2007). Concerns regarding a call for pluralism of information theory and hypothesis testing. *Journal of Applied Ecology*, 44, 456–460.

Stephens, P. A., Buskirk, S. W., Hayward, G. D., and Del Rio, C. M. (2005). Information theory and hypothesis testing: a call for pluralism. *Journal of Applied Ecology*, 42, 4–12.

Stephens, P. A., Buskirk, S. W., Hayward, G. D., and del Rio, C. M. (2007). A call for statistical pluralism answered. *Journal of Applied Ecology*, 44, 461–463.

White, G. C. and Burnham, K. P. (1999). Program MARK: survival estimation from populations of marked animals. *Bird Study*, 46 (Supplement), 120–138.

---

in conservation biology is obtaining measurements from as many representative and unbiased *units* (individuals, plots, habitats, ecosystems, etc.) as possible. The main reason for obtaining large sample sizes is that when one measures only a few units, the chance of obtaining a good estimate of the central tendency (e.g. mean or median), variance (i.e. the spread of true values), or distribution (i.e. shape of the frequency distribution of units such as Normal, binomial, log-Normal, etc. and extreme values which characterize the tails of distributions) of a parameter is low. Without good estimates of such parameters, the ability to tease pattern and noise apart becomes increasingly intractable.

There are no rules of thumb for 'adequate' sample sizes because they depend on the hypothesis being tested, the inherent variability of the measures chosen and the temporal or spatial scales examined. The most useful generalization is that there is no substitute for adequate sampling – more representative samples will inevitably provide more power to discern patterns (Caughley and Gunn 1996). While we generally recommend against using classic power tests (see Krebs 1999 for examples) because of their reliance on the NHT paradigm, there are techniques that can be applied to estimate adequate minimum sample size, and the sensitivity of information-theoretic and Bayesian methods (Boxes 16.3 and 16.4) to power can be evaluated in various ways. First, resampling can be used to assess to what extent sampling should continue, but this generally requires a moderately large initial sample. The basic approach is to resample (with replacement) observations from a distribution at incrementing subsample sizes (Manly 1997). The sample size at which the desired magnitude of effect can be detected then becomes the minimum target for future studies applying the same metric. These are typically known as *saturation* or *rarefaction* curves (Heck *et al.* 1975). Other

---

### Box 16.4 Bayesian inference
### Corey J. A. Bradshaw and Barry W. Brook

The most common statistical theory underpinning conservation (indeed, most ecological) research today is still *likelihood*-based; i.e. the likelihood of observing the data at hand based on the expected frequency (from a probability density function) that such data would be observed if the same procedure of data collection was repeated many times (McCarthy 2007). *Maximum likelihood* is therefore the optimisation process that chooses the model parameters that make the data the most likely relative to other parameter values. The process implicitly assumes no prior information on the relevant parameters, with the maximum likelihood estimate coinciding with the most probable values of that distribution. The approach essentially asks *what is the probability of observing the data given that the assumed model structure (hypothesis) is correct?*

An alternative approach is the Bayesian paradigm, which instead asks: *what is the probability the model/hypothesis is true given the data*? Bayes' theorem states that the probability of A occurring given that B has occurred is equal to the probability that both A and B occur divided by the probability of B occurring. Reframing A as a (or set of) parameter estimate θ and B as the data collected (x), then

$$P(\theta|x) = \frac{P(x|\theta)P(\theta)}{P(x)}$$

where $P(\theta|x)$ = the *posterior probability* of obtaining θ given x, and $P(\theta)$ = the *prior probability* of θ and $P(x)$ is the probability of the data – a scaling constant (usually derived numerically). Thus, $P(\theta)$ quantifies the available knowledge about θ prior to collecting x. This can often take the form of information collected during other studies that quantify the distribution (e.g. mean and standard deviation) of θ. Not only does the incorporation of prior information follow the spirit of scientific reasoning and logic (i.e. if A and B, then C) (McCarthy 2007), it generally provides higher certainty in parameter estimates because the model is not starting from scratch (no information). Other advantages of Bayesian approaches include: (i) errors are not assumed to follow any particular distribution, so departures from assumed data distributions are less problematic than in maximum likelihood-based models; (ii) Markov Chain Monte Carlo (MCMC) numerical optimisation (a computer-intensive method) is more flexible than maximum likelihood approaches because there is less of a tendency to become mired in local minima; and (iii) model parameters are assumed to be variable (i.e. a distribution), not fixed (a point value).

The most commonly used software to implement Bayesian models is the freely available WinBUGS (Windows Bayesian inference Using Gibbs Sampling – www.mrc-bsu.cam.ac.uk/bugs), which includes a friendly graphical user interface (GUI). While exceedingly popular, certain aspects of the software make it somewhat

*continues*

> **Box 16.4 (Continued)**
>
> cumbersome to implement, such as the requirement to re-initialise parameter settings whenever models are re-run. An alternative interface that is based on the same basic language is the BRugs library (R interface to R2WinBUGS) in the R programming language (R Development Core Team 2008 – also free, open source software). BRugs is a command-based, object-orientated implementation that can be re-run repeatedly without having to reset parameter values each time.
>
> **REFERENCES**
>
> McCarthy, M.A. (2007). *Bayesian methods for ecology*. Cambridge University Press, Cambridge, UK.
>
> R Development Core Team (2008). *R: A language and environment for statistical computing*. R Foundation for Statistical Computing, Vienna, Austria.

rules of thumb on sufficient sample sizes have emerged from the statistical literature based on assumptions regarding the underlying distribution of the observations (Krebs 1999), the width of Bayesian posterior credibility intervals compared to the prior distributions, or on experience from previous studies.

### 16.2.3 Replication and controls

One of the most common errors made when designing conservation studies is insufficient or biased *replication*. Replication essentially means *repetition* of the experiment (Krebs 1999) and is another type of sample size. Insufficient replication will inflate the estimates of error associated with any metric, so the statistical power to detect differences (or effects) even when present declines with reduced replication. Biased sampling will distort our ability to make inferences about population-level differences on the basis of finite samples. Replication is also essential to avoid the intrusion of chance events; for example, the comparison of only two sites experiencing different intensities of modification may be invalidated because some variable other than the one being tested (e.g. soil type instead of habitat quality) may drive the differences observed in, say, species richness. Only by replicating the sampling unit sufficiently will the chance of spurious events occurring be reduced.

It is important though to ensure that the appropriate statistical unit is replicated. In the above example, increasing the number of sub-samples in each of the two sites does not solve the problem of insufficient replication – the basic unit of comparison is still the 'site'. This is known as *pseudoreplication* because it may appear that increased effort leads to greater replication of the sampled unit, when in reality it is simply the reproduction of non-independent samples (see Hurlbert 1984; Underwood 1994; Krebs 1999). Without true independence among sampling units, estimates of variance, and hence, the power to detect differences (or effects), are downwardly biased, leading to higher probabilities of making Type II errors. Another form of pseudoreplication can occur when designs do not account for temporal autocorrelation among samples or repeat sampling of the same unit (e.g. multiple measures from the same animal that has been recaptured repeatedly). If sequential samples within plots are taken over time, there is a high probability that measures therein will be correlated. There are many experimental designs and statistical tests that can take temporal autocorrelation into account (e.g. Muller *et al.* 1992; Cnaan *et al.* 1997; Krebs 1999; Gueorguieva and Krystal 2004; Ryan 2007).

Another rule often broken by conservation biologists is the failure to incorporate some kind of *control* in their experimental (manipulative or mensurative) design. A control is an experimental unit that receives no direct treatment. In conservation terms, these could be, for example, sites that have not been changed (degraded) in a particular way, areas without invasive species (i.e. the 'treatment' being the presence of the invasive species), or sites where no re-introductions of native species have occurred. While gradient studies looking for correlations between well-known predictors of biodiversity patterns (e.g. forest fragment area explaining variation in species richness; Laurance et al. 2002) do not necessarily require 'controls' (e.g. contiguous forest patches of equivalent size) because the relationships are so well-established, any study attempting some form of manipulative or mensurative experimental inference MUST have controls (note that controls must also be replicated) (Krebs 1999). This applies particularly to the Before-After-Control-Impact (BACI) design – contemporaneous 'controls' are essential to be able to detect any differences (or effects) (Underwood 1994; Krebs 1999).

### 16.2.4 Random sampling

The complexities of experimental design cannot be treated sufficiently in this chapter; however, one last element that applies to all forms of experimental design is the concept of *randomization*. Randomization refers to the process of placing a random spatial or temporal order on the sampling design such that each unit measures statistically independent values. While complete randomization is not always possible (nor entirely desirable in cases of stratified random sampling – e.g. Krebs 1999) for many conservation studies, one should always strive to maximize sample randomization wherever and whenever possible. The key point is to ensure that your sample is representative of the population parameters about which you are trying to make inference – this is the fundamental theoretical tenet of statistical sampling theory.

## 16.3 Abundance Time Series

If species are the currency of biodiversity assessments, then counts of individuals represent the principal unit for population dynamics models used to assess conservation risk (see following section). The restrictions imposed on comprehensive biodiversity assessment by the sheer number of species on Earth (Chapter 2) also apply to the quantification of population dynamics for single species – there are simply too many species to be able to obtain detailed demographic data (e.g. survival, fertility, dispersal, etc.) for the majority of them to build population models (see following section). Therefore, many types of *phenomenological* model have been developed to deal with sequential censuses (time series) of absolute or relative population size. *Phenomenological* simply means that the dynamical properties these models emulate represent the end-point *phenomenon* of total population size (number of individuals at any given point in time), that is, the emergent property of various *mechanisms* such as birth, death, reproduction and dispersal. Therefore, phenomenological models applied to abundance time series are restricted in their capacity to explain ecological mechanisms, but they certainly provide fertile ground for testing broad hypotheses, describing gross population behavior, and making predictions about population change (provided mechanisms remain constant).

One of the commonest and simplest questions conservation biologists ask is whether a population is trending or stationary. Indeed, one of the main criteria used by the World Conservation Union (IUCN) to define a population or species as threatened (i.e. either *Vulnerable*, *Endangered* or *Critically Endangered*) on its *Red List* (www.iucnredlist.org) is its rate of decline. As such, reliably determining both the direction of the trend (i.e. if declining, to highlight conservation concern, or if increasing, to indicate successful recovery) and quantifying the rate of change, are central goals of conservation biology. While it may seem superficially straightforward to determine at least the direction of population's abundance trend, factors such as the difficulty in censusing the population (counting all individuals), measurement (observation) error, and the presence of high seasonal variance in abundance due to normal environmental stochasticity (variation), are common real-world challenges that can make conclusions of population trajectory uncertain.

Many statistical tools have been developed to deal with these problems, including traditional

NHT power analyses to detect trends (e.g. Gerrodette 1987; see also Gerrodette 1993 for associated software), nonlinear models (e.g. Fewster et al. 2000) and the simultaneous application of multiple time series models (Box 16.3) applied to relative abundance counts to determine the direction of trend and strength of feedbacks (e.g. McMahon et al. 2009). We certainly recommend the multiple working hypotheses approach (Box 16.3) when querying abundance time series, but argue that much more mathematical development and empirical testing is required on this topic.

Trending, or *nonstationary* populations may be driven by *exogenous* influences ("changes in the environment that affect population change, but are not themselves influenced by population numbers" – Turchin 2003) and/or by *endogenous* influences ("dynamical feedbacks affecting population numbers, possibly involving time lags" – Turchin 2003). It is of course important to determine the interplay between such drivers (Bradshaw 2008) because either may dominate at certain times or on certain stages of the population, or short-term trends may simply represent periods of re-equilibration of longer-term cycles that are not readily apparent when sampling over too few time intervals relative to the scale of disturbance or the species' generation length.

The development of population dynamics models in ecology dates back to the early 19th century (Pearl 1828; Verhulst 1838) and has developed in the intervening 180 years into an expansive discipline in its own right, dealing with the many and complex ways in which organisms interact within and among populations and species. We cannot possibly provide a summary of all the relevant components of time series analysis here (for an excellent overview with worked examples, see Turchin 2003), but we do highlight some of the essential basics.

An important component of extinction models is the presence of density feedback, because the strength and form of such endogenous influences can strongly affect predictions of extinction risk (see below) (Philippi et al. 1987; Ginzburg et al. 1990). In situations where detailed measurements of the ways in which population density modifies demographic processes are unavailable, phenomenological models applied to abundance time series can still provide some direction. The idea that populations tend to fluctuate around an equilibrium abundance, encapsulated by the general logistic (S-shaped curve) model (Turchin 2003), was generalized for time series by Ricker's model (Ricker 1954) where the rate of population change ($r$):

$$r = \log e\left(\frac{N_{t+1}}{N_t}\right)$$

($N$ is the discrete population size estimate at time $t$), can be expressed as a simple linear function of $N_t$ declining from an *intrinsic* (maximum) growth rate ($r_m$):

$$r = r_m\left(1 - \left(\frac{N_t}{K}\right)\right)$$

When $r$ is positive, the population is growing; above *carrying capacity* ($K$), the population declines. Here, the environment's $K$ is assumed to impose some upper limit to total abundance. There are many variants and complications of this basic model, and even more debates regarding its role in explaining complex population dynamics; however, we argue this basic model has been instrumental in defining some of the more important theoretical elements of population dynamics applied to questions of sustainable harvest and extinction risk. Indeed, Turchin (2003) goes as far as to call it a fundamental 'law' of population ecology.

In real-world situations, the negative influence of density on population rate of change is likely to apply mainly to the region around carrying capacity and be of less importance for small populations below their *minimum viable population size* (see below). For instance, as populations decline, individuals may lose average fitness due to phenomena such as inbreeding depression (see Genetic Tools section below), reduced cooperative anti-predator behavior (e.g. flocking or herding), reduced mate availability, and the loss or degradation of cooperative breeding effort (Courchamp et al. 2008). Thus, density feedback at these small population sizes can be *positive*, and this is generally known as an *Allee effect* (Allee 1931). Although the phenomenological evidence for Allee effects using abundance time series is sparse – mainly because obtaining observations at low densities is logistically challenging and observation error tends to be inflated when detection probabilities are low – there are some models that can be applied, such as the Ricker-Allee model:

$$r = r_m\left(1 - \frac{N_t}{K}\right)\left(\frac{N_t - A}{K}\right)$$

where $A$ represents the critical lower Allee threshold abundance below which positive feedback begins. For a comprehensive discussion of Allee effects, see Courchamp *et al.* (2008) and Berec *et al.* (2007).

## 16.4 Predicting Risk

A longstanding goal in conservation biology is predicting the risk a species, community or ecosystem faces when humans change the environment. Questions such as: *How many individuals are required for a population to have a high chance of persisting in the future? What species are most susceptible to human-induced changes to the environment? Are some species more likely to become invasive than others?* and *What types of species are required to maintain ecosystem function?* pervade the conservation literature from purely theoretical to highly applied perspectives. Not only do these questions require substantial data to provide realistic direction, the often arbitrary choice of the degree of risk (defined as a probability of, for example, becoming threatened, invasive, or falling below a predefined population size), can add subjectivity to the assessment.

### 16.4.1 Cross-taxa approaches

The ranking of species' life history traits (e.g. evolved characteristics such as generation time, mean body mass, reproductive potential; ecological attributes such as dispersal capacity, niche constraints) and environmental contexts, which together predict a species' response to environmental change, has received considerable attention in recent years (e.g. Bennett and Owens 1997; Owens and Bennett 2000; Purvis *et al.* 2000; Kolar and Lodge 2001; Heger and Trepl 2003; Brook *et al.* 2006; Pimm *et al.* 2006; Bielby *et al.* 2008; Bradshaw *et al.* 2008; Sodhi *et al.* 2008a, b, 2009). Determining which traits lead to higher extinction or invasion risk, for instance, is important for prioritizing management to eradicate harmful invasive species or recover threatened taxa (Bradshaw *et al.* 2008). Developing simple predictive generalizations ('rules') for categorizing poorly studied species into categories of relative risk (proneness) thus becomes a tool to assist in the efficient allocation of finite conservation resources.

There is now good correlative evidence that particular combinations of life history and ecological characteristics (e.g. organism size, dispersal capacity, geographic range, and other reproductive, dispersal, morphological and physiological attributes) influence a species' risk of becoming extinct or invasive, with the strength of effect depending on the spatial scale of measurement, environmental context, and rate of change of the forcing factor (e.g. deforestation or climate change) (Bradshaw *et al.* 2008). Much of this evidence is derived from three main types of models: generalized linear mixed-effects models (e.g. Brook *et al.* 2006; Bradshaw *et al.* 2008; Sodhi *et al.* 2008a, c), generalized estimating equations (Bielby *et al.* 2008) and phylogenetically independent contrasts (e.g. Bennett and Owens 1997; Owens and Bennett 2000; Purvis *et al.* 2000). The principal reason why these complex models must be used instead of simple correlations is because of the confounding effects of shared evolutionary traits when making cross-species comparisons (Felsenstein 1985). In other words, because species are related hierarchically according to their *phylogeny* (evolutionary relationships and common ancestry), they are not strictly independent statistical units, and so their relationships should be taken into account.

Linear mixed-effects models (Pinheiro and Bates 2000) take phylogeny inferred from Linnaean taxonomy into account by using a nested structure in the random effect component of the model (Blackburn and Duncan 2001); once the variance component due to correlated relationships is taken (partially) into account, the residual variation can be attributed to fixed effects (e.g. life history traits) of hypothetical interest. Generalized estimating equations are similar to mixed-effects models, but the parameters are estimated by taking correlations among observations into account (Paradis and Claude 2002). Phylogenetically independent contrasts (PIC) compute the differences in scores between sister clades and rescale the variance as a function of evolutionary branch length (Purvis 2008). The PIC approach (and its many variants – see Purvis *et al.* 2005; Purvis 2008) is useful, but has been criticized because of: (i) its sensitivity to errors in estimated phylogenetic distance (Ramon and Theodore 1998); (ii) incorrect treatment of extinction risk as an evolved trait (Putland 2005); (iii) overestimation of differences between closely related species (Ricklefs and Starck 1996); (iv) requirement of a complete phylogeny; (v) inability to deal with categorical variables; and (vi) its restriction of using the NHT framework (Blackburn and Duncan 2001; Bradshaw *et al.* 2008). Despite these criticisms, no one modeling approach is superior in all situations, so we recommend several techniques be applied where possible.

### 16.4.2 Population viability analyses

When the goal is to estimate risk to a single species or population instead of evolved life histories that may expose species to some undesirable state, then the more traditional approach is to do a *population viability analysis* (PVA). PVA broadly describes the use of quantitative methods to predict a population's extinction risk (Morris and Doak 2002). Its application is wide and varied, tackling everything from assessment of relative risk for alternative management options (e.g. Allendorf *et al.* 1997; Otway *et al.* 2004; Bradshaw *et al.* 2007), estimating minimum viable population sizes required for long-term persistence (e.g. Traill *et al.* 2007 and see section below), identifying the most important life stages or demographic processes to conserve or manipulate (e.g. Mollet and Cailliet 2002), setting adequate reserve sizes (e.g. Armbruster and Lande 1993), estimating the number of individuals required to establish viable re-introduced populations (e.g. South *et al.* 2000), setting harvest limits (e.g. Bradshaw *et al.* 2006), ranking potential management interventions (e.g. Bradshaw *et al.* in press), to determining the number and geographical structure of subpopulations required for a high probability of persistence (e.g. Lindenmayer and Possingham 1996).

The approaches available to do PVAs are as varied as their applications, but we define here the main categories and their most common uses: (i) count-based; (ii) demographic; (iii) metapopulation; and (iv) genetic. A previous section outlined the general approaches for the analysis of population dynamics and the uses of abundance time series in conservation biology; count-based PVAs are yet another application of basic abundance (either total or relative) surveys. Briefly, the distribution of population growth rates on the logarithmic scale, constructed from a (ideally) long time series (or multiple populations) of abundance estimates, provides an objective means of projecting long-term population trajectories (either declining, increasing, or stable) and their variances. The basic premise is that, given a particular current population size and a minimum acceptable value below which the population is deemed to have gone *quasi-extinct* (i.e. not completely extinct, but where generally too few individuals remain for the population to be considered viable in the long term), the mean long-term population growth rate and its associated variance enables the calculation of the probability of falling below the minimum threshold. While there are many complications to this basic approach (e.g. accounting for substantial measurement error, catastrophic die-offs, environmental autocorrelation, density feedback and demographic fluctuations (e.g. uneven sex ratio – for an overview, see Morris and Doak 2002), the method is a good first approximation if the only data available are abundance time series. A recent extension to the approach, based on the multiple working hypotheses paradigm (Box 16.3), has been applied to questions of sustainable harvest (Bradshaw *et al.* 2006).

A more biologically realistic, yet data-intensive approach, is the demographic PVA. Count-based PVAs essentially treat all individuals as equals – that is, equal probabilities of dying, reproducing and dispersing. In reality, because populations are usually structured into discernable and differentiated age, sex, reproductive and development stages (amongst others), demographic PVAs combine different measured (or assumed) *vital rates* that describe the probability of performing some demographic action (e.g. surviving, breeding, dispersing, growing, etc.). Vital rates are ideally estimated using capture-mark-recapture (CMR) models implemented in, for example, program MARK (White and Burnham 1999), but surrogate information from related species or allometry (body mass relationships) may also be used. The most common method of combining these different life stages' vital rates into a single model is the *population projection matrix*. While there are many complicated aspects to these, they allow for individuals in a population to advance through sequential life stages and perform their demographic actions at specified rates. Using matrix algebra (often via computer simulation), static, stochastic and/or density-modified matrices are multiplied by population vectors (stage-divided population abundance) to project the population into the future. The reader is referred to the comprehensive texts by Caswell (2001) and Morris and Doak (2002) for all the gory details. Freely or commercially available software packages such as VORTEX (www.vortex9.org) or RAMAS (www.ramas.com) can do such analyses.

*Metapopulations* are networks of spatially separated sub-populations of the same species that are connected by dispersal (see Chapter 5). A metapopulation can be thought of as a "population of populations" (Levins 1969) or a way of realistically representing patches of high habitat suitability within a continuous landscape. In ways that are analogous to the structuring of individuals within a single population, metapopulations 'structure' sub-populations according to habitat quality, patch size, isolation and various other measures. The mathematical and empirical development of metapopulation theory has burgeoned since the late 1990s (see Hanski 1999) and has been applied to assessments of regional extinction risk for many species (e.g. Carlson and Edenhamn 2000; Molofsky and Ferdy 2005; Bull *et al.* 2007). For a recent review of the application of metapopulation theory in large landscapes, see Akçakaya and Brook (2008).

Although genetic considerations are not nearly as common in PVAs as they perhaps should be (see more in the following section, and the book by Frankham *et al.* 2002 for a detailed overview), there is a growing body of evidence to suggest that the subtle determinants of extinction are strongly influenced by genetic deterioration once populations become small (Spielman *et al.* 2004; Courchamp *et al.* 2008). The most common application of genetics in risk assessment has been to estimate a *minimum viable population size* – the smallest number of individuals required for a demographically closed population to persist (at some predefined 'large' probability) for some (mainly arbitrary) time into the future (Shaffer 1981). In this context, genetic considerations are growing in perceived importance. Genetically viable populations are considered to be those large enough to avoid *inbreeding depression* (reduced fitness due to inheritance of deleterious alleles by descent), prevent the random accumulation or fixation of deleterious mutations (genetic drift and mutational meltdown), and maintain evolutionary potential (i.e. the ability to evolve when presented with changing environmental conditions; see following section). The MVP size required to retain evolutionary potential is the equilibrium population size where the loss of quantitative genetic variation due to small population size (genetic drift) is matched by increasing variation due to mutation (Franklin 1980). Expanded detail on the methods for calculating genetically effective population sizes and a review of the broad concepts involved in genetic stochasticity can be found in Frankham *et al.* (2002) and Traill *et al.* (2009). The next section gives more details.

## 16.5 Genetic Principles and Tools

The previous sections of this chapter have focused primarily on the organismic or higher taxonomic units of biodiversity, but ignored the sub-organism (molecular) processes on which

## Box 16.5 Functional genetics and genomics
### Noah K. Whiteman

Conservation genetics has influenced the field of conservation biology primarily by yielding insight into the provenance of individuals and the ecological and evolutionary relationships among populations of threatened species. As illuminated in the section on genetic diversity, conservation genetics studies rely primarily on genomic data obtained from regions of the genome that are neutral with respect to the force of natural selection (neutral markers). Conservation biologists are also interested in obtaining information on functional (adaptive) differences between individuals and populations, typically to ask whether there is evidence of local adaptation (Kohn et al. 2006). Adaptive differences are context-dependent fitness differences between individuals and are ultimately due to differences between individuals in gene variants (alleles) at one or multiple loci, resulting in differences in phenotype. These phenotypic differences are always the result of gene-environment interactions and can only be understood in that light. However, unraveling the association between particular nucleotide substitutions and phenotype is challenging even for scientists who study genetic model systems.

Adaptive differences between individuals and populations are difficult to identify at the molecular genetic level (see also Chapter 2). This is typically because genomic resources are not available for most species. However, with a set of unlinked molecular markers scattered throughout the genome, such as microsatellites, it is possible to identify candidate loci of adaptive significance that are physically linked to these markers. If the frequency of alleles at these loci is significantly greater or less than the expectation based on an equilibrium between migration and genetic drift, one can infer that this locus might have experienced the effects of natural selection. These analyses are often referred to as outlier analyses and aim to find genes linked to neutral markers that are more (or less) diverged between individuals and populations than the background (neutral) divergence (Beaumont 2005). Despite the immediate appeal of these studies, moving from identification of outlier loci to identification of the function of that locus and the individual nucleotide differences underlying that trait is a difficult task.

The genomics revolution is now enabling unprecedented insight into the molecular basis of fitness differences between individuals. Completed genome sequences of hundreds of plants and animals are available or in progress and next generation sequencing technology is rapidly increasing the number of species that will become genomically characterized. Massively parallel sequencing technology is enabling the rapid characterization of entire genomes and transcriptomes (all of the expressed genes in a genome) at relatively low cost. Currently, sequence reads from these technologies are, on average, <500 base pairs in length and so traditional Sanger sequencing still outperforms massively parallel technology at the level of the individual read. Digital gene expression (where all of the expressed genes are sequenced and counted; Torres et al. 2008) and microarray analysis allows one to study differences in global gene expression without a priori information on the identity of genes used in the analysis. Single nucleotide polymorphism (SNP) analysis is likely to be an effective tool in identifying loci and individual substitutions that are associated with differences in trait values between individuals, even when pedigree information and heritabilities of traits are not available, as is the case for most threatened species.

Although there is considerable debate over the relative importance of cis regulatory mutations (in non-coding sequences flanking protein-coding genes) versus structural mutations (in protein coding genes) in the molecular basis of phenotypic evolution across species, methods are best developed for detecting a signature of selection at codons within protein-coding genes. In this case, a conservation biologist may be interested in knowing what loci and what codons within that gene have experienced positive, adaptive selection. The redundancy of the DNA code

*continues*

> **Box 16.5 (Continued)**
>
> means that in protein-coding genes, nucleotide substitutions are either synonymous – the amino acid coded by the codon remains the same, or non-synonymous – the corresponding amino acid changes. Comparing the rates of non-synonymous/synonymous substitutions (the ω rate ratio) of a gene between species can provide evidence of whether that gene or locus is under selection (Yang 2003). A variety of methods are available to estimate ω ratios for a given gene tree. When ω <1, purifying selection is inferred because non-synonymous substitutions are deleterious with respect to fitness; when ω = 1, neutral evolution is inferred because there is no difference in fitness between non-synonymous and synonymous substitutions; and when ω >1, positive selection is inferred because non-synonymous substitutions are favored by natural selection. In their most general form, ω ratios are averaged across all nucleotide sites, but because non-synonymous rates are often quite variable across a gene, ω values can also be estimated for individual codons. While it is possible to test for significant differences among ω values, the most conservative interpretation holds that adaptive evolution has occurred only when ω values are >1. However, even when ω values are >1, demographic forces can elevate ω ratios if there is an imbalance between genetic drift and purifying selection. Because several non-mutually exclusive factors can affect ω ratios, comparisons using these data, which are always only correlative in nature, need to be interpreted with caution.
>
> The genomics research horizon is rapidly changing all areas of biology and conservation biology is no exception. A new arsenal of genomic and analytical tools is now available for conservation biologists interested in identifying adaptive differences between individuals and populations that will complement traditional neutral marker studies in managing wildlife populations.
>
> **REFERENCES**
>
> Beaumont, M. A. (2005). Adaptation and speciation: what can *Fst* tell us? *Trends in Ecology and Evolution*, 20, 435–440.
>
> Kohn, M. K., Murphy, W. J., Ostrander, E. A., and Wayne, R. K. (2006). Genomics and conservation genetics. *Trends in Ecology and Evolution*, 21, 629–637.
>
> Torres, T. T., Metta, M., Ottenwälder, B., and Schlötterer, C. (2008). Gene expression profiling by massively parallel sequencing. *Genome Research*, 18, 172–177.
>
> Yang, Z. (2003) Adaptive molecular evolution. In D. J. Balding, M. Bishop and C. Cannings, eds *Handbook of Statistical Genetics*, pp. 229–254, John Wiley and Sons, New York, NY.

evolution itself operates. As such, no review of the conservation biologist's toolbox would be complete without some reference to the huge array of molecular techniques now at our disposable used in "conservation genetics" (Box 16.5). Below is a brief primer of the major concepts.

Conservation genetics is the discipline dealing with the genetic factors that affect extinction risk and the methods one can employ to minimize these risks (Frankham *et al.* 2002). Frankham *et al.* (2002) outlined 11 major genetic issues that the discipline addresses: (i) inbreeding depression's negative effects on reducing reproduction and survival; (ii) loss of *genetic diversity*; (iii) reduction in *gene flow* among populations; (iv) *genetic drift*; (v) accumulation and purging of *deleterious mutations*; (vi) genetic adaptation to captivity and its implications for reintroductions; (vii) resolving uncertainties of taxonomic identification; (viii) defining *management units* based on genetic exchange; (ix) forensics (species identification and detection); (x) determining biological processes relevant to species management; and (xi) *outbreeding depression*. All these issues can be assessed by extracting genetic material [e.g. DNA (deoxyribonucleic acid), RNA (ribonucleic acid)] from tissue sampled from live or dead individuals (see Winchester and Wejksnora 1995 for a good introduction to the array of methods used to do this).

Of these 11 themes, the first three are perhaps the most widely applicable elements of conservation genetics, and so deserve special mention here. Inbreeding depression can be thought of as an Allee effect because it exacerbates reductions in average individual fitness as population size becomes small. Inbreeding is the production of offspring by related individuals resulting from self-fertilization (e.g. the extreme case of 'selfing' in plants) or by within-'family' (e.g. brother-sister, parent-offspring, etc.) matings. In these cases, the combination of related genomes during fertilization can result in reductions in reproduction and survival, and this is known as *inbreeding depression*. There are several ways to measure inbreeding: (i) the inbreeding coefficient ($F$) measures the degree of parent relatedness derived from a pedigree analysis (strictly – the probability that an allele is common among two breeding individuals by descent); (ii) the average inbreeding coefficient is the $F$ of all individuals in a population; and (iii) inbreeding relative to random breeding compares the average relatedness of parents to what one would expect if the population was breeding randomly.

The amount of *genetic diversity* is the extent of heritable variation available among all individuals in a population, species or group of species. *Heterozygosity* is the measure of the frequency of different of *alleles* [alternative forms of the same segment of DNA (locus) that differ in DNA base sequence] at the same gene locus among individuals and is one of the main ways genetic diversity is measured. Populations with few alleles have generally had their genetic diversity reduced by inbreeding as a result of recent population decline or historical bottlenecks. Populations or species with low genetic diversity therefore have a narrower genetic template from which to draw when environments change, and so their evolutionary capacity to adapt is generally lower than for those species with higher genetic variation.

Habitat fragmentation is the process of habitat loss (e.g. deforestation) and isolation of 'fragments', and is one of the most important direct drivers of extinction due to reductions in habitat area and quality (Chapter 5). Yet because fragmentation also leads to suitable habitats for particular species assemblages becoming isolated pockets embedded within (normally) inhospitable terrain (matrix), the exchange of individuals, and hence, the flow of their genetic material, is impeded. Thus, even though the entire population may encompass a large number of individuals, their genetic separation via fragmentation means that individuals tend to breed less randomly and more with related conspecifics, thus increasing the likelihood of inbreeding depression and loss of genetic diversity. For a more comprehensive technical demonstration and discussion of these issues, we recommend the reader refers to Frankham *et al.* (2002).

## 16.6 Concluding Remarks

The multidisciplinarity of conservation biology provides an expansive source of approaches, borrowed from many disciplines. As such, this integrative science can appear overwhelming or even intimidating to neophyte biologists, especially considering that each approach discussed here (and many more we simply did not have space to describe) is constantly being reworked, improved, debated and critiqued by specialists. But do not despair! The empirical principles of conservation biology (again, focusing here on the 'biology' aspect) can be broadly categorized into three major groups: (i) measuring species and abundance; (ii) correlating these to indices of environmental change; and (iii) estimating risk (e.g. of extinction). Almost all of the approaches described herein, and their myriad variants and complications, relate in some way to these aims. The specific details and choices depend on: (i) data quality; (ii) spatial and temporal scale; (iii) system variability; and (*iv*) nuance of the hypotheses being tested.

When it comes to the choice of a particular statistical paradigm in which to embed these techniques, whether it be null hypothesis testing or multiple working hypotheses (Box 16.3), likelihood-based or Bayesian inference (Box

> **Box 16.6 Useful Textbook Guides**
> **Corey J. A. Bradshaw and Barry W. Brook**
>
> It is not possible to provide in-depth mathematical, experimental or analytical detail for the approaches summarised in this chapter. So instead we provide here a list of important textbooks that do this job. The list is not exhaustive, but it will give emerging and established conservation biologists a solid quantitative background on the issues discussed in this chapter – as well as many more.
>
> **SUGGESTED READING**
>
> Bolker, B. M. (2008). *Ecological models and data in R*. Princeton University Press, Princeton, NJ.
> Burnham, K. P. and Anderson, D. R. (2002). *Model selection and multimodal inference: a practical information-theoretic approach*. 2nd edn. Springer-Verlag, New York, NY.
> Caswell, H. (2001). *Matrix population models: construction, analysis, and interpretation*. 2nd edn. Sinauer Associates, Inc., Sunderland, MA.
> Caughley, G. and Gunn, A. (1996). *Conservation biology in theory and practice*. Blackwell Science, Cambridge, MA.
> Clark, J. S. (2007). *Models for ecological data: an introduction*. Princeton University Press, Princeton, NJ.
> Ferson, S. and Burgman, M., eds (2002). *Quantitative methods for conservation biology*. Springer, New York, NY.
> Frankham, R., Ballou, J. D., and Briscoe, D. A. (2002). *Introduction to conservation genetics*. Cambridge University Press, Cambridge, UK.
> Krebs, C. J. (1999). *Ecological methodology*. 2nd edn. Benjamin Cummings, Upper Saddle River, NJ.
> Krebs, C. J. (2009). *Ecology: the experimental analysis of distribution and abundance*. 6th edn. Benjamin Cummings, San Francisco, CA.
> Lindenmayer, D. and Burgman, M. (2005). *Practical conservation biology*. CSIRO (Australian Commonwealth Scientific and Industrial Research Organization) Publishing, Collingwood, Australia.
> McCallum, H. (2000). *Population parameters: estimation for ecological models*. Blackwell Science, Oxford, UK.
> McCarthy, M. A. (2007). *Bayesian methods for ecology*. Cambridge University Press, Cambridge, UK.
> Millspaugh, J. J. and Thompson, F. R. I., eds (2008). *Models for planning wildlife conservation in large landscapes*. Elsevier, New York, NY.
> Morris, W. F. and Doak, D. F. (2002). *Quantitative conservation biology: theory and practice of population viability analysis*. Sinauer Associates, Sunderland, MA.
> Turchin, P. (2003). *Complex population dynamics: a theoretical/empirical synthesis*. Princeton University Press, Princeton, NJ.

16.4), is to some extent open to personal choice. We have been forthright regarding our particular preferences (we consider multiple working hypotheses to be generally superior to null hypothesis testing, and Bayesian outperforming likelihood-based inference), but there are no hard-and-fast rules. In general terms though, we recommend that conservation biologists must at least be aware of the following principles for any of their chosen analyses:

- Adequate and representative replication of the appropriate statistical unit of measure should be planned from the start.
- The high probability that results will vary depending on the spatial and temporal scale of investigation must be acknowledged.
- Choosing a single model to abstract the complexities of ecological systems is generally prone to oversimplification (and often error of interpretation).
- Formal incorporation of previous data is a good way of reducing uncertainty and building on past scientific effort in a field where data are inevitably challenging to obtain; and
- Multiple lines of evidence regarding a specific conclusion will always provide stronger inference, more certainty and better management and policy outcomes for the conservation of biodiversity.

This chapter represents the briefest of glimpses into the array of techniques at the disposal of conservation biologists. We have attempted to provide as much classic and recent literature to guide the reader toward more detailed information, and in this spirit have provided a list of what we consider to be some of the better textbook guides which provide an expanded treatment of the different techniques considered (Box 16.6). A parting recommendation – no matter how sophisticated the analysis, the collection of rigorous data using well-planned approaches will always provide the best scientific outcomes.

## Summary

- Conservation biology is a highly multidisciplinary science employing methods from ecology, Earth systems science, genetics, physiology, veterinary science, medicine, mathematics, climatology, anthropology, psychology, sociology, environmental policy, geography, political science, and resource management. Here we focus primarily on ecological methods and experimental design.
- It is impossible to census all species in an ecosystem, so many different measures exist to compare biodiversity: these include indices such as species richness, Simpson's diversity, Shannon's index and Brouillin's index. Many variants of these indices exist.
- The scale of biodiversity patterns is important to consider for biodiversity comparisons: $\alpha$ (local), $\beta$ (between-site), and $\gamma$ (regional or continental) diversity.
- Often surrogate species – the number, distribution or pattern of species in a particular taxon in a particular area thought to indicate a much wider array of taxa – are required to simplify biodiversity assessments.
- Many similarity, dissimilarity, clustering, and multivariate techniques are available to compare biodiversity indices among sites.
- Conservation biology rarely uses completely manipulative experimental designs (although there are exceptions), with mensurative (based on existing environmental gradients) and observational studies dominating.
- Two main statistical paradigms exist for comparing biodiversity: null hypothesis testing and multiple working hypotheses – the latter paradigm is more consistent with the constraints typical of conservation data and so should be invoked when possible. Bayesian inferential methods generally provide more certainty when prior data exist.
- Large sample sizes, appropriate replication and randomization are cornerstone concepts in all conservation experiments.
- Simple relative abundance time series (sequential counts of individuals) can be used to infer more complex ecological mechanisms that permit the estimation of extinction risk, population trends, and intrinsic feedbacks.
- The risk of a species going extinct or becoming invasive can be predicted using cross-taxonomic comparisons of life history traits.
- Population viability analyses are essential tools to estimate extinction risk over defined periods and under particular management interventions. Many methods exist to implement these, including count-based, demographic, metapopulation, and genetic.
- Many tools exist to examine how genetics affects extinction risk, of which perhaps the measurement of inbreeding depression, gene flow among populations, and the loss of genetic diversity with habitat degradation are the most important.

## Suggested reading

See Box 16.6.

## Relevant websites

- Analytical and educational software for risk assessment: www.ramas.com.
- Population viability analysis software: www.vortex9.org.
- *Ecological Methodology* software–Krebs (1999): www.exetersoftware.com/cat/ecometh/ecomethodology.html.
- Capture-mark-recapture analysis software: http://welcome.warnercnr.colostate.edu/gwhite/mark/mark.htm.
- Analysis of data from marked individuals: www.phidot.org.
- Open-source package for statistical computing: www.r-project.org.
- Open-source Bayesian analysis software: www.mrc-bsu.cam.ac.uk/bugs/.

## Acknowledgements

We thank T. Gardner, D. Bickford, C. Mellin and S. Herrando-Pérez for contributions and assistance.

## REFERENCES

Akçakaya, H. R. and Brook, B. W. (2008). Methods for determining viability of wildlife populations in large landscapes. In *Models for Planning Wildlife Conservation in Large Landscapes* (eds J. J. Millspaugh & F. R. I. Thompson), pp. 449–472. Elsevier, New York, NY.

Allee, W. C. (1931). *Animal Aggregations: A Study in General Sociology.* University of Chicago Press, Chicago, IL

Allendorf, F. W., Bayles, D., Bottom, D. L., *et al.* (1997). Prioritizing Pacific salmon stocks for conservation. *Conservation Biology*, **11**, 140–152.

Andersen, A. N. and Müller, W. J. (2000). Arthropod responses to experimental fire regimes in an Australian tropical savannah: ordinal-level analysis. *Austral Ecology*, **25**, 199–209.

Armbruster, P. and Lande, R. (1993). A population viability analysis for African elephant (*Loxodonta africana*): how big should reserves be? *Conservation Biology*, **7**, 602–610.

Belovsky, G. E., Mellison, C., Larson, C., and Van Zandt, P. A. (1999). Experimental studies of extinction dynamics. *Science*, **286**, 1175–1177.

Bennett, P. M. and Owens, I. P. F. (1997). Variation in extinction risk among birds: chance or evolutionary predisposition? *Proceedings of the Royal Society of London B*, **264**, 401–408.

Berec, L., Angulo, E., and Courchamp, F. (2007). Multiple Allee effects and population management. *Trends in Ecology and Evolution*, **22**, 185–191.

Bielby, J., Cooper, N., Cunningham, A. A., Garner, T. W. J., and Purvis, A. (2008). Predicting susceptibility to future declines in the world's frogs. *Conservation Letters*, **1**, 82–90.

Blackburn, T. M. and Duncan, R. P. (2001). Establishment patterns of exotic birds are constrained by non-random patterns in introduction. *Journal of Biogeography*, **28**, 927–939.

Bradshaw, C. J. A. (2008). Having your water and drinking it too – resource limitation modifies density regulation. *Journal of Animal Ecology*, **77**, 1–4.

Bradshaw, C. J. A., Fukuda, Y., Letnic, M. I., and Brook, B. W. (2006). Incorporating known sources of uncertainty to determine precautionary harvests of saltwater crocodiles. *Ecological Applications*, **16**, 1436–1448.

Bradshaw, C. J. A., Mollet, H. F., and Meekan, M. G. (2007). Inferring population trends for the world's largest fish from mark-recapture estimates of survival. *Journal of Animal Ecology*, **76**, 480–489.

Bradshaw, C. J. A., Giam, X., Tan, H. T. W., Brook, B. W., and Sodhi, N. S. (2008). Threat or invasive status in legumes is related to opposite extremes of the same ecological and life history attributes. *Journal of Ecology*, **96**, 869–883.

Bradshaw, C. J. A., Peddemors, V. M., McAuley, R. B., and Harcourt, R. G. in press. Population viability of Australian grey nurse sharks under fishing mitigation and climate change. *Marine Ecology Progress Series*.

Brook, B. W., Traill, L. W., and Bradshaw, C. J. A. (2006). Minimum viable population size and global extinction risk are unrelated. *Ecology Letters*, **9**, 375–382.

Brook, B. W., Sodhi, N. S., and Bradshaw, C. J. A. (2008). Synergies among extinction drivers under global change. *Trends in Ecology and Evolution*, **25**, 453–460.

Brosi, B. J., Armsworth, P. R., and Daily, G. C. (2008). Optimal design of agricultural landscapes for pollination services. *Conservation Letters*, **1**, 27–36.

Bull, J. C., Pickup, N. J., Pickett., B., Hassell, M. P., and Bonsall, M. B. (2007). Metapopulation extinction risk is increased by environmental stochasticity and assemblage complexity. *Proceedings of the Royal Society of London B*, **274**, 87–96.

Burnham, K. P. and Anderson, D. R. (2002). *Model Selection and Multimodal Inference: A Practical Information-Theoretic Approach.* 2nd edn Springer-Verlag, New York, NY.

Carlson, A. and Edenhamn, P. (2000). Extinction dynamics and the regional persistence of a tree frog metapopulation. *Proceedings of the Royal Society of London B*, **267**, 1311–1313.

Caswell, H. (2001). *Matrix Population Models: Construction, Analysis, and Interpretation.* 2nd edn Sinauer Associates, Inc., Sunderland, MA.

Caughley, G. and Gunn, A. (1996). *Conservation biology in theory and practice.* Blackwell Science, Cambridge, MA.

Cnaan, A., Laird, N. M., and Slasor, P. (1997). Using the general linear mixed model to analyse unbalanced repeated measures and longitudinal data. *Statistics in Medicine*, **16**, 2349–2380.

Courchamp, F., Luděk, B., and Gascoigne, B. (2008) *Allee effects in ecology and conservation.* Oxford University Press, Oxford, UK.

Drake, J. M. (2006). Extinction times in experimental populations. *Ecology*, **87**, 2215–2220.

Elliott, L. P. and Brook, B. W. (2007). Revisiting Chamberlain: multiple working hypotheses for the 21st Century. *BioScience*, **57**, 608–614.

Felsenstein, J. (1985). Phylogenies and the comparative method. *American Naturalist*, **70**, 115.

Fewster, R. M., Buckland, S. T., Siriwardena, G. M., Baillie, S. R., and Wilson, J. D. (2000). Analysis of population trends for farmland birds using generalized additive models. *Ecology*, **81**, 1970–1984.

Frankham, R., Ballou, J. D., and Briscoe, D. A. (2002). *Introduction to conservation genetics.* Cambridge University Press, Cambridge, UK.

Franklin, I. R. (1980). Evolutionary change in small populations. In *Conservation biology: An evolutionary-ecological perspective* (eds M.E. Soule & B.A. Wilcox), pp. 135–149. Sinauer, Sunderland, MA.

Fryxell, J. M., Lynn, D. H., and Chris, P. J. (2006). Harvest reserves reduce extinction risk in an experimental microcosm. *Ecology Letters*, **9**, 1025–1031.

Gaston, K. J. (2000). Global patterns in biodiversity. *Nature*, **405**, 220–227.

Gerrodette, T. (1987). A power analysis for detecting trends. *Ecology*, **68**, 1364–1372.

Gerrodette, T. (1993). TRENDS: Software for a power analysis of linear regression. *Wildlife Society Bulletin*, **21**, 515–516.

Ginzburg, L. R., Ferson, S., and Akçakaya, H. R. (1990). Reconstructibility of density dependence and the conservative assessment of extinction risks. *Conservation Biology*, **4**, 63–70.

Gueorguieva, R. and Krystal, J. H. (2004). Move over ANOVA: progress in analyzing repeated-measures data and its reflection in papers published in the Archives of General Psychiatry. *Archives of General Psychiatry*, **61**, 310–317.

Hanski, I. (1999). *Metapopulation Ecology.* Oxford University Press, Oxford, UK.

Hargrove, W. W. and Pickering, J. (1992). Pseudoreplication: a *sine qua non* for regional ecology. *Landscape Ecology*, **6**, 251–258.

Heck, K. L., van Belle, G., and Simberloff, D. (1975). Explicit calculation of the rarefaction diversity measurement and the determination of sufficient sample size. *Ecology*, **56**, 1459–1461.

Heger, T. and Trepl, L. (2003). Predicting biological invasions. *Biological Invasions*, **5**, 313–321.

Hilker, F. M. and Westerhoff, F. H. (2007). Preventing extinction and outbreaks in chaotic populations. *American Naturalist*, **170**, 232–241.

Hunter, M. (2002). *Fundamentals of Conservation Biology.* 2nd edn Blackwell Science, Malden, MA.

Hurlbert, S. H. (1984). Pseudoreplication and the design of ecological field experiments. *Ecological Monographs*, **54**, 187–211.

Hurlbert, S. H. (2004). On misinterpretations of pseudoreplication and related matters: a reply to Oksanen. *Oikos*, **104**, 591–597.

Jimenez, J. A., Hughes, K. A., Alaks, G., Graham, L., and Lacy, R. C. (1994). An experimental study of inbreeding depression in a natural habitat. *Science*, **266**, 271–273.

Kolar, C. S. and Lodge, D. M. (2001). Progress in invasion biology: predicting invaders. *Trends in Ecology and Evolution*, **16**, 199–204.

Koleff, P., Gaston, K. J., and Lennon, J. J. (2003). Measuring beta diversity for presence-absence data. *Journal of Animal Ecology*, **72**, 367–382.

Krebs, C.J. (1991). The experimental paradigm and long-term population studies. *Ibis*, **133**, 3–8.

Krebs, C. J. (1999). *Ecological Methodology.* 2nd edn Benjamin Cummings, Upper Saddle River, NJ.

Laurance, W. F., Lovejoy, T. E., Vasconcelos, H. L., et al. (2002). Ecosystem decay of Amazonian forest fragments: a 22-year investigation. *Conservation Biology*, **16**, 605–618.

Levins, R. (1969). Some demographic and genetic consequences of environmental heterogeneity for biological control." Bulletin of the Entomological Society of America. *Bulletin of the Ecological Society of America*, **15**, 237–240.

Lindenmayer, D. B. and Possingham, H. P. (1996). Ranking conservation and timber management options for leadbeater's possum in southeastern Australia using population viability analysis. *Conservation Biology*, **10**, 235–251.

Lukacs, P. M., Thompson, W. L., Kendall, W. L., et al. (2007). Concerns regarding a call for pluralism of information theory and hypothesis testing. *Journal of Applied Ecology*, **44**, 456–460.

MacKenzie, D. I., Nichols, J. D., Lachman, G. B., et al. (2002). Estimating site occupancy rates when detection probabilities are less than one. *Ecology*, **83**, 2248–2255.

Manly, B. F. J. (1997). *Randomization, Bootstrap and Monte Carlo Methods in Biology.* Chapman and Hall, London, UK.

Margules, C. R. and Pressey, R. L. (2000). Systematic conservation planning. *Nature*, **405**, 243–253.

Margules, C. R., Pressey, R. L., and Williams, P. H. (2002). Representing biodiversity: data and procedures for identifying priority areas for conservation. *Journal of Biosciences*, **27**, 309–326.

McMahon, C. R., Bester, M. N., Hindell, M. A., Brook, B. W., and Bradshaw, C. J. A. (2009). Shifting trends: detecting environmentally mediated regulation in long-lived marine vertebrates using time-series data. *Oecologia*, **159**, 69–82.

Mellin, C., Caley, M. J., Meekan, M. G., et al. (In review) Assessing the effectiveness of biological surrogates for measuring biodiversity patterns. *Ecological Applications*.

Mollet, H. F. and Cailliet, G. M. (2002). Comparative population demography of elasmobranchs using life history tables, Leslie matrices and stage-based matrix models. *Marine and Freshwater Research*, **53**, 503–516.

Molofsky, J. and Ferdy, J.-B. (2005). Extinction dynamics in experimental metapopulations. *Proceedings of the National Academy of Sciences of the United States of America*, **102**, 3726–3731.

Morris, W. F. and Doak, D. F. (2002). *Quantitative Conservation Biology: Theory and Practice of Population Viability Analysis*. Sinauer Associates, Sunderland, MA.

Muller, K. E., LaVange, L. M., Sharon Landesman, R., and Ramey, C. T. (1992). Power calculations for general linear multivariate models including repeated measures applications. *Journal of the American Statistical Association*, **87**, 1209–1226.

Oksanen, L. (2001). Logic of experiments in ecology: is pseudoreplication a pseudoissue? *Oikos*, **94**, 27–38.

Oksanen, L. (2004). The devil lies in details: reply to Stuart Hurlbert. *Oikos*, **104**, 598–605.

Osenberg, C. W., Schmitt, R. J., Holbrook, S. J., Abu-Saba, K. E., and Flegal, A. R. (1994). Detection of environmental impacts: natural variability, effect size, and power analysis. *Ecological Applications*, **4**, 16–30.

Otway, N. M., Bradshaw, C. J. A., and Harcourt, R. G. (2004). Estimating the rate of quasi-extinction of the Australian grey nurse shark (*Carcharias taurus*) population using deterministic age- and stage-classified models. *Biological Conservation*, **119**, 341–350.

Owens, I. P. F. and Bennett, P. M. (2000). Ecological basis of extinction risk in birds: habitat loss versus human persecution and introduced predators. *Proceedings of the National Academy of Sciences of the United States of America*, **97**, 12144–12148.

Paradis, E. and Claude, J. (2002). Analysis of comparative data using generalized estimating equations. *Journal of Theoretical Biology*, **218**, 175–185.

Pearl, R. (1828). *The rate of living, being an account of some experimental studies on the biology of life duration*. Alfred A. Knopf, New York, NY.

Pérez, S. H., Baratti, M., and Messana, G. (2008). Subterranean ecological research and multivariate statistics: a review (1945–2006). *Journal of Cave and Karst Studies*, **70**, in press.

Philippi, T. E., Carpenter, M. P., Case, T. J., and Gilpin, M. E. (1987). Drosophila population dynamics: chaos and extinction. *Ecology*, **68**, 154–159.

Pielou, E. C. (1966). The measurement of diversity in different types of biological collections. *Journal of Theoretical Biology*, **13**, 131–144.

Pimm, S., Raven, P., Peterson, A., Sekercioglu, C. H., and Ehrlich, P. R. (2006). Human impacts on the rates of recent, present, and future bird extinctions. *Proceedings of the National Academy of Sciences of the United States of America*, **103**, 10941–10946.

Pinheiro, J. C. and Bates, D. M. (2000). *Statistics and Computing. Mixed-Effects Models in S and S-Plus*. Springer-Verlag, New York, NY.

Popper, K. R. (1959). *The logic of scientific discovery*. Routledge, London, UK.

Purvis, A. (2008). Phylogenetic approaches to the study of extinction. *Annual Review of Ecology, Evolution, and Systematics*, **39**, 301–319.

Purvis, A., Gittleman, J. L., Cowlishaw, G., and Mace, G. M. (2000). Predicting extinction risk in declining species. *Proceedings of the Royal Society of London B*, **267**, 1947–1952.

Purvis, A., Gittleman, J. L., and Brooks, T., eds (2005) *Phylogeny and conservation*. Cambridge University Press, Cambridge, UK.

Putland, D. (2005). Problems of studying extinction risks. *Science*, **310**, 1277.

Ramon, D.-U. and Theodore, G., Jr. (1998). Effects of branch length errors on the performance of phylogenetically independent contrasts. *Systematic Biology*, **47**, 654–672.

Ricker, W. E. (1954). Stock and recruitment. *Journal of the Fisheries Research Board of Canada*, **11**, 559–623.

Ricklefs, R. E. and Starck, J. M. (1996). Applications of phylogenetically independent contrasts: a mixed progress report. *Oikos*, **77**, 167–172.

Romesburg, H. C. (1984). *Cluster Analysis for Researchers*. Lifetime Learning Publications, Belmont, CA.

Rothermel, B. B. and Semlitsch, R. D. (2002). An experimental Investigation of landscape resistance of forest versus old-field habitats to emigrating juvenile amphibians. *Conservation Biology*, **16**, 1324–1332.

Ryan, T. P. (2007). *Modern experimental design*. Wiley-Interscience, Hoboken, NJ.

Sarkar, S. and Margules, C. (2002). Operationalizing biodiversity for conservation planning. *Journal of Biosciences*, **27**, 299–308.

Shaffer, M. L. (1981). Minimum population sizes for species conservation. *BioScience*, **31**, 131–134.

Simpson, E. H. (1949). Measurement of diversity. *Nature*, **163**, 688.

Sodhi, N. S., Bickford, D., Diesmos, A. C., et al. (2008a). Measuring the meltdown: drivers of global amphibian extinction and decline. *PLoS One*, **3**, e1636.

Sodhi, N. S., Brook, B. W., and Bradshaw, C. J. A. (2009). Causes and consequences of species extinctions. In S. A. Levin, ed. *The Princeton Guide to Ecology*, pp. 514–520. Princeton University Press, Princeton, NJ.

Sodhi, N. S., Koh, L. P., Peh, K. S.-H., et al. (2008b). Correlates of extinction proneness in tropical angiosperms. *Diversity and Distributions*, **14**, 1–10.

South, A., Rushton, S., and Macdonald, D. (2000). Simulating the proposed reintroduction of the European beaver (*Castor fiber*) to Scotland. *Biological Conservation*, **93**, 103–116.

Spielman, D., Brook, B. W., and Frankham, R. (2004). Most species are not driven to extinction before genetic factors impact them. *Proceedings of the National Academy of Sciences of the United States of America*, **101**, 15261–15264.

Steidl, R. J., Hayes, J. P., and Schauber, E. (1997). Statistical power analysis in wildlife research. *Journal of Wildlife Management*, **61**, 270–279.

Thomas, L. (1997). Retrospective power analysis. *Conservation Biology*, **11**, 276–280.

Thomas, L. and Krebs, C. J. (1997). A review of statistical power analysis software. *Bulletin of the Ecological Society of America*, **78**, 128–139.

Thompson, P. M., Wilson, B., Grellier, K., and Hammond, P. S. (2000). Combining power analysis and population viability analysis to compare traditional and precautionary approaches to conservation of coastal cetaceans. *Conservation Biology*, **14**, 1253–1263.

Traill, L. W., Bradshaw, C. J. A., and Brook, B. W. (2007). Minimum viable population size: a meta-analysis of 30 years of published estimates. *Biological Conservation*, **139**, 159–166.

Traill, L. W., Brook, B. W., Frankham, R., and Bradshaw, C. J. A. (2009). Minimum population size targets for biodiversity conservation. *Biological Conservation*, in press.

Turchin, P. (2003). *Complex population dynamics: a theoretical/empirical synthesis.* Princeton University Press, Princeton, NJ.

Underwood, A. J. (1994). On beyond BACI: sampling designs that might reliably detect environmental disturbances. *Ecological Applications*, **4**, 3–15.

Verhulst, P. F. (1838). Notice sur la loi que la population poursuit dans son accroissement. *Correspondance mathématique et physique*, **10**, 113–121.

White, G. C. and Burnham, K. P. (1999). Program MARK: survival estimation from populations of marked animals. *Bird Study*, **46** (Supplement), 120–138.

Whittaker, R. H. (1972). Evolution and measurement of species diversity. *Taxon*, **21**, 213–251.

Winchester, A. M. and Wejksnora, P. J. (1995). *Laboratory Manual of Genetics.* 4th edn McGraw-Hill, New York, NY.

# Index*

*Intuitive topic coverage in chapters is not included here

*A. smithii*, 140
*A. superciliosa superciliosa*, 139
*A. tsugae*, 137
*A. undulata*, 140
*A. wyvilliana*, 140
*Acacia*, 147, 173
*Acacia cyclops*, 147
*Acer saccharum*, 155
*Achatina fulica*, 137
*Acridotheres tristis*, 141
adders-tongue fern, 28
*Adelges piceae*, 137
*Aedes albopictus*, 144
*Aegolius acadicus*, 225
*Aepyceros melampus*, 169
African buffalo, 169
African elephant, 233
African molassesgrass, 134
African mosquito, 147
*Agasicles hygrophila*, 137
Agrilus planipennis, 159
*Ailuropoda melanoleuca*, 233
Alagoas curassow, 184
*Alliaria petiolata*, 136
alligatorweed flea beetle, 137
*Alouatta seniculus*, 99
*Alternanthera philoxeroides*, 137
American ash tree, 159
American bison, 125
American chestnut, 135
American pika, 155
American tufted beardgrass, 134
*Anas platyrhynchos*, 139
*Aniba rosaeodora*, 113
*Anolis sagrei*, 143
*Anopheles darlingi*, 65
*Anopheles gambiae*, 147
*Anoplophora glabripennis*, 144
*Aphanomyces astaci*, 138
*Aphelinus semiflavus*, 139
*Arabidopsis thaliana*, 28
Arctic cod, 155, 158
Arctic hare, 158

*Arctogadus glacialis*, 155
*Areca catechu*, 245
arecanut, 245
Argentine ant, 135
*Arundo donax*, 143
Asian chestnut blight, 135
Asian parasitic tapeworm, 140
Asian tapeworm, 139
*Athrotaxis selaginoides*, 175
Australian eucalyptus trees, 134
Australian paperbark, 133, 134
Australian rooikrans tree, 147
avian influenza, 65

Bachman's warbler, 193
Baird's tapir, 305
bald eagle, 140, 225
balsam woolly adelgid, 137
Baltimore oriole, 155
Bay checkerspot butterfly, 96
*Bertholletia excelsa*, 113
*Bison bison*, 109
*Bithynia*, 141
*Bithynia tentaculata*, 139, 141
black and white colobus
    monkey, 94
black guillemot, 155
black rhinoceros, 124, 232, 234
blue monkey, 94
blue-breasted fairy-wren, 95
*Boiga irregularis*, 136, 194
*Bothriocephalus acheilognathi*, 139
Brazil nuts, 113
Brazilian pepper, 143
Brazilian *sardine*, 114
broadleaf mahogany, 110, 242
brown anole lizard, 143
brown tree snake, 136, 194
brown-headed cowbird, 99
*Bubalus bubalis*, 142
*Bufo houstonensis*, 232–233
*Bufo periglenes*, 156
bushmeat, 59, 100, 111, 225

*C. stoebe*, 135
C3 photosynthesis, 170, 176
C4 photosynthesis, 170
*Cactoblastis cactorum*, 137
cactus moth, 138, 148
*Caenorhabditis elegans*, 28
*Caesalpinia echinata*, 109
*Callitris intratropica*, 175
*Campephilus principalis*, 193
cape shoveller, 140
*Capra aegagrus hircus*, 137
*Carcharias taurus*, 115
*Carcinus maenas*, 143
*Caretta caretta*, 118
Carolina parakeet, 193
*carolinensis*, 193
Carson, Rachel, 11, 263
cassava mealybug, 137, 139
*Castanea dentata*, 135
*Castor canadensis*, 133
*Caulerpa taxifolia*, 134, 145
*Cebus capucinus*, 305
*Cenchrus echinatus*, 147
*Centaurea diffusa*, 135
*Cepphus grylle*, 155
*Ceratotherium simum*, 169
*Cercopithecus mitza*, 94
*Chaos chaos*, 28
chlorofluorocarbons, 153
chytrid fungus, 157
*Cirsium hygrophilum* var. *hygrophilum*,
    138
climate change, 16, 37, 47, 57, 153, 170,
    195, 206, 226, 274, 284
$CO_2$ (carbon dioxide), 47, 51, 153, 161,
    170, 178
cochineal bug, 137
*Colluricincla harmonica*, 94
*Colobus guereza*, 94
*Commidendrum robustum*, 148
common myna, 141
*Connochaetes* spp., 169
*Conus*, 192

Convention on International Trade in Endangered Species of Wild Fauna and Flora (CITES), 11, 124
coral reefs, 76, 157, 160, 192, 193
cordgrass, 134, 140
*Cordia interruptus*, 141
*Corvus corone*, 98
cowpeas, 51
crayfish plague, 138
*Cryphonectria parasitica*, 135
cryptic species, 30
crystalline ice plant, 135
*Ctenopharyngodon idella,.*
cutthroat trout, 140
*Cyathocotyle bushiensis*, 139
*Cynomys parvidens*, 233
cypress pine, 175

*Dactylopius ceylonicus*, 137
*Dalbergia melanoxylon*, 110
Darwin, Charles, 51, 163
DDT (dichloro-diphenyl-trichloroethane), 148, 263
*Dendroica chrysoparia*, 232
*Dermochelys coriacea*, 118
*Desmoncus*, 112
*Diceros bicornis*, 232
diffuse knapweed, 135
dipterocarp, 50
*Diuraphis noxia*, 137
diversity-stability hypothesis, 8
*Dreissena bugensis*, 142
*Dreissena polymorpha*, 135
*Drosophila melanogaster*, 28
dung beetles, 117

East African blackwood, 110
eastern yellow robin, 94
ecological diversity, 31, 33
economics of conservation, 15, 16, 52, 55, 61, 63, 112, 199, 233, 246, 256
economics of conservation, 55
ecoregions, 31, 32, 201, 240
*Ectopistes migratorius*, 193
Edith's checkerspot butterfly, 155
eel grass, 155
Ehrenfeld, David, 12
*Eichhornia crassipes*, 134
emerald ash borer, 159
Emerson, Ralph Waldo, 263
*Encephalitozoon intestinalis*, 28
Endangered Species Act (ESA), 11, 181, 220, 225
endemism, 1, 38, 201, 255
*Eopsaltria australis*, 94
*Epidinocarsis lopezi*, 139

*Equus* spp., 169
*Escherichia coli*, 28
ethics of conservation, 15, 21, 279
ethnobotany, 64
*Eucalyptus*, 164, 246
*Eucalyptus albens*, 97
*Euglandina rosea*, 137
*Euphydryas editha*, 155
*Euphydryas editha bayensis*, 96
Eurasian badger, 93
Eurasian weevil, 138
European green crab, 143
European mink, 140
European rabbit, 140
European rabbits, 137
Evolutionary-Ecological Land Ethic, 11
extinction, 60, 63, 95, 107, 137, 148, 156, 181, 225, 327
local, 107, 117, 123
mass, 1, 34, 35, 109, 183, 254

*F. microcarpa*, 141
*Falco femoralis septentrionalis*, 233
faucet snail, 139, 141
*Ficus* spp., 141
fig wasps, 143
figs, 142
fire ant, 135
firetree, 134, 142
Florida panther, 181
flying foxes (pteropodid fruit bats), 115
Franklin, Benjamin, 66
Fraser fir tree, 137
*Fraxinus americana*, 159

*G. amistadensis*, 139
*Gambusia affinis*, 137
*Gambusia amistadensis*, 139
garlic mustard, 136
genetic diversity, 28, 95, 181, 208, 333
*Genyornis newtoni*, 170
*Geophaps smithii*, 176
giant African snail, 137, 147
giant bluefin tuna, 124
giant panda, 233
giant reed, 143
Global Environment Facility (GEF), 199, 202
*Glossopsitta concinna*, 94
goats, 137
golden lion tamarin, 233
golden toad, 156
golden-cheeked warblers, 232
grass carp, 139
grey shrike-thrush, 94

grey-headed robin, 156
greynurse *sharks*, 115
grizzly bear, 141, 181
ground beetle, 246
groundsel, 140
*Grus americana*, 233
gumwood tree, 148
*Gyps* vulture, 63
gypsy moth, 137

*Haemophilus influenzae*, 28
*Haliaeetus leucocephalus*, 140, 225
Hawaiian duck, 140
Hawaiian honeycreepers, 159
*Hemidactylus frenatus*, 135
hemlock woolly adelgid, 137
*Heodes tityrus*, 155
*Herpestes auropunctatus*, 136
*Hesperia comma*, 95
heterogeneity, 32, 33
*Heteromyias albispecularis*, 156
*Hibiscus tiliaceus*, 49
HIV, 64
*Holcaspis brevicula*, 246
homonymy, 30
hooded crow, 98
hotspots (biodiversity), 77, 194, 200, 203, 204
house gecko, 135
Houston toads, 232
howler monkeys, 99
*humile*, 135
*Hydrilla verticillata*, 147
*Hydrodamalis gigas*, 191
*Hyperaspis pantherina*, 148

*Icerya purchasi*, 137
*Icterus galbula*, 155
*Iguana iguana*, 99
iguanas, 99
impala, 169
implementing policy, 17, 19
Indian house crow, 147
Indian mongoose, 136
Intergovernmental Panel on Climate Change (IPCC), 16, 47, 153, 158, 161
International Union for Conservation of Nature (IUCN), 225, 263
redlist, 185, 192, 204, 226, 234, 326
irreplaceability, 200, 201
island biogeography, 12, 14, 88, 186, 204
ivory-billed woodpecker, 193

Jacques-Yves Cousteau, 11
Japanese white-eye, 142

Joshua trees, 159
jumper ant, 28

kangaroo, 170
key deer, 157
keystone species, 57
King Billy pine, 175
*Kochia scoparia*, 147
kokanee salmon, 140

lady beetle, 137, 148
*Lagorchestes hirsutus*, 176
Lake Erie water snake, 143
landscape ecology, 210
*Lantana camara*, 142
large blue butterfly, 140
*Lates niloticus*, 136
latitudinal species gradient, 39
leaf monkeys, 304
leaf-cutter ants, 99
leatherback turtle, 118
*Leontopithacus rosalia*, 233
*Leopardus pardalis*, 305
Leopold, Aldo, 3, 4, 10, 181, 263, 279
*Lepus arcticus*, 158
*Leyogonimus polyoon*, 141
*Ligustrum robustrum*, 141
Linnaean taxonomy, 328
Linnaeus, Carl, 184, 185
lion, 233
lodgepole pine, 166
loggerhead turtle, 118
longhorned beetle, 144
Lousiana crayfish, 138
lowland tapir, 160, 213
*Loxodonta africana*, 233
*Lymantria dispar*, 137

*M. vison*, 140
*Maculina arion*, 140
maize, 51
malaria, 65, 147
  avian, 138, 139, 159
*Malurus pulcherrimus*, 95
mangrove trees, 49
Mangroves, 78
*Manorina melanocephala*, 99
marine conservation, 18, 203, 264
Marsh, George Perkins, 9, 262
masked palm civet, 65
Mayr, Ernst, 8
Mediterranean salt cedars, 134
*Melaleuca quinquenervia*, 133
*Melamprosops phaeosoma*, 184
*Meles meles*, 93

*Melinis minutiflora*, 134
*Mesembryanthemum crystallinum*, 135
mesic spruce-fir, 166
mesquite, 147
metacommunity, 253
metapopulation, 95, 253, 330
methane, 47, 153
Michael Soulé, 12
Millennium Ecosystem Assessment, 45
*Mimosa pigra*, 142
*Mitu mitu*, 184
*Molothrus ater*, 99
Monterrey pine, 135
*Morella faya*, 134, 142
mosquito fish, 137, 148
Muir, John, 263, 269
*Mus musculus*, 28
musk lorikeet, 94
*Mustela erminea*, 136
*Myriophyllum spicatum*, 142
*Myrmecia pilosula*, 28
*Myrmica sabuleti*, 140
*Mysis relicta*, 140
myxoma virus, 139, 140
*Myxosoma cerebralis*, 139

*N. bruchi*, 137, 148
National Environmental Policy Act, 11
*Neochetina eichhorniae*, 137, 148
*Neogobius melanostomus*, 142
*Nerodia sipedon insularum*, 143
Network of Conservation Educators and Practitioners (NCEP), 303
New Zealand grey duck, 139
Nile perch, 136
nitrogen, 47, 51, 134, 142, 176
nitrous oxide, 47, 153
noisy miner, 99
North America mink, 140
North American beaver, 59, 117, 133
North American buffalo, 109
North American gray squirrel, 135
North American mallard, 139
northern aplomado falcons, 233
northern saw-whet owl, 225
northern spotted owl, 15, 233
Norway rats, 136, 147
*Nothofagus* spp, 133
*Notropis lutrensis*, 139
*Nyctereuteus procyonoides*, 65

*O. clarki*, 140
*O. corallicola*, 138
*O. jamaicensis*, 140
oaks, 135

ocelot, 305
*Ochotona princeps*, 155
*Odocoileus virginianus*, 123
*Odocoileus virginianus clavium*, 157
oil palm, 50, 239
*Oncorhynchus mykiss*, 139
*Oncorhynchus nerka*, 140
*Ophioglossum reticulatum*, 28
opossum shrimp, 140
*Opuntia* spp., 137
*Opuntia vulgaris*, 137
orangutan, 243
organismal diversity, 28, 31, 33
*Orthezia insignis*, 148
*Oryctolagus cuniculus*, 137
Oxford ragwort, 140
*Oxyura leucocephala*, 140

*Pacifastacus lenusculus*, 138
Pacific rat, 136
*Paguma larvata*, 65
*Panthera leo*, 233
parasitic wasp, 139
parasitic witchweed, 139
partridge pigeon, 176
passenger pigeon, 193
Pau-Brasil legume tree, 109
Pau-Rosa, 113
pet trade, 124
*Phacochoerus africanus*, 169
pharmaceuticals, 64
*Pharomachrus mocinno*, 305
*Phenacoccus manihoti*, 137
Philippine cockatoo, 305
philosophy of conservation, 148
phosphorous, 47, 63
phylogenetic irreplaceability, 206
*Phytophthora pinifolia*, 135
*Picea engelmannii*, 166
*Picoides borealis*, 232
Pinchot, Gifford, 263
*Pinus contorta*, 166
*Pinus radiata*, 135
*Plagopterus argentissimus*, 139, 140
*Plasmodium relictum capistranoae*, 138
po'o uli, 184
polar bear, 155
pollinators, 56, 60, 98, 115
*Pongo borneo*, 243
population biology, 10
*Praon palitans*, 139
*Presbytis* sp., 304
prickly pear cactus, 137
*Procambarus clarkii*, 139
*Prosopis* spp., 147
*Pseudocheirus peregrinus*, 156
pteropods, 158

puma, 305
*Puma concolor*, 305
*Puma concolor coryi*, 181
*Pycnonotus jocosus*, 141

quagga mussel, 142
*Quercus* spp., 135

*R. exulans*, 136
*R. norvegicus*, 136
raccoon dog, 65
rain forest conservation, 36
rainbow trout, 139
Rare, 305
    Pride campaign, 305
*Rattus norvegicus*, 28
*Rattus rattus*, 136
red shiner, 139
red signal crayfish, 138
red squirrel, 135
red-cockaded woodpecker, 232
reduced impact logging (RIL), 116, 242
red-whiskered bulbul, 141
resplendent quetzal, 306
*Rhinocyllus conicus*, 138
*Rhizobium*, 51
rinderpest, 138
ringtail possum, 156
*Rodolia cardinalis*, 137
Roosevelt, Theodore, 10
rosy wolf snail, 137
round goby, 142
*Rubus alceifolius*, 141
ruddy duck, 140
rufous hare-wallaby, 176

*S. alterniflora*, 143
*S. cambrensis*, 140
*S. squalidus*, 140
*Saccharomyces cerevisiae*, 28
saiga antelope, 125
*Saiga tatarica*, 125
Saint Lucia parrot, 305
sand bur, 147
Sardinella brasiliensis, 114
SARS, 65
scale insect, 148
Scarabaeinae, 117
*Schinus terebinthifolius*, 143
*Schizachyrium condensatum*, 134
*Sciurus carolinensis*, 135

*Sciurus vulgaris*, 135
*Sciurus. vulgaris*, 135
semaphore cactus, 138
*Senecio*, 140
*Senecio squalidus*, 140
Shannon's Index, 315
ship rat, 136
silver-spotted skipper, 95
skipper butterflies, 30, 33
smooth prickly pear, 137
Society for Conservation Biology (SCB), 14
*Solenopsis invicta*, 135
sooty copper, 155
Soulé, Michael E., 12
South American water hyacinth, 134
South American weevils, 137
southern beech, 133
*Spartina*, 140
*Spartina anglica*, 134, 143
spatial conservation planning, 31
species complex, 142
species richness, 28, 30, 33, 57, 62, 201, 314, 317
spotted knapweed, 135
Steller's sea cow, 191
stoat, 136
*Stochastic processes*, 95
*Striga asiatica*, 139
*Strix aluco*, 93
*Strix occidentalis caurina*, 15, 233
sugar maple, 155
*Suidae*, 116
Suisun thistle, 138
sulfur, 47
surrogacy, 203, 314, 318
sustained yield, 8
swallowtail butterflies, 33
*Swietenia macrophylla*, 110, 242
*Syncerus caffer*, 169
synonymy, 30

*Tachycineta bicolor*, 155
*Tamarix*, 231
*Tamarix* spp., 134
*Taprius terrestris*, 94
*Tapirus bairdii*, 305
tawny owl, 93
*Tayassu pecari*, 94
temperate forest conservation, 78, 176
The Nature Conservancy, 12

*Therioaphis trifolii*, 139
Thoreau, Henry David, 263
*Thunnus thynnus*, 124
tiger mosquito, 144
tragedy of the commons, 65, 124
tree swallows, 155
trematode, 139, 141
*Trioxys utilis*, 139
tropical forest conservation, 13, 47, 49, 73, 76, 78, 82, 111, 116, 156, 185, 254, 284, 297

urban planning, 18, 253
*Ursus arctos horribilis*, 140, 181
*Ursus maritimus*, 155
Utah prairie dogs, 233

*Vermivora bachmanii*, 193

Wallace, Alfred Russel, 9, 187
warthog, 169
waru trees, 49
water buffalo, 142
water hyacinth, 134, 147, 148
watermilfoil, 142
weevils, 148
Welsh groundsel, 140
wheat, 51
wheat aphid, 137
whirling disease, 139
white box tree, 97
white rhino, 169
white-faced capuchin monkey, 305
white-headed duck, 140
white-lipped peccary, 94
white-tailed deer, 109, 123
whooping crane, 233
Wilcox, Bruce A., 7, 18
wild pigs, 116
wildebeest, 169
wilderness conservation, 8, 110, 208, 263, 266, 268
woundfin minnow, 140

yellow clover aphid, 139
yellowbilled duck, 140
*Yucca brevifolia*, 159

zebra, 169
zebra mussel, 135, 142
*Zostera marina*, 155
*Zosterops japonicus*, 142